AIR CONDITIONING AND REFRIGERATION FOR THE PROFESSIONAL

AIR CONDITIONING AND REFRIGERATION FOR THE PROFESSIONAL

ROBERT CHATENEVER

Oxnard College
Oxnard, Ca.

JOHN WILEY & SONS

New York Chichester Brisbane Toronto Singapore

Library of Congress Cataloging in Publication Data:

Chatenever, Robert.
  Air conditioning and refrigeration for the
professional

  Bibliography: p.
  Includes indexes.
    1. Air conditioning—Equipment and supplies—
Maintenance and repair. 2. Refrigeration and refrig-
erating machinery—Maintenance and repair. I. Title.
TH7687.5.C48   1988   621.5′6   87-21650
ISBN 0-471-83045-3

Printed in the United States of America

10 9 8 7 6 5 4 3 2 1

PREFACE

This textbook covers HVAC and refrigeration system concepts and equipment, with the emphasis on practical instruction in procedures and techniques for the installation, maintenance, and repair of these systems. Written for the student training to become a field service technician, it is appropriate for both technical school and community-college programs, and as a reference for the apprentice in the field.

My primary aim has been to provide instruction that will have a direct bearing on the work a student can expect to be doing after graduation. For this reason, I have stressed widely applied, modern equipment and systems. Specific, detailed chapters on psychrometrics, microprocessor control systems, and electronic ignition heating systems have been included. Practical hands-on hints, common pitfalls, as well as data on specific pressures and temperatures, pneumatic controls, electrical troubleshooting approaches, and practical uses for psychrometrics also make this text unique. Nearly 1000 illustrations and photographs are included.

The four-part, thirty-three chapter division of this book corresponds to the typical course sequence of a classroom-laboratory approach. Parts One and Two introduce the student to specifics of refrigeration and heating. Parts Three and Four cover particulars of airside sytems and electricity for HVAC applications. Learning objectives are at the beginning of each part; each chapter is followed by a Key Terms and Concepts list and Study Questions.

Part One, REFRIGERATION, describes the mechanical equipment that is used to produce cooling for a wide variety of field applications. An operating description of the refrigeration cycle and each of the four major components common to every mechanical refrigeration system is included. Estimated operating pressures and temperatures are included wherever possible. The chapters in Part One also cover soldering and brazing, service procedures, applications, and troubleshooting.

Part Two, HEATING, covers heating systems, with separate chapters on gas fired, oil fired, electric, and solar heating systems. Boilers and piping are also covered. Extensive treatment is given to all operating controls for these systems.

Part Three, AIRSIDE SYSTEMS, builds on the two previous sections, describing how heating and refrigeration equipment are used to condition air for personal comfort. Included are practical applications of psychrometrics, load calculation methods, and diagrams of air distribution systems used in building air conditioning, and the control and zoning of these systems. Particular emphasis is placed on special air measuring tools and techniques used in air balancing.

Part Four, ELECTRICITY, covers basic electricity and the specific components that are most common in the refrigeration, heating, and air conditioning industry. The chapter on troubleshooting details the basic procedures for using voltmeters and ohmmeters to identify failed switches and defective loads. The circuits described are divided into cooling applications and heating applications, and the unique aspects of each wiring diagram are explained fully.

I hope that students, as they go through their apprenticeship, will carry this text into the field with them. I have sought to emphasize the practical, and have brought to the writing of the text all my experience as a service technician, a designer, a professional engineer, a facilities manager, and a community college vocational education instructor.

v

Acknowledgments

The author and publisher wish to thank the following companies for their assistance in obtaining technical information and illustrations:

A.W. Sperry Instruments
Aeroquip Corp.
Airserco Manufacturing Co.
Alco Controls, Division of
  Emerson Electric Co.
Anamet, Inc.
ASHRAE
Aurora Pump
Bacharach Instruments
Barry Blower
Biddle Instruments
Bussman Division Cooper
  Industries
Carrier Air Conditioning
Champion Cooler Corp.
Check-It Electronics Corp.
Cleaver-Brooks, Division of
  Aqua Chem, Inc.

Coleman Co., Inc.
Copeland Corp.
Dayton Electric Mfg. Co.
Dwyer Instruments
E.I. du Pont DeNemours & Co.
Ebco Manufacturing Co.
Environmental Products Co.
Furnas Electric Co.
Hart & Cooley
Herrmidifier Co., Inc.
Highside Chemicals
Honeywell
ITT Fluid Handling Division
J.W. Harris Co., Inc.
Johnson Controls, Inc.
Joslyn Clark Controls, Inc.
Joy Mfg. Co.

March Manufacturing, Inc.
Mc-Graw Hill
Myron L Co.
Paragon Electric Co., Inc.
Parker Hannifin Corp.
R.W. Beckett Corp.
Research Products Corp.
Robinair
Russell Coil Co.
Selkirk Metalbestos
Servel Gas Air Conditioning Co.
Sporlan Valve Co.
The Trane Co.
Tutco, Inc.
Uniweld Products, Inc.
Watts Regulator Co.
Weil-McLain Co.

A NOTE ON SAFETY

In this text you will learn practical HVAC service techniques, ones actually used by service technicians in the field. Where they are necessary, safety warnings have been provided, and these safety notes should be read and taken seriously. When you perform for the first time each of the service procedures that carries a safety warning, you should be under the supervision of an instructor or another individual with HVAC service experience.

In general, you should be careful to follow the following safety warnings:

1. Do not attempt any service procedure that will cause any pressure-containing component to exceed its rated pressure.

2. Do not attempt any procedure that would cause either a short-circuit, or a condition that will present a shock hazard to the technician.

3. Do not attempt any procedure that can result in the uncontrolled release to the atmosphere of liquid refrigerant, steam, or hot water.

4. Do not attempt to solder or braze where this could cause a fire.

CONTENTS

PART TWO

HEATING

AIR CONDITIONING AND REFRIGERATION FOR THE PROFESSIONAL

PART ONE

REFRIGERATION

Part One describes the mechanical equipment that is used to produce cooling for a variety of applications in the fields of refrigeration and air conditioning. In addition to an operating description of the refrigeration cycle, a separate chapter is devoted to each of the four major components that are common to every mechanical refrigeration system. The devices that are used to control and regulate the system are grouped together in the chapter on accessories. One chapter is provided to describe the many applications, including heat pumps. Estimated operating pressures and temperatures are given wherever possible. Mechanical field procedures that are common to all phases of installation and service of cooling equipment are covered in the chapters on soldering and brazing, service procedures, and troubleshooting.

OBJECTIVES

After completing Part One, the student should be able to do the following:

1. Describe the function of all the major refrigeration components and system accessories in an operating system.
2. Identify mechanical malfunctions using manifold gages and physical observations.
3. Perform tube working and brazing operations.
4. Use specialized refrigeration tools.
5. Set regulating valves and pressure switches.
6. Evacuate and recharge systems.

CHAPTER 1

DEFINITIONS

In order for us to be able to say what we mean, and have you understand what we say in the remainder of this text, we need to define some words. Some will be common words, such as temperature and heat; others will be new to you. In either event, it is important that you clearly understand the language of the heating, air conditioning, and refrigeration industry.

Temperature and Heat

You know that **temperature** and **heat** are somehow related, but what is the difference between them? To illustrate that there is a significant difference (Figure 1-1), imagine that the room in which you are sitting is totally sealed and well-insulated. You light a match, and then watch a thermometer located some distance away. Even though the match flame may be at a temperature of over 2000°F, it does not do any measurable amount of heating to the air in the room.

Next, let's bring in a block of steel that weighs 5000 lb, and is at a temperature of 400°F. Within a short period of time, you will not need the thermometer to tell you that the air in the room is being heated by the block of steel. Even though the match produced a higher temperature, the block of steel contained much more heat.

Heat is a form of energy. Some other forms of energy are electrical, chemical, kinetic (such as the energy available from fast-moving water), light, sound, and others. These various forms of energy may be converted from one form into another (Figure 1-2), as illustrated in the following examples:

1. The utility company burns coal (chemical energy) in a boiler to produce steam (heat energy).
2. Steam is used to create a high velocity (kinetic energy) through a steam turbine to create mechanical rotation.
3. The mechanical energy of the turbine is used to drive a generator and produce electrical energy.
4. The electrical energy is used by you to energize a light bulb that produces light and heat.

Temperature can be described as the intensity level of the heat. Hot, cold, and warm are relative terms. For example, a 65 degree room would be considered by most people to be cold, but a 65 degree refrigerator would be quite warm. In many cases, as heat is added to a substance, its temperature increases. As heat is removed, the temperature of the substance may decrease.

Temperature is measured in degrees. The most commonly used scale is the **Fahrenheit** scale. A Fahrenheit thermometer may be created as illustrated in Figure 1-3. When it is inserted into boiling water, the fluid inside rises. A mark is made on the thermometer and labeled 212°F. When it is inserted into water that was formed by melting ice, another mark is made, and labeled 32°F. One hundred and eighty spaces are then made between 32°F and 212°F, and each division represents one degree Fahrenheit.

The **Celsius** scale is not yet commonly used by refrigeration technicians or designers. The Celsius scale is created in a similar fashion to the Fahrenheit scale, except the level of the fluid inside is marked as 100°C in boiling water, and 0°C in melting ice.

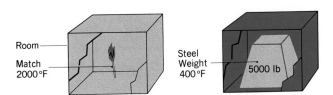

Figure 1-1 The match is at a higher temperature than the block of steel. The block of steel, however, contains more heat, and will do more heating of the room air.

It can be easily seen that a change in temperature of 100°C represents the same as a temperature change of 180°F. Or, a change of 9°F is equivalent to 5°C. In order to find the Fahrenheit equivalent of a known Celsius temperature, multiply the Celsius temperature by $\frac{9}{5}$, which gives you the number of Fahrenheit degrees above freezing. Then add 32°F, which is the Fahrenheit freezing temperature, and you have the Fahrenheit equivalent temperature. The general formulas to convert between Fahrenheit and Celsius are:

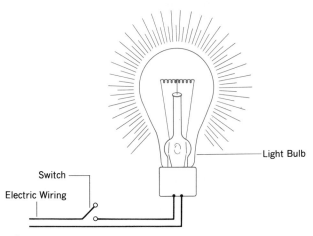

Figure 1-2 Energy may be converted from one form to another. A light bulb converts electrical energy to light and heat energy.

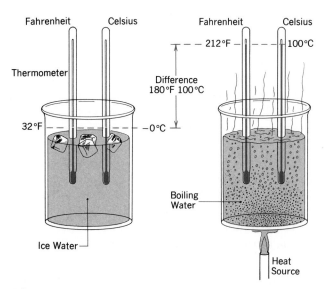

Figure 1-3 The Fahrenheit and Celsius thermometers are identical in every detail, except for the numbers imprinted on the scales.

$$°F = \frac{9}{5} °C + 32$$

and

$$°C = \frac{5}{9} \times (°F - 32)$$

Figure 1-4 gives tabular conversion information for Fahrenheit and Celsius.

There is a limit to the coldest temperature that can be reached. Scientists explain that as the temperature of a substance is reduced, the amount of movement of the atoms within the substance actually slows down. At the point where the movement of the atoms has stopped, no further cooling can take place. The temperature at which this happens is called **absolute zero.** On the Fahrenheit scale, it occurs at −460°F, and on the Celsius scale, it occurs at −273°C. Scientists have attained temperatures within a few tenths of a degree of absolute zero.

While the units of measurement for temperature are probably familiar to you, the units of measurement for heat are not as common. The conventional unit of heat used in the heating and air conditioning and refrigeration industry is the **Btu (British thermal unit).** This unit is defined as the amount of heat required to raise the temperature of one pound of water by one degree Fahrenheit (Figure 1-5). There are other units of heat, such as the calorie and the joule, but they are rarely used by the designers and technicians in refrigeration and air conditioning.

The Btu is a relatively small unit of heat. The heat liberated from the burning of one match is approximately 1 Btu. A common domestic water heater or outdoor gas barbecue grille can supply 30,000 Btu/hr, and a medium-size window air conditioner can remove 6000 Btu/hr of heat from the air.

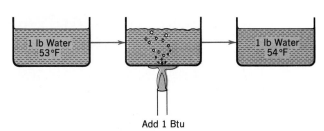

Figure 1-5 An amount of heat required to change the temperature of one pound of water one degree Fahrenheit is defined as a Btu.

Celsius	C or F	Fahr	Celsius	C or F	Fahr	Celsius	C or F	Fahr	Celsius	C or F	Fahr
−40.0	**−40**	−40.0	−6.7	**+20**	+68.0	+26.7	**+80**	+176.0	+60.0	**+140**	+284.0
−39.4	**−39**	−38.2	−6.1	**+21**	+69.8	+27.2	**+81**	+177.8	+60.6	**+141**	+285.8
−38.9	**−38**	−36.4	−5.5	**+22**	+71.6	+27.8	**+82**	+179.6	+61.1	**+142**	+287.6
−38.3	**−37**	−34.6	−5.0	**+23**	+73.4	+28.3	**+83**	+181.4	+61.7	**+143**	+289.4
−37.8	**−36**	−32.8	−4.4	**+24**	+75.2	+28.9	**+84**	+183.2	+62.2	**+144**	+291.2
−37.2	**−35**	−31.0	−3.9	**+25**	+77.0	+29.4	**+85**	+185.0	+62.8	**+145**	+293.0
−36.7	**−34**	−29.2	−3.3	**+26**	+78.8	+30.0	**+86**	+186.8	+63.3	**+146**	+294.8
−36.1	**−33**	−27.4	−2.8	**+27**	+80.6	+30.6	**+87**	+188.6	+63.9	**+147**	+296.6
−35.6	**−32**	−25.6	−2.2	**+28**	+82.4	+31.1	**+88**	+190.4	+64.4	**+148**	+298.4
−35.0	**−31**	−23.8	−1.7	**+29**	+84.2	+31.7	**+89**	+192.2	+65.0	**+149**	+300.2
−34.4	**−30**	−22.0	−1.1	**+30**	+86.0	+32.2	**+90**	+194.0	+65.6	**+150**	+302.0
−33.9	**−29**	−20.2	−0.6	**+31**	+87.8	+32.8	**+91**	+195.8	+66.1	**+151**	+303.8
−33.3	**−28**	−18.4	0	**+32**	+89.6	+33.3	**+92**	+197.6	+66.7	**+152**	+305.6
−32.8	**−27**	−16.6	+0.6	**+33**	+91.4	+33.9	**+93**	+199.4	+67.2	**+153**	+307.4
−32.2	**−26**	−14.8	+1.1	**+34**	+93.2	+34.4	**+94**	+201.2	+67.8	**+154**	+309.2
−31.7	**−25**	−13.0	+1.7	**+35**	+95.0	+35.0	**+95**	+203.0	+68.3	**+155**	+311.0
−31.1	**−24**	−11.2	+2.2	**+36**	+96.8	+35.6	**+96**	+204.8	+68.9	**+156**	+312.8
−30.6	**−23**	−9.4	+2.8	**+37**	+98.6	+36.1	**+97**	+206.6	+69.4	**+157**	+314.6
−30.0	**−22**	−7.6	+3.3	**+38**	+100.4	+36.7	**+98**	+208.4	+70.0	**+158**	+316.4
−29.4	**−21**	−5.8	+3.9	**+39**	+102.2	+37.2	**+99**	+210.2	+70.6	**+159**	+318.2
−28.9	**−20**	−4.0	+4.4	**+40**	+104.0	+37.8	**+100**	+212.0	+71.1	**+160**	+320.0
−28.3	**−19**	−2.2	+5.0	**+41**	+105.8	+38.3	**+101**	+213.8	+71.7	**+161**	+321.8
−27.8	**−18**	−0.4	+5.5	**+42**	+107.6	+38.9	**+102**	+315.6	+72.2	**+162**	+323.6
−27.2	**−17**	+1.4	+6.1	**+43**	+109.4	+39.4	**+103**	+217.4	+72.8	**+163**	+325.4
−26.7	**−16**	+3.2	+6.7	**+44**	+111.2	+40.0	**+104**	+219.2	+73.3	**+164**	+327.2
−26.1	**−15**	+5.0	+7.2	**+45**	+113.0	+40.6	**+105**	+221.0	+73.9	**+165**	+329.0
−25.6	**−14**	+6.8	+7.8	**+46**	+114.8	+41.1	**+106**	+222.8	+74.4	**+166**	+330.8
−25.0	**−13**	+8.6	+8.3	**+47**	+116.6	+41.7	**+107**	+224.6	+75.0	**+167**	+332.6
−24.4	**−12**	+10.4	+8.9	**+48**	+118.4	+42.2	**+108**	+226.4	+75.6	**+168**	+334.4
−23.9	**−11**	+12.2	+9.4	**+49**	+120.2	+42.8	**+109**	+228.2	+76.1	**+169**	+336.2
−23.3	**−10**	+14.0	+10.0	**+50**	+122.0	+43.3	**+110**	+230.0	+76.7	**+170**	+338.0
−22.8	**−9**	+15.8	+10.6	**+51**	+123.8	+43.9	**+111**	+231.8	+77.2	**+171**	+339.8
−22.2	**−8**	+17.6	+11.1	**+52**	+125.6	+44.4	**+112**	+233.6	+77.8	**+172**	+341.6
−21.7	**−7**	+19.4	+11.7	**+53**	+127.4	+45.0	**+113**	+235.4	+78.3	**+173**	+343.4
−21.1	**−6**	+21.2	+12.2	**+54**	+129.2	+45.6	**+114**	+237.2	+78.9	**+174**	+345.2
−20.6	**−5**	+23.0	+12.8	**+55**	+131.0	+46.1	**+115**	+239.0	+79.4	**+175**	+347.0
−20.0	**−4**	+24.8	+13.3	**+56**	+132.8	+46.7	**+116**	+240.8	+80.0	**+176**	+348.8
−19.4	**−3**	+26.6	+13.9	**+57**	+134.6	+47.2	**+117**	+242.6	+80.6	**+177**	+350.6
−18.9	**−2**	+28.4	+14.4	**+58**	+136.4	+47.8	**+118**	+244.4	+81.1	**+178**	+352.4
−18.3	**−1**	+30.2	+15.0	**+59**	+138.2	+48.3	**+119**	+246.2	+81.7	**+179**	+354.2
−17.8	**0**	+32.0	+15.6	**+60**	+140.0	+48.9	**+120**	+248.0	+82.2	**+180**	+356.0
−17.2	**+1**	+33.8	+16.1	**+61**	+141.8	+49.4	**+121**	+249.8	+82.8	**+181**	+357.8
−16.7	**+2**	+35.6	+16.7	**+62**	+143.6	+50.0	**+122**	+251.6	+83.3	**+182**	+359.6
−16.1	**+3**	+37.4	+17.2	**+63**	+145.4	+50.6	**+123**	+253.4	+83.9	**+183**	+361.4
−15.6	**+4**	+39.2	+17.8	**+64**	+147.2	+51.1	**+124**	+255.2	+84.4	**+184**	+363.2
−15.0	**+5**	+41.0	+18.3	**+65**	+149.0	+51.7	**+125**	+257.0	+85.0	**+185**	+365.0
−14.4	**+6**	+42.8	+18.9	**+66**	+150.8	+52.2	**+126**	+258.8	+85.6	**+186**	+366.8
−13.9	**+7**	+44.6	+19.4	**+67**	+152.6	+52.8	**+127**	+260.6	+86.1	**+187**	+368.6
−13.3	**+8**	+46.4	+20.0	**+68**	+154.4	+53.3	**+128**	+262.4	+86.7	**+188**	+370.4
−12.8	**+9**	+48.2	+20.6	**+69**	+156.2	+53.9	**+129**	+264.2	+87.2	**+189**	+372.2
−12.2	**+10**	+50.0	+21.1	**+70**	+158.0	+54.4	**+130**	+266.0	+87.8	**+190**	+374.0
−11.7	**+11**	+51.8	+21.7	**+71**	+159.8	+55.0	**+131**	+267.8	+88.3	**+191**	+375.8
−11.1	**+12**	+53.6	+22.2	**+72**	+161.6	+55.6	**+132**	+269.6	+88.9	**+192**	+377.6
−10.6	**+13**	+55.4	+22.8	**+73**	+163.4	+56.1	**+133**	+271.4	+89.4	**+193**	+379.4
−10.0	**+14**	+57.2	+23.3	**+74**	+165.2	+56.7	**+134**	+273.2	+90.0	**+194**	+381.2
−9.4	**+15**	+59.0	+23.9	**+75**	+167.0	+57.2	**+135**	+275.0	+90.6	**+195**	+383.0
−8.9	**+16**	+60.8	+24.4	**+76**	+168.8	+57.8	**+136**	+276.8	+91.1	**+196**	+384.8
−8.3	**+17**	+62.6	+25.0	**+77**	+170.6	+58.3	**+137**	+278.6	+91.7	**+197**	+386.6
−7.8	**+18**	+64.4	+25.6	**+78**	+172.4	+58.9	**+138**	+280.4	+92.2	**+198**	+388.4
−7.2	**+19**	+66.2	+26.1	**+79**	+174.2	+59.4	**+139**	+282.2	+92.8	**+199**	+390.2

The numbers in boldface in the center column refer to the temperature, either in Celsius or Fahrenheit which is to be converted to the other scale. If converting Fahrenheit to Celsius the equivalent temperature will be found in the left column. If converting Celsius to Fahrenheit, the equivalent temperature will be found in the column on the right.

Figure 1-4 Fahrenheit–Celsius temperature conversion chart. (Courtesy Russell Coil Company)

Sensible and Latent Heat

Consider the pail of water at 50°F shown in Figure 1-6. As heat is applied to the 1 lb of water, its temperature begins to rise. As a matter of fact, for every Btu of heat added, the temperature of the water rises 1°F. This may be clearly seen by watching the temperature of the water rise. But when the water temperature rises to 212°F, something different happens. The water begins to boil, and the temperature of the water stops rising. If we add heat at a faster rate, it only serves to make the water boil faster. Once the water has reached its boiling temperature, the thermometer can no longer sense that we are continuing to add heat.

When the heat that is being added results in a temperature rise that can be sensed with a thermometer, we are said to be adding **sensible heat.** When the heat input cannot be sensed with a thermometer, we are said to be adding **latent heat.** The word latent means hidden. In this context, the fact that we are adding heat to the boiling water is hidden from the thermometer.

The same effect can be seen when removing heat from water. The temperature continues to fall until it reaches 32°F. Then, as heat continues to be removed, the water begins turning into ice. The temperature remains at 32°F until all the water has turned to ice.

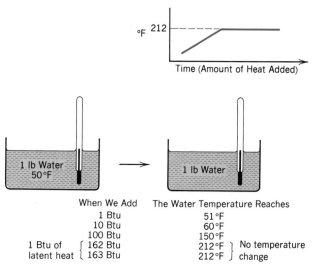

When We Add	The Water Temperature Reaches	
1 Btu	51°F	
10 Btu	60°F	
100 Btu	150°F	
1 Btu of ⎰ 162 Btu	212°F ⎱	No temperature
latent heat ⎱ 163 Btu	212°F ⎰	change

Figure 1-6   Adding heat causes the water temperature to rise (sensible heat) until the water reaches its boiling point.

Change of State			Latent Heat of
Liquid	$\xrightarrow{\text{(boiling)}}$	Vapor	Vaporization
Vapor	$\xrightarrow{\text{(condensing)}}$	Liquid	Condensation
Liquid	$\xrightarrow{\text{(freezing)}}$	Solid	Fusion
Solid	$\xrightarrow{\text{(melting)}}$	Liquid	Melting

Figure 1-7   For any substance, the latent heat of vaporization equals the latent heat of condensation, and the latent heat of fusion equals the latent heat of melting.

Change of State

When we add latent heat to the boiling water, the effect of the heat is to cause a **change of state.** That is, the heat causes the water to change from a liquid state to a vapor state. There are four different changes of state that can occur (Figure 1-7).

1. The amount of heat that must be added to 1 lb of liquid at its boiling point to change it into 1 lb of vapor is called the **latent heat of vaporization.** For water, the latent heat of vaporization at normal pressure is 970 Btu/lb.

2. The heat that must be removed from a vapor to cause it to condense into a liquid is called the **latent heat of condensation.** Its value is the same as the latent heat of vaporization. If 970 Btu are removed from one pound of water vapor at atmospheric pressure, we will once again have 1 lb of liquid.

3. The heat that must be removed from 1 lb of liquid at its freezing temperature in order to turn it into a solid is called the **latent heat of fusion.** The latent heat of fusion for water is 144 Btu/lb.

4. The heat required to turn 1 lb of a solid into 1 lb of liquid is called the **latent heat of melting.** When 1 lb of ice at 32°F melts, it will absorb 144 Btu.

Refrigeration Capacity

Refrigeration capacity of a system refers to the rate at which the system can remove Btus from air or from any other material. If we had 1000 lb of water that we wanted to cool from 54°F to 42°F (Figure 1-8) we would have to remove a quantity of heat equal to

$$Q = w \times (T_2 - T_1)$$
$$Q = 1000 \times (54 - 42)$$
$$Q = 12{,}000 \text{ Btu}$$

This formula is usually written as

$$Q = w \times \Delta T$$

where the term ΔT (delta T) is used to represent the change in temperature.

If we were willing to wait for one hour for a refrigeration system to accomplish this task, we would need a system with a capacity to remove 12,000 Btu/hr. If we were willing to wait 2 hr for this change to be accomplished, however, the refrigeration system would only need to have a capacity of 6000 Btu/hr.

Although Btu per hour is a very exact way to state the capacity of a refrigeration or air conditioning system, a **ton** is another commonly used term. In the early 1900s, people used ice blocks in an ice box to store their food (Figure 1-9). The ice was manufactured in an ice plant and delivered daily. When people referred to the capacity of the plant that manufactured the ice, they referred to how many tons of ice that plant could produce each day. The terminology describing the ice production rate came to be used as a description of the capacity of the refrigeration system that produced the ice. Let's see how much heat must be removed from 1 ton of water in order to form 1 ton of ice in a day.

Each pound of water has a latent heat of fusion of 144 Btu. Therefore, in order to remove the latent heat of fusion from each of 2000 lb of water, we must remove (2000 × 144) or 288,000 Btu each day, or 12,000 Btu each hour (Figure 1-10). Today, we refer to **one ton of refrigerating effect** as the equivalent cooling that would be done by one ton of ice melting in a day, or 12,000 Btu/hr.

Example

How many tons of refrigeration would be required to cool a stream of water of 500 lb/min from 56°F to 44°F?

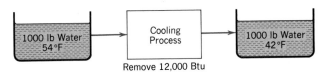

Figure 1-8 A cooling process of any size could remove 12,000 Btu, but to do the job in 1 hr, a 1-ton cooling process is required.

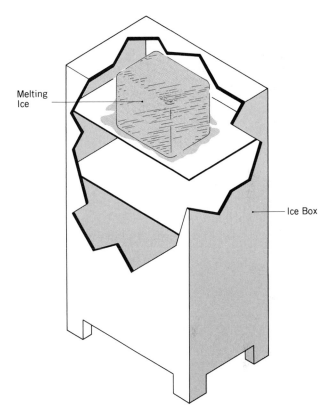

Figure 1-9 In early ice boxes the block of ice would absorb its latent heat of melting from the surrounding air. The ice melted and the air became cool.

Pounds of Water Being Cooled (W)	Number of Degrees of Cooling (ΔT)	Btus to Be Removed (Q)
1	1	1
1	2	2
2	1	2
2	2	4
1000	1	1000
1000	12	12,000

Figure 1-10 The change in heat content (Q) for water equals the weight of water (lb) times its change in temperature (°F).

Solution

$$\frac{500 \text{ lb}}{\text{min}} \times \frac{60 \text{ min}}{\text{hr}} \times \frac{1 \text{ Btu}}{\text{lb-°F}}$$

$$\times (56 - 44)°F \times \frac{\text{ton}}{12,000 \text{ Btu/hr}} = 30 \text{ tons}$$

Specific Heat

When sensible heat is added to a substance, its temperature rises. Not all material, however, will experience the same amount of temperature rise when the same amount of heat is added. One Btu of heat added to a pound of water will produce a temperature rise of 1°F; 1 Btu of heat added to a pound of iron will cause a temperature rise of 8.5°F; and 1 Btu added to a pound of aluminum will cause a temperature rise of 4.6°F. The **specific heat** of a substance is defined as the amount of heat (Btus) that would be required to raise the temperature of 1 lb of that substance by 1°F. Therefore, the iron has a specific heat of 1/8.5, or 0.118 Btu/lb-°F, and the aluminum has a specific heat of 0.215. When water changes its state from water to either a vapor or a solid, it becomes a new substance with a new specific heat. Ice has a specific heat of 0.5, water vapor (steam) has a specific heat of 0.4, and air has a specific heat of 0.24 (Figure 1-11).

Refrigeration and heating equipment is usually used to cool or heat a continuously moving stream of air or water. With the specific heat known, and if the density of the material is known, some handy formulas can be derived. For water, with a specific heat of 1.0 Btu/lb and a density of 8.3 lb/gal, we can figure the Btus that must be added or removed from a known gpm (gallons per minute) as follows

Substance	Temperature Rise Caused by Adding 1 Btu	Specific Heat	
Water	1.0°F	1.0	Btu/lb-°F
Ice	2.0°F	0.5	Btu/lb-°F
Steam	2.5°F	0.4	Btu/lb-°F
Iron	8.5°F	0.118	Btu/lb-°F
Aluminum	4.6°F	0.215	Btu/lb-°F
Air	4.17°F	0.24	Btu/lb-°F

Figure 1-11 Specific heats of common substances.

$$Q = \frac{\text{gal}}{\text{min}} \times \frac{60 \text{ min}}{\text{hr}} \times \frac{1 \text{ Btu}}{\text{lb-°F}} \times \frac{8.3 \text{ lb}}{\text{gal}} \times \Delta T$$

or

$$Q = 500 \times \text{gpm} \times \Delta T$$

Example

It is desired to cool 15 gpm of water from 56°F to 44°F. How much refrigeration capacity will be required?

Solution

$$Q = 500 \times \text{gpm} \times \Delta T$$
$$Q = 500 \times 15 \times (56 - 44)$$
$$Q = 90,000 \text{ Btu/hr}$$

If the previous formula is divided by 12,000, which is the number of Btu/hr in 1 ton, the required capacity in tons can be figured as

$$Q \text{ (tons)} = \frac{\text{gpm} \times \Delta T}{24}$$

For air that has a specific heat of 0.24 and a density at normal conditions of 0.075 lb/cu ft, a similar formula can be derived. For air, flow rates are normally known in cfm (cubic feet per minute). When heating or cooling air (with no addition or removal of moisture)

$$Q = \frac{\text{ft}^3}{\text{min}} \times \frac{60 \text{ min}}{\text{hr}} \times \frac{0.075 \text{ lb}}{\text{ft}^3} \times \frac{0.24 \text{ Btu}}{\text{lb-°F}}$$

or

$$Q = 1.08 \times \text{cfm} \times \Delta T$$

Example

How much heat is being absorbed by air that enters a room at 58°F and leaves at 76°F? The airflow rate is 1200 cfm.

Solution

$$Q = 1.08 \times \text{cfm} \times \Delta T$$
$$Q = 1.08 \times 1200 \times (76 - 58)$$
$$Q = 23,328 \text{ Btu/hr}$$

The formulas just given for water and air are the most commonly used. For any other liquid or gas, however, other formulas can be derived if you know the density and the specific heat of the material being heated or cooled.

Note that the formula given for air only applies if there is no change in moisture content for the air. Where humidification or dehumidification of the air is involved, other formulas are presented in the chapter on psychrometrics.

Heat Transfer

If there are two materials at different temperatures, there exists the potential to transfer heat from the warmer to the cooler material. There are three separate mechanisms through which this heat transfer can potentially take place: **Conduction, Convection, Radiation.**

Of these three mechanisms, conduction will be of the most interest to us. Conduction may be best illustrated as in Figure 1-12. If one end of a metal bar is heated, the other end will very soon experience a temperature rise. The heat travels easily through the metal bar because it is a good conductor. Generally, materials that are good conductors of electricity are also good conductors of heat. Materials such as fiberglass and Styrofoam that do not readily conduct heat are called insulators.

If there is a wall separating two spaces that are being held at two different temperatures, such as the refrigerator wall shown in Figure 1-13, heat

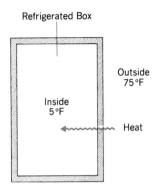

Figure 1-13 Heat travels through a wall by means of conduction. The heat flows from the warm side to the cooler side.

will flow through the wall from the warm air on the outside to the cooler space on the inside. The rate at which the heat flow will occur will depend upon three factors:

1. The temperature difference across the wall.
2. The number of square feet of area over which the heat transfer is being allowed to take place.
3. The material of construction of the wall.

The relationship of these factors is as follows:
$$Q = U \times A \times \Delta T$$
where

Q = heat flow, Btu/hr.

U = heat transfer coefficient, Btu/hr-sq ft.

A = the area through which the heat transfer is taking place, sq ft.

ΔT = temperature difference.

All materials have a U factor (heat transmission coefficient). A very high U factor would be slightly over 1.0 (single-pane glass is 1.13). A very low U factor would be 0.05 (a well-insulated ceiling).

Example
How much heat will be lost through the ceiling of a house if the ceiling area is 1200 sq ft, the U factor is 0.08, the room temperature is 74°F, and the attic temperature is 40°F?

Solution
$$Q = U \times A \times \Delta T$$
$$Q = 0.08 \times 1200 \times (74 - 40)$$
$$Q = 3264 \text{ Btu/hr}$$

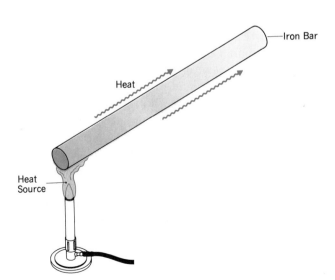

Figure 1-12 When one end of an iron bar is heated the heat travels quickly to the other end by means of conduction.

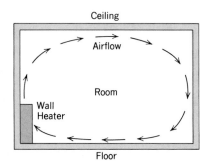

Figure 1-14 The heat from the heater is carried to the room through convection.

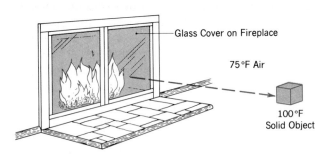

Figure 1-15 An object that can "see" the flame from the fireplaces warms to 100°F by radiant heat transfer. The air between the fireplace and the object doe not absorb the radiant energy.

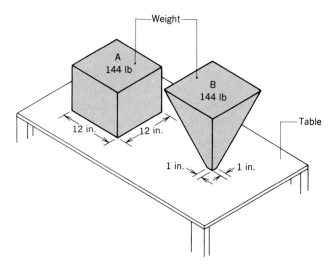

Figure 1-16 Each block exerts a force of 144 1b on the table. A exerts a pressure of 1 psi, while B exerts a pressure of 144 psi.

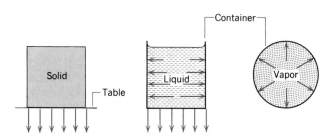

Figure 1-17 Pressures exerted for solids, liquids, and vapor.

Convection, shown in Figure 1-14, is the second method of heat transfer. Energy from a hot radiator warms the surrounding air, which then moves through the room to spread the heat. Convection may be natural where the air moves due to differences in density, or it may be forced by a fan.

Radiation is a means of heat transfer that occurs from a warm body to a cooler body without heating any medium in between. When you stand in front of a fireplace, the portion of your body facing the flame feels warm, evey though the air temperature between the fire and your body is not warm (Figure 1-15). The earth is warmed by radiation from the sun, which is transferred through the vacuum of outer space.

Pressure

Pressure is defined as a force divided over the area over which the force is applied. The difference between force and pressure can best be illustrated by example. If a 100-lb person is wearing shoes with an area of 50 sq in. of sole area, a pressure of 2 pounds per square inch (psi) is being exerted on the ground. If, however, the sole area of the shoes that is carrying the weight is only 1 sq in. (as might be the case with high-heeled shoes), then the pressure on the ground would be 100 psi. Even though the force is the same in both cases (100 lb), the pressure is much higher when the force is applied over a smaller area (Figure 1-16). Figure 1-17 illustrates the direction in which pressures may be exerted. A solid exerts pressure only in a downward direction; a liquid exerts pressure that is dependent upon the depth of the liquid on the sides of its container; a vapor exerts the same pressure in all directions.

There is a great deal of air pressure being exerted in all directions, all around you. This can be simply illustrated by sucking water out of a glass

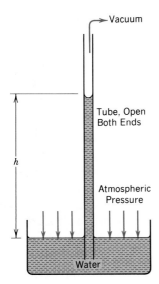

Figure 1-18 As the vacuum becomes deeper, the water rises further, until the column reaches 33.4 ft at a perfect vacuum.

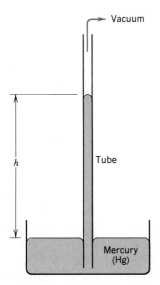

Figure 1-19 When a perfect vacuum is pulled on the mercury, its maximum height will be 29.92 in. Hg.

as shown in Figure 1-18. What makes the water rise up the straw? When you suck the air out of the straw, the pressure inside the straw is reduced. The remaining pressure on the surface of the liquid outside the straw pushes the water up the straw. If every molecule of air were to be sucked out of the straw, we would find that the water would rise 33.4 ft. That is the limit of the height of water that can be supported by atmospheric pressure. If we were to calculate the

weight of water contained in that 33.4 ft column, and divide it by the cross-sectional area of the column, we would find that the pressure being exerted at the bottom of the column would be 14.7 psi. It is left as an interesting student exercise to prove that this is true, regardless of the diameter of the column of water being supported.

There is one further measure we can use to describe pressure. If we were to replace the water with mercury (Figure 1-19), the same 14.7 psi of atmospheric pressure would only be able to lift the column of mercury 29.92 in. (commonly rounded to 30 in.). This can be explained by the difference in density between water and mercury (the abbreviation for mercury is Hg). Because mercury has a density of 13.5 times the density of water, a column of mercury 29.92 in. high will exert the same pressure as a 33.4 ft column of water. As atmospheric pressure changes, its ability to lift a column of mercury will also change. When the weather forecaster reports that the barometer is 29.92 in. Hg, the atmospheric pressure is being described in terms of mercury that can be supported by the atmospheric pressure of the moment.

When measuring refrigerant pressures, the refrigeration technician uses the units of **psi** for pressures above atmospheric, and **in. Hg** for describing pressures below atmospheric **(vacuum).** One in. Hg would be a very slight vacuum, while 30 in. Hg would be a perfect vacuum. When measuring air pressures or gas pressures, the most commonly used units are in. w.c. (water column) for pressure or vacuum.

Absolute and Gauge Pressure

We have just finished describing that atmospheric pressure is 14.7 psi, or 33.4 ft of water column. Why, then, if we take a perfectly good pressure gauge out of a box, will it register zero pressure? The answer is that it is convenient in most applications to measure pressure in terms of the number of psi above atmospheric pressure. For example, a tire pressure on an automobile might register 30 psi, but that actually means 30 psi above atmospheric. In order to know whether we are speaking of actual absolute measurement of pressure or simply what the pressure gauge is

	Gage Pressure	Absolute Pressure
A perfect vacuum	30 in. Hg	0 psia
Atmospheric pressure	0 psig	14.7 psia
Automobile tire	30 psig	44.7 psia

Figure 1-20 Relationship between gauge pressure and absolute pressure.

reading, we say **psia** or **psig.** This refers to psi absolute or psi gauge. Figure 1-20 shows the relationship between psig and psia for several common pressures. A perfect vacuum is zero psia, or 29.92 in. Hg (say, 30 in. Hg). Normal atmospheric pressure is 14.7 psia, or zero psig. An automobile tire would be inflated to 30 psig, or about 45 psia. A gauge that can measure pressure above atmospheric or a vacuum is called a **compound gauge.**

Density and Specific Gravity

Density is a physical property of all substances. Simply stated, it is the weight of the substance per unit of volume (Figure 1-21). Some commonly used densities are

1. Water has a density of 62.4 lb/cu ft, or 8.33 lb/gal.
2. Air has a density of 0.075 lb/cu ft.

Other materials have different densities. For liquids, we can calculate **specific gravity** as the density of that material divided by the density of

water, while for gasses, we can calculate specific gravity as the density of that gas divided by the density of air. By definition, the specific gravity of water and of air are both equal to 1.0.

Saturation Temperature

In a previous section, we discussed the fact that water boils at a temperature of 212°F. But this is not always true. For example, the water in a pressure cooker can become heated to almost 250°F if its pressure is allowed to build up to 15 psig. The water in an automobile radiator will normally be higher than 212°F (even without antifreeze) because the cooling system runs at a pressure of 7 psi. The dependence of boiling point on temperature also works for pressures below 14.7 psia. If you go into the high mountains, atmospheric pressure is lower than 14.7 psia, and the temperature at which water will boil is less than 212°F. The **saturation temperature** is another word for boiling point. It is the temperature at which a material will boil, which corresponds to a pressure. For example, the saturation temperature for water is 250°F at a pressure of 15 psig, and 212°F when the pressure is normal atmospheric 0 psig (Figure 1-22).

Other liquids have boiling points different from water, but each has a saturation temperature that changes with the pressure.

When a liquid has been heated to its saturation temperature, it is called a saturated liquid, and a vapor at the saturation temperature is

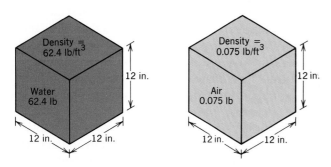

Figure 1-21 Density is the weight of a substance per unit of volume.

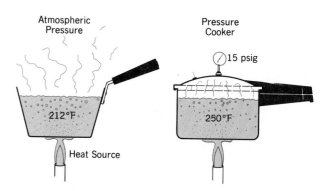

Figure 1-22 The boiling temperature of a liquid depends upon the pressure. Another name for boiling point is saturation temperature.

called a saturated vapor. A liquid may exist at any temperature below or up to the saturation temperature, and a vapor may exist at any temperature at or above the saturation temperature. The only temperature at which both can exist together is the saturation temperature.

Subcooling and Superheat

Whenever we have a liquid whose temperature is lower than the saturation temperature, we say that the liquid is **subcooled.** By this definition, if we had water in a pot at 200°F and at atmospheric pressure, it would have 12°F of subcooling. It is 12°F cooler than the temperature at which it would boil.

Whenever we have a vapor whose temperature is higher than the saturation temperature, we say that the vapor is **superheated.** If we allowed water to boil at 212°F, and then collected the vapor and heated it still further, its temperature would rise above 212°F. If it is heated to a tem-

perature of 220°F (still at atmospheric pressure), we would say that it has 8° of superheat.

The example in Figure 1-23 ties together the concepts of saturation temperature, superheat, and subcooling. Consider a closed-loop system that is being maintained at approximately atmospheric pressure throughout. The pump is supplying water at 190°F, and the water is at a subcooled condition. As we add heat, the water temperature rises to 212°F (sensible heat). As we continue to add heat, the water remains at 212°F (its saturation temperature) while it is boiling into a vapor (latent heat). When all the water has been boiled into a vapor, the continued addition of heat will cause the vapor temperature to rise again (sensible heat), resulting in a superheated vapor. If next we begin to remove heat by the use of cooling fans, the superheated vapor will first be cooled to its saturation temperature (sensible cooling). The continued removal of heat will result in the vapor condensing back into a liquid at the saturation temperature (removing the latent heat of vaporization). Then, after all vapor has been condensed, the continued removal of heat will result in subcooling of the liquid.

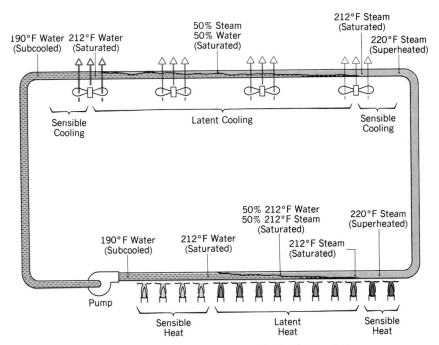

Figure 1-23 A fluid (water in this case) can be boiled and then condensed repeatedly as heat is added or removed.

KEY TERMS AND CONCEPTS

Temperature

Heat

Fahrenheit

Celsius

Absolute Zero

Btu

Sensible Heat

Latent Heat

Change of State

Latent Heat of
  Vaporization

Latent Heat of
  Condensation

Latent Heat of Fusion

Latent Heat of Melting

Ton

Specific Heat

Conduction

Convection

Radiation

In. Hg

Vacuum

Psia

Psig

Compound Gauge

Density

Specific Gravity

Saturation
  Temperature

Subcooled

Superheated

QUESTIONS

1. For each of the following changes at atmospheric pressure, state whether it involves sensible heating, sensible cooling, latent heating, or latent cooling:

   (a) Change water at 120°F to water at 130°F.

   (b) Change steam at 250°F to steam at 220°F.

   (c) Change water at 32°F to ice at 32°F.

   (d) Change water at 200°F to steam at 212°F.

2. How much heat must be added to 5 lb of water at 212°F to create 5 lb of steam at 212°F?

3. When 100 Btu of heat is added to 10 lb of a certain material, its temperature increases from 80°F to 130°F. What is the specific heat of this material?

4. What is the Fahrenheit equivalent of 5°C?

5. How much cooling will be done by 3 lb of ice that is brought into a 70°F room, and allowed to melt? (Assume the final water temperature is the same as the room temperature.)

6. How much ice must be allowed to melt if we want it to absorb 6000 Btu?

7. How much heat must be added to 5 lb of ice at −10°F in order to change it into 5 lb of steam at 250°F?

8. How many tons of refrigeration capacity would be required to cool a stream of water of 15 lb/min from 52°F to 42°F?

9. Six gpm of water is being cooled from 55°F to 43°F through a water chiller. What is the capacity of this water chiller in Btus per hour? In tons?

10. How much heat must be added to 1000 cfm of air in order to heat it from 70°F to 140°F?

11. What are the factors that will affect the rate at which heat will flow through a material due to conduction?

12. A wall with a U factor of 0.15 is 20 ft long and 8 ft high. If the indoor temperature is 72°F and the outdoor temperature is 30°F, how fast will heat flow out of the room through this wall?

13. On a gauge, what would it read if sensing atmospheric pressure?

14. On a vacuum gauge, what would it read if sensing a very deep vacuum?

15. What is a compound gauge?

16. What is the relationship between the pressure on a liquid and the temperature at which it will boil?

17. Can a liquid be superheated?

18. Liquid and vapor are together in a tube. Are they subcooled, saturated, or superheated?

19. What is the difference between heat and temperature?

20. One cubic foot of a material A is at a temperature of 50°F. One cu ft of another material B is at a temperature of 100°F. What other information would you need in order to know which material contained more heat?

CHAPTER 2

THE REFRIGERATION CYCLE

In this chapter we will deal with the mechanical vapor compression cycle. Vapor compression simply means that equipment that creates a mechanical force is used to compress a vapor. This is the most commonly used type of refrigeration and cooling system. There are other, less common ways of producing a cooling effect that will not be included in this chapter. These specialized type systems are discussed in Chapter 15.

Evaporation

The concept of **evaporation** is the key to producing a cooling effect. In order to demonstrate the cooling that can take place due to evaporation, consider the following example. Your body (which is at about 98°F or slightly less) is lying on a lounge chair next to a pool. It is 98°F outside with a warm breeze, so you feel quite warm. You jump into the pool, which also happens to be at 98°F, but you don't even feel cooled off because the pool is so warm. When you get out of the pool, however, you feel quite cool because the warm breeze is causing the 98°F water on your skin to evaporate. In order for this evaporation to take place, the water absorbs its latent heat of vaporization before it turns into a vapor. The evaporating water draws its latent heat of vaporization from your body, and it makes you feel cool.

There are many other examples of evaporation causing cooling, (Figure 2-1). When you are perspiring, standing in front of a fan will make you feel cooler. The fan does not actually make the air cooler. Rather, the increased air movement increases the rate of evaporation of the perspiration from your body, causing the perspiration to absorb its latent heat of vaporization from your body at a faster rate.

Flashing

In Chapter 1, it was explained that a liquid cannot exist at a temperature higher than its boiling point or saturation temperature. We can have 230°F water in an automobile radiator only if the pressure in the radiator system is higher than atmospheric. When at atmospheric pressure, the water cannot exist as a liquid at any temperature higher than 212°F. What happens if we suddenly remove the pressure cap from an automobile radiator that contains water as a liquid at 230°F? Those who have done it are not likely to forget the experience. The 230°F liquid immediately begins to boil rapidly. As the steam is formed, it absorbs its latent heat of vaporization from the rest of the water, which cools rapidly. This boiling con-

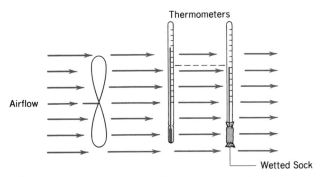

Figure 2-1 Evaporation of water causes the thermometer to be cooled. The evaporating water absorbs its latent heat of vaporization from the thermometer.

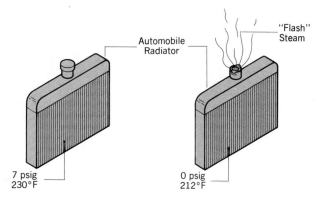

Figure 2-2 When the radiator cap is removed, steam is formed through *flashing*, and the remaining liquid immediately cools to its new saturation temperature.

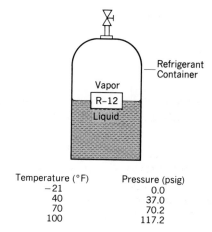

Temperature (°F)	Pressure (psig)
-21	0.0
40	37.0
70	70.2
100	117.2

Figure 2-3 Boiling (saturation) temperatures for R-12 at selected pressures.

tinues until sufficient vapor has formed to cool the remaining water to 212°F (Figure 2-2). Of course, when so much steam is formed in the radiator, it comes out of the fill cap with dramatic force, and can cause serious burns.

When the water was pressurized as a 230°F liquid, it was at a saturated condition. When the pressure was reduced, boiling and cooling took place. When rapid boiling of a saturated liquid is caused by a reduction in pressure, the liquid is said to be **flashing** into a vapor. The cooling caused by a saturated liquid flashing to a vapor is the same phenomenon that causes cooling to happen in a refrigeration or air conditioning system.

Refrigerants

A **refrigerant** is the fluid that is used inside the refrigeration system to transfer heat. The refrigerants are either natural fluids, such as ammonia, or manufactured chemicals. By far, the two most popular refrigerants in use today are dichlorodifluoromethane and monochlorodifluoromethane, both manufactured substances. Their names are so difficult to say that they are commonly referred to as R-12 and R-22, respectively, or sometimes as Freon (a registered trademark of E. I. DuPont, one of several manufacturers of refrigerants).

Refrigerants are similar to water in many respects. The temperature at which the refrigerant will boil depends upon its pressure (**pressure–temperature relationship**), and it may be found in either vapor or liquid form. The liquid refrigerant at its saturation temperature (boiling point that corresponds to a specific pressure) will flash when exposed to a lower pressure. When it flashes, it will cool to a new temperature that corresponds to the saturation temperature at its new pressure. When liquid refrigerant at its saturation temperature is exposed to a higher temperature, it will absorb the heat and boil away into a vapor.

The major difference between water and the refrigerants used in vapor compression systems is the difference in boiling temperatures. At atmospheric pressure, water boils at 212°F, while R-12 boils at -21°F. If you were outside on a cold winter day of -22°F, you could carry around some R-12 in a cup. It would be at a subcooled condition (1°F lower than its saturation temperature at 0 psig), and in fact, it would look very much like water. But if the R-12 at atmospheric pressure is exposed to a temperature warmer than -21°F, it begins to absorb heat and the liquid boils off into a vapor. If you put the R-12 into a closed container so that the vapor cannot escape as the container absorbs heat, its temperature and pressure would both increase. Figure 2-3 shows what the pressure in the closed container would reach for several different temperatures.

These are the saturation conditions for R-12. Remember, for any pressure, the liquid cannot exist at a temperature above its corresponding saturation temperature. We will now describe how we can take advantage of the properties of the refrigerant to cause a cooling effect.

Expendable Refrigerant System

Consider the tank of R-12 shown in Figure 2-4. It has been sitting in a 100°F room for several hours, so the liquid and vapor inside have reached a temperature of 100°F. According to the chart in Figure 2-3, when the liquid and vapor are at 100°F, the pressure inside the tank is 117.2 psi. In fact, if you measure the pressure on a can of R-12 at 100°F, you will find its pressure to be 117.2 psi regardless of the size of the can or how much liquid it contains.

Now, let us bleed off some pressure through the valve so that the tank pressure is maintained at 37 psi instead of 117.2 psi. The 100°F liquid refrigerant is at a temperature well above the new boiling point of 40°F that corresponds to 37 psi, so it flashes into a vapor. As it flashes, it absorbs its latent heat of vaporization from the remaining liquid, and the liquid that remains becomes cooled to 40°F. The cold R-12 liquid is now able to absorb heat from its surroundings, and as it does, it continues to boil at a temperature of 40°F. If we put this cold tank inside an insulated box, we will have a refrigerated box. If we use a fan to blow air across this tank, the air will be

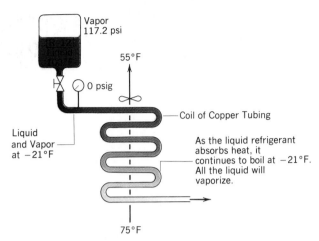

Figure 2-5 An expendable refrigerant system.

cooled, and we would have a crude form of air conditioner.

We would be all done with the explanation of the refrigeration cycle at this point, except that as the liquid absorbs heat and boils, the level of liquid in the tank is continually going down. It will not be long before we run out of liquid.

In an **expendable refrigerant system** (Figure 2-5), we merely replace the refrigerant tank with a new one. For operation on a continuous basis, however, this would become quite expensive. The rest of the refrigeration cycle is provided for only one reason: to recover the cool vapor at 40°F that we allowed to escape in the previous example and reconstitute it to a liquid form at 100°F.

There are expendable refrigerant systems in use today, but instead of using R-12 as the refrigerant, they use liquid nitrogen or some other liquid that is less expensive than R-12. These systems are used in truck trailer refrigeration and on other systems that are used only infrequently. For those systems, the high operating cost is not significant compared to the costs that are avoided in installing a **vapor-compression refrigeration system.**

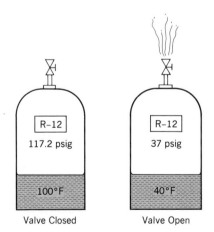

Figure 2-4 When pressure is bled out of a can of R-12, flash gas is formed. The remaining liquid cools to the new saturation temperature corresponding to the new pressure.

Mechanical Refrigeration Cycle

The coil of copper tubing shown in Figure 2-5 is called an **evaporator** because the refrigerant inside is evaporating as it absorbs heat. Let's take the cool refrigerant vapor being released from the evaporator and allow it to flow into a **compressor,** as shown in Figure 2-6. The compressor squeezes (compresses) the refrigerant vapor so that its pressure increases. For the sake of discussion, let's say that this compressor compresses the vapor to a pressure of 117.2 psi. When we compress any gas, we are doing a lot of work on it, and as a result, its temperature goes up. The refrigerant leaving the compressor will therefore not only be at a high pressure, it will also be at a high temperature. For discussion, let's say that the refrigerant leaves the compressor as a superheated vapor at 130°F. Note that this is not the saturation temperature.

Next, let's run the hot gas through a tube to a coil located outdoors. The coil is very similar to an automobile radiator. As the high pressure, high temperature gas passes through this coil, we blow outside air over it. If the outside air is at 90°F, it will remove heat from the 130°F refrigerant, and the refrigerant gas will cool. When the hot gas cools to 100°F, an interesting phenomenon occurs. Remember that the saturation temperature of 117.2 psi refrigerant is 100°F. Once the vapor is cooled to 100°F, any further cooling done to the gas will cause some of the gas to condense into a liquid. As we continue to allow

the outdoor coil to remove heat from the refrigerant, the refrigerant continues to condense until there is only liquid and no vapor left. This outdoor coil is called the **condenser,** because it removes heat from the refrigerant and causes it to condense. We now have a liquid refrigerant at 100°F, and we can pass it back into the refrigerant tank to replenish the refrigerant supply.

In the completed refrigeration cycle shown in Figure 2-7, we have eliminated the refrigerant tank. The tank valve has been replaced with a **metering device** that meters the amount of liquid refrigerant that is allowed to enter the evaporator. Metering device is actually a general term that is used to describe any of a number of different types of valves or orifices (an orifice is a small opening). All that remains in our refrigeration system is the compressor, condenser, metering device, and evaporator coil. In the simplest type of cooling systems (refrigerators, window air conditioners), these four components very often make up the entire system. Some systems are far more complex, but they still have at least these same basic four components. A line is shown in Figure 2-7 that divides the system pressures and temperatures into a **high side** and a **low side.** On the high side, both the pressure and the temperature are high. On the low side, we have only low pressure refrigerant at a relatively low temperature.

The refrigeration system does not actually

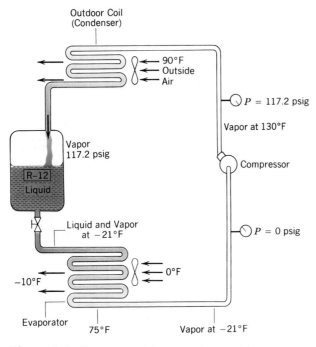

Figure 2-6 Recovering the vapor leaving the evaporator.

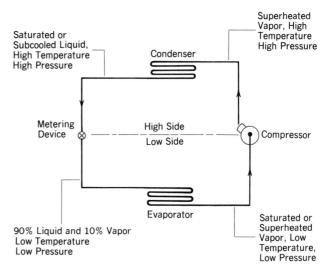

Figure 2-7 A simple complete refrigeration system.

eliminate heat. It only moves it around. In the previous example, the refrigerant in the evaporator absorbs heat from whatever it is that we are trying to cool. This heat is then carried around the system by the vaporized refrigerant to a point where we can get rid of it (condenser). The condenser is located in a place where the release of this heat will not be objectionable, such as outdoors. Thus, the point of heat discharge is isolated from the point of heat pickup (cooling).

If you understand the basic cycle described in this chapter, you have the basis for understanding most cooling systems, including refrigerators, air conditioners, process cooling, water coolers, freezers, and automobile air conditioners. There are many differences in these various applications, but they all use the same principles. Every system uses a refrigerant that is compressed, condensed, allowed to flash, and evaporates while it absorbs heat. The differences between systems include the following:

1. Type of refrigerant.
2. Operating pressures.
3. Physical characteristics of the four major components.
4. Physical location of the four major components.
5. Addition of accessories to regulate the operation.

Arrangement of Components

It is convenient to think of the refrigeration cycle in terms of the high side and the low side. The refrigerant is at a high pressure from the time it leaves the compressor until the point at which it flashes in the metering device. It is at a low pressure from the metering device all the way back to the compressor.

In many systems, the compressor and condenser are located together in a location where the waste heat will be rejected. The metering device and the evaporator are then located together at the point where the useful cooling is being done. Industry technicians refer to a low side, meaning the metering device and the evaporator, or to a high side, meaning the compressor and condenser. Another very common name for a

compressor and condenser mounted together on a common base with electrical controls is **condensing unit** (Figure 2-8).

Figure 2-9 shows a common household refrigerator in which the compressor is located un-

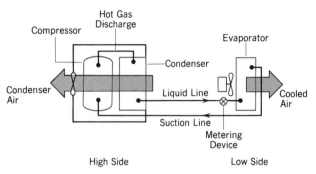

Figure 2-8 The high side consists of the compressor and condenser (condensing unit). The low side consists of the metering device and evaporator coil.

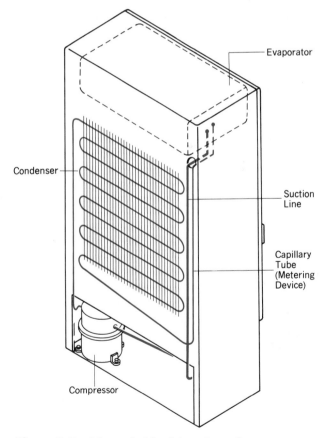

Figure 2-9 A household refrigerator—the compressor is located below, the condenser is mounted on the back, the metering device consists of a long, small-diameter tube, and the evaporator is located inside the refrigerated space.

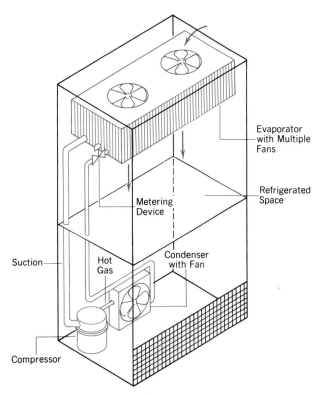

Figure 2-10 A commercial refrigerator with the evaporator and metering device exposed for service inside the box.

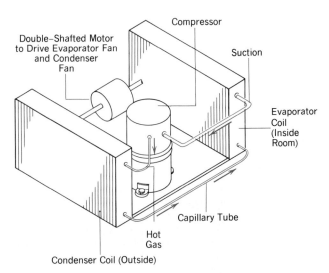

Figure 2-11 The refrigeration circuit for a window air conditioner.

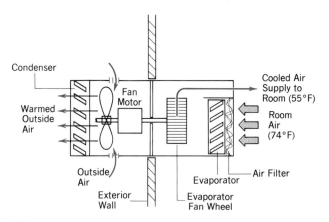

Figure 2-12 Airflow through a window (through-the-wall) air conditioner.

der the refrigerated compartment. The condenser coil is mounted on the back of the box. The metering device is a long tube of small diameter (capillary tube) that runs from the condenser to the evaporator, which is located inside the box.

Figure 2-10 shows a commercial refrigerator that has a condensing unit located under the refrigerated section. The box contains the evaporator coil along with a metering valve mounted just at the inlet to the coil. In this application, both the condenser and the evaporator use fans to blow the air across the coil to improve heat transfer. One or both of these fans may also be found on many household refrigerators.

Figures 2-11 and 2-12 show a common window air conditioner. While all the components are located within one box, the box is divided into an outdoor section and an indoor section. One motor is used to drive both the outdoor condenser fan and the indoor evaporator fan. The outdoor fan draws air through the condenser, where it picks up heat from the condensing re-

frigerant. The outdoor air is warmed in the process, and is discharged back outside the building. The indoor fan draws room air through the evaporator coil, where it is made 18–20°F cooler before being discharged back to the room.

Central air conditioning also uses the same four basic components (Figures 2-13 and 2-14). The condensing unit (compressor plus condenser) are located outdoors, while the evaporator and metering device are closer to the area to be air conditioned. And there is some distance physically separating the high side from the low side of the system. This arrangement is commonly called a split system.

The number of different ways in which the refrigeration components may be designed and

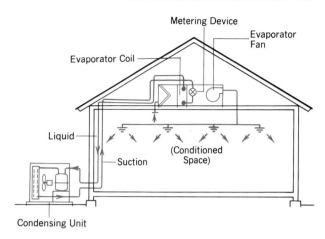

Figure 2-14 A split-system air conditioning system in a house without forced-air heating.

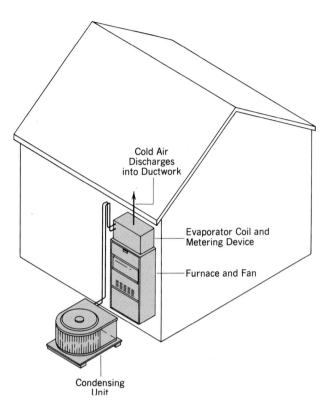

Figure 2-13 A residential split-system air conditioner. The same fan that provides forced air heating also serves as the evaporator fan.

matched with each other is almost limitless. In Chapter 10, many of the specific applications of the mechanical refrigeration cycle are explored further. But no matter how complex the systems may become, they all maintain the same common bond of a refrigerant being circulated through a compressor, condenser, metering device, and evaporator.

KEY TERMS AND CONCEPTS

Evaporation
Flashing
Refrigerant
Pressure–Temperature
 Relationship
Expendable
 Refrigerant System

Vapor-Compression
 Refrigeration
 System
Evaporator
Compressor
Condenser
Metering Device

High Side
Low Side
Condensing Unit

QUESTIONS

1. Name the four basic components that are common to every mechanical refrigeration system.

2. For each of the four components, state whether the refrigerant leaving is a liquid or a vapor or both liquid and vapor.

3. For each of the four components, state whether the refrigerant leaving is at a high pressure or a low pressure.

4. For each of the four components, state whether the refrigerant leaving is at a high temperature or low temperature.

5. For each of the four components, state whether the refrigerant absorbs heat, loses heat, or has no change in its heat content.

6. In Question 5, for each component where the refrigerant absorbed heat, state where the heat came from. For each component where the refrigerant lost heat, state where the heat went.

7. In which three of the components does the refrigerant experience a change of state?

8. Which component is usually located physically closest to the metering device?

9. Why do we compress the refrigerant to a high pressure and temperature?

10. In which two components might the refrigerant absorb sensible heat?

11. In which component does the refrigerant lose sensible heat?

12. In which component does the refrigerant absorb its latent heat of vaporization?

13. In which component does the refrigerant lose latent heat?

14. What happens to the pressure of a refrigerant vapor when it is compressed?

15. What happens to the temperature of a refrigerant vapor when it is compressed?

16. What happens to the heat content of a refrigerant vapor when it is compressed?

17. Refrigerant liquid and vapor are flowing together through a tube. Are they at their saturation temperature? How do you know?

18. Refrigerant liquid and vapor are flowing together through an evaporator. As the refrigerant absorbs heat, what happens to its pressure?

19. A properly operating refrigeration system has superheated refrigerant leaving two of the four major components. Which ones?

20. A properly operating refrigeration system has subcooled refrigerant leaving one of the components. Which one?

21. A properly operating refrigeration system has superheated refrigerant entering and leaving one of the components. Which one?

22. In which two components does the refrigerant experience a significant change in its pressure?

CHAPTER 3

COMPRESSORS

The compressor is the heart of the mechanical refrigeration system. It circulates the refrigerant around the system, and creates the pressure differential required for system operation. This chapter describes the major types of compressors, their uses, operation and common maintenance procedures.

Principles of Vapor Compression

Figure 3-1 shows a simple type of compressor—a bicycle tire pump. There is a chamber called a **cylinder,** and a plunger that you can pump up and down called a **piston.** As you lift the piston, air at atmospheric pressure is admitted into the cylinder. As you push down on the piston, the volume that the air is allowed to occupy is reduced, and the pressure of the air is increased. When the pressure of the air in the cylinder is higher than the pressure of air in the bicycle tire tube, the valve at the tire tube is forced open, and you are able to push the air into the tire tube. Do this for five minutes and you will make two observations (besides the tire pressure getting high). The first is that the cylinder is getting warm. The second is that you are also getting very warm.

The reason that you are getting warm is that it takes a lot of work to compress a vapor. As a matter of fact, for an air conditioning application, the motor size required to drive a compressor is one horsepower for each ton of refrigeration capacity. For refrigeration and low temperature applications, the compressor horsepower is even higher (Figure 3-2). The reason that the bicycle pump cylinder gets warm is not, as you might suspect, due to the friction between the piston and the cylinder. The increase in temperature is due to the work being done on the

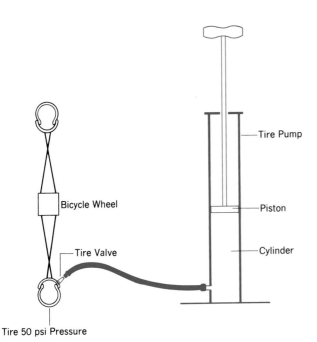

Tire Pump

Bicycle Wheel

Piston

Tire Valve

Cylinder

Tire 50 psi Pressure

Figure 3-1 A simple compressor. As the piston is pushed down, the air in the cylinder is compressed. When the pressure in the cylinder exceeds the pressure in the tire, the air is forced into the tire.

Application	Evaporator Temperature	Compressor hp/ton
Air Conditioning	40°F	1.0
Refrigeration	25°F	1.25
Freezing	−5°F	2.0
Freezing	−35°F	3.0
Process	−65°F	4.5
Process	−95°F	6.0

Figure 3-2 Approximate horsepower requirements to drive a compressor.

vapor appearing as heat due to the friction between the molecules of the gas. The more the vapor is compressed, the higher will be the outlet temperature and pressure.

Compressor Types

There are four common compressor designs in use today. They are **reciprocating, rotary, screw, and centrifugal.**

The reciprocating compressor (Figure 3-3) is very much like the bicycle pump. It may have from one to eight separate cylinders. The larger compressors look somewhat similar to an automobile engine, without the ignition, carburetion, and fuel systems. Automobile engines use power from gasoline explosions to move pistons that, in turn, rotate a crankshaft. The compressor uses a source of power to drive a crankshaft that, in turn, causes the pistons to compress a vapor. Reciprocating compressors are available in sizes from one quarter of a ton to 100 tons.

Figure 3-3 will help you identify the parts of a reciprocating compressor. Although a single-piston compressor is shown, reciprocating compressors may have as many as eight cylinders,

each pumping in sequence. When the piston moves downward, a suction is created in the cylinder. Refrigerant vapor is drawn through the **suction valve** into the cylinder. When the piston moves upward, the pressure is increased until the refrigerant is forced past the **discharge valve** to the condenser. The piston is connected to the **crankshaft** by the **connecting rod** and the **piston pin** or **wrist pin.** Figure 3-3 shows the open suction valve admitting refrigerant into the cyclinder on the piston downstroke. Figure 3-4 shows the discharge of the compressed refrigerant.

Reciprocating compressors are by far the most common type of compressor used in refrigeration and air conditioning.

The rotary compressor works on the same principle as the rotary automobile engine. Figure 3-5 shows a sequence that describes how a rotor moves in a continuous motion inside a cylinder. The blades (or vanes) rotate with the rotor, moving in and out of the slots in the rotor. Figure 3-6 shows a rotary compressor with multiple blades. The trapped vapor is forced into a smaller area, and then released to the condenser. The concept of a rotary compressor is attractive because the continuous direction of rotation eliminates the back and forth motion of the reciprocating compressors. For equal capacities, the rotary com-

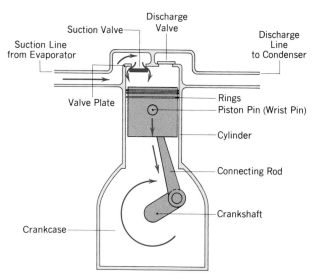

Figure 3-3　Reciprocating compressor. As the piston moves downward, the suction valve opens, pulling refrigerant in from the evaporator.

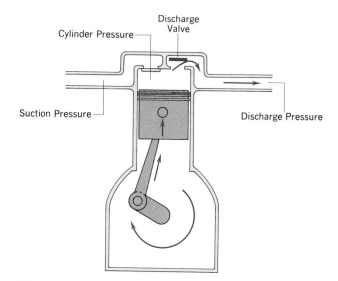

Figure 3-4　As the piston moves upward, pressure inside the cylinder increases until it forces the discharge valve off its seat. The cylinder pressure pushes the suction valve closed.

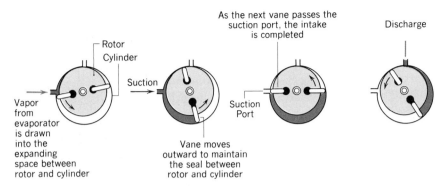

As the next vane passes the suction port, the intake is completed

Discharge

Rotor

Cylinder

Suction

Vapor from evaporator is drawn into the expanding space between rotor and cylinder

Suction Port

Vane moves outward to maintain the seal between rotor and cylinder

Figure 3-5 A rotary compressor.

pressor is much smaller and is therefore gaining popularity in the smaller size applications now dominated by the reciprocating compressors. Figure 3-7 shows another arrangement for the rotary compressor in which a single nonrotating vane maintains a seal between the cylinder and an eccentric motor.

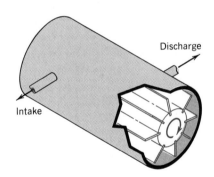

Discharge

Intake

Figure 3-6 Cutaway showing the rotor inside the cylinder of a rotary compressor.

The screw type of compressor makes use of a driven helical gear that mates with and drives a second helical gear (Figure 3-8). The vapor is trapped between the two gears and forced into a smaller volume as the gears rotate. Screw compressors are quite small and quiet for the volume of refrigerant they are able to move. Screw compressors are popular in equipment in the 100- to 600-ton range.

The most popular type of compressor for large tonnage refrigeration systems is the centrifugal compressor, which uses centrifugal force. Figures 3-9 and 3-10 shows a centrifugal compressor **impeller.** A motor causes the impeller to rotate, forcing vapor to the outer perimeter of the wheel where the volume is smaller than at the center.

Centrifugal compressors cannot produce the high discharge pressures achieved with the other more positive displacement types. But this type

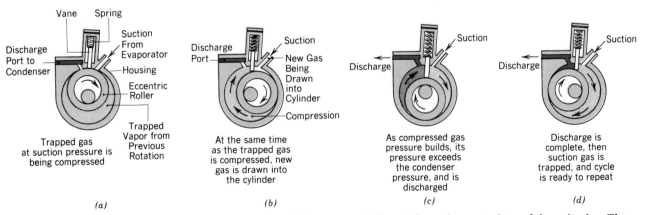

Vane   Spring

Discharge Port to Condenser

Suction From Evaporator

Housing

Eccentric Roller

Trapped Vapor from Previous Rotation

Trapped gas at suction pressure is being compressed

(a)

Discharge Port

Suction

New Gas Being Drawn into Cylinder

Compression

At the same time as the trapped gas is compressed, new gas is drawn into the cylinder

(b)

Suction

Discharge

As compressed gas pressure builds, its pressure exceeds the condenser pressure, and is discharged

(c)

Suction

Discharge

Discharge is complete, then suction gas is trapped, and cycle is ready to repeat

(d)

Figure 3-7 A rotary compressor. The centerline of the rotor is different from the centerline of the cylinder. The vanes move in and out of the slots in the rotor, filling the constantly changing gap.

of compressor can move tremendously large volumes of vapor. These characteristics make centrifugal compressors compatible with R-11, which condenses at a low pressure, but requires large volumes of refrigerant to be circulated. Centrifugal compression may be accomplished by using an impeller that rotates at 6000–8000 rpm. An alternate method uses two impellers in series. They rotate at 3600 rpm, driven directly from a 3600 rpm motor, and they do not use a gearbox to attain high rpm. Centrifugal compressors are simple in operation, as there are no pistons or valves. Figure 3-10 shows a cutaway of a hermetic two-stage centrifugal compressor.

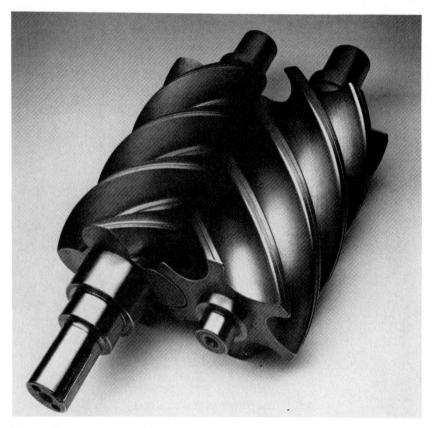

Figure 3-8 Screw compressor. (*Courtesy Sullair Corporation*)

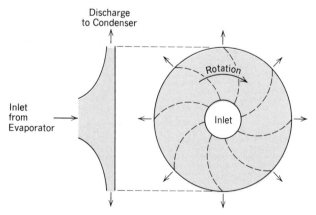

Figure 3-9 Schematic of the impeller for a centrifugal compressor.

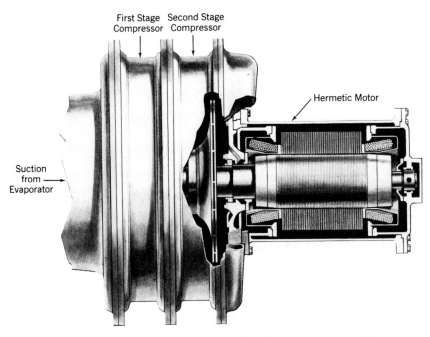

First Stage Compressor
Second Stage Compressor
Hermetic Motor
Suction from Evaporator

Figure 3-10 The impellers of this centrifugal compressor are driven by a 3600 rpm motor. Impeller and motor are both housed within the refrigeration system. (*Courtesy The Trane Company*)

Reciprocating Compressors

The three general classifications of reciprocating compressors are **Welded hermetic, Accessible hermetic,** and **Open.** The welded hermetic compressor is shown in Figures 3-11 and 3-12. The motor and compressor are housed in the same casing, which is welded closed, making it a sealed system. They are built for lowest first cost, high reliability, and are not field repairable. If the motor burns out, or even if a valve seat becomes spoiled, the compressor is discarded. There are some companies that will use parts from ruined compressors to remanufacture new compressors to original specifications. Field repairs to hermetic compressors should not be attempted, however.

The suction line from the evaporator penetrates the shell and ends just inside the shell. The entire contents of the shell (including the motor) are cooled by the cool refrigerant vapor. The compressor picks up cool vapor from inside the shell and compresses it. The discharge line which is normally a smaller diameter than the suction line, leads from the compressor through the shell and on to the condenser.

Many hermetic compressors have a third line emerging from the shell. It is usually less than 12 in. in length, $\frac{1}{4}$ in. in diameter, and its end is crimped and brazed closed. It is called a **purge tube** or **processs tube,** and it also ends just inside the shell of the compressor. After the manufacturer has installed the compressor in the refrigerator or air conditioner, the system is charged through the purge tube, which is sealed after the correct charge has been added. In this way, the manufacturer leaves no valves where potential leaks can develop.

Some small compressors also have two additional connections for an **oil cooler,** making a total of five connections to the shell. The function of an oil cooler is simple, but widely misunderstood. Figure 3-13 shows that the oil cooler connections are opposite ends of a coil that lies in the pool of oil in the bottom of the compressor (the function of the oil will be explained later). The hot gas discharge from the compressor goes to a small auxiliary condenser that cools the hot gas. The gas passes through the coil in the compressor bottom, where it removes heat from the oil. The gas then proceeds to the main condenser, where the cycle proceeds in the normal fashion. The auxiliary condenser is referred to as

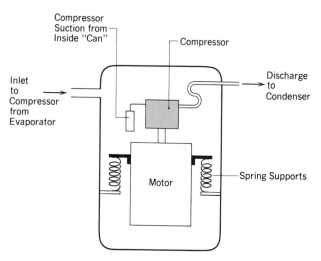

Figure 3-11 A reciprocating compressor, supported inside a "can" by springs.

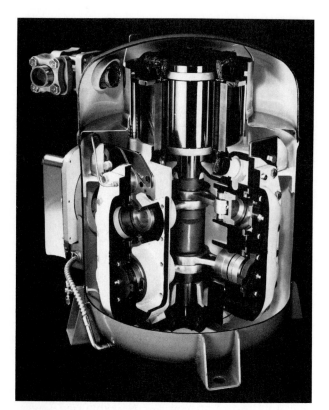

Figure 3-12 Cutaway of a welded hermetic 4-cylinder compressor. (*Courtesy The Trane Company*)

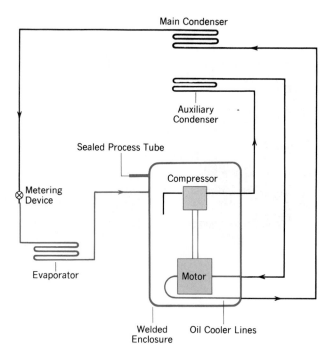

Figure 3-13 Welded hermetic compressor with an oil cooler.

the oil cooler, even though there is no oil circulating through the coil. When a compressor with an oil cooler is replaced with a compressor with only an inlet and discharge connection, the auxiliary condenser may be repiped so that it is in series with the main condenser.

Depending upon the manufacturer, a replacement compressor with an oil cooler may be used in a system without an auxiliary condenser if the condenser air is drawn over the compressor shell for cooling. In that case, the oil cooler connections may be left unconnected.

The second major classification of reciprocating compressors is the accessible hermetic (Figure 3-14). It is also sometimes called **semi-hermetic** or **serviceable hermetic.** The concept of the motor and compressor being enclosed inside the refrigerant system is the same for the accessible hermetic as it is for the welded hermetic compressor. The difference is that the accessible hermetic compressor can be completely disassembled and rebuilt (i.e., is field repairable). The motor section is flanged to the compressor section and can be easily replaced. There are valves (called service valves) at the suction and

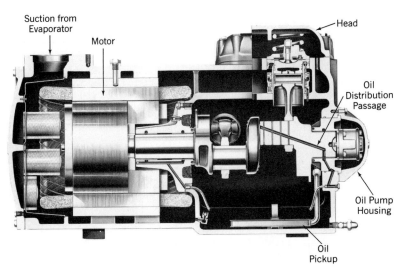

Figure 3-14 Cutaway of an accessible hermetic compressor. (*Courtesy The Trane Company*)

discharge connections to the compressor. A sight glass in the crankcase indicating oil level is an accessory that is not found on the welded hermetic. The cylinder head has pressure ports for attachment of safety controls, and the oil pump on the end of the crankcase can be replaced. If the valves should fail, the cylinder head may be removed, providing access to the valve plate, which is easily replaced.

Being bigger, heavier, and more expensive than a welded hermetic unit of equal capacity, the accessible hermetic compressor is designed for more severe applications. Still, accessible hermetics compete with welded hermetics from three tons to twenty tons. Accessible hermetics are rarely used below 3 tons, and welded hermetics are uncommon above 20 tons.

The third major classification of reciprocating compressors is the open compressor or external drive compressor (Figure 3-15). The distinguishing feature of the open compressor is that the motor is located outside of the refrigerant system. The compressor itself has no integral drive, and may be driven by a standard-type electric motor, a gas engine, a steam turbine, or any other drive. The compressor shaft must pass through the casing of the compressor in order to be attached to the external drive. Therefore, open compressors require a **shaft seal,** which is not required on either of the hermetic types of com-

pressor. The crankshaft seal (Figure 3-16) consists of two surfaces, one that is stationary and the other that rotates with the compressor shaft. These two surfaces are pushed against each other to form a seal. The force is provided by either a diaphragm, bellows, or spring, so that as the material wears, the seal is maintained. The common materials used for sealing surfaces are polished steel, carbon, bronze, ceramic, and Teflon. When crankshaft seals do begin to leak, they are replaced rather than adjusted or repaired. When installing a new crankshaft seal, it is important to recognize how critical the cleanliness of the parts and close dimensional tolerances are to maintaining a good seal. Several other types of seals are also shown in Figure 3-16.

Open compressors are available over the entire size range covered by the welded and accessible hermetic compressors. Their major advantage is their ability to operate when use of an electric motor is not practical. Another advantage is their ability to pump refrigerants (ammonia) that would chemically attack motor materials of construction (copper). All automobile air conditioning compressors are of the open type, and are driven by a belt that is, in turn, driven by the automobile engine. The final advantage of the open compressor is that motor heat is not absorbed into the refrigerant. Therefore, given an

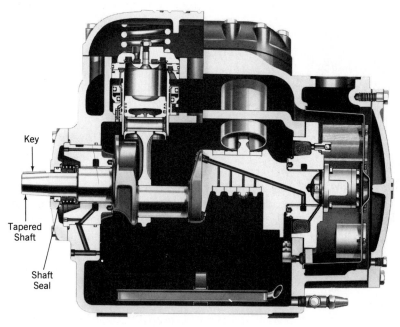

Figure 3-15 Cutaway of an open (external drive) compressor. (*Courtesy The Trane Company*)

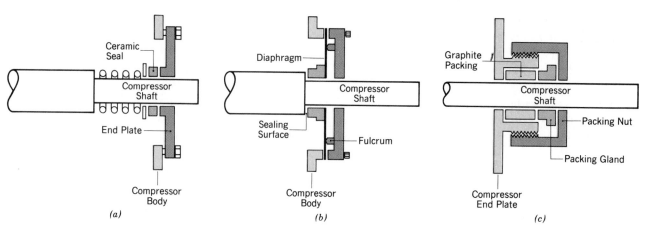

Figure 3-16 (*a*) Rotary seal. The ceramic seal rotates with the crankshaft. Pressure against the end plate is maintained by a spring. (*b*) Diaphragm seal. (*c*) Packing gland.

open and a hermetic compressor of equal size, the former would be capable of handling a larger evaporator load than the latter.

There are, however, two disadvantages to the open compressor. The components needed to make an open compressor system, such as an external drive, belts, flexible couplings, and guards, make it more expensive than a hermetic system. The second disadvantage is the increased maintenance associated with the shaft seal and the drives. If a standard electric motor drive on an open compressor needs to be replaced, however, the replacement parts are far more readily available and less expensive than comparable motors for accessible hermetic units.

Cylinder Pressure—Reciprocating Compressors

The compressor produces the pressure differential required by the system. It must draw in vapor from the evaporator and discharge it to the condenser at a much higher pressure. When the piston is at the bottom of its stroke, the cylinder is full of refrigerant at a pressure equal to the system suction pressure. As the piston moves up, the cylinder volume is decreased and the pressure inside the cylinder is increased (Figure 3-17). The pressure continues to build until the cylinder pressure is high enough to overcome the high-side pressure that is holding the discharge valve closed. Then, further movement of the piston toward the top of the cylinder causes the discharge valve to open because of the high cylinder pressure. The refrigerant is discharged to the condenser until the piston reaches the top of its stroke.

At the top of the piston stroke, there is still a small **clearance volume** left between the top of the piston and the top of the cylinder. When the piston starts on its way down (Figure 3-18), the high pressure refrigerant that was trapped in this clearance volume reexpands to fill the increasing cylinder volume, and the pressure inside the cylinder falls. When the cylinder pressure drops below the suction pressure, the pressure differential causes the suction valve to open. Further movement of the piston toward the bottom of its stroke causes refrigerant vapor to be drawn from the evaporator into the cylinder. The four parts of the compression cycle just described are called **compression, discharge, reexpansion,** and **suction** or intake.

Theoretically, the compressor discharges a volume of refrigerant equal to the total volume displaced each stroke by the piston. The **piston displacement** may be calculated as

$$d = \frac{\pi b^2 l}{4}$$

where

    d = piston displacement.
    b = bore (diameter) of cylinder.
    l = length of piston stroke.

The displacement of the compressor may be calculated as

$$D = \frac{\pi b^2 l}{4} \times n \times \text{rpm}$$

where

    D = **compressor displacement.**
    b = bore.
    l = stroke.
    n = number of cylinders.
  rpm = compressor speed.

The units of piston displacement will be cubic feet (cu ft), and the units of compressor displacement will be cubic feet per minute (cu ft/min). Figure 3-19 shows what volume of refrigerant must be pumped in order to produce a ton of refrigeration capacity for several applications.

Theoretically, the amount of refrigerant pumped equals the compressor displacement. But part of the downstroke is wasted in allowing reexpansion rather than allowing fresh vapor to

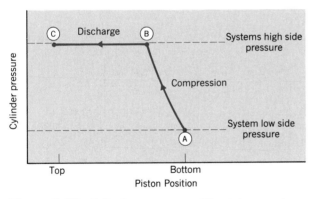

Figure 3-17 Cylinder pressure with piston moving toward the top of its stroke.

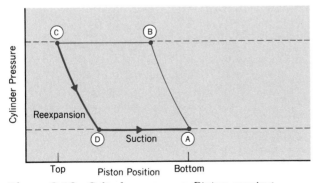

Figure 3-18 Cylinder pressure. Piston moving toward the bottom of its stroke.

R-12	11.2 ft³/ton
R-22	6.9 ft³/ton
R-502	6.8 ft³/min

Figure 3-19 Volume of refrigerant which must be circulated for each ton of refrigeration capacity.

be drawn into the cylinder. The ratio of the actual volume of vapor pumped divided by the calculated volume is called the **volumetric efficiency** of the compressor.

Figure 3-20 shows why it is important to keep the clearance space as small as possible. The dotted lines show how the pressure inside the cylinder will be changed if the clearance volume is allowed to be greater than indicated. Starting at the top of the stroke, there is an increased amount of refrigerant trapped in the clearance volume. A greater portion of the downstroke is then devoted to reexpansion, and the resulting suction portion of the stroke from D to A is shortened. A reduced volume of refrigerant enters the

cylinder, and the compressor (and system) capacity is reduced. An extra gasket left under a valve plate that is being replaced can increase the clearance volume significantly and have a devastating effect on compressor capacity.

Figure 3-21 shows another situation that can reduce compressor capacity. A malfunctioning condenser causes the high-side pressure to be too high. Again, the compressor will use more of its stroke for the compression and less for the discharge of refrigerant.

Compressor Lubrication

The compressor has the following mating surfaces that move in relation to each other and require lubrication.

1. Piston to cylinder wall.

2. Connecting rod to piston pin.

3. Connecting rod to crankshaft.

4. Crankshaft to bearings.

Figure 3-20 Capacity reduction caused by a too large clearance volume.

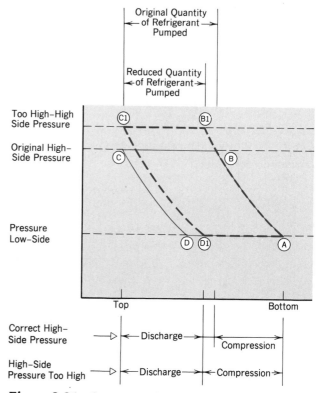

Figure 3-21 Capacity reduction caused by a high-side pressure that is higher than normal.

The refrigerant oil reservoir is the bottom of the crankcase. There are two methods used to move this oil from the reservoir up to the parts requiring lubrication: (1) splash lubrication and (2) pressure lubrication.

The **splash lubrication** system is quite simple in operation (Figure 3-22). With each rotation, the crankshaft or a small paddle attached to a connecting rod splashes into the reservoir of oil. The resulting spray is distributed throughout the inside of the compressor, with some finding its way to the places where lubrication is required. This is a common system in the smaller size compressors. Its advantage is simplicity and reliability. Its disadvantage is that in order to allow room for the oil to get to the mating surfaces, clearances between mating parts must be relatively large, which results in increased compressor noise.

The **pressure lubrication** system uses an oil pump (Figure 3-23) that is mounted on the end of the compressor and driven by the crankshaft (Figure 3-24). The oil pump draws oil from the crankcase and delivers it to a hollowed passageway running through the center of the crankshaft (Figure 3-24). There are radial holes

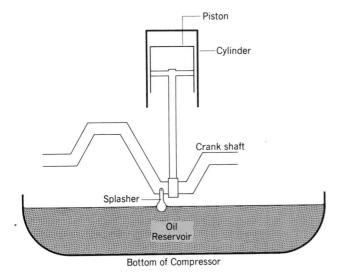

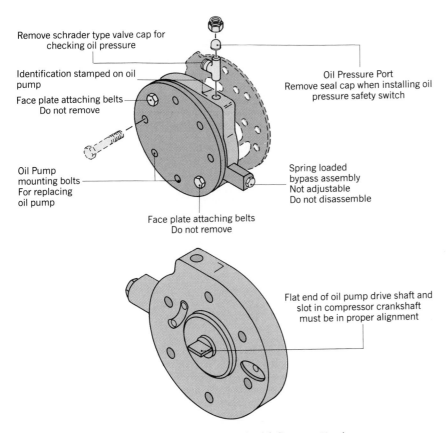

Figure 3-23 Oil pump. (*Courtesy Copeland Corporation*)

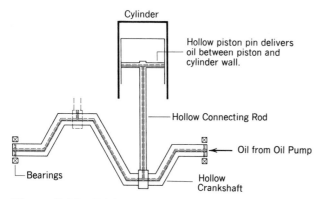

Figure 3-24   Oil from the pump is discharged through the hollow crankshaft.

in the crankshaft at each of the connecting rods to allow oil to lubricate this surface. Additionally, the connecting rod may also have a passage that allows oil to move through the connecting rod, through a hollow piston pin, and directly into the piston/cylinder surface. The crankshaft bearings are lubricated similarly to the connecting rods. The oil, after squeezing through the lubrication points, falls back into the reservoir.

In later chapters, we will discuss oil pressure safety controls. It will be important to know that the oil pump suction pressure is equal to the crankcase pressure, which is also equal to suction pressure. The oil pump adds pressure to the oil. The **net oil pressure** available for forcing the oil through its normal route is equal to the difference between the oil discharge pressure and the crankcase pressure.

Oil pumps will commonly produce a net oil pressure of 20 to 50 psi. If pressure exceeds 60 psi, it will be relieved internally by the pump discharge to the crankcase. If net oil pressure falls below 10 to 15 psi, the oil pump assembly needs to be replaced (it is usually available as a bolt-on kit).

Electrical Connections

With the electric motor located inside the refrigerant circuit, the power wiring must pass through the shell of the compressor (Figure 3-25). This is effected by three pins, to which the electrical motor wiring connects, passing through a sealed opening.

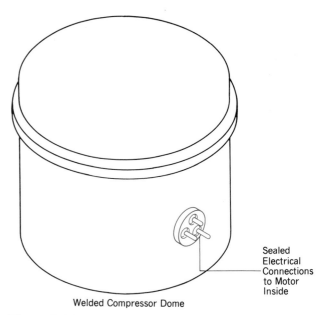

Figure 3-25   Electrical terminals for the windings in the compressor motor.

There may be two additional pins, which, if present, are connected to a switch inside the motor windings that opens if the motor windings overheat. Compressors that have this motor winding thermostat are sometimes referred to as having **inherent overload protection.**

Swash Plate Compressors

For the most part, automotive compressors are simple open-type compressors of the kind previously described. There is, however, one unique

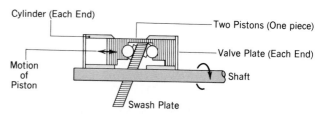

Figure 3-26   Partial cross section of a swash plate compressor. As the plate rotates with the shaft, the piston is driven from side to side. Each stroke is a discharge for a cylinder on one side and a suction for the opposing cylinder. Used on General Motors automobiles.

design used by General Motors that deserves special note because its design is quite compact and efficient. Figure 3-26 shows the construction of a double-action piston operated by a rotating **swash plate.** Double action means that as the piston moves to the left, the cylinder on the left is in the compression or discharge stroke while, at the same time, the cylinder on the right is in the suction or reexpansion portion of its stroke. As the swash plate rotates, the piston is moved back and forth. There are multiple pairs of cylinders around the crankshaft.

Compression Ratio

The **compression ratio** of a compressor is defined as the absolute discharge pressure divided by the absolute suction pressure. Systems should not be designed that require a compression ratio higher than 10:1. Higher compression ratios cause too much work to be done on the refrigerant, and high refrigerant temperature will result. This may cause equipment failure due to breakdown of the lubricating oil or fatigue of the compressor parts due to high temperature.

Example
Calculate the compression ratio required if you want a system to operate with a head pressure of 150 psi and a suction pressure of 10 psi.

Solution
The compression ratio is not 15:1. The operating pressures are given in psi gauge, and the calculation of compression ratio requires the use of psi absolute.

Absolute head pressure = 150 + 15 = 165 psi

Absolute suction pressure = 10 + 15 = 25 psi

$$\text{Compression ratio} = \frac{165}{25} = 6.6{:}1$$

When the suction pressure is given in inches of mercury vacuum, it must first be converted into psia. As 30 in. Hg is the equivalent of 15 psi below atmospheric pressure, we can say that each 2 in. Hg of vacuum equals 1 psi below atmospheric pressure.

Example
Calculate the compression ratio for a system operating at 10 in. Hg suction, and 150 psi discharge.

Solution

Absolute head pressure = 150 + 15 = 165

$$\text{Absolute suction pressure} = 15 - \frac{10}{2} = 10 \text{ psi}$$

$$\text{Compression ratio} = \frac{165}{10} = 16.5{:}1$$

If in fact the system required operation at this suction and discharge pressure, the use of a single compressor would probably cause a premature compressor failure. In Chapter 15, which is on special systems, a compound system is described that accomplishes the compression in two steps.

Factors Affecting Compressor Capacity

When a manufacturer designs a compressor, the capacity is determined by the volume of refrigerant pumped. This capacity will be determined by the piston and cylinder dimensions, the length of the stroke, the number of cylinders, and the speed at which the compressor rotates. After the compressor leaves the factory, there are a number of application factors that will determine the system capacity. Two identical compressors installed in different systems, in different applications, may produce different capacities.

For example, the type of refrigerant used will have a tremendous **effect on compressor capacity.** A compressor pumping a dense refrigerant will produce more capacity than one pumping a lighter refrigerant. Also, refrigerants that have a higher latent heat of vaporization will have more capacity. With the combination of these two factors, a system using R-22 will have about twice the capacity of a system using R-12 at the same temperatures and using the same compressors.

A second application factor that affects compressor capacity is suction pressure. A compressor operating at 40 psi suction pressure will have a higher capacity than the same compressor operating at a 20 psi suction pressure. This occurs because when the cylinder fills with refrigerant at the system suction pressure, the 40 psi system fills the cylinder with denser refrigerant, and therefore more weight. When more pounds of refrigerant are circulated, and each pound of refrigerant absorbs its latent heat of vaporization in the evaporator, more capacity is produced.

Head pressure affects compressor capacity, but to a lesser extent than suction pressure. As the head pressure increases, the compressor capacity decreases. The piston, on its upstroke, must travel further before producing sufficient pressure to overcome the head pressure. Less of the stroke can then be devoted to discharging the refrigerant into the condenser (Figure 3-21). And, of course, for an open compressor, increasing the rpm at which the compressor operates will increase the compressor capacity by an equal percentage.

Figure 3-27 shows a rating table for a typical compressor. The column labeled bhp stands for brake horsepower, which is a term used in rating the output power of motors. The compressor rating table shows the amount of horsepower that must be available from a motor in order to drive the compressor at the indicated operating conditions. You will note that as suction pressure increases, so does the bhp. The increase in power consumption, however, is smaller than the increase in capacity. We can say that even though horsepower goes up with suction pressure, the bhp/ton is reduced and lower operating costs will result.

Capacity Control

A refrigeration or air conditioning system is not called upon to deliver its maximum cooling at all times. A means of capacity control must therefore be provided for all systems, so that we can reduce the amount of cooling when the space temperature is low enough. The simplest method is called compressor cycling, or **on–off control.** Refrigerators, window air conditioners, and most other small applications have on–off compressor cycling for capacity control. A thermostat senses the space temperature. When the thermostat is satisfied, its electrical contacts open and the compressor stops. The space then warms. Sometime later, the thermostat once again calls for cooling, so its contacts close and the compressor starts again. With on–off control, the temperature within the cooled space is constantly changing, as shown in Figure 3-28. If closer control of temperature is desired, the compressor would have to be cycled on and off more frequently (Figure 3-29). But this is not practical.

Condensing Temperature	Saturated Suction Temperature	400 rpm		500 rpm		600 rpm	
		tons	kW	tons	kW	tons	kW
110°	20	9.1	12.4	11.4	15.5	13.6	18.6
	30	11.7	13.5	14.6	16.9	17.6	20.3
	40	14.6	14.2	18.2	17.7	21.9	21.3
	50	18.1	14.6	22.6	18.2	27.1	21.9
120°	20	8.3	12.9	10.4	16.1	12.4	19.4
	30	10.7	14.3	13.4	17.9	16.1	21.5
	40	13.5	15.3	16.9	19.1	20.2	22.9
	50	16.7	16.1	20.9	20.1	25.1	24.1
130°	20	7.5	13.4	9.4	16.8	10.7	20.1
	30	9.7	14.9	12.1	18.6	14.6	22.4
	40	12.4	16.3	15.5	20.4	18.6	36.7
	50	15.3	17.6	19.1	22.0	22.9	26.4

Figure 3-27 Compressor capacity rating table.

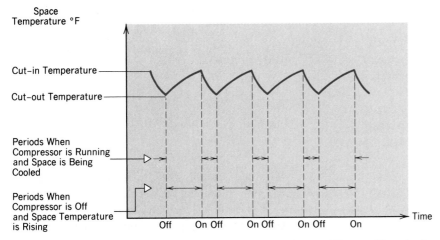

Figure 3-28 Constantly changing space temperature with on–off control.

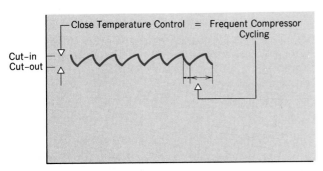

Figure 3-29 To reduce the space temperature variation with on–off control, the frequency of compressor cycling must increase.

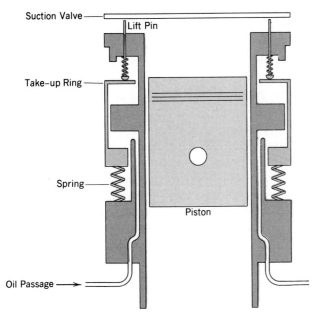

Figure 3-30 Compressor unloading mechanism.

When compressors start, there is an inrush of current to the motor called locked rotor amps. As the compressor runs, the heat produced by this inrush of current is dissipated. But if the compressor is forced to start too frequently, another inrush of current will be added before the compressor has cooled from the previous start. On larger compressors, the motor is not allowed to start more often than once every 30 min, or even once each hour, so other means must be provided to reduce compressor output capacity without turning off the motor.

For compressors with more than one cylinder, capacity control may be accomplished by the use of **cylinder unloaders.** A cylinder unloader is a device that can mechanically hold open the suction valve on a cylinder (Figure 3-30). An unloaded cylinder does no compression, supplies no refrigerant to the condenser, and requires very little power from the motor to move up and down. A two-cylinder compressor can have just one step of unloading (50 percent). A four-cylinder compressor can unload to 75, 50, and 25 percent. When a compressor is loading and unloading a cylinder, you will hear a distinctly different noise each time the compressor load changes.

Figure 3-31 shows a different method of accomplishing unloading without holding open the suction valve. A bypass circuit is provided for each cylinder. When the solenoid valve is deenergized, the discharge pressure holds the bypass valve closed and the cylinder operates normally. When the solenoid is energized, the high-side

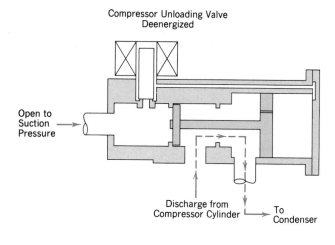

Compressor Unloading Valve
Deenergized

Open to
Suction
Pressure →

Discharge from
Compressor Cylinder

To
Condenser

Energized

To Compressor
Suction
Passage ←

Discharge from
Compressor Cylinder

Open to
Condenser
Pressure

Figure 3-31 Compressor unloading accomplished by using cylinder bypass. (*Courtesy Copeland Corporation*)

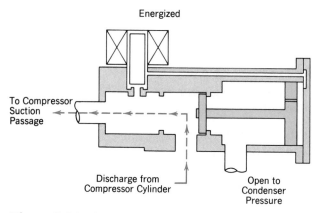

Condenser

Hot
Gas

Metering
Device

Hot Gas
Bypass
Valve

Compressor

Pressure
Regulator

Evaporator

Figure 3-32 Hot gas bypass to reduce system capacity. The bypass valve may be electrically operated from low-side pressure or from a space temperature. Or it may be a pressure-controlled valve maintaining constant pressure between the metering device and the evaporator.

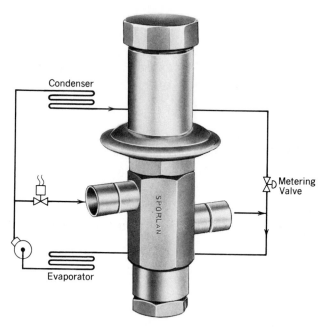

Condenser

Metering
Valve

Evaporator

Figure 3-33 Discharge bypass valve responds to changes in suction pressure. As suction pressure drops below a minimum set point, hot gas is bypassed to the evaporator inlet. (*Courtesy Sporlan Valve Company*)

pressure on the right side of the valve is bled off and the valve moves to the right. The cylinder discharge then returns to the suction, and the cylinder has effectively caused no compression.

Another method of compressor capacity control is called **hot gas bypass** (Figure 3-32). With this method, very close control of temperature is possible. When the hot gas bypass valve is opened, a portion of the compressor discharge gas is routed directly to the inlet of the evaporator, which reduces the cooling done in the space. Close control is possible because there is almost no limit to the frequency at which the hot gas bypass valve may be opened and closed. The disadvantage of the hot gas bypass system is that the compressor sees the equivalent of full load all the time, so the operating costs for compressor motor electricity are quite high. This method is therefore only used in applications where the close control of temperature is essential.

The hot gas bypass valve can also be a pressure-operated valve that modulates to maintain a constant downstream pressure. As the evaporator load drops, the low-side pressure also tends to drop, causing the hot gas bypass valve to open (Figure 3-33).

For open compressors and some limited applications of hermetic compressors, two-speed motors may be used for capacity control. At full load the compressor operates at high speed. As the load diminishes, however, the compressor switches to low speed instead of turning off. The compressor turns off only if the evaporator load continues to fall. Two-speed operation allows closer temperature control without frequent compressor starting.

KEY TERMS AND CONCEPTS

Cylinder

Piston

Reciprocating Compressor

Rotary Compressor

Screw Compressor

Centrifugal Compressor

Suction Valve

Discharge Valve

Crankshaft

Connecting Rod

Piston Pin

Wrist Pin

Impeller

Welded Hermetic Compressor

Accessible Hermetic Compressor

Open Compressor

Purge Tube

Process Tube

Oil Cooler

Semihermetic

Serviceable Hermetic

Shaft Seal

Clearance Volume

Compression Stroke

Discharge Stroke

Reexpansion Stroke

Suction Stroke

Piston Displacement

Compressor Displacement

Volumetric Efficiency

Splash Lubrication

Pressure Lubrication

Net Oil Pressure

Inherent Overload Protection

Swash Plate

Compression Ratio

Effects on Compressor Capacity

On-off Control

Cylinder Unloaders

Hot Gas Bypass

QUESTIONS

1. What are the three types of reciprocating compressors? What are the unique features of each?

2. What types of compressors might you find on a single-compressor 300-ton refrigeration system?

3. Which type of compressor is used in automobile air conditioners?

4. Which type of compressor is used in a window air conditioner?

5. What connects the crankshaft to the piston?

6. A compressor is operating at a suction pressure of 10 psi and the oil pump discharge pressure is 40 psi. How much net oil pressure is available?

7. A compressor is operating at a suction pressure of 10 in. Hg, and the oil pump discharge pressure is 30 psi. How much net oil pressure is available?

8. Where is the oil pump located, and what makes it turn?

9. How can you tell if the oil level is correct in a semihermetic compressor? A welded hermetic compressor?

10. A compressor has four cylinders and a 1750 rpm motor. Each piston is 2 in. in diameter and 1.5 in. long. The stroke is 2.5 in. What is the piston displacement? What is the compressor displacement?

11. A compressor is operating at a suction pressure of 2 psi and a discharge pressure of 120 psi. What is the compression ratio?

12. A compressor is operating at a suction pressure of 2 in. Hg and a discharge pressure of 120 psi. What is the compression ratio?

13. A compressor is operating at a condensing temperature of 110°F and a saturated suction temperature of 40°F. What is the compression ratio if R-12 is used?

14. For the compressor in Question 13, what is the compression ratio if R-22 is used?

15. What is volumetric efficiency?

16. What type of compressor will use a crankshaft seal?

17. How is the motor of a hermetic compressor cooled?

18. What enters the oil cooler of a small hermetic compressor? What is connected to the outlet of the oil cooler?

19. What type of compressor is used in an ammonia refrigerant system? Why?

20. What are the steps of unloading available with a three-cylinder compressor?

CHAPTER 4

CONDENSERS AND RECEIVERS

The condenser is the exit door for the heat that the refrigerant has absorbed in the evaporator and compressor. This chapter describes the different types of condensers, how to estimate the correct head pressure for systems using each type, and what malfunctions are most common.

Heat Rejection

Figure 4-1 shows how the quantity of heat rejected in the condenser balances the heat inputs from an air conditioner. Assuming a 1-ton air conditioner, the refrigerant absorbs 12,000 Btu/hr in the evaporator. The compressor in an air conditioning system usually draws slightly under 1 kilowatt (kW) of electricity for each ton of evaporator capacity. This electricity is all converted to heat when it pumps the refrigerant. A heat input

of 3000 Btu/hr due to compressor work is assumed in Figure 4-1. The heat rejected in the condenser must exactly equal the total amount of heat absorbed by the refrigerant in the evaporator and compressor. For systems operating at lower suction temperatures, the heat of compression increases per ton of refrigeration. The heat rejected also is increased at lower suction temperatures to match the increased input. Compared to a 40°F evaporator, a 25°F evaporator adds 25 percent more compressor heat. A −5°F evaporator adds 100 percent more compressor heat.

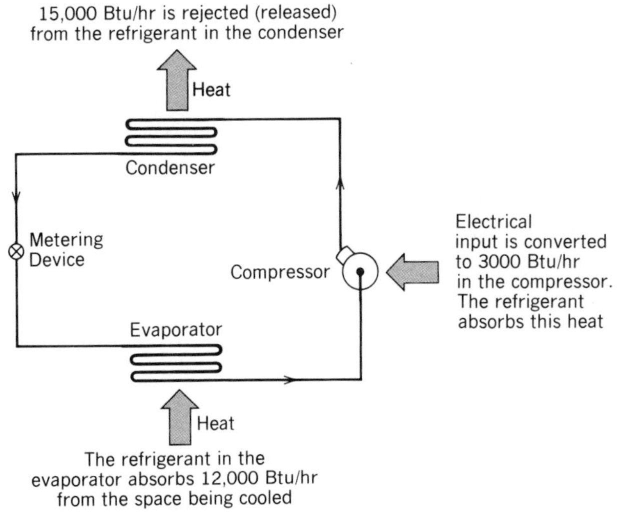

There are two general types of condensers, air-cooled and water-cooled. **Air-cooled condensers** use outside air as the condensing medium, while **water-cooled condensers** use a source of water between 60°F and 90°F to absorb the heat from the condensing refrigerant. Evaporative condensers are a third type that is actually a combination of an air-cooled and a water-cooled condenser.

Air-Cooled Condensers

There are two types of air-cooled condensers, **natural draft** and **forced draft.** Natural-draft condensers depend on natural air convection currents to carry heat away from the refrigerant. These condensers are very reliable because there are no moving parts to fail. Because of the limited air velocity, however, they usually require a relatively large surface area to accomplish the required heat transfer. The most common application for natural-draft condensers is found on the household refrigerator (Figure 4-2).

A forced-draft condenser uses a fan to force air through a finned tube condenser. Figure 4-3 shows a very popular type of forced-draft con-

Figure 4-1 The heat rejected in the condenser equals the sum of the heat absorbed in the evaporator and the compressor.

(Figure labels:)
15,000 Btu/hr is rejected (released) from the refrigerant in the condenser

Heat

Condenser

Metering Device

Compressor

Electrical input is converted to 3000 Btu/hr in the compressor. The refrigerant absorbs this heat

Evaporator

Heat

The refrigerant in the evaporator absorbs 12,000 Btu/hr from the space being cooled

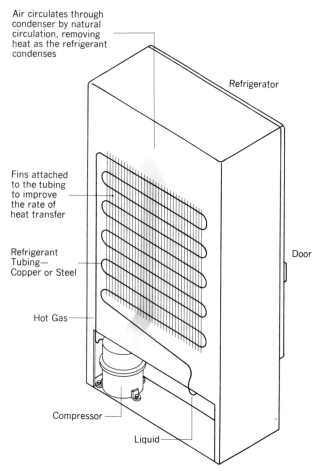

Air circulates through condenser by natural circulation, removing heat as the refrigerant condenses

Refrigerator

Fins attached to the tubing to improve the rate of heat transfer

Refrigerant Tubing— Copper or Steel

Door

Hot Gas

Compressor

Liquid

Figure 4-2 A natural-draft condenser mounted on the back of a domestic refrigerator.

denser used on a small condensing unit (a compressor and condenser together with electric controls). This arrangement is used in many types of refrigerated boxes, ice makers, freezers, and other applications usually not exceeding 3 tons.

Forced air-cooled condensers are used in applications up to 120 tons in a single unit. Figures 4-4 and 4-5 show two configurations for these

Condenser Fan (3)

Condensing Coil

Figure 4-4 A large air-cooled condenser with multiple condenser fans. (*Courtesy Russell Coil Company*)

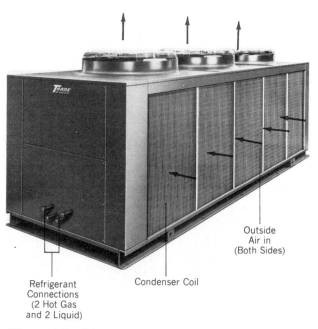

Outside Air in (Both Sides)

Refrigerant Connections (2 Hot Gas and 2 Liquid)

Condenser Coil

Figure 4-5 A large air-cooled condenser with three fans and two separate refrigeration circuits. (*Courtesy The Trane Company*)

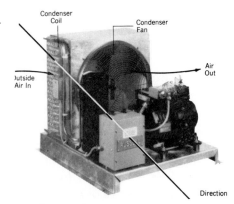

Condenser Coil

Condenser Fan

Air Out

Outside Air In

Direction

Figure 4-3 Small condensing unit using a forced-draft condenser. (*Courtesy Russell Coil Company*)

large air-cooled condensers. In the larger sizes, multiple fans are used to draw outside air over the condenser coil. Air-cooled condensers may also be located inside a building if a centrifugal fan is used to duct the outside air to and from the condenser coil (Figure 4-6).

Air-cooled condensers have the peculiar distinction of being able to deliver increased capacity whenever it is not required! This is to say, an air-cooled condenser is sized sufficiently large to reject the required amount of heat when the outside air temperature is 95°F (or the actual outside air design temperature wherever the condenser is actually installed). When the outside air temperature drops, the condenser becomes able to reject more heat, even though the cooling load has been reduced. This can create a problem on systems requiring the refrigeration system to run even when the outside air temperature has dropped significantly. When the condenser operates at outside air temperatures of less than 60°F, the increased condensing capacity causes unacceptably low head pressure, which decreases the flow of refrigerant across the metering device. Also, the decreased pressure drop across the metering device can cause erratic control of refrigerant flow. There needs to be a method to reduce the capacity of the air-cooled condenser when ambient (outdoor) temperatures are low.

Capacity Control

Three methods are commonly used to control the capacity of air-cooled condensers: (1) **fan cycling,** (2) **shutters,** and (3) **flooding.**

Fan cycling is by far the most common method of what is commonly referred to as **head-pressure control.** As outside air temperature drops, the fans are cycled off in sequence (one variation on this scheme is to use variable-speed motors to reduce airflow as air temperature drops). A sensing device may sense outside air temperature or head pressure. For example, consider a condenser with two fans (Figure 4-7). One fan would automatically cycle off when the outside air temperature dropped to 60°F, the second when the outside air temperature dropped to 50°F. With all fans turned off, the condenser can operate with satisfactory head pressures down to about 40°F outside air temperature. When it becomes colder than that, the natural convection of air across the condenser coil causes the head pressure to be too low.

Where operation at outside air temperatures below 40°F (down to 0°F) is required, capacity is further reduced by the use of shutters or condenser flooding. Shutters are modulating dampers (Figure 4-8) that prevent even natural convection of air around the condenser coil. The

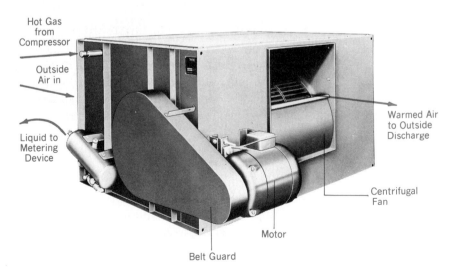

Figure 4-6 Centrigugal-fan air-cooled condenser. *(Courtesy The Trane Company)*

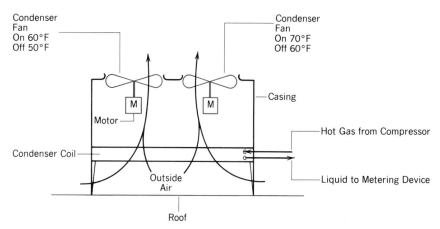

Figure 4-7 An air-cooled condenser with fan-cycling head-pressure control.

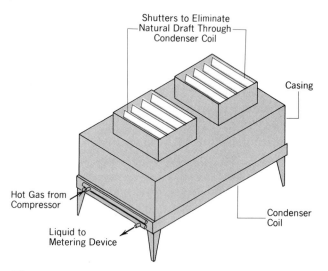

Figure 4-8 Air-cooled condenser using shutters in addition to fan cycling for head-pressure control.

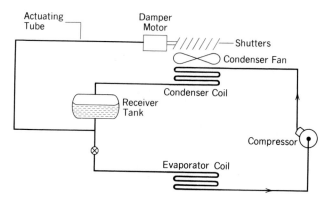

Figure 4-9 Shutters operated directly from refrigerant pressure. As the high-side pressure drops, a spring in the damper motor causes the shutters to close.

dampers are actuated by either a pneumatic or electric damper motor, or directly from refrigerant pressure (Figure 4-9).

Flooding is a refrigerant-side method of controlling condenser capacity. It is also suitable to control head pressures with outside air temperatures down to 0°F. A variety of valving arrangements are available to bypass hot gas around the condenser (Figure 4-10). In one arrangement the bypass valve is preset for a minimum allowable head pressure. Whenever the head pressure is higher than this setting, the valve allows unrestricted flow from the condenser

to the receiver. When head pressure fails, however, the valve restricts the flow from the condenser, causing the tubes to fill with liquid refrigerant (flooding). Hot gas from the compressor is allowed to mix with a reduced quantity of liquid from the condenser to maintain the required head pressure. The bypass valve modulates the mixture of liquid and hot gas. Figure 10a shows a photo of this bypass valve.

The check valve shown in Figure 4-10 is required to prevent liquid from migrating from the receiver to the condenser during the off-cycles. Without the check valve, refrigerant migration to the cold condenser will occur, resulting in a starved evaporator on start-up which causes a nuisance trip out on the low pressure cutout.

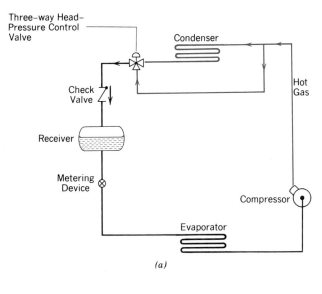

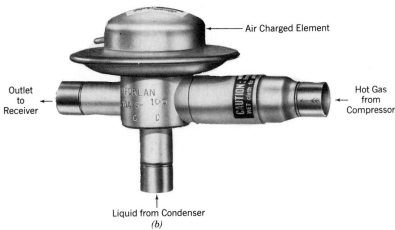

Figure 4-10 (a) Head-pressure control accomplished by flooding the condenser. (b) A nonadjustable head-pressure control valve that limits the flow of liquid refrigerant from the condenser. As the receiver pressure decreases, the flow of liquid is reduced, and the flow of hot gas from the compressor is increased. (*Courtesy Sporlan*)

Physical Arrangements

The most common physical arrangement for an air-cooled condenser is the vertical updraft configuration shown in Figures 4-4 and 4-5. It is the preferred configuration because the airflow is relatively unaffected by wind direction. Where an overhead obstruction exists, however, it may be preferable to have a vertical coil with a horizontal discharge (Figure 4-11). The direction of discharge should be carefully chosen to take advantage of the prevailing wind direction during the hottest summer months. If the direction of air discharge from the fan is against the prevailing wind direction, the condenser may not be able to produce the required condensing capacity.

Air-Cooled Condenser Problems

The operational problems with air-cooled condensers are most commonly related to reduced airflow or heat transfer. The symptom of a reduced condenser capacity problem is a high head

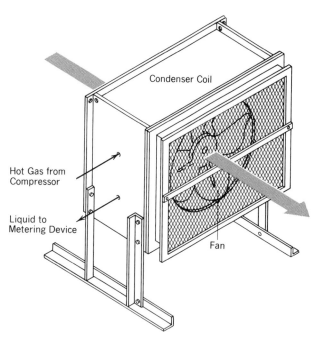

Figure 4-11 Air-cooled condenser with horizontal air discharge.

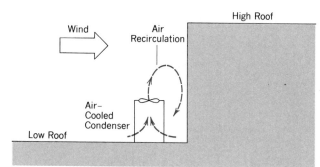

Figure 4-12 Air recirculation caused by a combination of wind direction and physical obstruction. High head pressure and loss of capacity will result.

pressure. The condensing temperature rises, increasing the temperature difference between the refrigerant and the outside air. It will continue rising until the increased temperature is sufficient to transfer a quantity that will match the input from the evaporator and compressor.

The following are the most commonly found causes of reduced condenser capacity (and high head pressure).

1. Fan not rotating. This can be caused by a bad motor or controls, a broken or loose belt for belt drive fans, or a fan slipping on the motor shaft for direct-drive fans.

2. Airflow obstructed. Physical obstructions can be caused by leaves, dirt (especially in agricultural or industrial areas), grass clippings, debris, or any other airborne contaminants. Bent condenser fins may also obstruct air flow. Corrosion of fins is common in areas where aluminum fins are exposed to salt air (near the ocean), or to fertilizers containing ammonia. When the corrosion problem is severe, air-cooled condenser coils may be provided with copper fins.

3. Air recirculation. Figure 4-12 shows how physical obstructions can cause the warmed air

from the condenser outlet to recirculate to the air inlet. This recirculation will simulate operation of the condenser at an artificially high ambient temperature, resulting in reduced capacity and high operating cost. Air-cooled condensers need room to breathe.

In a properly designed refrigeration or air conditioning system, the condensing temperature may reasonably be expected to be 25°F to 35°F higher than the entering condenser air temperature. If a high-side pressure reading shows a higher condensing temperature, check for a condenser problem.

Example
What head pressure would you reasonably expect to find on an air-cooled air conditioner (R-22) operating with 95°F outside air entering the condenser?

Solution
At 95°F outside air temperature, we expect a condensing temperature of 95°F + 30°F = 125°F. The R-22 pressure that allows condensing to take place at 125°F is 278 psi. A range of 260 to 297 psi would be considered normal.

Water-Cooled Condensers

Water-cooled condensers reject heat from the refrigerant to a source of water. The three most

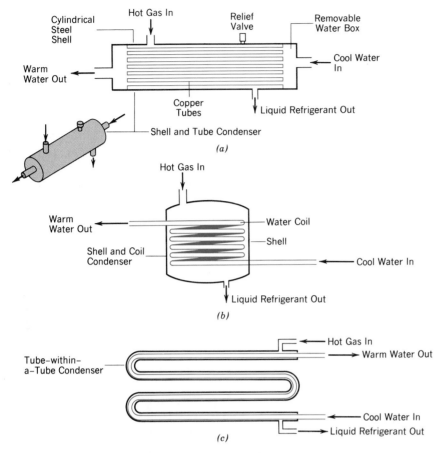

Figure 4-13 Three types of water-cooled condensers. (*a*) Shell and tube. (*b*) Shell and coil. (*c*) Tube within a tube.

popular configurations for water-cooled condensers are (1) **shell and tube,** (2) **shell and coil,** and (3) **tube within a tube,** and are shown in Figure 4-13. Operationally, they work on the same heat transfer principles. Water enters the tube side of the heat exchanger and the hot refrigerant gas enters the shell. As the refrigerant loses heat to the water, the refrigerant condenses and the water gets warm.

The main advantage water-cooled condensers have over air-cooled condensers is they create a lower head pressure. This lower pressure means the compressor uses less electrical power and costs less to operate. As a rule of thumb, the condensing temperature will be about 10°F warmer than the leaving condenser water temperature, and the leaving condenser water temperature will be about 10°F warmer than the entering water temperature.

Example

What head pressure would you expect to find on a water-cooled air conditioner (R-22) operating with condenser water entering at 85°F and leaving at 95°F?

Solution

With 95°F leaving water temperature, we would expect a condensing temperature of 95°F + 10°F = 105°F. The corresponding R-22 pressure is 213 psi.

The reduction in head pressure compared to an air-cooled unit is dramatic. But there are also some offsetting costs associated with water-cooled condensers, such as higher first cost for installation, higher water costs, and higher maintenance costs.

Water Sources

One method of providing water to the condenser is to use a once-through system (Figure 4-14). That is, water may be taken from a city water supply, a well, or a process, and supplied to the condenser. After absorbing heat from the refrigerant, the water is not reused in the refrigeration system. With a once-through system, it is important that we supply only enough water to keep the head pressure reasonable. If more than the minimum required water flow is used, the water cost may more than offset the reduced compressor operating cost.

The device used to adjust the water flow through the condenser is called a **condenser water regulating valve** (see Figures 4-15 and 4-16). It is a self-contained valve, with pressure from the high side of the refrigerant system opposing a spring pressure. Head pressure may want to increase, either due to an increasing refrigeration load or an increase in the temperature of the entering condenser water. In either case, the increased head pressure will overcome the spring pressure, and the regulating valve will be moved to a more open position (Figure 4-17). The increased water flow allowed through the valve will tend to offset the tendency of the head pressure to increase. Relatively constant head pressure will thus be maintained.

The absolute value of head pressure being maintained is field adjustable. The adjusting screw on the top of the regulating valve can be used to change the spring tension, thereby changing the head pressure required to open the valve.

Example

You encounter a water-cooled system (R-12) operating with 75°F entering water temperature, 90°F leaving water temperature, and a head pressure of 140 psi. Would you adjust the water regulating valve for more flow, less flow, or leave it as it is running?

Solution

With 90°F leaving water temperature, the condensing temperature should be 100°F and the condensing pressure should be 121 psi. The existing head pressure is too high. The water regulating valve should be readjusted to a lower spring pressure and more water flow.

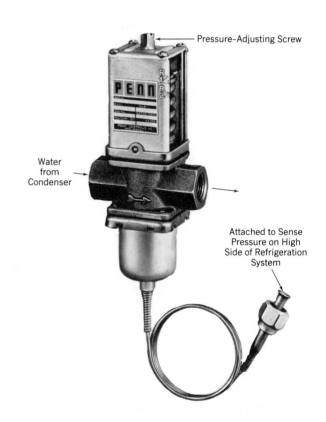

Figure 4-15 Two-way condenser water regulating valve, pressure actuated. An increase in head pressure will cause the valve to open, allowing increased condenser water flow. *(Courtesy Johnson Controls, Inc.)*

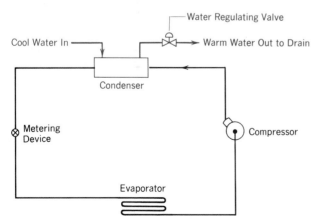

Figure 4-14 Refrigeration system with a once-through water-cooled condenser.

A normal flow rate for condenser water is 3.0 gpm/ton, which results in a 10°F water temperature rise through the condenser. For a 20-ton system, this translates into a flow rate of 60 gpm! In order to give you a reference point, a good strong water flow from a ¾-in. garden hose is only 10–15 gpm. You can readily see that unless a very inexpensive source of water is available, the water cost for a 20-ton water-cooled system would be quite high. The answer to high water cost is to recool and reuse the water leaving the condenser.

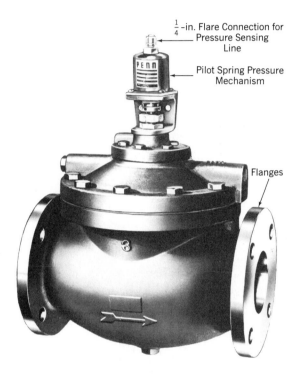

Figure 4-16 Pressure-actuated valve for a condenser requiring a high-flow capacity. Valve is pilot operated. (*Courtesy Johnson Controls, Inc.*)

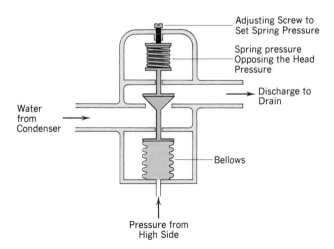

Figure 4-17 Schematic of a condenser water-pressure regulating valve.

Cooling Towers

The device used to cool the condenser water is called a **cooling tower** (Figures 4-18 and 4-19). It works on the principle that when warm water is allowed to evaporate, it absorbs its latent heat of vaporization, thus cooling. In the cooling tower, the warm water from the condenser is pumped through spray nozzles. The small droplets thus produced provide an increase in the amount of water surface area, allowing evaporation to take place easily. The droplets fall over the tower fill material, keeping the water broken up into fine droplets. The water in the bottom of the trough has been cooled to within 5°F to 10°F of the outside air wet bulb temperature (a measure of humidity), and is ready for reuse in the condenser.

Make-up water (Figure 4-20) is required for the cooling tower to replace the water lost to evaporation, **windage,** and **bleed-off.** For every 1050 Btu of heat rejected in the cooling tower, 1 lb of water is evaporated (1050 Btu/lb is the latent heat of vaporization for water at 80°F). Windage is the term used to describe the water loss from the cooling tower due to wind. That is, some water is lost to the atmosphere without first having evaporated. The third term, bleed-off, is the water that we intentionally drain from the cooling tower sump. Without this bleed-off, the evaporating water would leave behind an ever-increasing concentration of solids, minerals, and other impurities in the cooling tower water. In order to maintain the concentration of solids at an acceptably low level, we allow a continuous bleed-off rate of 10–20 percent of the water circulation rate.

The bleed-off from a cooling tower can still represent a significant operating cost to the owner. In recent years, automatic bleed-off controllers have become available (Figure 4-21). They are designed to bleed off only the minimum

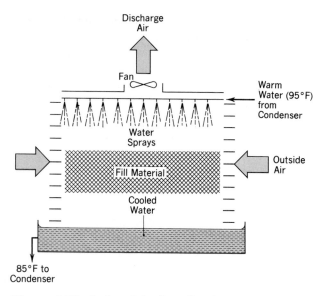

Figure 4-18 Induced-draft cooling tower.

quantity of water required to keep solids below a predetermined concentration. The controller senses the concentration of impurities in the water by continually measuring the electrical resistance of the water. The controller actually reads out in **micromhos** (pronounced micro-moze), representing the inverse value of resistance (mho spelled backwards, is ohm, the conventional measure of electrical resistance). The controller pays for itself by reducing make-up water costs, without allowing concentration of impurities to exceed acceptable levels.

As evaporation takes place cooling tower water tends to become alkaline and requires a continuous addition of acid. Figure 4-22 shows a **pH** controller (pH is a measurement of acidity or alkalinity) that automatically adds the required amount of acid or other chemicals. On a scale of 1

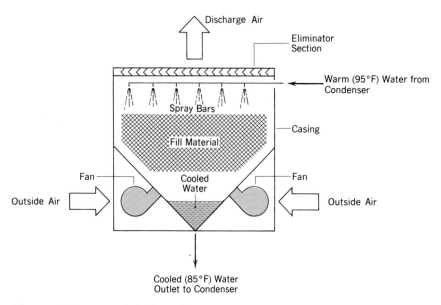

Figure 4-19 Forced-draft cooling tower.

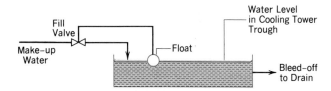

Figure 4-20 Make-up water is supplied through a float-controlled fill valve to replenish water lost through evaporation, windage, and bleed-off.

to 14, a pH below 7.0 is acidic, while a pH above 7.0 is alkaline. In addition to bleed-off and acid treatment, cooling towers also require the occasional addition of a biocide to stop the growth of algae.

The tower in Figure 4-18, called an induced-draft cooling tower, has a fan that draws outside air through the fill, helping evaporate the water. Figure 4-19 shows a different sort of tower arrangement called forced-draft or blow-through, because the fans blow the air through the fill instead of drawing it through.

Most air conditioning and refrigeration applications use cooling towers that have fans. There are some though that are natural-draft towers. Figure 4-23 shows a large natural-draft cooling tower used in some large power generating stations. The draft through the tower is produced by natural convection.

As outside air temperature drops, the temperature of the water leaving the cooling tower also drops. The same problems that were described for air-cooled condensers (due to low head pressure) apply to water-cooled condensers. As water temperature drops to 75°F, the capacity

of the cooling tower must be reduced. The first step of capacity reduction is accomplished by fan cycling. The next step is cooling tower bypass, using either a two-way valve (Figure 4-24) or a three-way valve (Figures 4-25 and 4-26). In each of these schemes, the water temperature leaving the cooling tower is sensed. If the temperature drops below the set point (usually 60°F to 70°F), the action of the tower bypass valve causes the

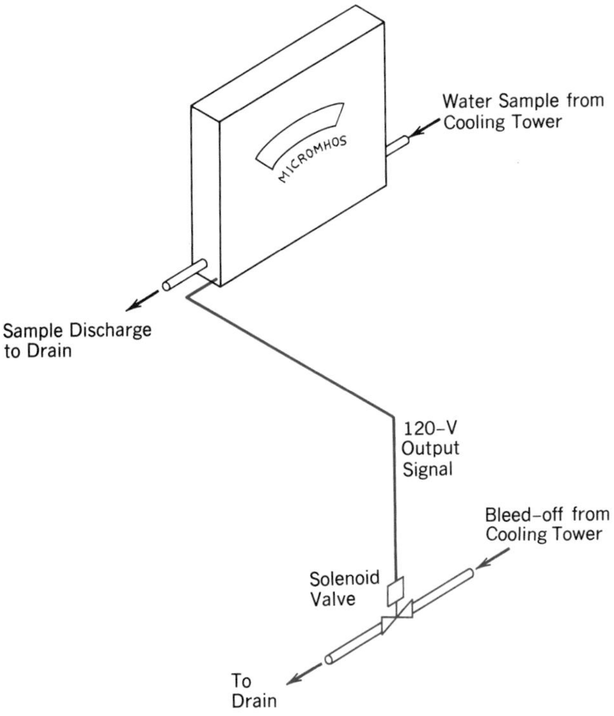

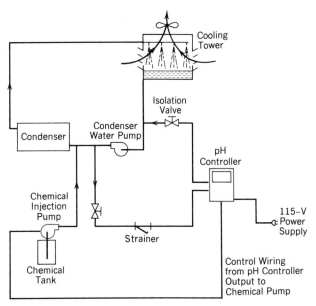

Figure 4-22 A pH controller senses the pH of the circulating condenser water. It automatically starts the chemical injection pump (acid or alkaline) when the sensed pH is not within desired limits.

Figure 4-21 Automatic bleed-off controller.

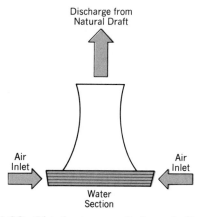

Figure 4-23 This large-capacity hyperbolic cooling tower uses no fan.

condenser water to bypass the cooling effect of the tower.

When cooling towers are required to operate where freezing outside air conditions may be encountered, some means of freeze protection is required. This is usually accomplished by the

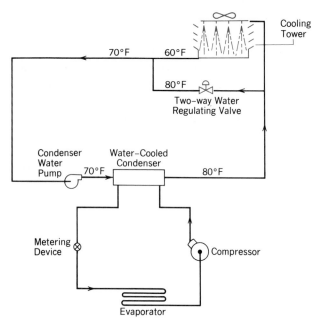

Figure 4-24 Cooling tower water temperature controlled by a two-way bypass valve.

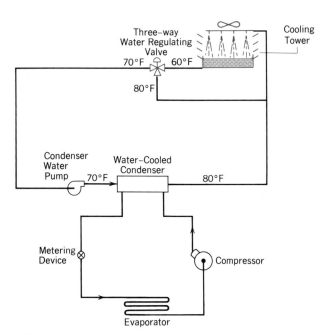

Figure 4-25 Cooling tower water temperature controlled from a three-way mixing valve.

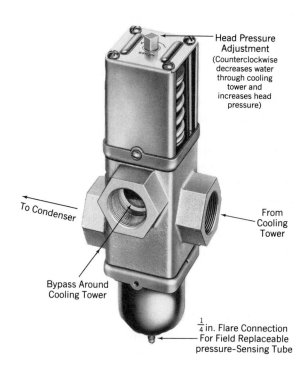

Figure 4-26 Three-way water regulating valve used on condensers that use water from a cooling tower. *(Courtesy Johnson Controls, Inc.)*

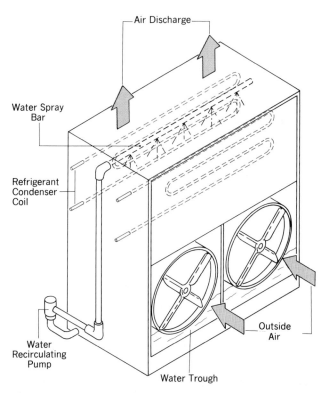

Figure 4-27 Evaporative condenser.

addition of either electric resistance heaters or steam injection into the cooling tower sump as required.

Evaporative Condensers

The **evaporative condenser** shown in Figure 4-27 contains some elements of both the air-cooled and the water-cooled condensers. The hot gas from the compressor discharge is piped to the evaporative condenser through a closed refrigerant circuit. Water is constantly recirculated over the surface of the refrigerant piping, and a fan forces outside air over the wetted refrigerant tubes. The evaporation of the water from the tube removes heat from the refrigerant, causing it to condense.

The evaporative condenser requires the same water treatment and bleed-off as the cooling tower in order to prevent the accumulation of solids. Protection against freezing must also be provided for units that operate during the winter months.

The evaporative condenser may operate with the fan system only (no pump, no water spray) until the outside air temperature rises to 70°F–80°F. At that point an outside air thermostat activates the water system.

KEY TERMS AND CONCEPTS

Air-cooled Condenser	Head-pressure Control	Cooling Tower
Water-cooled Condenser	Estimating Head Pressures	Windage
		Bleed-off
Natural Draft	Shell and Tube	Micromhos
Forced Draft	Shell and Coil	pH
Fan Cycling	Tube within a Tube	Cooling Tower Control
Shutters	Condenser Water Regulating Valve	Evaporative Condenser
Flooding		

QUESTIONS

1. What happens to the temperature of a refrigerant as it passes through a condenser?

2. What happens to the temperature of air as it passes through an air-cooled condenser?

3. What happens to the temperature of water as it passes through a water-cooled condenser?

4. An R-12 water-cooled system is operating with water entering at 75°F and leaving at 85°F. What head pressure would you consider to be normal?

5. An R-22 water-cooled system is operating with water entering at 75°F and leaving at 88°F. What head pressure would you consider to be normal?

6. An R-12 air-cooled system has 92°F air entering the condenser. What is the range of head pressures you would consider to be normal?

7. An R-500 air-cooled system has 100°F air entering the condenser. What is the range of head pressures you would consider to be normal?

8. What will happen to the head pressure of an air-cooled system if the outside air temperature falls?

9. What will happen to the head pressure of a water-cooled system if the condenser water flow is increased?

10. What are the advantages of an air-cooled system compared with a water-cooled system?

11. What are the advantages of a water-cooled system compared with an air-cooled system?

12. What are the consequences of having a condenser water flow that is less than required?

13. What are the consequences of having a condenser water flow that is more than required?

14. What operational problems will result when an air-cooled condenser has very cold entering air temperature?

15. Name three methods used to control the capacity of an air-cooled condenser when the outside air temperature falls.

16. How is the capacity of a water-cooled condenser controlled?

17. What causes water to become cooled as it passes through a cooling tower?

18. How is the capacity of a cooling tower controlled?

19. Why is bleed-off used on a cooling tower system?

20. How much water would normally be supplied to a 15-ton water-cooled condenser?

21. How much water would evaporate to the atmosphere in a 15-ton cooling tower? (One gallon of water weighs 8.3 lb.)

22. What does an automatic bleed-off controller measure in order to determine how much water to bleed?

CHAPTER 5

EVAPORATORS

The evaporator is the reason that we have a refrigeration system. In the evaporator, the cold refrigerant is allowed to absorb heat from the warmer material that needs to be cooled. The evaporator may be used to cool air, water, or any other gas or liquid.

Direct Expansion Evaporators

Evaporators that are supplied with a cold liquid and vapor mixture from the metering device and carry this mixture through tubing are called **direct expansion (D-X) evaporators** (Figure 5-1). As the refrigerant moves through the tubing, it absorbs heat from the air or water that it is trying to cool. This absorption of heat causes more of the liquid to turn into vapor, but the temperature of the refrigerant does not rise. As long as there remains some unboiled liquid, the temperature at which the boiling takes place remains constant. If the evaporator coil has been sized correctly, there will be liquid present through 90 percent of the coil. Then in the last 10 percent of

the coil, with only cool vapor remaining inside the tubing, the continued addition of heat will cause the temperature of the refrigerant to rise. It will be superheated.

Types of D-X Evaporators

The most common type of evaporator is called a finned-tube coil (Figure 5-2). Refrigerant is carried inside the tube, and the aluminum fins are mechanically attached to the tube to increase heat transfer area (Figure 5-3). Except for size,

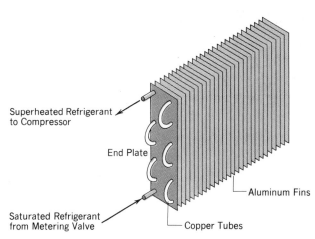

Figure 5-2 Finned-tube evaporator coil.

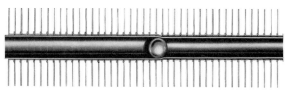

Figure 5-3 The aluminum evaporator fins are attached to the copper tube at the factory by forcing a ball through the tube. (*Courtesy The Trane Company*)

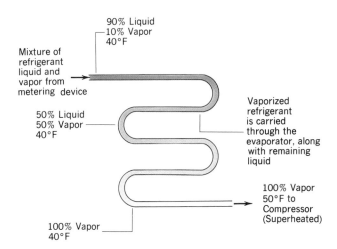

Figure 5-1 Direct expansion (D-X) evaporator.

the finned-tube evaporator is very similar to some condensers described in Chapter 4.

Evaporators may be **static** or **forced-draft.** Finned-tube evaporators are usually forced-draft. This is the type of evaporator used in air conditioners and refrigerators that use evaporator fans. Using a forced-draft evaporator makes it possible to have the evaporator coil in a different location than the area to be cooled. The air is then forced through a duct to an air conditioned space that might be quite far from the refrigeration system.

Figure 5-4 shows a type of evaporator used in residential air conditioners. The condensing unit is outside, and the evaporator sits on top of the furnace. It is a forced-draft evaporator, called an **A coil,** that uses the same fan for both heating and air conditioning.

The number of fins on finned-tube evaporators will be between 7 and 14 fins per inch. The higher **fin densities** provide more efficient heat transfer, but have the disadvantage of a higher air pressure drop and a tendency to become more easily plugged if the air is not well filtered.

Plate-type coils have no fins and can be shaped into a variety of physical shapes. Figure

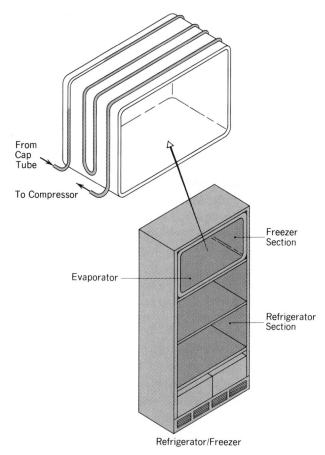

Figure 5-5   Plate evaporator formed into a freezer compartment.

Figure 5-6   Cross section of a plate evaporator made from two pieces of formed aluminum, matched to provide a path for the refrigerant.

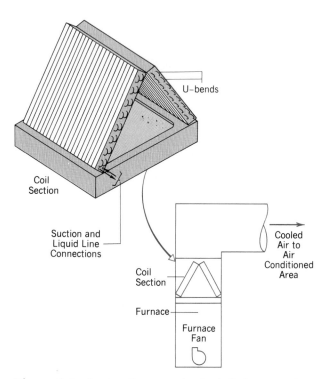

Figure 5-4   An A coil evaporator installed on a residential furnace.

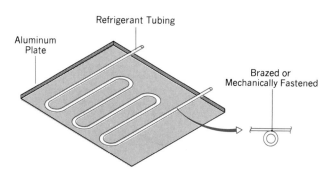

Figure 5-7   Plate evaporator formed by attaching tubing to a plate.

5-5 shows a freezer section inside a refrigerator that is nothing more than a plate-type evaporator formed into the shape of a box, while Figure 5-6 shows two separate embossed aluminum plates brazed together to form a plate with integral passages for the refrigerant. Figure 5-7 shows another common kind of construction, aluminum tubing brazed to an aluminum plate. Several of these may be used as the shelves in a freezer.

Plate-type evaporators may take on complex shapes, such as the one in Figure 5-8. In machines that make shaved or flaked ice, the cylindrical-shaped evaporator is submerged in a tank of water. The water freezes on the outside surface of the evaporator, after which an auger inside the water tank scrapes off a thin layer of ice, then moves it into a storage bin. Evaporators used for making ice are frequently constructed from stainless steel in order to prevent contamination due to corrosion.

Figure 5-9 shows another type of evaporator used for making ice. The plate is shaped into the form of an inverted ice tray. Water is then sprayed onto the surface, gradually building up the thickness of ice. When the ice is ready, the evaporator is warmed, and the ice cubes fall out.

Many chest freezers use plate-type evaporators to form the entire inside of the box, as in Figure 5-10. There are industrial applications of vapor degreasers where a solvent is used inside a tank to remove grease from parts. In order to prevent

the evaporating solvent inside from escaping to the room, a plate-type coil may be installed around the tank perimeter. When the solvent vapor contacts the cold evaporator, it condenses and falls back into the tank (see Figure 5-11).

Plate-type evaporators are even used in frozen-food processing plants (see Figure 5-12). Vegetables are packaged into rectangular boxes. The boxes are then placed between two plate-type evaporators that hold the sides of the box perfectly flat while the food inside freezes.

The tubing attached to a tank shown in Figure 5-13 is cooling the water for a drinking fountain. The drinking water is held within the tank while

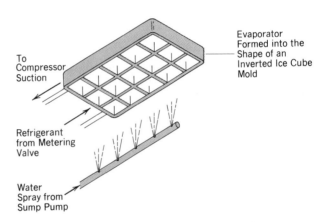

Figure 5-9 Evaporator for forming ice cubes.

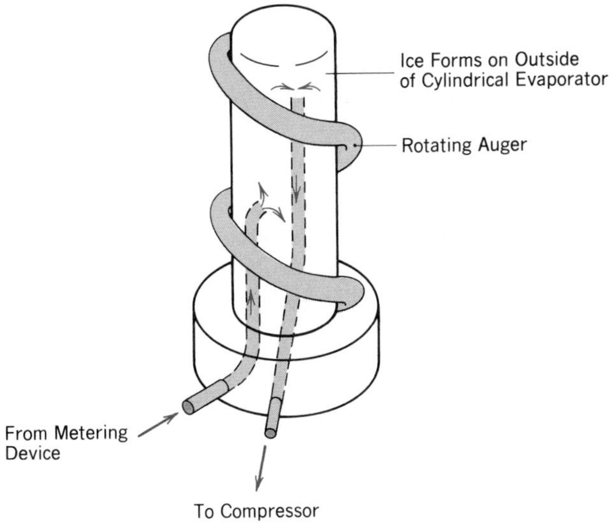

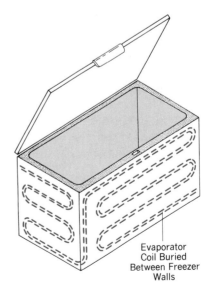

Figure 5-10 Evaporator imbedded between the walls of a chest freezer.

Figure 5-8 Evaporator used for making flaked ice.

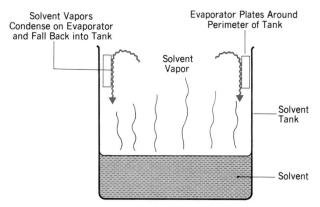

Figure 5-11 Evaporator installed on a vapor-degreasing tank to prevent escape of solvent vapors into occupied areas.

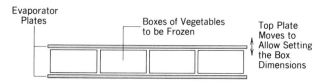

Figure 5-12 Plate evaporator used in the processing of frozen foods.

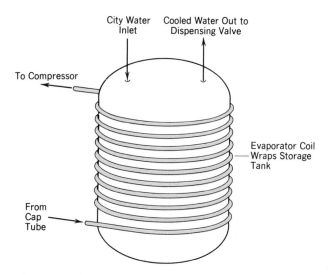

Figure 5-13 Evaporator used to cool water in a drinking fountain.

the evaporator coil attached to the outside causes the water to be cooled. This whole assembly is housed within an insulated compartment, and is usually not visible without major disassembly.

Another method of cooling liquids is to immerse a **bare tube evaporator** coil into the liquid

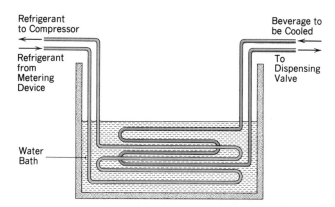

Figure 5-14 Evaporator chills a water bath which, in turn, cools a beverage coil.

(Figure 5-14). This is the common approach for beverage cooling.

Flooded Evaporators

Flooded evaporators use a bath of cold liquid refrigerant that has been supplied from the metering device. The bath contains a water coil, as in Figure 5-15. As the liquid refrigerant absorbs heat from the water, it vaporizes and is returned to the compressor. A more common use of the flooded evaporator is in water chillers. In large building air conditioning systems, it is not practical to run refrigerant piping around the whole building in order to distribute the refrigerant to many users. Instead, we use a refrigeration system that cools water to 44°F, after which it is pumped around the building to many air conditioning units (Figure 5-16). The building water returns from the building at 54°F and is then passed through the flooded evaporator, where it cools to 44°F. The refrigerant that boils is drawn to the compressor. The eliminator section is provided above the level of refrigerant to prevent the boiling liquid from becoming entrained with the vapor and passing along to the compressor.

Condensation

If you have a glass of ice water sitting on a table, it won't be long until the outside of the glass

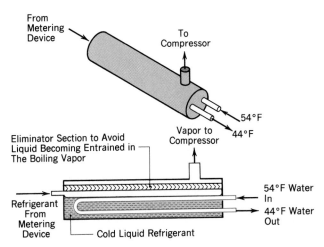

Figure 5-15 A flooded evaporator coil used to produce chilled water.

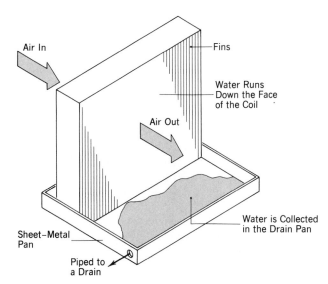

Figure 5-17 Condensate that forms on the coil is collected in a drain pan.

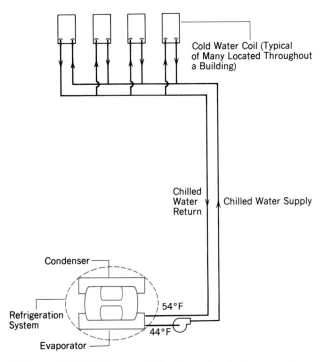

Figure 5-16 Water is chilled in a flooded evaporator and pumped to various parts of the building for use in air conditioning.

becomes wet. The water came from the water vapor that is in the air. As the air around the glass of ice water became cooled, the water vapor in the air condensed onto the outside surface of the glass. Condensation of water vapor on a cold surface is sometimes referred to as **sweating.**

When air is cooled by an evaporator, conden-

sation usually occurs. If the evaporator surface happens to be at a temperature below freezing, the water vapor that condenses on the surface also freezes. With static plate coils, this ice can be allowed to build up to a considerable thickness before it must be manually removed. On forced-draft finned-tube evaporators operating at subfreezing temperatures, however, the evaporator must be **defrosted** several times a day. Automatic defrost is discussed later in the chapter.

For evaporators operating at refrigerant temperatures between 32°F and 50°F, the water that condenses out of the air runs down the face of the coil and into a drain pan (Figure 5-17). A 3-ton evaporator coil used to provide the cooling for an 1800-sq-ft house could easily produce up to 2 gal of condensate in 1 hour! The condensate is usually carried away by gravity into a drain.

In some applications, there is not a drain available near the evaporator coil. The condensate must be pumped up, perhaps into an attic space, and then to a drain or to the outside (Figures 5-18 and 5-19). The **condensate pump** and tank unit is controlled by a level switch. When the level of condensate in the tank reaches a predetermined level, the pump starts. If the pump fails, or if the discharge line becomes clogged so that the pump cannot remove the condensate, a high level switch will open and cause the refrigeration system to stop. This prevents the possibility of a flood.

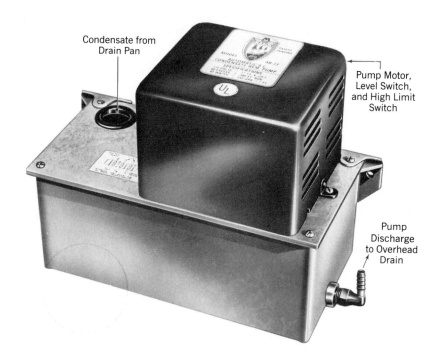

Figure 5-18 Condensate pump. *(Courtesy March Manufacturing, Inc.)*

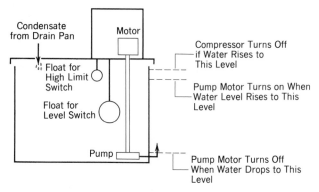

Figure 5-19 Operation of condensate pump.

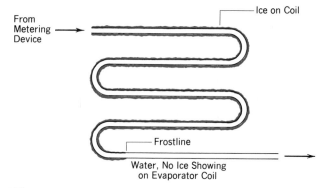

Figure 5-20 The frostline occurs very shortly after the last drops of liquid refrigerant have been vaporized.

Frostline

The **frostline** in an evaporator refers to the point at which the liquid inside the evaporator has been expended. In the remainder of the evaporator the refrigerant becomes superheated. The location of the frostline can be observed from outside the evaporator. By observing where the outside temperature of the evaporator has changed, you know that the frostline was only a short distance upstream. Evaporator temperature change may be noted as the point where the condensation on the evaporator surface no longer freezes (Figure 5-20). With evaporators operating at temperatures above freezing, the frostline may be observed by noting the point at which condensation no longer occurs. For evaporators without such obvious indicators, frostline may only be determined by taking temperature readings with a thermometer.

Evaporator Operating Pressure

The temperature of the refrigerant inside the evaporator is determined by the pressure inside the evaporator. Where evaporator temperatures are below 26°F to 30°F, the surface temperature will be cold enough to freeze the condensate. Figure 5-21 gives approximate evaporator operating pressures for a number of applications. For applications in which the correct suction pressure is unknown, it may be reasonably estimated. Consider a box that needs to be maintained at −20°F using an R-12 system. In order to maintain −20°F in the box, the cold air discharging into the box will have to be colder than the box. Estimate the discharge air 10°F colder than the box, or −30°F. Now, in order to cool the air to −30°F, the refrigerant will need to be 10°F lower than −30°F. Estimate the required refrigerant temperature at −40°F. This corresponds to a suction pressure of 11 in. Hg for R-12.

It is possible for a system to operate with evaporator temperatures well below 32°F and not require defrosting. The capacity of the system must be large enough that the box temperatures can be maintained while the refrigeration system need only run about half the time. Also, the box temperature must be at 35°F or higher. In this way, when the refrigeration system operates, a slight amount of frost builds up on the coil, but during the off cycle, it melts, leaving the coil clean before the next cycle.

Application	Refrigerant	Normal Suction Pressure
Domestic refrigerator/ freezer	12	4 in. Hg–2 psi
Air conditioning	22	60–75 psi
Ice maker	502	25–35 psi
Water chiller (44°F leaving water)	11	16 in. Hg
Auto air conditioner	12	30 psi

Figure 5-21 Approximate evaporator operating pressures.

Evaporator Capacities

Evaporator capacity is actually system capacity. Calculation of the capacity required is a lengthy and complex process that is described in Chapter 26. Some applications are so common, however, that there are some rules of thumb can be used with a fair degree of accuracy.

Figure 5-22 gives evaporator capacities required for walk-in boxes. These are figured for boxes constructed with 3-in. foam insulation, normal product load, and a design temperature 55°F lower than the surrounding temperature. For boxes subject to heavy traffic, different design temperature, or of different construction, a load calculation should be done.

For comfort air conditioning, the required evaporator capacity is normally within the range of 300 to 600 sq ft of conditioned area for each ton of refrigeration capacity. Offices and commercial occupancies will tend to be closer to the 300 figure, while residences are closer to the 600 figure. The major impacts on these loads are number, size, and direction of windows, lighting densities for office applications, type of construction (insulating values), and type of occupancy. With so many variables, these guides should be used only for the roughest type of estimates and not as a substitute for accurate load calculations.

Defrosting Evaporators

For evaporators that operate at temperatures low enough to maintain space temperatures at less than 32°F, the condensate on the evaporator surface will freeze. This is commonly the case in domestic as well as walk-in-type freezers. If the evaporator is a plate type, it may be feasible to operate for days or even weeks at a time without defrosting. If a finned-tube coil with an evaporator fan is being used, however, the formation of ice will very quickly cause problems. The space for air passage between fins is relatively narrow, on the order of $\frac{1}{8}$ in. or less. When any thickness of frost is allowed to build up on the fins, the airflow is restricted. The build-up reduces the evaporator load, causes suction pressure to drop lower, and increases the rate of frost formation. It

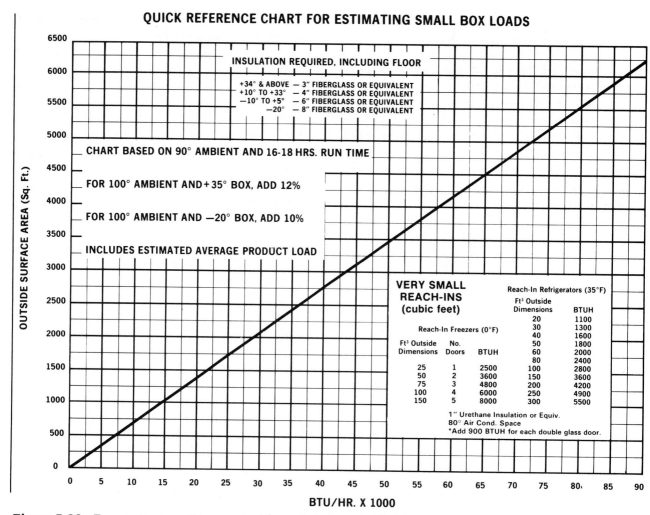

QUICK REFERENCE CHART FOR ESTIMATING SMALL BOX LOADS

INSULATION REQUIRED, INCLUDING FLOOR

+34° & ABOVE — 3" FIBERGLASS OR EQUIVALENT
+10° TO +33° — 4" FIBERGLASS OR EQUIVALENT
−10° TO +5° — 6" FIBERGLASS OR EQUIVALENT
−20° — 8" FIBERGLASS OR EQUIVALENT

CHART BASED ON 90° AMBIENT AND 16-18 HRS. RUN TIME

FOR 100° AMBIENT AND +35° BOX, ADD 12%

FOR 100° AMBIENT AND −20° BOX, ADD 10%

INCLUDES ESTIMATED AVERAGE PRODUCT LOAD

VERY SMALL REACH-INS (cubic feet)

Reach-In Freezers (0°F)

Ft³ Outside Dimensions	No. Doors	BTUH
25	1	2500
50	2	3600
75	3	4800
100	4	6000
150	5	8000

Reach-In Refrigerators (35°F)

Ft³ Outside Dimensions	BTUH
20	1100
30	1300
40	1600
50	1800
60	2000
80	2400
100	2800
150	3600
200	4200
250	4900
300	5500

1" Urethane Insulation or Equiv.
80° Air Cond. Space
*Add 900 BTUH for each double glass door.

Y-axis: OUTSIDE SURFACE AREA (Sq. Ft.) — 0 to 6500
X-axis: BTU/HR. X 1000 — 0 to 90

Figure 5-22 Evaporator capacities required for walk-in boxes with 3-in. foam insulation, average product load, and a temperature 55°F lower than the surroundings. *(Courtesy Russell Coil Company)*

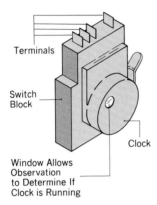

Figure 5-23 Electric defrost heater installed on the face of an evaporator coil.

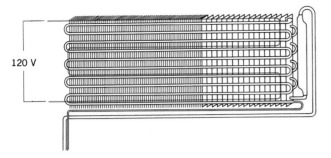

Figure 5-24 Defrost timer used in refrigerator-freezers.

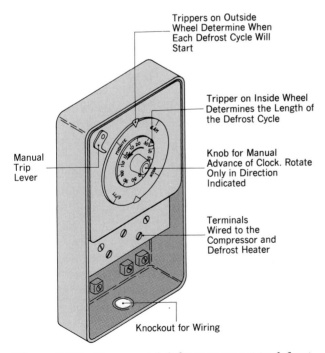

Figure 5-25 Commercial defrost timer set to defrost twice daily for 35-min duration.

will not take long before we find our evaporator coil encased in a solid block of ice!

For this type of system, an automatic defrost is provided. Figure 5-23 shows an evaporator coil with an electric heater element. A timer such as that shown in Figures 5-24 and 5-25 causes the compressor to stop, and the heater to be energized. The defrost period will be between 30 min and 1 hr. During this period, the condenser and evaporator fans also remain off. After that time has elapsed, the timer motor allows the system to return to normal operation.

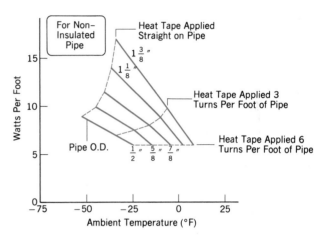

Figure 5-26 Electrical heating required to prevent pipe freezing. Turns per foot of pipe applies only to heat tape that has a capacity of 6 watts/ft.

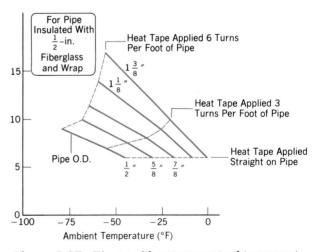

Figure 5-27 Electrical heating required to prevent pipe freezing for insulated pipe and thermostatic control. Turns per foot applies only to heat tape that has a capacity of 6 watts/ft.

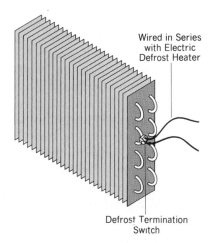

Figure 5-28   The defrost termination switch senses when the evaporator surface temperature rises to 40°F and turns off the electric defrost element.

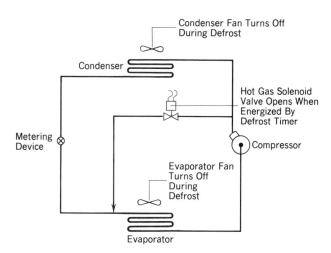

Figure 5-29   Hot gas defrost system.

When the ice melts, the water drains out of the coil section. Drain heaters are provided to prevent blocking this drain. For field-installed evaporators, heat tape is wrapped around the drain. The number of turns recommended is shown in Figures 5-26 and 5-27. If the heat tape is not thermostatically controlled, it should not be insulated, otherwise it may melt.

The thermostat shown in Figure 5-28 is called a termination thermostat. It senses the temperature of the evaporator coil surface. If the heater is able to defrost the evaporator before the timer says its time to resume normal operation, the termination thermostat will turn off the heater for the remainder of the defrost period. This saves energy in two ways. Less energy is consumed by the heater, and less energy will need to be expended by the refrigeration system to remove the heat that the electric heater has added to the space.

A second method of defrosting the evaporator called **hot gas defrost** is shown in Figure 5-29. With this system, the timer energizes a valve that allows the hot gas from the compressor discharge to be routed directly to the evaporator. The compressor continues to operate during the hot gas defrost cycle.

Defrost timers for domestic freezers are normally set for either two, three, or four defrosts per 24 hr. Depending on how the timer is wired into the circuit, this may be 24 hr of clock time or 24 hr of running time. For commercial timers, mov-

able pins are used to set both the number of defrosts per 24 hr and the duration of each defrost.

Circuiting

The evaporator coil shown in Figure 5-1 has only a single tube running back and forth in the coil. All the refrigerant follows the same path. This coil is said to have only one **circuit.** In larger coils this circuit can get to be quite long, and the refrigerant would experience an unacceptable pressure drop. In Figure 5-30, the coil has been circuited so that there are two tubes being fed with refrigerant. There are two independent parallel paths for the refrigerant through the coil. There may be as many as 15 or 20 circuits provided in large evaporator coils.

With large, multicircuited coils, a **distributor** such as that shown in Figure 5-31 is used. It receives the mixture of liquid and vapor from the metering device, and assures that it is distributed evenly among the different circuits. The odd arrangement of tubing that connects the distributor to the evaporator is constructed in such a fashion that the pressure drop through each tube is the same. This tangle of tubing is often referred to as spaghetti (Figure 5-32).

In some cases, as in Figure 5-33, the evaporator coil is piped so that it functions as two separate evaporators. Each section may even be piped to a separate refrigeration system. The

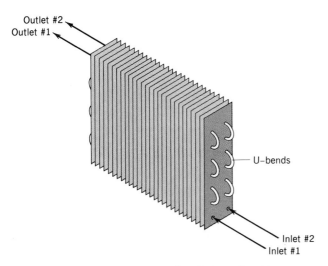

Outlet #2
Outlet #1

U–bends

Inlet #2
Inlet #1

Figure 5-30 Evaporator coil with two circuits.

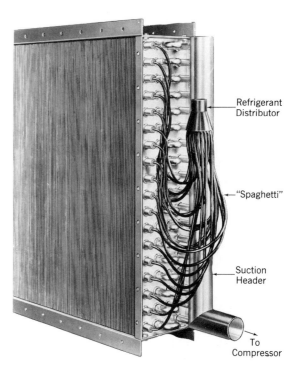

Refrigerant
Distributor

"Spaghetti"

Suction
Header

To
Compressor

Figure 5-32 A refrigerant distributor supplying 18 separate refrigerant circuits. *(Courtesy Sporlan Valve Company)*

evaporator may be split either vertically or horizontally. When air passing the coil is split horizontally, half the air passing through the coil will not be cooled at all if only half of the coil is working. The vertically split coil will cool all of the air, but only half as much as if the entire coil is in operation.

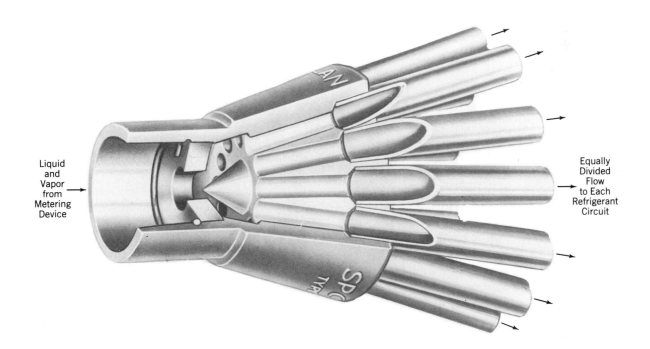

Liquid
and
Vapor
from
Metering
Device

Equally
Divided
Flow
to Each
Refrigerant
Circuit

Figure 5-31 Refrigerant distributor. *(Courtesy Sporlan Valve Company)*

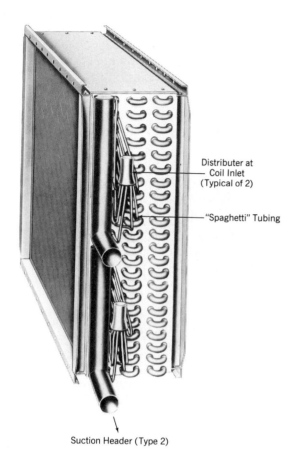

Figure 5-33 Evaporator coil six rows deep with multiple refrigerant circuits fed from two distributors. The coil circuiting is split horizontally. (Courtesy The Trane Company)

Flow Direction

There have been an embarrassingly high number of new installations where the system had inadequate cooling capacity, and the culprit turned out to be a backwards evaporator coil! It is an easy mistake to make, as the entering side and the leaving side of the evaporator coil look very much alike. The trick in getting it installed properly is to remember that the coldest refrigerant should contact the coldest air. Figures 5-34 and 5-35 show both the correct and incorrect piping arrangements. Figure 5-34 shows the incorrect method. With an evaporator operating with a 40°F saturated suction temperature, and 10°F of

superheat, the coldest that the air could be leaving the coil is 50°F. Unless the coil surface area were extremely large, the air temperature leaving the coil would probably not be lower than 55°F.

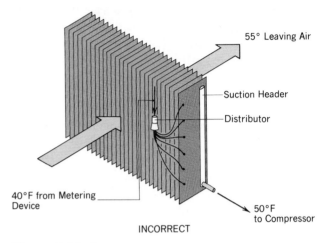

Figure 5-34 Incorrect airflow direction. The leaving air contacts the warmest refrigerant.

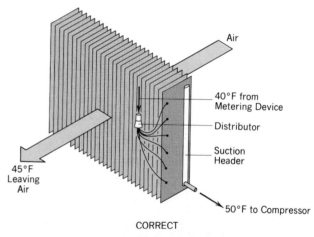

Figure 5-35 The correct airflow direction allows the coldest air to contact the coldest refrigerant.

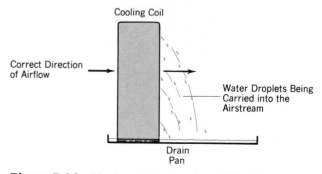

Figure 5-36 The longest extension of the drain pan is on the leaving air side of the coil.

Compare that to the correct arrangement in Figure 5-35. With the leaving air able to see the 40°F entering refrigerant, we could reasonably expect to be able to cool the air to 45°F.

There is another clue that you might look for in determining the correct **flow direction.** If the coil is supplied with a drain pan to catch the condensate, the drain pan will extend well beyond the leaving air side of the coil (see Figure 5-36).

Water Coils

Strictly speaking, an evaporator is a device where refrigerant evaporates. Coils that are supplied with cold water instead of refrigerant, however, are also referred to as evaporator coils. Figures 5-37 and 5-38 show a four-row **water coil.** Water chilled to a temperature between 40°F and 45°F is supplied to a header that supplies water to all the

tubes in one row. This type of circuiting is called single-row serpentine. Different circuiting arrangements are used for higher or lower water flows. It is desirable to have water velocities

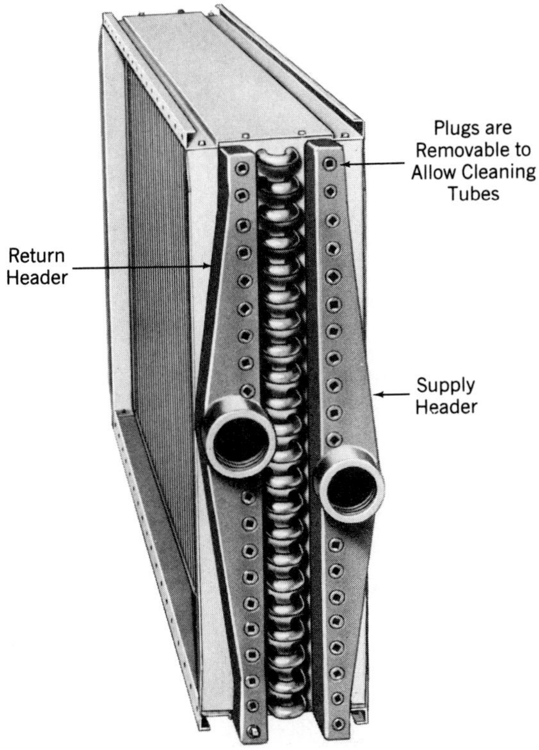

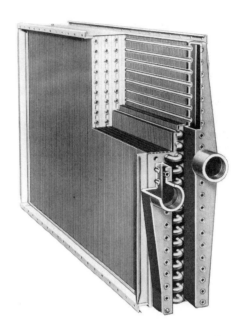

Figure 5-38 Cutaway of a cleanable chilled water coil. *(Courtesy The Trane Company)*

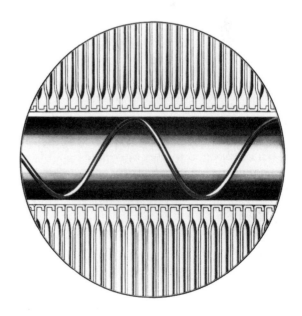

Figure 5-39 A spring device inside a water coil tube increases turbulence and heat transfer when water velocities are low. *(Courtesy The Trane Company)*

Plugs are Removable to Allow Cleaning Tubes

Return Header

Supply Header

Figure 5-37 Chilled water coil. *(Courtesy The Trane Company)*

through the tubes of 3 to 6 ft/sec. Velocities that are too high can cause tube erosion, while velocities that are too low will prevent proper heat transfer from taking place.

When very limited flow rates of water are available, devices to cause increased turbulence (and improved heat transfer) are used (Figure 5-39). These are normally factory installed.

KEY TERMS AND CONCEPTS

D-X Evaporator

Static Evaporator

Forced-draft Evaporator

A Coil

Fin Density

Plate Coil

Bare Tube Evaporator

Flooded Evaporator

Sweating

Defrosting the Evaporator

Condensate Pump

Frostline

Estimating Correct Evaporator Pressure

Load Estimating

Electric Defrost

Hot Gas Defrost

Circuiting

Distributor

Flow Direction

Water Coils

QUESTIONS

1. What is the evaporator capacity (Btu/hr) of a 5-ton air conditioning system?

2. At what point in the evaporator would you find superheated refrigerant?

3. What happens to the refrigerant temperature inside the evaporator as the liquid/vapor combination is absorbing heat?

4. Where would you find the evaporator coil in a residential central air conditioning system?

5. What pressure would you expect in the evaporator of a window air conditioner using R-22 refrigerant?

6. What pressure would you expect in the evaporator of a refrigerator/freezer using R-12 refrigerant?

7. What happens to the condensate that forms on the evaporator of an air conditioner?

8. Name two sources of heat that may be used to automatically defrost an evaporator.

9. What evaporator capacity would be required for an average walk-in cooler 10 ft by 10 ft by 10 ft?

10. Estimate the evaporator capacity required for an average residential air conditioner for an 1800-sq-ft house.

11. Estimate the evaporator capacity required for a 5000-sq-ft office area.

12. How often will an automatic-defrost household freezer defrost itself?

13. When is it necessary to use a condensate pump?

14. If you have two identical-size coils with the same number of refrigerant tubes, which will have lower pressure drop on the refrigerant side, the one with four circuits or the one with six circuits?

15. What is the correct direction of airflow through the evaporator coil relative to the direction of refrigerant flow?

16. What is a flooded evaporator?

17. How is shaved ice formed?

18. How can you identify the point in the evaporator where all of the liquid has become vaporized?

19. Why must some evaporators be defrosted?

20. What function is served by a defrost termination thermostat?

CHAPTER 6

METERING DEVICES

The **metering device** is the part of the system that causes the pressure drop as refrigerant flows from the high pressure side of the system to the low side. Metering device is a general term that is used to describe any of a number of different devices that are explained in this chapter.

Metering Device Theory

Refrigerant enters the metering device from the condenser as either a saturated or a subcooled liquid. The metering device presents a significant obstruction to the flow of this liquid and causes a pressure drop. Some of the liquid flashes, causing a reduction in temperature. No heat is gained or lost by the refrigerant as it passes through the metering device, even though a significant reduction in temperature occurs.

Figure 6-1 illustrates how the temperature can drop with no accompanying loss in heat content. Suppose the condenser is supplying 10 lb/min of R-12 at a saturated liquid condition at 100°F. Each pound of liquid contains 31.1 Btu, and the total amount of heat in the refrigerant entering the metering device is 311 Btu/min. When the refrigerant's pressure drops to 37 psig, we have only 7.85 lb/min of liquid, and 2.15 lb/min of gas that has been produced by the flashing of the liquid. Each pound of the liquid now contains only 17.3 Btu, but each pound of vapor contains 81.4 Btu. A little arithmetic will show that the total heat content of the liquid plus the vapor leaving the metering device is equal to the 311 Btu/min that was entering.

In the previous example, most of the refrigerant leaving the metering device is liquid. In fact, from 75 to 90 percent liquid leaving the metering device would be considered normal. But, if you have an opportunity to see the refrigerant at this point in a trainer with transparent tubing, it will appear as if just the opposite is true. That is, it will appear as if there is primarily vapor leaving the metering device, because each pound of vapor occupies a space hundreds of times larger than a pound of liquid.

Capillary Tube

The **capillary tube,** called **cap tube** for short, shown in Figure 6-2 is the most commonly used metering device on small refrigeration and air conditioning systems. It is used on virtually all domestic refrigerators and window air conditioners. The capillary tube consists of a long small-diameter copper tube. As the liquid from the condenser is pushed through such a small passageway, the friction between the refrigerant

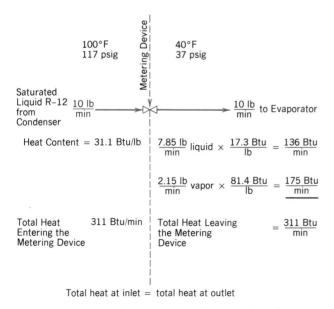

Figure 6-1 Heat balance around a metering device.

and the tube causes a pressure drop. When this pressure drop causes flashing of the liquid to occur, the additional space occupied by the flash gas causes the rate of pressure drop to increase rapidly. If the capillary tube is properly sized, the point of very rapid pressure drop will occur during the last quarter of the length of the capillary tube.

Correctly sizing the capillary tube depends on two factors: the inside diameter of the tube and the length of the tube. The same job may be performed by a short capillary tube of small diameter, or a longer tube with a larger diameter. From the standpoint of reliability, the larger diameter is preferred, as it is less prone to becoming restricted.

Even large-diameter capillary tubes must be protected from becoming clogged by particles that may have been left inside the system during manufacture. Filter driers such as those shown in Figure 6-3 are normally installed in front of the cap tube. Whenever a cap tube system has been opened, the filter drier should be replaced.

As a further note to avoid future confusion, a capillary tube as described in this section pertains to a tube that carries a flow of refrigerant and causes a pressure drop. Capillary tubes are also widely used for other services, such as to sense pressure. Not all cap tubes you find are being used as metering devices.

When cutting a capillary tube to the proper length, do not use a tube cutter. A tube cutter deforms (squeezes) the tube, thus slightly reducing the inside diameter (Figure 6-4). The correct way to cut a cap tube is to score the outside with a knife or a file. Bending back and forth a few times will cause the tube to break, leaving a full dimension inside diameter (Figure 6-5).

Cap tubes are frequently soldered or brazed into larger size tubing. Figure 6-6 shows how the larger tube is crimped onto the cap tube. It is important that the cap tube be inserted 1 to $1\frac{1}{2}$ in. into the larger tube to prevent solder from accidently flowing into the small opening of the cap tube.

The cap tube is commonly much longer than the distance from the condenser to the evaporator, the excess length accommodated by rolling the cap tube into a coil. Extreme care must be taken not to kink the capillary tube. This may be avoided by using any solid cylindrical shape as a

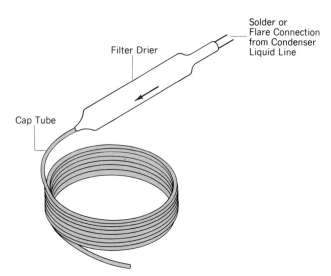

Figure 6-3   Cap tube protected by a filter drier.

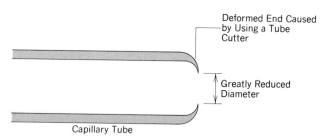

Figure 6-4   Enlarged view of a cap tube that has been improperly cut to length with a tube cutter.

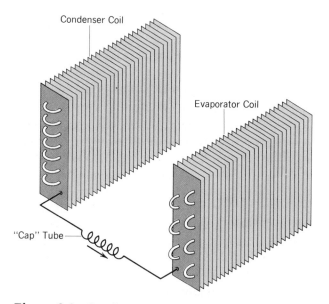

Figure 6-2   Capillary tube used as a metering device on a window air conditioner.

form to wrap the capillary tube around. Anything from a D-size battery to a tin can will work adequately as a form.

The obvious advantage of a capillary tube as a metering device is that it is inexpensive and has no moving parts. Because it cannot change in order to match the different amounts of refrigerant that may be flowing through the system, however, its use is restricted to those systems that have a relatively constant load.

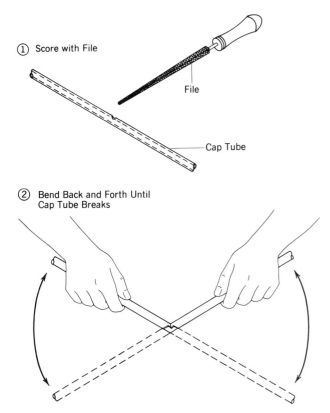

① Score with File

File

Cap Tube

② Bend Back and Forth Until Cap Tube Breaks

Figure 6-5 Correct method for cutting a cap tube to the correct length.

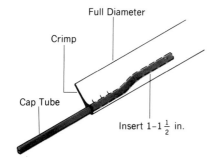

Full Diameter

Crimp

Cap Tube

Insert 1–1½ in.

Figure 6-6 Cap tube crimped into a larger diameter tube.

When a capillary tube is replaced, the length and diameter of the original should be matched. Where this is not possible, Figures 6–7 and 6–8 may be used as a sizing guide.

Thermostatic Expansion Valves

The **thermostatic expansion valve (TXV)** is another very popular type of metering device (Figure 6-9). It is actually a valve that modulates in order to allow the correct amount of refrigerant to enter the evaporator. In the chapter on evaporators (Chapter 5), it was explained that the liquid refrigerant will all be converted to vapor at a point near the end of the coil. If the metering device supplies too much refrigerant to the evaporator, we have a possibility of getting liquid back to the compressor (Figure 6-10), which can result in a ruined compressor. If the metering device does not supply enough refrigerant to the evaporator, there will not be enough liquid to absorb all the evaporator load, and the vapor leaving the evaporator will be very superheated (Figure 6-11).

The thermostatic expansion valve (commonly abbreviated TXV or TEV) is a very intelligent device. Its job is to sense the pressure and the temperature of the refrigerant leaving the evaporator, and determine the number of degrees of superheat. If there is too much superheat, more refrigerant will be allowed to flow. Not enough superheat will cause the TXV to close down and reduce the refrigerant flow. Let's see how the TXV accomplishes its job.

The schematic of the TXV in Figure 6-12 shows three different pressures acting on a diaphragm. A temperature-sensing bulb containing the same refrigerant that is in the system senses the temperature of the refrigerant leaving the evaporator. The pressure produced by the refrigerant in the bulb is transmitted to the top of the diaphragm. A pressure-sensing line senses the pressure at the evaporator outlet and transmits that pressure to the bottom of the diaphragm. The third pressure is produced by a spring and it acts on the bottom of the diaphragm. The amount of spring pressure can be set manually by an adjusting screw. Some typical pressures are shown in Figure 6-12 for an R-12 system. This system is operating at a suction

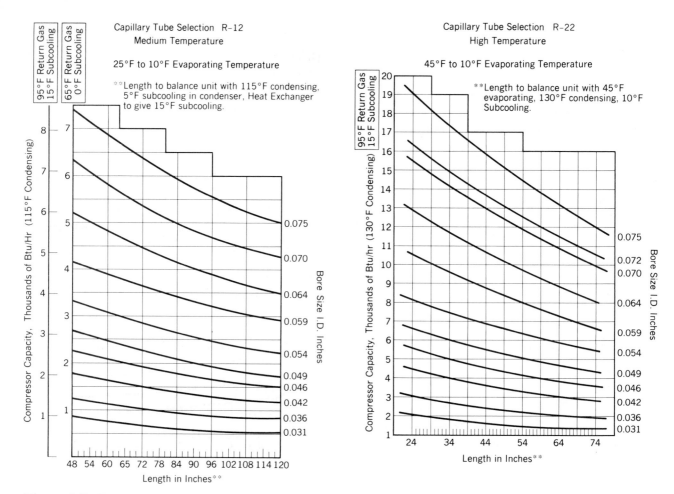

Figure 6-7 Tonnage capacities of cap tubes. *(Courtesy Copeland Corporation)*

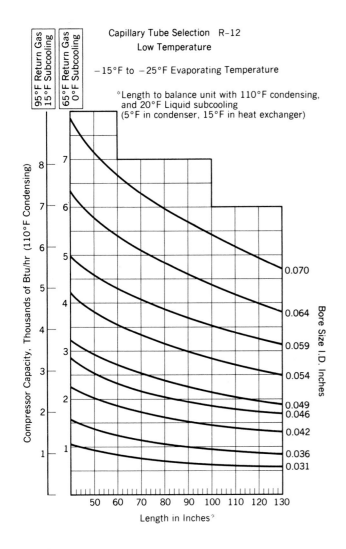

Capillary Tube Selection R-12
Low Temperature

−15°F to −25°F Evaporating Temperature

*Length to balance unit with 110°F condensing,
and 20°F Liquid subcooling
(5°F in condenser, 15°F in heat exchanger)

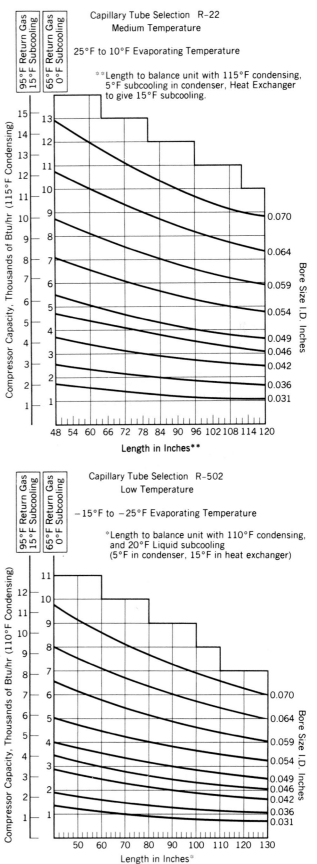

Capillary Tube Selection R-22
Medium Temperature

25°F to 10°F Evaporating Temperature

**Length to balance unit with 115°F condensing,
5°F subcooling in condenser, Heat Exchanger
to give 15°F subcooling.

Capillary Tube Selection R-502
Low Temperature

−15°F to −25°F Evaporating Temperature

*Length to balance unit with 110°F condensing,
and 20°F Liquid subcooling
(5°F in condenser, 15°F in heat exchanger)

Original Inside Diameter	Inside Diameter of Replacement					
	0.031	0.036	0.044	0.050	0.064	0.070
0.028	1.59					
0.030	1.16					
0.031	1.00					
0.032	0.86					
0.033	0.75	1.54				
0.034	0.65	1.35				
0.035	0.58	1.16				
0.036	0.50	1.00				
0.037		0.90				
0.038		0.80				
0.039		0.71				
0.040		0.62	1.55			
0.041		0.56	1.38			
0.042		0.50	1.24			
0.043			1.11			
0.044			1.00			
0.045			0.90			
0.046			0.82	1.47		
0.047			0.74	1.31		
0.048			0.67	1.20		
0.049			0.61	1.09		
0.050			0.56	1.00	1.56	
0.051			0.51	0.93	1.44	
0.052				0.85	1.32	
0.053				0.78	1.20	
0.054				0.70	1.09	
0.055				0.64	1.00	
0.056				0.60	0.94	
0.057				0.55	0.87	
0.058				0.51	0.80	1.50
0.059					0.73	1.00
0.060					0.67	0.73
0.064					0.50	0.54
0.070						
0.075						
0.080						
0.085						
0.090						

Multiply length of original cap tube by the factor in this table to find length of replacement cap tube. If less than 3 ft, select a smaller diameter replacement

Figure 6-8 Equivalent lengths of different diameter cap tubes.

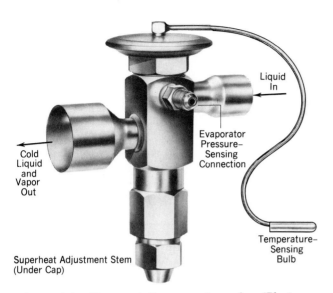

Figure 6-9 Thermostatic expansion valve. *(Photo Courtesy Sporlan Valve Company)*

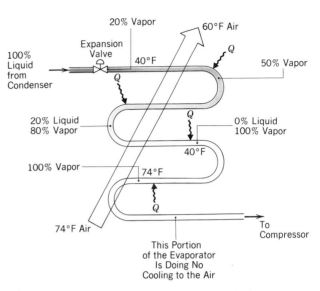

Figure 6-11 This evaporator is "starved" for refrigerant. All the liquid has been vaporized at a point only halfway through the coil. The cooling effect has been reduced.

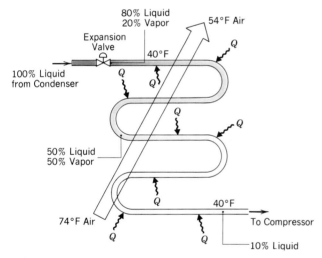

Figure 6-10 This evaporator is being fed too much refrigerant by the expansion valve, resulting in a flooded coil. The heat (Q) being absorbed by the cold refrigerant is not enough to vaporize all the liquid. Slugging of the compressor with liquid may result.

pressure of 37 psig, a saturated suction temperature of 40°F, and the refrigerant leaving the evaporator is at 50°F. It has 10°F of superheat. The temperature-sensing bulb senses the 50°F, and sends a pressure to the TXV of 46.7 psi. The spring pressure is 9.7 psi. The pressure on the top of the diaphragm balances the pressures on the bottom of the diaphragm. The operation continues with these pressures and no change

in refrigerant flow occurs until an imbalance occurs.

Let's say the load on the evaporator has now increased. Initially, this additional load on the same flow of refrigerant will cause the temperature of the refrigerant leaving the evaporator to rise. This will cause the bulb pressure to increase on the diaphragm, and force the valve farther from its seat (Figure 6-13). This allows more refrigerant to flow through the valve. The suction pressure increases, the saturated suction temperature increases, and the capacity of the evaporator increases to match the load. The only value that remained constant when the load changed was the number of degrees of superheat in the refrigerant leaving the evaporator. With a decrease in load, the bulb pressure will decrease, the valve will modulate closed, and the quantity of refrigerant admitted to the evaporator will be decreased.

The TXV just described is called an **externally equalized** valve. An **internally equalized** valve is shown in Figures 6-14, 6-15, and 6-16. The internally equalized valve may be used only with evaporators that have a low pressure drop. Typically, this would mean less than a 2 psi drop from the evaporator inlet to the evaporator outlet. In that case, the TXV senses the evaporator pressure internally, where the refrigerant is leaving the valve. The principle of operation is the same, ex-

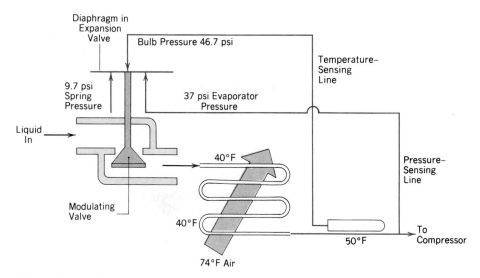

Figure 6-12 Pressures acting upon the diaphragm of a thermostatic expansion valve.

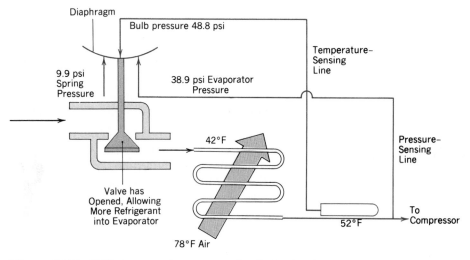

Figure 6-13 With increased evaporator load, the diaphragm is pushed down, opening the valve. The same approximate superheat setting is maintained.

cept that the TXV outlet pressure is used as an approximation of the evaporator outlet pressure. There is no operational advantage to using an internally equalized thermostatic expansion valve. It is used only to reduce the cost of the system and still provide control comparable to the externally equalized valve.

The mounting of the temperature-sensing bulb is important for accurate operation of the TXV (Figure 6-17). On horizontal lines, it is mounted at 4 o'clock or 8 o'clock. It should be mounted low so that it will sense any low temperature liquid coming from the evaporator. We don't want it on the bottom of the tube because there may be an oil film inside the tube that would act as an insulator. The bulb must be tightly strapped to the suction line, and well insulated to prevent the bulb from being affected by ambient air temperature. On vertical suction lines, the bulb is mounted so that the pressure-

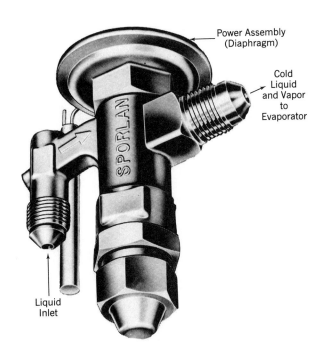

Power Assembly (Diaphragm)

Cold Liquid and Vapor to Evaporator

Liquid Inlet

Figure 6-14 Internally equalized TXV. There is no external pressure connection to sense evaporator pressure. *(Courtesy Sporlan Valve Company)*

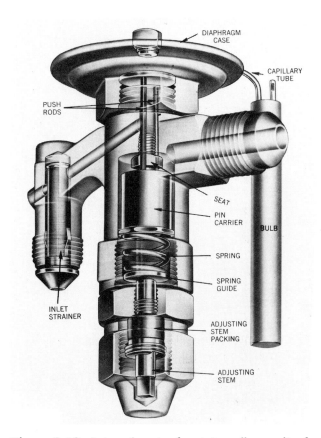

DIAPHRAGM CASE

CAPILLARY TUBE

PUSH RODS

SEAT

PIN CARRIER

BULB

SPRING

SPRING GUIDE

ADJUSTING STEM PACKING

INLET STRAINER

ADJUSTING STEM

Figure 6-16 Internal parts of an internally equalized TXV. The pushrods are triangular in cross section, passing through round openings. The TXV outlet pressure leaks around the pushrods to the bottom side of the diaphragm. *(Courtesy Sporlan Valve Company)*

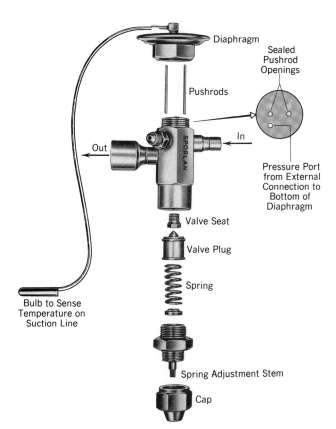

Diaphragm

Sealed Pushrod Openings

Pushrods

Out

In

Pressure Port from External Connection to Bottom of Diaphragm

Valve Seat

Valve Plug

Spring

Bulb to Sense Temperature on Suction Line

Spring Adjustment Stem

Cap

Figure 6-15 Internal parts of an externally equalized TXV. *(Photo © Sporlan Valve Company)*

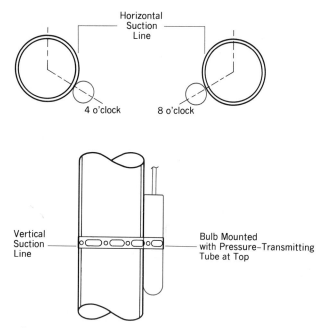

Figure 6-17 Recommended mounting location for TXV temperature-sensing bulb.

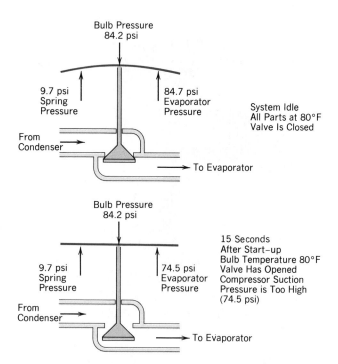

Figure 6-18 Compressor overload caused by start-up under warm conditions.

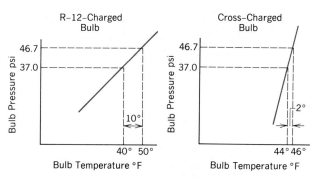

Figure 6-19 A cross-charged bulb will produce a faster pressure change response to a change in temperature.

transmitting tube emerges from the top of the bulb.

There are some variations on the materials that are used inside the temperature-sensing bulbs. So far, we have only described a **liquid-charged bulb** with a refrigerant matching the system refrigerant. There is a potential problem with this bulb that can occur on start-up of a system. Consider the system shown in Figure 6-18. It has been shut down for some period of time. All the components in the system have reached a temperature of 80°F. The bulb pressure is 84.2 psi, the evaporator pressure is 84.2 psi, and the spring pressure is 9.7 psi. The valve is being held in the closed position because the pressure on the bottom of the diaphragm is higher than the pressure on the top. Now the compressor is called upon to start. The TXV will remain closed until the suction pressure drops below 74.5 psi. At that point, the TXV will begin to open, admitting new refrigerant into the evaporator. After some period of operation the system pressures will drop to the normal 37 psi previously described. In the interim period, however, the compressor has operated with a high suction pressure. This could cause the compressor to overload before normal operating pressures are reached. The problem can be solved with a **lim-**

ited charge bulb. A limited charge bulb would not be able to produce the 84.2 psi, because it would run out of liquid at perhaps 60°F (57.7 psi). The valve would then be held closed until the evaporator pressure was pulled down to 48 psi. This would be accomplished very quickly with no additional refrigerant being admitted to the evaporator. Normal operating pressures are accomplished very quickly with the limited charge bulb.

Another variation on the material used in the temperature-sensing element is the **cross-charged bulb.** This type of bulb uses either a different refrigerant or a combination of re-

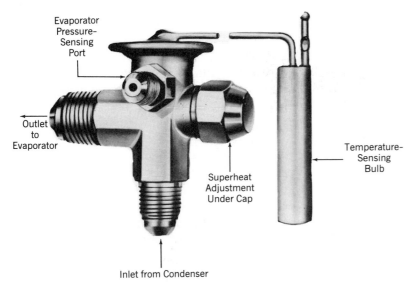

Figure 6-20 Externally equalized thermostatic expansion valve. *(Courtesy Alco Controls, Division Emerson Electric Co.)*

Figure 6-15 Internal parts of an externally equalized TXV. *(Photo Courtesy © Sporlan Valve Company)*

frigerants that will have a greater change in pressure for a given change in temperature (see Figure 6-19). This makes the response time for the valve faster, and closer control of superheat is accomplished.

The physical configuration of expansion valves can be quite varied. Arrangements used by another major manufacturer are shown in Figures 6-20 and 6-21. Although the methods of adjusting the spring pressure and replacement of the

power elements are different, their principles of operation are identical. Figures 6-21 and 6-22 show **pilot-operated valves.** These would be used on large systems where the force required to operate the valve would otherwise require the use of a very large diaphragm. The pilot-operated expansion valve uses a small diaphragm that monitors the use of the refrigerant being pumped through the system. The pressure of the system refrigerant is then used to move the large TXV.

The expansion valve must be properly sized to match the evaporator load. If a too large valve is used, small corrections in the valve position will overcompensate for changes in the load. This is referred to as hunting. The valve alternately supplies too much, then too little refrigerant, swinging back and forth until the correct flow is finally attained. Aside from wearing out the valve, this situation creates other problems. When the refrigerant flow is too low, the evaporator is starved and capacity is lost. When the refrigerant flow is too high, the potential exists for allowing liquid refrigerant droplets into the compressor.

The expansion valves are rated according to tons of evaporator capacity required. The valve seats may be changed so that one size valve body may be used for different refrigerants and different capacities. The power elements, however, must be matched to a single refrigerant. They are

Figure 6-22 Thermostatic expansion valve used on medium- to large-capacity systems. Bolted flange construction with replaceable power element. *(Courtesy Parker Hannifin Corporation)*

normally color-coded according to the refrigerant.

Automatic Expansion Valve

The valve shown in Figure 6-23 is an **automatic expansion valve** (AEV). Its job is to sense pressure at the inlet to the evaporator, and open or close as required in order to keep the evaporator at a constant pressure. The pressure at the valve outlet (which is the same as the evaporator inlet) acts upon a diaphragm and is opposed by a spring. If the spring pressure is increased, it will tend to open the valve and increase the quantity of refrigerant being admitted to the evaporator. Thus, the new evaporator pressure will be controlled at a higher set point.

There is an odd characteristic about the automatic expansion valve. If the evaporator load were to increase, the valve would initially sense an increase in evaporator pressure. But because the job of the AEV is to maintain constant evaporator pressure, the valve will close! The evaporator will become starved, and even though the load has increased, the system capacity will be reduced. The automatic expansion valve is generally used when it is desired to keep the evaporator coil at a constant temperature.

Orifice Tube

Figure 6-24 shows a **fixed orifice tube.** This is a fixed restriction, very similar in concept to the capillary tube. It is found on some air conditioning systems, and quite commonly on automobile air conditioning systems. The hole in the orifice tube must be quite small in order to accomplish the pressure reduction in such a short distance.

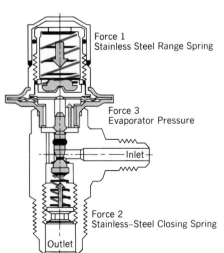

How the Valve Works

Three Forces Control the Operation of the Valve

Force 1 the adjustable range spring above the diaphragm. It moves the diaphragm down, opening the valve.

Force 2 the closing spring beneath the diaphragm. It moves the pushrod and ball assembly up, closing the valve.

Force 3 the outlet pressure acting under diaphragm. This is the pressure valve controls when the spring force, F1, is equal to the sum of forces F2 and F3.

Figure 6-23 Automatic expansion valve. *(Courtesy Parker Hannifin Corporation)*

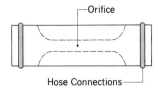

Figure 6-24 Orifice tube used as a metering device on automobile air conditioning.

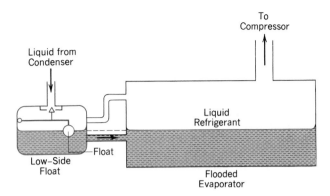

Figure 6-25 Low-side-float metering device.

The orifice tube may be more prone to clogging than a capillary tube of a larger diameter.

Low-Side Float

The **low-side float** shown in Figure 6-25 is used in some large-capacity systems. The float tank is attached to a flooded evaporator. The level of the liquid in the float chamber will be equal to the level of liquid in the evaporator shell. If the evaporator load increases, the level of refrigerant will tend to go down, thus increasing the amount of liquid that is then allowed to enter from the condenser.

High-Side Float

Figure 6-26 shows a metering device called a **high-side float.** The refrigerant reservoir is located in the high pressure side of the system. Whereas the low-side float modulated refrigerant flow to match the rate of evaporation in the evaporator, the high-side float matches the condenser capacity. As the rate of condensing increases, the level in the tank increases, and the float allows more refrigerant to pass on to the evaporator. It

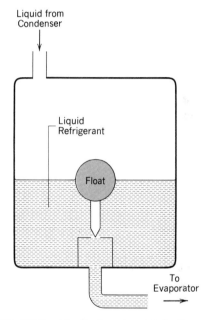

Figure 6-26 High-side-float metering device.

is important with this type of system that the evaporator be sized large enough to prevent liquid from being pumped into the compressor. Also, an overcharge of refrigerant with this system may easily cause compressor slugging.

KEY TERMS AND CONCEPTS

Metering Device	Limited Charge Bulb	Orifice Tube
Cap Tube	Cross-charged Bulb	Low-side Float
TXV	Pilot-operated Valve	High-side Float
Externally Equalized	Automatic Expansion Valve	
Internally Equalized		
Liquid-charged Bulb		

QUESTIONS

1. As refrigerant passes through the metering device, what happens to
   (a) its pressure?
   (b) its temperature?
   (c) its heat content?

2. Where might you find each of the following refrigerant conditions: at the metering device inlet, outlet, neither, or both:
   (a) subcooled liquid?
   (b) saturated liquid?
   (c) all vapor?
   (d) 90 percent liquid, 10 percent vapor?
   (e) warm liquid?
   (f) warm vapor?

3. What type of metering device is used in domestic refrigerators?

4. If a capillary tube of 0.028 in. inside diameter is replaced with a capillary tube of 0.031 in. inside diameter, will the new cap tube be longer than or shorter than the original? Why?

5. What is the correct method to cut a capillary tube to length?

6. When crimping a larger tube onto a capillary tube, how far should the capillary tube be inserted? Why?

7. When starting up from a warm condition, will the capillary tube tend to starve the evaporator or flood the evaporator? How do you know?

8. What three pressures act upon the diaphragm of the thermostatic expansion valve?

9. What does the thermostatic expansion valve hold constant?

10. For a TXV controlling an R-12 evaporator at 34°F saturated suction temperature and 12°F superheat leaving the evaporator:
    (a) What is the bulb pressure?
    (b) What is the evaporator pressure?
    (c) What is the spring pressure?

11. For a TXV controlling an R-22 evaporator at 34°F saturated suction temperature and 12°F superheat leaving the evaporator:
    (a) What is the bulb pressure?
    (b) What is the evaporator pressure?
    (c) What is the spring pressure?

12. For each of the following, state whether the TXV will overfeed, underfeed, or supply the correct amount of refrigerant to the evaporator:
    (a) The spring is broken.
    (b) The diaphragm has been punctured.
    (c) The refrigerant has leaked out from the sensing bulb.
    (d) The sensing bulb is making poor contact with the suction line.

(e) The insulation around the sensing bulb has been stripped away.

(f) An R-22 valve has been mistakenly installed on an R-12 system.

13. What is the advantage to using a limited charge temperature sensing bulb?

14. If you replace an internally equalized TXV with an externally equalized value, will it work OK? Why?

15. If you replace an externally equalized TXV with an internally equalized valve, will it work OK? Why?

16. Why is a cross-charged bulb sometimes used?

17. What is hunting? What causes it? What is bad about it?

18. How can you tell just by looking at a thermostatic expansion valve which refrigerant it is supposed to be used with?

19. What tools would you need in order to check how much superheat the thermostatic expansion valve is set to hold? How would you do it?

20. Where is the filter screen located on the thermostatic expansion valve?

CHAPTER 7

ACCESSORIES

The basic components described up to this point are the only parts that are common to all systems. All mechanical refrigeration systems have a compressor, condenser, metering device, and evaporator. There are a number of devices, however, that are used on systems to regulate operation, protect the basic components from failure, and simplify service for the technician. This chapter describes the purpose and function of these accessories. There probably is no single refrigeration system that uses all of these accessories, but they are all found in common usage.

Filter Driers

Filter driers serve a dual purpose. First, they act to strain out any particles that may be in the system. Most commonly, these particles might be oxidation that was formed on the inside of brazed tubing, which breaks loose during operation of the system. The second function of a filter drier is to dry the refrigerant. That does not mean that it removes liquid, but that it absorbs and holds water that may have not been properly removed when the system was put together. Some filters also serve a third purpose by allowing the refrigerant to flow through a filter bed that absorbs acids that may have been formed as a result of a compressor motor burnout.

Figure 7-1 shows an assortment of small filter driers used on small refrigeration systems. The filter drier is installed in the liquid line, which connects the condenser with the cap tube or other metering device. This type of filter may be bidirectional; if it is not, the correct flow direction will be noted by an arrow on the filter body. The connections are $\frac{1}{4}$-in. tubing that may either be brazed into the system or flared. This type of filter may also be supplied with an integral access valve fitting.

Figure 7-2 shows a larger liquid line filter drier that is used in larger systems. A single filter drier can be used on a system of up to 15 tons. The filter driers may be used on systems using any of the popular refrigerants, but the capacities will vary. For example, a filter drier that has a 15-ton capacity on an R-22 system may only be suitable for use on a 10-ton R-12 system. These filters are available in either male flare or solder fitting connections. The inside of the filter drier contains a coarse and fine mesh screen and a dessicant material to absorb water (Figure 7-3).

For larger systems, or for systems that operate frequently under vacuum, **replaceable core filter driers** are sometimes the best choice (Figure 7-4). When installed with valves on the inlet and outlet

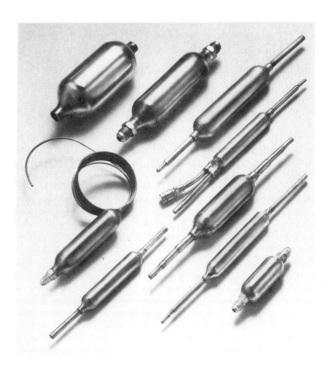

Figure 7-1 Filter driers used on small refrigeration and air conditioning systems. *(Courtesy Parker Hannifin Corporation)*

Figure 7-2 Liquid line filter drier used in the range of 2 to 15 tons of capacity. (*Courtesy Parker Hannifin Corporation*)

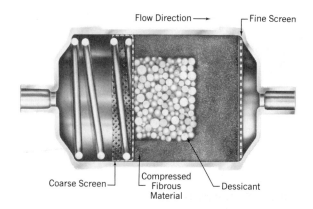

Figure 7-3 Cutaway of a liquid line filter drier. (*Courtesy Parker Hannifin Corporation*)

Figure 7-4 Liquid line filter drier with a replaceable core. (*Courtesy Parker Hannifin Corporation*)

sides, the filter may be disassembled without loss of refrigerant. The old filter element is removed and replaced with a new one (Figure 7-5).

Where maximum cleanliness and compressor protection are desired, there may also be a suction line drier installed between the evaporator and the compressor. They are used less frequently than liquid line driers because they are far more expensive. In order not to impose too much pressure drop while handling the full refrigerant flow in its vapor form, the suction line drier is physically much larger than a liquid line drier of equal tonnage capacity.

Troubleshooting the liquid line filter drier can be as simple as holding your hands on the connections to the drier. If the outlet temperature is cooler than the inlet temperature, it is an indication that there is sufficient pressure drop in the clogged filter to cause the entering liquid to flash and cool.

Sight Glass

The **slight glass** (Figure 7-6) is your window to see what is going on inside the refrigerating system. When furnished, it is an excellent aid for the service technician. It is installed in the liquid line, just before the metering device and after any other accessories that may be in the liquid line. In a properly operating system, you will have a solid column of liquid in the sight glass. To the inexperienced technician, the line may appear to be totally empty instead of totally full, as the refrigerant is clear like water. If the refrigerant in the sight glass shows bubbles or a foamy white appearance (also due to bubbles), it is a sure indication of trouble. The trouble might be an undercharge of refrigerant (some has leaked out), excessive pressure drop in the filter or other liquid line accessories (causing flash gas), or a condenser problem that is not allowing the condensation of all the gas being supplied to the condenser. Large systems will have a liquid line that is larger than the largest available sight glasses. For these applications, a sight glass may be in-

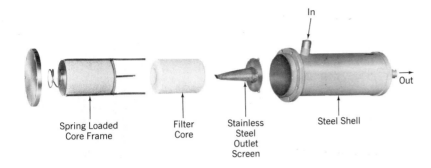

Figure 7-5 Internals of a replaceable core filter drier. *(Courtesy Parker Hannifin Corporation)*

stalled in parallel with the liquid line, as in Figure 7-7.

Many sight glasses are also supplied with a moisture indicator. In the presence of water in the refrigerant, the moisture indicator will change color. By matching the color on the indicator to the WET or DRY key on the casing, you can be assured that you have a dry system. Some indicators, however, once they have changed to the WET color, will not change back to the DRY color after the moisture has been removed.

Suction Line Accumulator

Figure 7-8 shows the installation of a **suction line accumulator** between the evaporator and the compressor. The purpose of the accumulator is to intercept any liquid that may get through the evaporator without being vaporized. This could happen upon a sudden drop in evaporator load, before the metering device has had a chance to react and reduce the refrigerant flow. In that case, the liquid will drop out in the accumulator, and gradually vaporize while absorbing heat

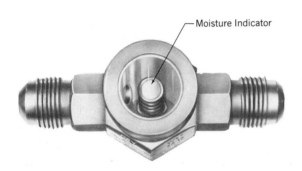

Figure 7-6 Sight glass. *(Courtesy Sporlan Valve Company)*

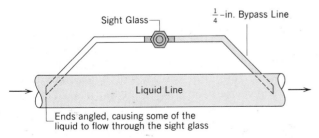

Figure 7-7 Sight glass installed in parallel with a large-diameter liquid line.

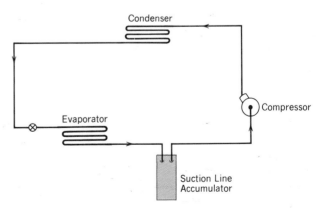

Figure 7-8 Suction line accumulator is installed to prevent temporary surges of liquid refrigerant from reaching the compressor.

Figure 7-9 Suction line accumulator. *(Courtesy Parker Hannifin Corporation)*

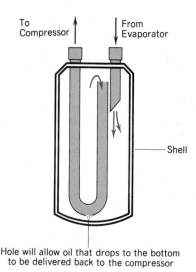

Hole will allow oil that drops to the bottom
to be delivered back to the compressor

Figure 7-10 Internal piping of a suction line accumulator. Any liquid that enters from the evaporator will fall to the bottom. The pickup to the compressor is located at the top.

Note that system problems that continuously allow liquid at the evaporator outlet will not be solved by the use of a suction line accumulator. Once the accumulator fills with liquid, it can no longer prevent liquid from entering the compressor suction.

Oil Separator

Chapter 3 discussed the function of oil in the compressor. Unfortunately, as the compressor pumps refrigerant, it also pumps a small quantity of the oil. To keep the compressor from running out of oil, the system is piped in such a fashion that all the oil that leaves the compressor will eventually return after traveling through the condenser, metering device, and evaporator. The oil is carried through the refrigeration tubing along the inside surface of the tube. There are two disadvantages to having the oil circulate through the system. First, the oil acts as an insulator, reducing the rate of heat transfer in both the condenser and the evaporator. The second disadvantage is limited to low temperature applications, but can be quite serious. All refrigerant oils contain some degree of wax dissolved in the oil. At very low temperature (below −40°F), the wax begins to precipitate out of the

through the shell. Some installations run a small liquid line from the condenser outlet around the shell of the accumulator to vaporize the liquid in the accumulator more quickly. Subcooling the liquid line at the same time causes the refrigeration system to operate more efficiently.

The photo and a cutaway view of the accumulator in Figures 7-9 and 7-10 show the internal piping. The same expansion that causes the liquid to drop out will also cause oil to drop out. A small hole in the bottom of the U-tube causes the oil to be added to the flow back to the compressor. A small quantity of liquid refrigerant may also find its way back to the compressor through this hole, but the quantity will not be sufficient to cause any damage.

Suction line accumulators are instrumental in reducing the chance of a compressor failure due to liquid entering from the suction line. As a matter of fact, eliminating liquid slugging of the compressor is so important that some manufacturers have even incorporated a suction line accumulator as an integral part of the compressor.

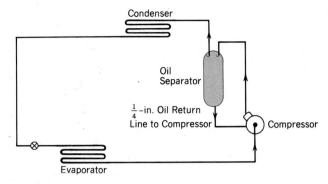

Figure 7-11 The oil separator is installed in the hot gas line to prevent oil from being circulated through the system.

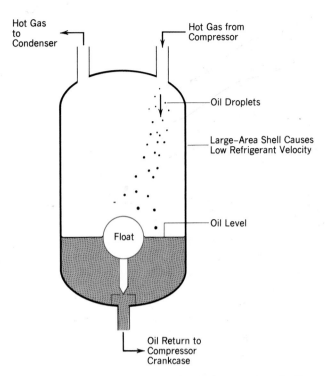

Figure 7-12 Internal view of an oil separator. As the oil level rises, the float will open the oil return valve.

oil. This free wax can clog the metering device and render the entire system inoperable until the oil is allowed to warm.

The **oil separator** is installed between the compressor and the condenser, in the hot gas line (Figure 7-11). It separates most of the oil from the hot gas and returns it directly back to the crankcase of the compressor. Figure 7-12 shows how the hot gas velocity is reduced in a chamber, causing the oil droplets to fall to the bottom. When the oil builds up higher than the allowable level, it lifts the float, which in turn opens the valve back to the compressor. The oil is forced through the valve, as the tank is under high pressure, while the crankcase is at the low-side pressure.

Inspection of the cutaway of the oil separator makes it apparent that this device must be installed in the upright position. Any position other than upright will not allow the float valve to operate properly.

Muffler

Mufflers are sometimes installed in refrigeration systems for the same reason they are installed in automobiles—to reduce noise. The muffler may be externally mounted immediately at the compressor outlet, or it may be incorporated into the discharge line inside the compressor shell by the manufacturer. Figure 7-13 shows the inside of a discharge muffler. As the hot gas passes through, it undergoes a series of expansions to reduce the noise level. The discharge muffler

must be correctly installed to prevent trapping large quantities of oil as the hot gas velocity is reduced. The inlet and outlet connections are off-center for a very good reason. When installed in a horizontal line, the inlet and outlet are positioned at the bottom of the muffler. For installation in vertical lines, the direction of flow through the muffler must be downward.

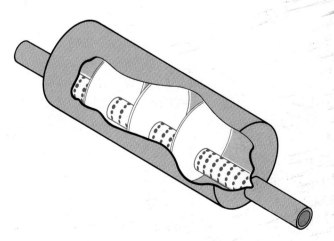

Figure 7-13 Cutaway view of a muffler.

Suction/Liquid Heat Exchanger

The efficiency of the refrigeration system is increased when we allow heat exchange to take place from the liquid line to the suction line. In small systems, this effect is often achieved as shown in Figure 7-14. The capillary tube is soldered to the suction line, thus allowing the liquid refrigerant to become subcooled (temperature lower than the boiling point), while the refrigerant in the suction line picks up a few extra degrees of superheat. Some systems will actually run the cap tube inside the suction line for even better heat transfer.

Larger systems may use a fabricated heat exchanger, as shown in Figure 7-15.

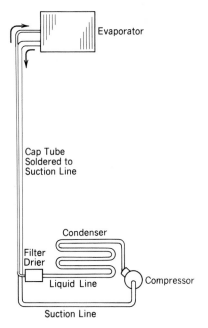

Figure 7-14 The cap tube is soldered to the suction line to act as a suction/liquid heat exchanger.

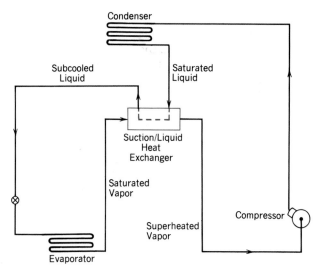

Figure 7-15 The suction/liquid heat exchanger subcools the refrigerant in the liquid line and superheats the refrigerant in the suction line.

Crankcase Heater

The **crankcase heater** is an electrical heater element that is used to heat the crankcase of the compressor. Its purpose is to maintain a high enough temperature to prevent significant quantities of refrigerant from dissolving in the oil. If the compressor is located in a colder area than the evaporator coil, there will be a tendency for the refrigerant to migrate to the compressor.

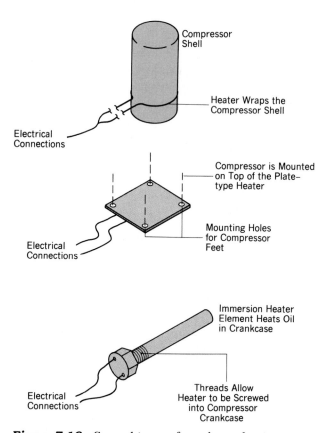

Figure 7-16 Several types of crankcase heaters.

Then when the compressor starts up, and there is no crankcase heater, the liquid refrigerant that has dissolved in the oil flashes. The resulting oil/refrigerant mixture foams, and much of the oil will be carried out of the compressor with the refrigerant.

Several different types of crankcase heaters are shown in Figure 7-16. The heater may be bolted to the compressor, wrapped around the compressor shell, or it may be located inside the compressor for direct heating of the oil in the crankcase. The controls of the system will function to cause the crankcase heater to be energized whenever the compressor is off, and to be deenergized when the compressor is running.

Manufacturers will often place warnings on their units that advise against allowing the compressor to start if the power has not been on for the preceding 24 hours. They want to make sure that the crankcase heater has been on for a sufficiently long time to assure than there is not a lot of refrigerant still dissolved in the oil.

Solenoid Valves

A solenoid is a coil of wire that, when supplied with an electrical voltage, creates a magnetic field. When a core of metal is located inside the coil of wire, it tends to move when the coil is energized (Figure 7-17). If the core of metal is actually the stem of a valve as shown in Figure 7-18, we have a solenoid valve. The valve may be a normally closed valve, which opens when the coil is energized (as in Figure 7-18), or it may be normally open (as in Figure 7-19), closing when the solenoid is energized. From looking at the outside of the solenoid valve, it is usually not possible to determine whether it is normally open or normally closed (Figures 7-20 and 7-21).

Probably the most common location for a solenoid valve in a refrigeration system is between the receiver and the metering device. When it is in this position, it is known as a **liquid solenoid valve** or LSV. This is a common location because many systems make use of an **automatic pump-down** sequence each time the compressor turns off. It works as follows (see Figure 7-22).

1. A thermostat has sensed that the space or box has become cool enough. The thermostat opens, deenergizing a normally closed liquid solenoid valve.
2. The solenoid valve closes. As the compressor continues to run, the pressure in the low side of the system is reduced, and all remaining liquid flashes to vapor.
3. When the low-side pressure has dropped to almost zero, a pressure-sensing switch opens and turns off the compressor.

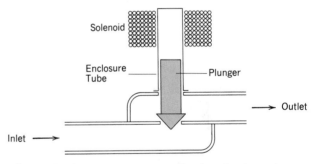

Figure 7-18 Two-way, normally closed solenoid valve. The valve opens when the coil is energized. Note that the plunger moves inside an enclosure tube, which allows removal of the coil without releasing the contents of the system.

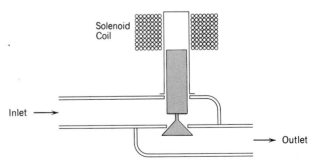

Figure 7-19 This normally open valve will close when the solenoid coil is energized.

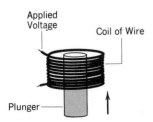

Figure 7-17 When voltage is applied to the coil the magnetic field produced by the coil will pull the plunger up.

This pumpdown sequence serves two purposes. It removes the refrigerant from the low side, making it impossible for the oil to become diluted with liquid refrigerant, and it prevents the compressor from having to start up with a high suction pressure that could overload the compressor motor.

A second common use for the solenoid valve is on systems that use a hot gas bypass for capacity control. A solenoid valve is placed in the bypass line that connects the hot gas to the evaporator inlet or compressor inlet. When applied in this application, it is called a **hot gas bypass solenoid** valve (Figure 7-23). A similar application for a hot gas solenoid valve is to use hot gas to defrost an evaporator.

It is easy to tell whether or not a solenoid valve is energized. An energized, properly operating solenoid will have a magnetic field around it. If you hold a light screwdriver near the top of the solenoid, you should be able to feel the effect of the magnetic field on the screwdriver. If you discover a faulty solenoid, you can remove and replace the

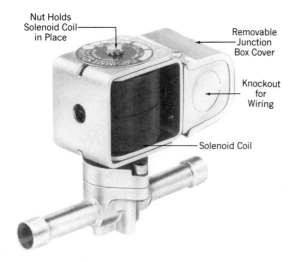

Figure 7-20 Solenoid valve. *(Courtesy Parker Hannifin Corporation)*

Figure 7-21 Solenoid valve that can be manually operated. *(Courtesy Parker Hannifin Corporation)*

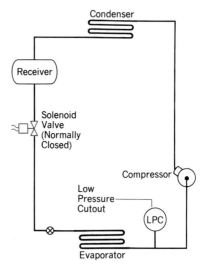

Figure 7-22 Liquid line solenoid valve used on a system that includes automatic pumpdown.

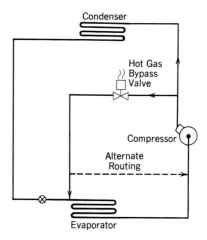

Figure 7-23 Solenoid valve used to activate a hot gas bypass line for capacity control.

Direct Operated Solenoid Valves

Are used in systems requiring low flow capacities or in applications with low pressure differentials across the valve orifice. The sealing surface that opens and closes the main valve orifice is connected to the solenoid plunger. The valve operates from zero pressure differential to maximum rated pressure differential (MOPD) regardless of line pressure. Pressure drop across the valve is not required to hold the valve open.

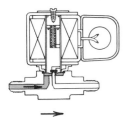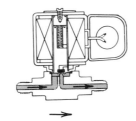

Pilot-Operated Valves (Normally Closed)

Open when electrical current is applied. They are the most widely used solenoid valves. The pilot orifice is much larger than the equalizer orifice. When the coil is energized, the plunger lifts off the pilot orifice. The pressure above the diaphragm is reduced to the outlet pressure of the valve. The resulting pressure differential across the diaphragm creates a force that lifts the diaphragm off the main port, opening the valve.

When the coil is deenergized, the pilot orifice is closed. The inlet pressure through the equalizer orifice equalizes the pressure above and below the diaphragm. The diaphragm resets, closing the valve.

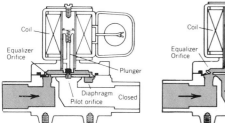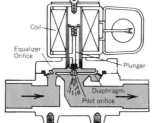

Figure 7-24 Operational difference between directly operated solenoid valves and pilot-operated solenoid valves. *(Courtesy Parker Hannifin Corporation)*

Pilot-Operated Valves (Normally Open)

Close when electrical current is applied. They are used in applications in which the valve remains open most of the time or where it is required to open in case of an electrical malfunction. Operation is the reverse of the normally closed valve.

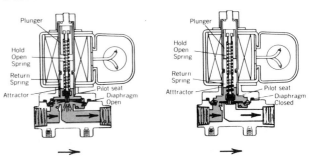

Figure 7-25 Normally open pilot-operated solenoid valve. *(Courtesy Parker Hannifin Corporation)*

Service Valves

Service valves function as manual valves that include a connection for the purpose of attaching gauges to read system pressures. Typical service valves are shown in Figure 7-26. The most common locations for service valves are at the compressor suction, at the compressor discharge, and at the receiver outlet.

The normal operating position for a service valve is for the stem to be turned counterclockwise, as far as it will go. In this position (Figure 7-27), the flow of refrigerant through the valve is unimpeded. The pressure port is closed off. The cap may be removed from the $\frac{1}{4}$-in. service port without loss of refrigerant.

After pressure gauges have been attached, the stem may be turned one-half turn or less, allowing the pressure in the system to bleed out to the pressure port connection. In this position, the valve is said to be **cracked off back seat** (Figure 7-28).

If the stem is turned clockwise until it stops, the valve is **front seated** (Figure 7-29). The flow of refrigerant through the valve has been shut off. If the service valves on both sides of the compressor have been front seated, the compressor has been isolated. It may be opened up (if it is a semihermetic or open compressor) or removed entirely from the system, without loss of the refrigerant charge.

solenoid portion only. It will not be necessary to replace the entire solenoid valve.

The solenoid valves described to this point use the magnetic field to provide all the power that is needed to move the valve stem. For larger size applications, a **pilot-operated solenoid** valve can be used (Figures 7-24 and 7-25). A pilot-operated valve uses a small solenoid to bleed the system pressure from one side of a piston. The unbalanced system pressure remaining on the other side of the piston is the motivating force that causes the main valve to open or close.

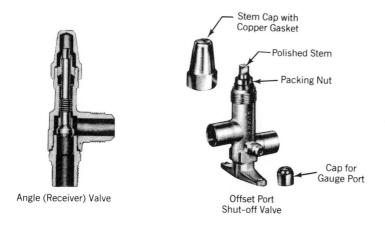

Angle (Receiver) Valve

Offset Port Shut-off Valve

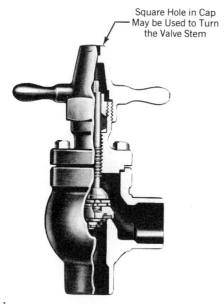

Wing Cap Shut-off Valve

Figure 7-26 Service valves. *(Courtesy Henry Valve Company)*

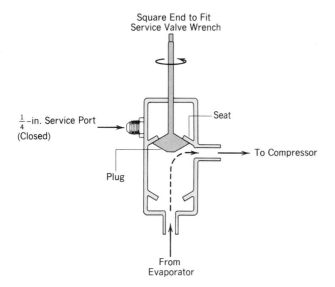

Figure 7-27 A service valve in the backseated position. Full refrigerant flow is available and the service port is shut.

Several precautions about service valves need to be mentioned. The stem material is polished so that it forms a tight seal with the packing as it passes through the valve. If it is dirty or scratched as you turn it through the packing, the packing will be ruined.

The square top on the end of the valve stem is easily ruined by service technicians or others who use an adjustable wrench to turn the valve stem. Use a service valve wrench that is specially designed to fit the square top on the valve stem. If the stem does not turn easily with the service valve wrench, loosen the packing nut one-half turn, then retighten when service is completed.

In severe cases, the valve plug may be frozen to the seat from a long period of nonuse. It is permissible to apply heat gently to the valve body with a torch in order to cause expansion of the valve. Too much heat, however, may cause the valve to warp.

The cap that fits on top of the valve stem should always be replaced when service is completed. It protects the valve stem from dirt and damage. It also prevents losing the refrigerant charge if the valve packing is not perfectly tight. The cap should be snugged down with a wrench approximately one-quarter of a turn past finger tight. There is a soft copper gasket that forms a seal between the cap and the valve body.

Check Valves

A **check valve** is a valve that will allow flow in one direction only. An arrow on the casing of the valve indicates the direction of flow. They may be either a swing-type check valve (Figure 7-30) or a spring loaded check valve (Figure 7-31). The swing-type has a flapper that merely swings out of the way when there is flow in the proper direction. The spring loaded check valve is sometimes

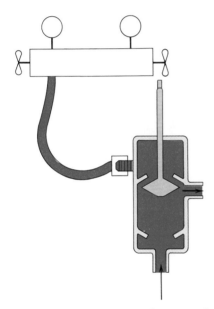

Figure 7-28 When the service valve is cracked off back seat, the flow of refrigerant is still unrestricted, but the pressure inside the valve can be sensed at the service port.

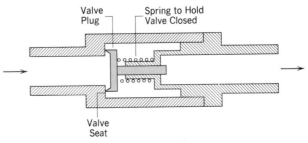

Figure 7-30 A swing-check valve only allows flow in one direction.

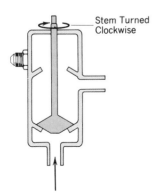

Figure 7-29 When front seated the service valve acts as a shut-off valve.

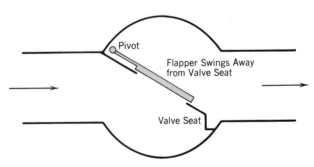

Figure 7-31 Spring loaded check valve.

called a positive closing check valve. This type would be suitable for installing in a line where the correct direction of flow is vertically down, whereas a swing check valve would not be suitable for that application. The spring loaded check valve, however, will present more of a resistance to flow than will the swing check valve.

Relief Valves

Relief valves or **safety valves** (Figure 7-32) are provided to protect vessels from pressure that exceeds the rating for which they were designed. The use of these valves is mandated by law. The governing boards that approve these valves are ASME (American Society of Mechanical Engineers) and the National Board (N.B.). The importance of the correct use of relief valves cannot be overemphasized. A condenser shell that is designed for 300 psi and is subjected to a high enough pressure to cause it to rupture will explode with no less force than if a bomb had exploded.

The usual location for a relief valve is on the condenser shell. Under no circumstances may a valve be installed between the relief valve and the vessel that it is designed to protect. Figure 7-33 shows an arrangement that is sometimes used to enable a relief valve to be changed or tested without losing the refrigerant charge. The dual relief valves are sized so that either one is sufficiently large to relieve the required refrigerant flow. When the valve is front seated, one of the valves may be removed; when it is back seated, the other relief valve may be removed. The system remains protected as long as one of the relief valves is open to the system. Since the manual valve cannot be front seated and back seated at the same time, this arrangement is safe. When one of the relief valves is removed, the piping connection must remain open. Pipe plugs or caps should never be used.

Piping of the discharge of the relief valve is also critical. The relief valve outlet may remain unpiped, or it may be piped to allow relief of the refrigerant to a safer area (Figure 7-34). No valves are allowed in the discharge relief piping. Safety should be a major consideration in the location of the discharge of the relief valve. If the discharge

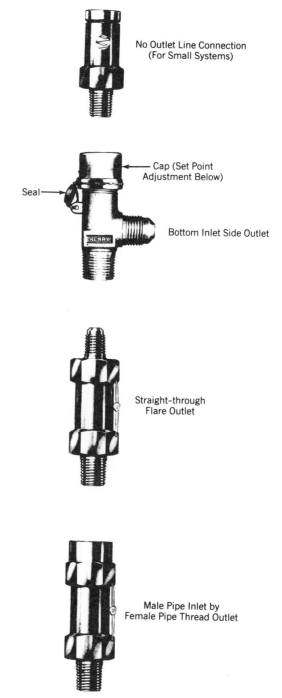

No Outlet Line Connection (For Small Systems)

Cap (Set Point Adjustment Below)

Seal

Bottom Inlet Side Outlet

Straight-through Flare Outlet

Male Pipe Inlet by Female Pipe Thread Outlet

Figure 7-32 Relief valves used to protect condenser or receiver.

line is piped through the roof, a small outlet should be provided at the bottom of the stack to eliminate the possibility of any accumulation of rain water or condensation. A column of water in the relief valve discharge can be dangerous for three reasons:

1. The weight of water on the valve outlet will make the effective relieving pressure higher than design.

2. The constant presence of water can cause the parts of the relief valve to rust or stick, and thus not operate at the required relief pressure.

3. If a column of water is present when the valve relieves, the water may be moved through the pipe with velocities high enough to cause the pipe to rupture when the column of water reaches an elbow.

Fusible Plug

The **fusible plug** (Figure 7-35) is another safety device normally found on the condenser or receiver. It looks like a standard plug, except that the center has been drilled out and filled with a solder material. Standard temperatures for the solder plug to melt out are 162°F, 212°F, and 275°F. Fusible plugs relieve pressure in the event that temperature gets too high. Even though the condenser may not see a pressure higher than its rated pressure, it may fail because the materials of construction weaken as they are subjected to elevated temperatures. Fusible plugs should never be replaced with nonfusible plugs.

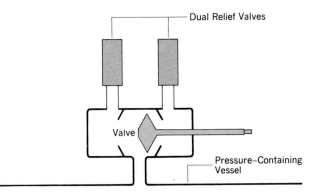

Figure 7-33 With dual relief valve, one can be valved off for service or testing while the other maintains protection for the system.

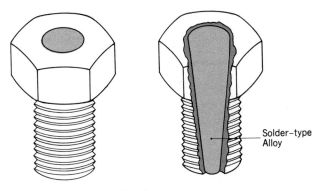

Figure 7-35 Fusible plug.

Rupture Discs

Rupture discs are assemblies that provide an area of thin metal that will rupture at a predetermined pressure. A rupture disc (Figure 7-36) is sometimes used in place of a relief valve. It is less expensive, and it can never fail in an unsafe mode, because it cannot stick closed as a relief valve can. There are also several disadvantages. Once the rupture disc ruptures, the refrigerant charge is completely lost. It cannot reseat itself once the overpressure condition has passed. The

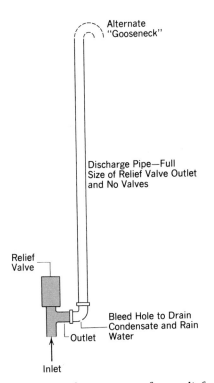

Figure 7-34 Discharge piping for a relief valve.

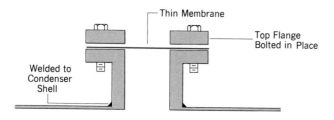

second disadvantage is that rupture discs cannot be installed where they will experience pressure pulsations, such as on the compressor discharge. The constant flexing of the disc due to these pulsations will cause a fatigue failure. The disc will then rupture and the refrigerant will be lost, even though no overpressure condition existed.

Evaporator Pressure Regulator

The **evaporator pressure regulator (EPR)** is a self-contained pressure-regulating valve that is installed between the evaporator outlet and the compressor inlet (Figures 7-37 and 7-38). It

senses pressure in the evaporator. If the evaporator pressure is higher than the set point of the EPR, the valve remains open and has no effect on the system. If the evaporator pressure starts to drop lower than the setting of the EPR, however, the valve will close to prevent the evaporator pressure from dropping too low. Two common applications for the EPR valve are on automotive systems and on systems that use more than one evaporator.

Automotive systems make use of an EPR because the compressor capacity varies so dramatically. The compressor speed varies with the engine speed, and it is not unusual for the com-

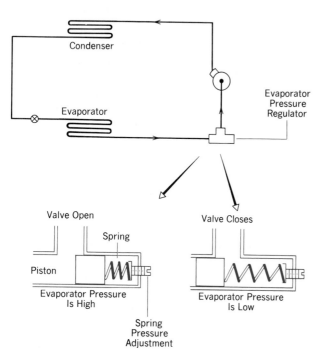

Figure 7-37 Evaporator pressure regulator closes when the evaporator load drops, preventing evaporator pressure from falling below a set pressure.

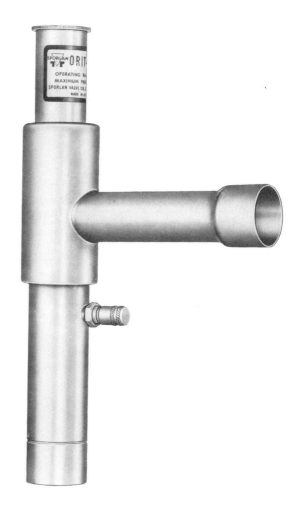

Figure 7-38 Evaporator pressure-regulating valve. ORIT in the model number indicates that the valve opens on a rise in temperature. *(Courtesy Sporlan Valve Company)*

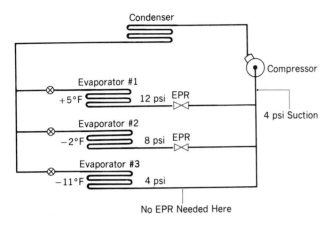

Figure 7-39 EPRs used on a multiple evaporator system to provide different evaporator temperatures from the same compressor.

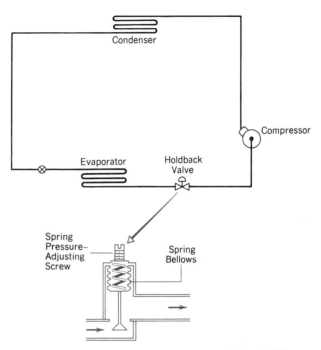

Figure 7-40 Crankcase pressure regulator (holdback valve) used to prevent compressor crankcase pressure from getting too high.

pressor to have a higher capacity than required. The automotive evaporator must be held at a temperature higher than 32°F. There is no provision for defrosting the evaporator, so the condensate cannot be allowed to freeze. The EPR on this R-12 system is set for 28–30 psig. On automotive applications the EPR is sometimes called a suction throttling valve (STV).

Multiple evaporator systems (Figure 7-39) use EPR valves to allow each evaporator to operate at a different saturated suction temperature. The evaporator that is designed to operate at the lowest pressure will see the compressor suction pressure directly, but each of the other evaporators will be held at some higher pressure and temperature by the EPR valve.

Holdback Valve

A **holdback valve** (Figures 7-40 and 7-41), or **crankcase pressure-regulating valve,** is designed to prevent the pressure at the compressor suction from getting too high. This would normally only be a problem on start-up or upon coming out of a defrost cycle. When bringing the system down to normal operating temperatures and pressures, the evaporator load is high. The suction pressure during this pull-down period will also be high. The compressor, seeing a high suction pressure also sees more dense refrigerant vapor, and as a result, handles more pounds of refrigerant. This causes high current draw by the

compressor motor, and can result in a motor burn-out.

The holdback valve is located between the evaporator outlet and the compressor inlet. It senses pressure entering the compressor, and closes as required to prevent the suction pressure from overloading the compressor motor. Once the suction pressure in the evaporator has dropped to normal, the holdback valve remains open and has no effect on the system operation.

Hot Gas Bypass Regulator

The EPR valve sets a limit on how low a pressure it would allow, and the holdback valve sets a limit on the maximum allowable pressure. The **hot gas bypass regulator** has a set-point pressure that it will maintain downstream from the valve—not higher, not lower. The hot gas bypass valve is located in the hot gas bypass line (Figure 7-42). It regulates how much hot gas is allowed to bypass from the compressor discharge into the evaporator. It should be set to a pressure that will match the suction pressure just before the sys-

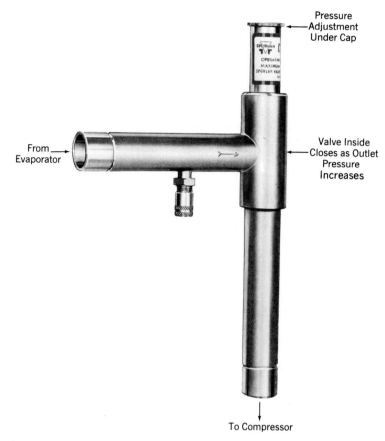

Figure 7-41   Crankcase pressure-regulating valve.
(Courtesy Sporlan Valve Company)

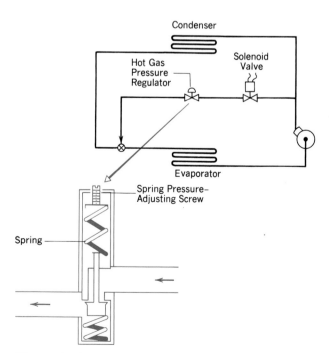

Figure 7-42   Hot gas bypass regulator.

tem has gone into the hot gas bypass mode of operation.

Condenser Water Regulator

Figure 7-43 shows a **condenser water-regulating valve,** which regulates the quantity of condenser water that enters a water-cooled condenser. It may sense either temperature or pressure in the condenser. An increase in temperature or pressure overcomes the spring pressure, and allows the valve to move to a more open position. Increasing the spring pressure will cause the valve to maintain control at a higher condensing temperature and pressure. This saves water, but increases the electrical consumption of the compressor motor. This valve is normally set to maintain a condensing temperature 10°F higher than the temperature of the water leaving the con-

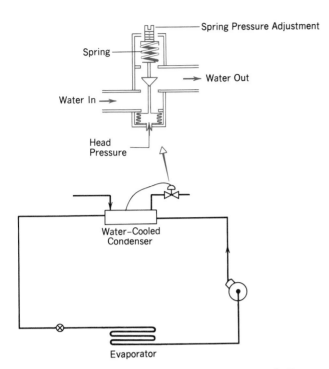

Figure 7-43 Condenser-water pressure-regulating valve.

denser. This has been found to be the optimum balance between water and electrical operating cost for many cases. In cases where the conservation of either water or electrical power is especially valuable, the setting of the condenser water-regulating valve may be different.

Hot Gas Defrost Valve

The system in Figure 7-44 uses hot gas to defrost the evaporator. When defrost is initiated, hot gas from the compressor is allowed to flow through the **defrost valve,** and backwards through the evaporator. The gas then flows through the bypass around the metering valve, then through the condenser, and back to the compressor.

Heat Reclaim Valve

The **heat reclaim valve** in Figure 7-45 is used when it is desirable to use the hot gas from the compressor for building heating, instead of dis-

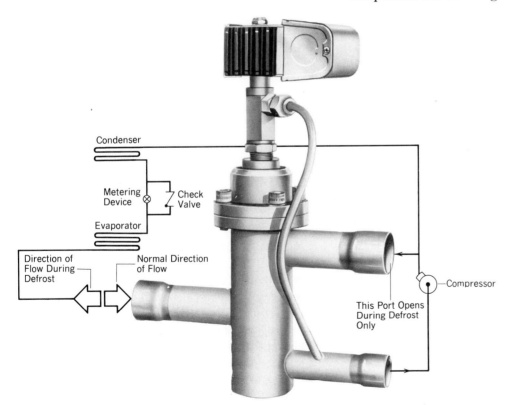

Figure 7-44 Three-way hot gas defrost valve. When the solenoid is energized hot gas flows backwards into the evaporator. *(Photo Courtesy Sporlan Valve Company)*

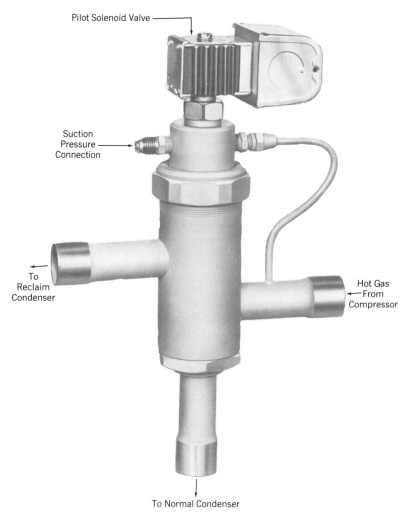

Pilot Solenoid Valve

Suction
Pressure
Connection

To
Reclaim
Condenser

Hot Gas
From
Compressor

To Normal Condenser

Figure 7-45 Three-way heat reclaim valve. When the solenoid valve is energized the hot gas from the compressor is routed to the reclaim condenser.

carding the heat in the condenser. Supermarkets
have refrigeration systems that provide hot gas
year round. When heating in the store is required, the solenoid valve is energized, sending
hot gas to a heating coil (reclaim condenser) in
the building heating system.

Thermostats

Thermostats are devices that sense temperature
in a refrigerated or air conditioned space, and tell
the cooling system when to run and when to shut
off. The thermostat is a switch that is caused
to open or close according to the temperature
sensed. The temperature-sensing mechanism
may be either a (1) **bimetal element,** (2) **gas-
charged bulb,** or a (3) **liquid-charged element.**

Figure 7-46 shows how a bimetal sensing element works. All metals will expand when they are
heated, and different metals will expand at different rates. When two metal strips with different
coefficients of expansion are fastened together
and then heated, the element will bend away
from the material with the higher coefficient of

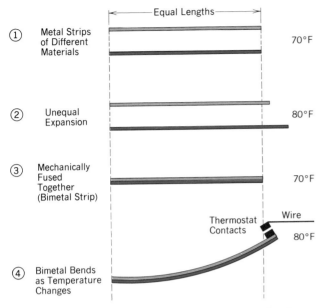

Figure 7-46 The bimetal thermostat principle of operation.

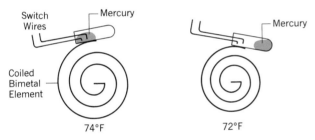

Figure 7-47 A coiled bimetal cooling thermostat.

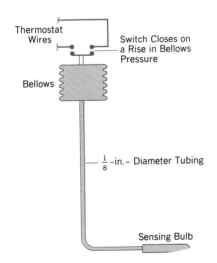

Figure 7-48 Principle of operation for a gas- or liquid-charged remote sensing thermostat.

expansion. This bending action may be used to open or close a switch. Sometimes, in order to get a lot of movement, a long bimetal element is used. It is arranged in a coil as shown in Figure 7-47 so that it will fit into a casing of reasonably small size. This type of thermostat must be mounted within the space where the temperature is being sensed.

Figure 7-48 shows a thermostat with a remote sensing bulb. The bulb is filled with a gas that expands or contracts, depending upon the temperature sensed. With increasing temperature, the expanding gas causes the pressure within the bellows to increase, moving a mechanism that will close a switch and cause the cooling system to start up. The element may also be charged with a liquid, with the pressure also acting upon a diaphragm to open or close the switch. With this type of thermostat, the switching portion of the thermostat may be located outside of the space being cooled.

Thermostats may operate on either line voltage or control voltage. Line voltage thermostats are wired directly in series with a compressor motor, and carry the full current of the compressor. This is the common arrangement for small systems such as household refrigerators. For larger systems, however, it is not practical to have a thermostat carrying a very large current. The thermostat then closes a switch in a lower current circuit (control voltage). The closing of the thermostat causes a small current to flow through a magnetic switch that can then pass a very high current to the compressor.

Thermostats have a **set point** and a **differential.** A thermostat that is set at 40°F might not turn on the compressor until the sensed temperature reached 44°F, and then would not turn the compressor off until the sensed temperature fell to 36°F. This thermostat would have a set point of 40°F, and a differential of 8°F between its cut-in and cut-out temperatures.

Some thermostats have a bulb that senses the temperature of the evaporator coil instead of air temperature. In this application, the thermostat may have a differential of 25°F or more. Soon after the compressor starts, the evaporator temperature drops 15°F to 20°F. Then, after 5 to 10 minutes of operation, the coil temperature gradually falls to the cut-out temperature.

High Pressure Cutout

The **high pressure cutout (HPC)** in Figure 7-49 senses pressure on the high pressure side of the system, between the compressor outlet and the metering device. Its job is to turn off the compressor if the high-side pressure ever approaches a temperature that would be high enough to physically damage the compressor or the condenser. The sensed pressure acts directly upon a diaphragm, in a manner similar to the thermostat.

The high pressure cutout is commonly (although not always) a **manual reset** device. That is, once it cuts the compressor off, it will not turn the compressor back on, even though the high-side pressure has returned to a safe level. The reset button on the switch itself must be pressed in order to restart the system.

Low Pressure Cutout

The **low pressure cutout (LPC)** in Figures 7-50 and 7-51 is different from the high pressure cut-

out in that its switch contacts will open on a fall in pressure, rather than with a rise in pressure, as does the high pressure cutout. Whereas the high pressure cutout is a safety control, the low pressure cutout may be used as either a safety control or an operating control **(automatic reset).** When used as an operating control, the set point and the differential are both adjustable. As the refrigeration system runs and the space cools, the lowside pressure will gradually drop. When the pressure drops to the cut-out setting, the compressor will turn off. Then, as the space warms, the lowside pressure increases, and the

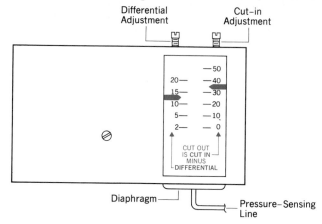

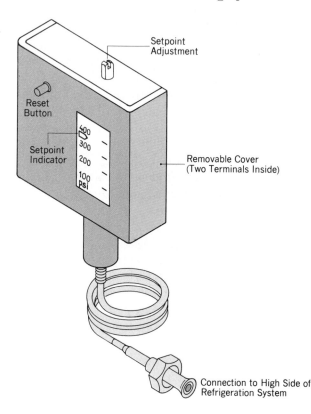

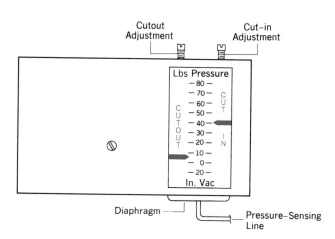

Figure 7-50 Two arrangements for a low pressure cutout (automatic reset).

Figure 7-49 Manual reset high pressure cutout.

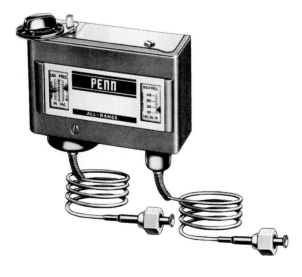

Figure 7-51 Low pressure cutout. The knob allows a limited range of temperature control by the operator. *(Courtesy Johnson Controls, Inc.)*

Figure 7-52 Combination high/low pressure cutout. *(Courtesy Johnson Controls, Inc.)*

LPC will turn the compressor on again once the pressure rises to the cut-in pressure. When the LPC is used as an operating control, a thermostat is not used. Figure 7-52 shows a combination high/low pressure cutout. The knob allows temperature adjustments by the customer within a limited range.

Oil Pressure Cutout

Figure 7-53 shows an **oil pressure cutout (OPC)** that is designed to turn off the compressor if the oil pump is not providing sufficient pressure for lubrication. There are two pressure-sensing ele-

Figure 7-53 Oil pressure cutout with adjustable cutout and fixed differential. *(Courtesy Johnson Controls, Inc.)*

ments. One, which senses the oil pressure as it enters the oil pump, is connected to the compressor crankcase. The other pressure-sensing line connects to the discharge of the oil pump. It senses how much higher the pump discharge pressure is than the crankcase pressure. The pressures oppose each other inside the controller, resulting in a measurement of net oil pressure (Figure 7-54). There is a time delay built into this device so that the compressor will be allowed to start. After the compressor starts, the OPC will allow it to run for 60 to 90 sec. If, after that period, the correct oil pressure has not been established, the OPC will open its contacts and cause the compressor to stop. The OPC is almost always a manual reset device. Otherwise, it would keep attempting to restart, running 60 sec each time without proper lubrication.

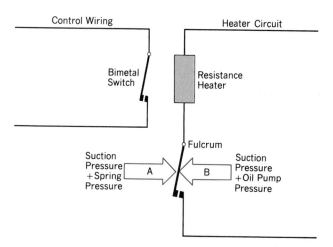

Figure 7-54 Schematic of oil pump operation. Oil pressure must overcome spring pressure and break the heater circuit before the heat causes the bimetal switch to open.

KEY TERMS AND CONCEPTS

Filter Drier

Replaceable Core Filter Driers

Diagnosing a Clogged Drier

Sight Glass

Suction Line Accumulator

Oil Separator

Muffler

Suction/Liquid Heat Exchanger

Crankcase Heater

Liquid Migration

Liquid Solenoid Valve

Automatic Pumpdown

Hot Gas Bypass Solenoid

Pilot-Operated Solenoid

Service Valves

Back Seated

Front seated

Check Valve

Safety Valve

Fusible Plug

Rupture Disc

EPR

Holdback Valve

Hot Gas Bypass Regulator

Condenser Water Regulating Valve

Hot Gas Defrost Valve

Heat Reclaim Valve

Thermostat

Bimetal Element

Gas-Charged Bulb

Liquid-Charged Element

Set Point

Differential

HPC

Manual Reset

LPC

Automatic Reset

OPC

QUESTIONS

1. Between which two major components (compressor, condenser, metering device, evaporator) would you find a filter drier? sight glass? accumulator? moisture indicator? oil separator? solenoid valve? EPR valve? holdback valve?

2. Using whichever of the following terms apply, describe the sequence that happens during automatic pumpdown shutdown (not all terms will be used): thermostat opens, thermostat closes, LPC opens, LPC closes, low-side pressure goes up, low-side pressure goes down, compressor starts, compressor stops.

3. What troubles besides low charge can be indicated by bubbles in the sight glass?

4. What does the suction line accumulator prevent from happening?

5. What does the EPR prevent from happening?

6. What does the holdback valve prevent from happening?

7. Why are oil separators found more commonly on low temperature systems?

8. Why are capillary tubes often soldered to or run inside suction lines?

9. When does the crankcase heater become energized?

10. What does the crankcase heater prevent from happening?

11. What is a quick and easy way to determine whether a solenoid valve is energized?

12. In what positions may the discharge line muffler be installed?

13. In what positions may the oil separator be installed?

14. A filter drier feels warm at the inlet and cold at the outlet. What conclusions can you draw?

15. Where would you find a fusible plug?

16. Where would you find a relief valve?

17. What two control devices would you find in a hot gas bypass line?

18. What is sensed by the condenser water regulating valve?

CHAPTER 8

REFRIGERANTS, OILS, AND REFRIGERANT PIPING

The purpose of this chapter is to describe some of the important properties of the fluids that circulate in the refrigeration system. You are aware by this point that different refrigerants are used in different applications. Isn't there one refrigerant that is best? Evidently not. When we understand the differences in behavior of the refrigerants and oils that are available, we will understand why the choice depends on the particular application.

The sections of this chapter describe the pipe sizing, arrangements, and routing required in order to keep all of the fluids where they belong within the system.

Common Refrigerants

One class of refrigerants in common use today are called **halogens,** or **chlorinated fluorocarbons.** This is a fancy phrase used to describe a group of manufactured refrigerants that have as their primary components chlorine, fluorine, and carbon atoms. Their individual chemical names are quite long, so we have given each a number for identification. The ones we will discuss are R-11, R-12, R-13, R-22, R-113, R-114, R-500, and R-502.

There are other refrigerants that are materials more commonly used in other applications. When used as a refrigerant, these are also given identification numbers. The most commonly used of these refrigerants is ammonia, which is R-717. A number of other materials have also been used for refrigerants in the past. Included

in this group would be methane, ethane, propane, sulfur dioxide, and others.

Refrigerant Physical Properties

The refrigerants in use today are, for the most part, relatively safe to use. The chlorinated fluorocarbon refrigerants are nonflammable, nonexplosive, and nontoxic. That is not to imply that they cannot be hazardous if you do not understand their properties.

For example, even though they are not flammable, when these refrigerants are heated by a flame, they produce a toxic gas called **phosgene.** This is similar to the gas that has been used during wartime as nerve gas.

And even though the material is classified as not toxic, when large quantities are released, the refrigerant displaces the air, and people have been suffocated. A classic situation for you to avoid would be entering a basement or a tank that might be filled with one of these refrigerants. There would be very minimal consolation for you in knowing that you were suffocated and not poisoned by a toxic refrigerant!

The **miscibility** of the refrigerant with refrigeration oil is an important physical factor. When reciprocating compressors discharge refrigerant, they also discharge small quantities of the oil that belong in the compressor crankcase. If the refrigerant and oil are quite miscible (mix together well), the natural tendency will be for the oil to circulate through the system with the refrigerant, and return to the compressor. The miscibility of the various refrigerants with oil are shown in Figure 8-1. In very low temperature applications, sometimes a small quantity of R-12 will be mixed with a different refrigerant because of its good miscibility with the oil. Note that ammonia (R-717) has almost no miscibility with oil. In ammonia systems, special provisions must be made to collect the oil in the system, and pump it back to the compressor.

Toxicity of the manufactured refrigerants (R-11, R-12, R-22, etc.) is not a problem for the service technician. Caution must be exercised, however, if you enter a tank that may contain refrigerant. The refrigerant, being heavier than air, will displace the air from the tank. Suffocation can result. The only commonly used toxic refrigerant is R-717 (ammonia). It is toxic in concentrations of 0.5 percent (by volume) or higher. Its presence, however, is obvious by its odor at much lower concentrations.

Figure 8-2 shows the molecular weights of some refrigerants. These are given to indicate the relative tendency of a refrigerant to leak. Refrigerants with high molecular weights have relatively large molecules, and so will be less likely to leak than a refrigerant with smaller molecules.

Physical properties of refrigerants as they pertain to the detection of leaks are important to the service technician. All of the refrigerant systems in use may be tested by use of the bubble method. That is, the system to be tested may be pressurized with either air, nitrogen, or the refrigerant to be used, to at least 20 psi. When a viscous (thick) solution of soap or detergent and water is applied to all suspected potential leaks, leakage will be revealed by the appearance of bubbles.

An **electronic leak detector** may be used for any of the halogen refrigerants. It is the most sensitive of all the leak detection methods, but it will not work on ammonia refrigerant. It may also be unreliable when other vapors such as alcohol or carbon monoxide are present.

The **halide torch** may also be used on the halogen refrigerants only. When the halogen refrigerants are present in a flame and in the presence of a heated copper element, the refrigerant decomposes and causes the flame to turn bluish green in color.

Ammonia leak detection is a different story. When a **sulfur candle** is burned in the area where a leak is suspected, a white cloud of smoke indicates that an ammonia leak is present.

A second method of testing for ammonia leakage is the use of **litmus paper.** Litmus paper can be used to determine the pH (acidity or alkalinity) of various substances. Ammonia vapor has a pH of less than 7, and may be detected by a suitable litmus paper that changes its color in response to the pH being measured.

The corrosivity of the halogen refrigerants is usually not a factor when used with materials such as steel, brass, copper, tin, lead, and aluminum. Ammonia, however, is not suitable for use with any metals that contain copper or brass.

Refrigerant	Miscibility
R-11	Good
R-12	Good
R-22	Good above −10°F
R-500	Good
R-502	Good
R-717 (ammonia)	Poor

Figure 8-1 Miscibility of refrigerants with oil.

Refrigerant	Molecular Weight	Leakage Tendency
R-113	187	Lowest
R-11	137	
R-12	121	
R-502	112	
R-500	99	
R-22	86	Highest

Figure 8-2 Leakage tendency of refrigerants.

Thermal Properties of Refrigerants

Thermal properties of refrigerants are those characteristics that determine how well they work as refrigerants. Thermal properties include pressure, temperature, density, volume, and latent heat. The selection of a refrigerant by a manufacturer usually represents a trade-off in these properties.

The **pressure–temperature (P–T) relationship** of several refrigerants is presented in Figure 8-3. We can draw a number of conclusions upon examining these data. For example, if we were to

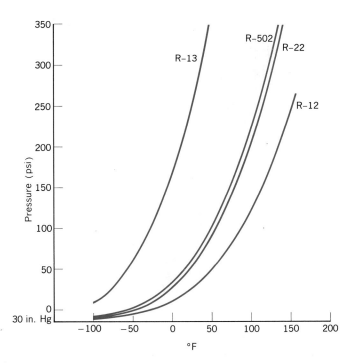

Figure 8-3 Pressure–temperature graph.

design a system that was using water at 85°F as the condensing medium, and assuming a condensing temperature of 105°F, an R-22 system would need to have a high-side pressure of 211 psi. An R-12 system, however, could be constructed more cheaply, as the high-side pressure would only be 127 psi. An R-11 system high side, on the other hand, would only have to withstand a pressure of 11 psi.

The **density** (or specific volume) of a refrigerant is important because it affects the size of the compressor required. Density is defined as the weight in pounds of a cubic foot of refrigerant, and will quite obviously change, depending upon pressure. Specific volume is the inverse of density; that is, it is the volume, in cubic feet, that will be occupied by 1 lb of refrigerant at any pressure.

Figure 8-4 shows how several refrigerants require different volumes of refrigerant to be circulated in order to obtain 1 ton of refrigerating effect in the evaporator. We can see that even though we needed to build the high side of the R-22 system stronger than that of an R-12 system, we can use a smaller compressor with the R-22 system. Which is better? The design engineer must make that choice by deciding which factor affects the cost of the system to a greater

extent. But how about R-11? It looked so good from a pressure–temperature standpoint, but we must circulate tremendous volumes of R-11 in order to produce a ton of refrigeration. We therefore find that the use of R-11 is well-suited for centrifugal compressors that can handle large flow volumes, but that cannot produce the high pressures that can be attained with reciprocating compressors.

The last factor that affected the cfm per ton comparisons in Figure 8-4 is the **latent heat of vaporization** of the different refrigerants. Remember, the refrigerating capacity produced by

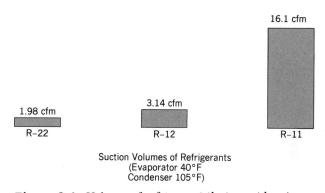

Suction Volumes of Refrigerants
(Evaporator 40°F
Condenser 105°F)

Figure 8-4 Volume of refrigerant that must be circulated by the compressor to produce 1 ton of refrigeration.

the evaporator in a system can be calculated as the number of pounds per hour that is circulating multiplied by the latent heat of vaporization for each pound. While we have shown some typical figures for a single set of operating conditions, the whole situation changes when different suction and discharge temperatures are required.

The purpose in presenting the refrigerant information in this chapter is not to enable you to design a system. That is the subject of a far more involved text. Many service technicians, however, have wondered why various applications require the use of different refrigerants, and why systems are designed the way they are. Some of the common applications for the various refrigerants are shown in Figure 8-5, and an overall comparison of refrigerants is shown in Figure 8-6. A P–T chart, such as the one in Figure 8-7, is available in pocket form for use by service technicians.

Refrigerant	Application
R-11, R-113	Low pressure, high volume refrigerants that are used with centrifugal compressors. Size applications are from 100 to 7000 tons in a single unit. These systems are used for chilling water to 44°F, for building air conditioning. They may also be used for industrial process cooling. Sometimes used as a solvent (especially R-113)
R-12	Household refrigerators, automotive air conditioners, walk-in refrigerators and freezers, display cases, drinking fountains, many applications requiring evaporator temperatures between −30°F and +40°F
R-22	Virtually all packaged air conditioning systems, including window units, central air conditioning, and heat pumps. Some use in refrigeration where evaporator temperatures above −10°F are adequate high stage of cascade systems
R-502	Refrigeration where evaporator temperatures below 0°F are required—frozen-food cases, refrigerated warehouses
R-13	Very low temperature applications, to −100°F. Low stage of cascade systems

Figure 8-5   Common refrigerant usage.

Refrigerant Color Codes

In order to aid the service technician in using the correct refrigerant, a system of color codes has been adopted (Figure 8-8). These colors are used on refrigerant containers and on refrigerant components, such as thermostatic expansion valves that are designed for use with only one specific refrigerant.

Refrigerant Oils

The properties required of oil used in a refrigeration system are so critical that conventional oils are not suitable for use. Special refrigeration oils that are highly refined and produced under strict quality-control methods are required. While the oil in an automobile can be changed every few

	Comparative Refrigerant Performance (per ton)			
Refrigerant	Low-side Pressure	High-Side Pressure	Compressor Displacement (cfm)	hp
R-11	23.9 in. Hg	3.5 psig	36.5	0.94
R-12	11.8 psig	93.3 psig	5.83	1.00
R-13*	7.6 psig	90.9 psig	6.74	1.13
R-113	27.9 in. Hg	13.9 in. Hg	102	0.97
R-114	16.1 in. Hg	22.0 psig	20.1	1.05
R-22	28.2 psig	158 psig	3.55	1.01
R-22*	25.0 in. Hg	4.9 psig	40.6	1.07
R-500	16.4 psig	113 psig	4.95	1.01
R-502	36.0 psig	175 psig	3.61	1.08
Ammonia	19.6 psig	154 psig	3.44	0.99

Note: Comparison of performance of refrigerants based on 5°F evaporating temperature and 86°F condensing temperature (except for *, which indicates performance based on −100°F evaporating temperature and −30°F condensing temperature).

Figure 8-6   Comparison of performance of refrigerants based on a 5°F evaporating temperature and an 86°F condensing temperature (except R-13, which is based on a −100°F evaporating temperature and a −30°F condensing temperature).

°F	R-11	R-12	R-13	R-22	R-113	R-500	R-502	R-717 Ammonia
−100	*29.8*	*27.0*	7.5	*25.0*		*26.4*	*23.3*	*27.4*
−95	*29.8*	*26.4*	10.9	*24.1*		*25.7*	*22.1*	*26.8*
−90	*29.7*	*25.8*	14.2	*23.0*		*24.9*	*20.7*	*26.1*
−85	*29.7*	*25.0*	18.2	*21.7*		*24.0*	*19.0*	*25.3*
−80	*29.6*	*24.1*	22.3	*20.2*		*22.9*	*17.1*	*24.3*
−75	*29.5*	*23.0*	27.1	*18.5*		*21.7*	*15.0*	*23.2*
−70	*29.4*	*21.9*	32.0	*16.6*		*20.3*	*12.6*	*21.9*
−65	*29.3*	*20.5*	37.7	*14.4*		*18.8*	*10.0*	*20.4*
−60	*29.2*	*19.0*	43.5	*12.0*		*17.0*	*7.0*	*18.6*
−55	*29.0*	*17.3*	50.0	*9.2*		*15.0*	*3.6*	*16.6*
−50	*28.8*	*15.4*	57.0	*6.2*		*12.8*	*0.0*	*14.3*
−45	*28.6*	*13.3*	64.6	*2.7*		*10.4*	2.1	*11.7*
−40	*28.4*	*11.0*	72.7	0.5		*7.6*	4.3	*8.7*
−35	*28.1*	*8.4*	81.5	2.6		*4.6*	6.7	*5.4*
−30	*27.8*	*5.5*	90.9	4.9	*29.3*	*1.2*	9.4	*1.6*
−28	*27.7*	*4.3*	94.9	5.9	*29.3*	0.1	10.5	0.0
−26	*27.5*	*3.0*	98.9	6.9	*29.2*	0.9	11.7	0.8
−24	*27.3*	*1.6*	103.0	7.9	*29.1*	1.6	13.0	1.7
−22	*27.2*	*0.3*	107.3	9.0	*29.1*	2.4	14.2	2.6
−20	*27.0*	0.6	111.7	10.2	*29.0*	3.2	15.5	3.6
−18	*26.8*	1.3	116.2	11.3	*29.0*	4.1	16.9	4.6
−16	*26.6*	2.1	120.8	12.5	*28.9*	5.0	18.3	5.6
−14	*26.4*	2.8	125.7	13.8	*28.8*	5.9	19.7	6.7
−12	*26.2*	3.7	130.5	15.1	*28.8*	6.8	21.2	7.9
−10	*26.0*	4.5	135.4	16.5	*28.7*	7.8	22.8	9.0
−8	*25.8*	5.4	140.5	17.9	*28.6*	8.8	24.4	10.3
−6	*25.1*	6.3	145.7	19.3	*28.5*	9.9	26.0	11.6
−4	*25.3*	7.2	151.1	20.8	*28.4*	11.0	27.7	12.9
−2	*25.0*	8.2	156.5	22.4	*28.3*	12.1	29.4	14.3
0	*24.7*	9.2	162.1	24.0	*28.2*	13.3	31.2	15.7
2	*24.4*	10.2	167.9	25.6	*28.1*	14.5	33.1	17.2
4	*24.1*	11.2	173.7	27.3	*28.0*	15.7	35.0	18.8
6	*23.8*	12.3	179.8	29.1	*27.9*	17.0	37.0	20.4
8	*23.4*	13.5	185.9	30.9	*27.7*	18.4	39.0	22.1
10	*23.1*	14.6	192.1	32.8	*27.6*	19.7	41.1	23.8
12	*22.7*	15.8	198.6	34.7	*27.5*	21.2	43.2	25.6
14	*22.3*	17.1	205.2	36.7	*27.3*	22.6	45.5	27.5
16	*21.9*	18.4	211.9	38.7	*27.1*	24.1	47.7	29.4
18	*21.6*	19.7	218.8	40.9	*27.0*	25.7	50.1	31.1
20	*21.1*	21.0	225.7	43.0	*26.8*	27.3	52.5	33.5
22	*20.6*	22.4	233.0	45.3	*26.6*	28.9	54.9	35.7
24	*20.1*	23.9	240.3	47.6	*26.4*	30.6	57.4	37.9
26	*19.7*	25.4	247.8	49.9	*26.2*	32.4	60.0	40.2
28	*19.1*	26.9	255.5	52.4	*26.0*	34.2	62.7	42.6
30	*18.6*	28.5	263.2	54.9	*25.8*	36.0	65.4	45.0
32	*18.1*	30.1	271.3	57.5	*25.6*	37.9	68.2	47.6
34	*17.5*	31.7	279.5	60.1	*25.3*	39.9	71.1	50.2
36	*16.9*	33.4	287.8	62.8	*25.1*	41.9	74.1	52.9
38	*16.3*	35.2	296.3	65.6	*24.8*	43.9	77.1	55.7
40	*15.6*	37.0	304.9	68.5	*24.5*	46.1	80.2	58.6
45	*13.9*	41.7	327.5	76.0	*23.7*	51.6	88.3	66.3
50	*12.0*	46.7	351.2	84.0	*22.9*	57.6	96.9	74.5
55	*10.0*	52.0	376.1	92.6	*22.0*	63.9	106.0	83.4
60	*7.8*	57.7	402.3	101.6	*21.0*	70.6	115.6	92.9
65	*5.4*	63.8	429.8	111.2	*19.9*	77.8	125.8	103.1
70	*2.8*	70.2	458.7	121.4	*18.7*	85.4	136.6	114.1
75	0.0	77.0	489.0	132.2	*17.3*	93.5	148.0	125.8
80	1.5	84.2	520.8	143.6	*15.9*	102.0	159.9	138.3
85	3.1	91.8	—	155.7	*14.1*	111.0	172.5	151.7
90	4.9	99.8	—	168.4	*12.3*	120.6	185.8	165.9
95	6.8	108.3	—	181.8	*10.6*	130.6	199.7	181.1
100	8.8	117.2	—	195.9	*8.6*	141.2	214.4	197.2
105	10.9	126.6	—	210.8	*6.4*	152.4	229.7	214.2
110	13.2	136.4	—	226.4	*4.0*	164.1	245.8	232.3
115	15.6	146.8	—	242.7	*1.4*	176.5	262.8	251.5
120	18.2	157.7	—	259.9	0.7	189.4	280.3	271.7
125	21.0	169.1	—	277.9	2.2	203.0	298.7	293.1
130	24.0	181.0	—	296.8	3.7	217.2	318.0	315.0
135	27.1	193.5	—	316.6	5.4	232.1	338.1	335.0
140	30.4	206.6	—	337.3	7.2	247.7	359.1	365.0
145	33.9	220.3	—	358.9	9.2	266.1	381.1	390.0
150	37.7	234.6	—	381.5	11.2	281.1	403.9	420.0
155	41.6	249.5	—	405.1	13.4	298.9	427.8	450.0
160	45.8	265.1	—	429.8	15.7	317.4	452.6	490.0

Italic figures = vacuum

Figure 8-7 Pressure–temperature chart.

Refrigerant	Color Code
R-11	Orange
R-12	White
R-13	Light blue
R-22	Green
R-113	Purple
R-500	Yellow
R-502	Orchid
R-717 (ammonia)	Silver

Figure 8-8 Color codes for refrigerants.

thousand miles, the oil in a refrigeration system may last for 20 years or more.

Viscosity

The primary purpose for having oil in the system is to provide lubrication to the moving and rubbing parts. In order to provide effective lubrication, the oil must be sufficiently thin to be able to get between parts that fit together with a close tolarance. If the oil is too thin, however, there is not enough body to provide the oil layer of separa-

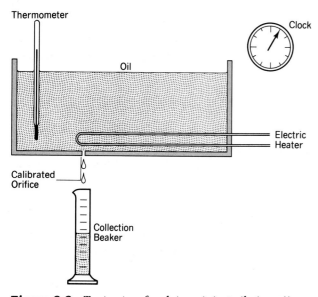

Figure 8-9 Test setup for determining oil viscosity.

tion between the moving parts. **Viscosity** is the measure of how easily or poorly the oil is able to flow. Conventional machinery oils are rated according to a scale in which 10W oil is relatively thin, while 90W oil is heavy enough to be used in gear casings. Refrigeration oil viscosity is measured in terms of Saybolt Seconds Universal (SSU). The setup for this test is shown in Figure 8-9. The oil is allowed to flow through a carefully calibrated orifice. The number of seconds required for a standard volume of oil to pass through the orifice is the SSU rating of that oil. The viscosity of the oil is affected by the temperature of the oil, as well as the percentage of refrigerant that has been mixed with the oil. For most applications, an oil viscosity of 150 SSU at 100°F is suitable. In automotive air conditioning systems, an oil of 500 SSU is common, while water chillers will use oils with a viscosity of 300 SSU.

Pour Point

The **pour point** of a refrigerant oil is the lowest temperature at which it can be poured out of an open container (Figure 8-10). If the oil is cooled to a temperature lower than its pour point, it will not move freely through the evaporator, and will not be returned to the compressor. The pour point is not just dependent upon the viscosity of the oil, but also depends upon the amount of wax contained in the oil.

Floc Point

The **floc point** of an oil is a measure of the amount of wax contained. All oils contain some

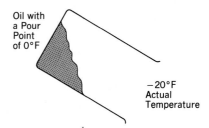

Figure 8-10 Oil that has been cooled to a temperature lower than its pour point.

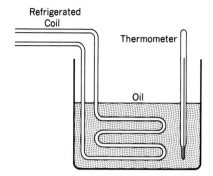

Figure 8-11 Test setup for determining the floc point. The temperature of the oil is lowered until wax just becomes visible.

degree of wax, although refrigerant oils have had far more of the wax removed during the refining process than conventional oils. At normal temperatures, the wax remains dissolved in the oil, but as the temperature of the oil is lowered, its ability to hold the wax in solution decreases. In the test shown in Figure 8-11, the temperature of the oil is slowly reduced until the first appearance of wax particles precipitating out of solution. The temperature at which this occurs is called the floc point.

The refrigerant oil used should have a floc point lower than the lowest temperature expected to be encountered in the system. Otherwise, the formation of wax will form a restriction, usually in the metering device.

Moisture

Moisture from any source can be the ruination of a refrigeration system. When moisture and oil are heated to the temperatures normally encountered in the compressor high side, they will form acid and sludge. The sludge can clog the oil passageways. The acids can cause pitting of the polished surfaces in the compressor, and can also attack the insulation on the motor windings for hermetic compressors. All refrigeration oils have had sufficient water removed so that it should not be a factor in the operation of any refrigeration system. When using refrigerant oils, it is important that you do not leave the containers open, because the oil will quickly absorb moisture from the atmosphere, and be rendered useless.

Testing the Oil

There are three tests available to the service technician to determine the condition of the oil in a system. The first is odor. When a system with oil in good condition is opened, there should be no unpleasant smells. The presence of such an odor would indicate that the oil contains impurities and should be changed.

A second indicator of the condition of the oil is its color. A sample of oil may be placed in a glass container, where it should appear very light brownish. A contaminated oil will be much darker.

The third method for testing the oil is with an acid test kit. An oil sample is mixed with a solution furnished with the kit, and the acid content is determined by comparing the mixture's color with a standard chart.

Refrigerant Piping

The design and installation of refrigerant piping is important to the proper operation of every refrigeration system. The problems that can be caused by poor refrigerant piping are as follow:

1. Compressor failure due to lack of oil.
2. Nuisance compressor tripouts due to lack of oil.
3. Compressor failure due to liquid refrigerant entering the crankcase.
4. High operating cost and/or low system capacity due to excessive pressure drop in the lines.

The theory of proper refrigerant piping is simple. Piping must be large enough to prevent excessive pressure drop while being small enough to prevent oil from being trapped in the system, prevent loss of oil from the compressor on startup, and not allow liquid refrigerant to enter the compressor.

Hot Gas from Compressor to Condenser

The **hot gas line** between the compressor and the condenser will contain refrigerant vapor and oil

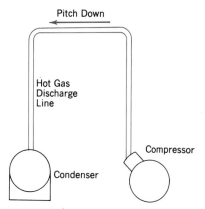

Figure 8-12 Hot gas discharge line with the condenser at an elevation either the same as or lower than the compressor.

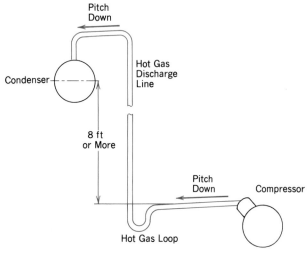

Figure 8-13 Hot gas discharge line with condenser more than 8 ft above the compressor. The hot gas loop prevents liquid that may form during the off-cycle from draining back to the head of the compressor.

that has escaped from the compressor. The oil is carried through the line around the inside of the tube wall. The oil and the refrigerant vapor do not mix very well, so it is necessary that the line be sized small enough so that the refrigerant velocity is sufficient to move the oil along the tube wall. The oil must eventually be allowed to return to the compressor. If the hot gas line is too small, there will be excessive pressure drop, and the compressor will produce an abnormally high discharge pressure.

Installation of the hot gas line should not allow any traps in which the oil can accumulate. If a muffler is installed in the hot gas line, the piping

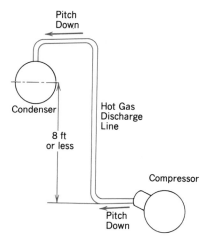

Figure 8-14 Hot gas discharge line with condenser less than 8 ft above the compressor.

should be arranged so that the flow is vertically downward. If an oil separator is installed, the oil that collects in the tank must be piped back to the compressor crankcase. Figures 8-12 through 8-14 show several arrangements for hot gas piping.

Liquid Line

The miscibility of the oil with the refrigerant in the liquid state is much better than with refrigerant in the vapor state. The oil will move freely through the **liquid line,** regardless of the pipe size chosen. If the pipe size is excessively small, there will be high pressure drop, high operating cost, and loss of system capacity. If the liquid line produces so much pressure drop that some of the liquid flashes, then drastic reduction in system capacity will result. The flash vapor will compete with the liquid trying to move through the metering device, and the evaporator will be starved.

If the liquid line is sized too large, it will affect the original cost of the system (higher piping cost and refrigerant cost), but there will not be any operational problems. Figure 8-15 shows liquid line piping to a vented receiver. The vent prevents liquid from being trapped in the condenser, but may be omitted when the receiver is less than 6 ft from the condenser. Figure 8-16 shows a special liquid line loop used when the evaporator is located below the condenser.

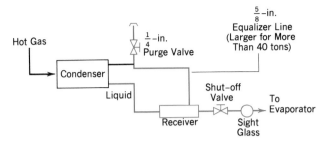

Figure 8-15 Liquid line piping with a receiver vent line.

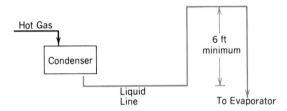

Figure 8-16 Liquid line loop to prevent liquid from siphoning into the evaporator during shutdown.

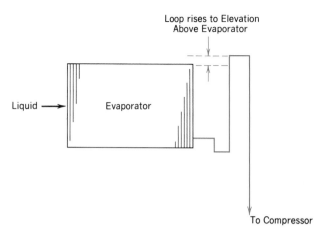

Figure 8-17 Suction line loop used to prevent liquid from draining into the compressor located below the evaporator.

Suction Line from Evaporator to Compressor

The **suction line** is the most critical and potentially troublesome of all the refrigerant piping. The effect of pressure drop in the suction line is much greater on system capacity and compressor horsepower than any other line. The oil does not mix well with the refrigerant vapor, so refrigerant velocity must remain high enough to entrain the oil. Properly sized suction lines may be run level or slightly pitched down toward the compressor. For installations where the evaporator is located at a higher elevation than the compressor, piping must be provided that will prevent the liquid refrigerant from draining into the compressor on shutdown of the system (suction line loop). For installations where the evaporator is located at a lower elevation than the compressor, special consideration must be given to sizing the portion of the line that must carry the oil vertically upward. And where compressors have unloading capability, suction line risers are provided to provide sufficient velocity at low loads, yet not impose excessive pressure drop when the compressor operates fully loaded.

Suction Line Loop

When some systems shut down, all of the liquid refrigerant in the evaporator boils to a vapor. Other systems have liquid refrigerant remaining, even after the high- and low-side pressures equalize. The liquid will be located where it is normally found during operation, that is, in the liquid line and in the evaporator. When the evaporator is located above the compressor, the liquid might run downhill into the compressor, which can cause liquid slugging upon start-up. It can also dilute the oil. Upon start-up, the liquid refrigerant mixed with the oil will flash, and a large quantity of oil will be pumped out of the compressor with the refrigerant.

The **suction line loop** shown in Figure 8-17 is designed to prevent these problems. The loop works because the liquid refrigerant will not flow uphill out of the evaporator coil. The loop must rise to an elevation higher than the liquid in the coil in order to prevent siphoning of the liquid into the compressor.

Double Suction Riser

The **double suction riser** shown in Figure 8-18 is a method for carrying a lot of suction gas with minimum pressure drop, and for carrying a much lower flow of suction gas without losing the

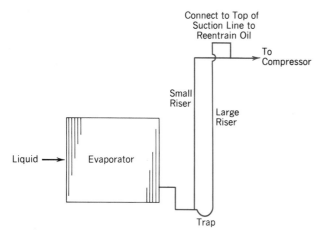

Figure 8-18 Double suction riser for evaporator located below the compressor.

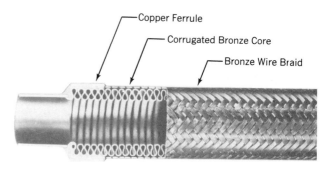

Figure 8-19 Vibration eliminator. *(Courtesy Anamet Inc.)*

required oil-carrying velocity. During full load operation, both risers carry refrigerant back to the compressor at sufficient velocity to carry the oil. As the compressor unloads and the quantity of suction gas is reduced, the oil will tend to collect in the trap at the bottom of the risers. When the trap fills, suction gas will be drawn only through the remaining smaller riser.

The larger riser ties into the top of the elevated line to the compressor. This is done for two reasons.

1. When the system is operating for a long period of time at partial load, oil is prevented from dripping into the larger riser, causing it to totally fill with oil.

2. When the system returns to full load operation, the oil trap is blown out. By causing it to enter the pipe from the top, it will become distributed around the inside of the pipe more easily than if it came up from the bottom.

The length of the trap at the bottom of the suction line loop must be kept short, otherwise too much oil will be allowed to accumulate during part load operation.

Vibration Eliminators

The **vibration eliminators** in Figures 8-19 and 8-20 are designed to prevent compressor vibration from being transmitted through the refrigerant piping. The benefits include noise

Figure 8-20 Compressors installed with vibration eliminators on the suction and discharge lines. *(Courtesy Anamet Inc.)*

reduction and a lower likelihood of developing refrigerant leaks due to vibration failure of joints. On low temperature applications where the vibration eliminator is installed vertically, some service technicians wrap the lower copper ferrule with tape to prevent condensation collection during the Off cycle, as it will freeze and expand during the On cycle. This prevents frost damage from occurring.

Precharged Piping

Quick connect couplings (Figure 8-21) are used by manufacturers of split systems in order to allow factory precharging of the outdoor and indoor units and interconnecting piping. This can

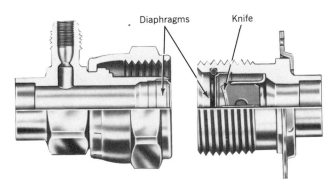

Figure 8-21 Quick connect coupling used on precharged systems. *(Courtesy Aeroquip Corporation)*

result in cleaner field installations. Precharging reduces improper field joints, leaks, and system contamination, and failure due to improper charging. The connection sequence is shown in Figures 8-22 and 8-23. On initial connection, a rubber seal prevents loss of refrigerant. Tightening the union nut draws the coupling halves together, piercing and folding both metal diaphragms back, opening the refrigerant passage. When fully coupled, the sealing surfaces in the two brass halves are mated, creating a permanent metal-to-metal seal.

Pipe Sizing

The charts at the end of this chapter (Figures 8-24 through 8-38) give you sizing information for each of the refrigerant lines, as well as for several common refrigerants. They represent an optimum compromise between first cost and op-

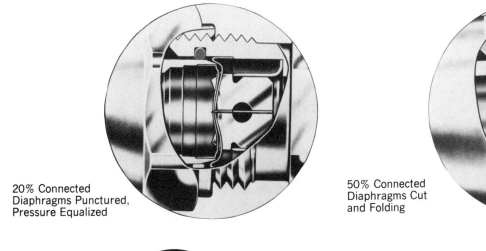

20% Connected
Diaphragms Punctured,
Pressure Equalized

50% Connected
Diaphragms Cut
and Folding

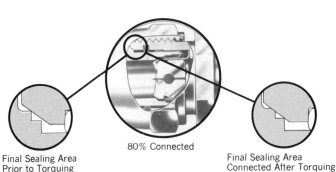

Final Sealing Area
Prior to Torquing

80% Connected

Final Sealing Area
Connected After Torquing

Figure 8-22 Connection sequence for quick connect couplings on precharged tubing installations. *(Courtesy Aeroquip Corporation)*

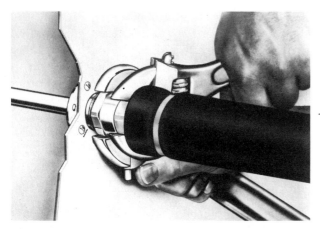

Figure 8-23 Tightening quick connect couplings using a back-up wrench. *(Courtesy Aeroquip Corporation)*

erating cost. Note that in all cases, for long runs of refrigerant piping, the size of the line should be increased. In order to determine the **equivalent length,** first determine the actual length of the line to be sized, then add 50 percent to this length to account for the added friction through fittings, valves, and accessories. Depending upon the complexity of the run, you may choose to add anything between 25 and 75 percent to the actual length. Enter the chart at the equivalent length and at the capacity of refrigeration and select the closest pipe size. If you select the higher pipe size, the pressure drop will be lower than allowable, but if the smaller size is chosen, the operating costs will be slightly higher than optimum.

RECOMMENDED LIQUID LINE SIZES

Capacity BTU/hr	R-12					R-22					R-502				
	Condenser to Receiver	Receiver to Evaporator Equivalent Length, Ft.				Condenser to Receiver	Receiver to Evaporator Equivalent Length, Ft.				Condenser to Receiver	Receiver to Evaporator Equivalent Length, Ft.			
		50	100	150	200		50	100	150	200		50	100	150	200
6,000	⅜	⅜	⅜	⅜	⅜	⅜	¼	⅜	⅜	⅜	⅜	¼	⅜	⅜	⅜
12,000	½	⅜	⅜	½	½	½	⅜	⅜	⅜	⅜	½	⅜	½	½	½
18,000	½	½	½	½	½	½	⅜	⅜	½	½	⅝	½	½	½	½
24,000	⅝	½	½	½	⅝	⅝	⅜	½	½	½	⅝	½	⅝	⅝	⅝
36,000	⅝	½	⅝	⅝	⅝	⅝	½	½	½	½	⅞	½	⅝	⅝	⅝
48,000	⅞	½	⅝	⅝	⅞	⅞	½	⅝	⅝	⅝	⅞	⅝	⅝	⅝	⅞
60,000	⅞	⅝	⅝	⅞	⅞	⅞	½	⅝	⅝	⅝	⅞	⅝	⅞	⅞	⅞
75,000	⅞	⅝	⅞	⅞	⅞	⅞	½	⅝	⅝	⅝	⅞	⅝	⅞	⅞	⅞
100,000	1⅛	⅞	⅞	⅞	⅞	⅞	⅝	⅞	⅞	⅞	1⅛	⅞	⅞	⅞	⅞
150,000	1⅛	⅞	⅞	1⅛	1⅛	1⅛	⅞	⅞	⅞	⅞	1⅜	⅞	⅞	1⅛	1⅛
200,000	1⅜	⅞	1⅛	1⅛	1⅛	1⅛	⅞	⅞	1⅛	1⅛	1⅜	1⅛	1⅛	1⅛	1⅛
300,000	1⅝	1⅛	1⅛	1⅜	1⅜	1⅜	1⅛	1⅛	1⅛	1⅛	1⅝	1⅜	1⅜	1⅜	1⅜
400,000	1⅝	1⅜	1⅜	1⅜	1⅜	1⅝	1⅛	1⅛	1⅜	1⅜	1⅝	1⅜	1⅜	1⅜	1⅝
500,000	1⅝	1⅜	1⅜	1⅝	1⅝	1⅝	1⅛	1⅜	1⅜	1⅜	2⅛	1⅜	1⅜	1⅝	1⅝
600,000	2⅛	1⅝	1⅝	1⅝	1⅝	1⅝	1⅜	1⅜	1⅜	1⅝	2⅛	1⅝	1⅝	1⅝	1⅝
750,000	2⅛	1⅝	1⅝	1⅝	2⅛	2⅛	1⅝	1⅝	1⅝	1⅝	2⅛	2⅛	2⅛	2⅛	2⅛

Recommended sizes are applicable with evaporating temperatures from -40° F. to 45° F. and condensing temperatures from 80° F. to 130° F.

Figure 8-24 Recommended liquid line sizes. *(Courtesy Copeland Corporation)*

RECOMMENDED DISCHARGE LINE SIZES

Capacity BTU/hr	Light Load Capacity Reduction	R-12 Equivalent Length, Ft.				R-22 Equivalent Length, Ft.				R-502 Equivalent Length, Ft.			
		50	100	150	200	50	100	150	200	50	100	150	200
6,000	0	½	½	½	⅝*	⅜	½	½	½	½	½	½	⅝*
12,000	0	⅝	⅝	⅝	⅞	½	½	⅝	⅝	⅝	⅝	⅝	⅞
18,000	0	⅝	⅞	⅞	⅞	⅝	⅝	⅝	⅞	⅝	⅞	⅞	⅞
24,000	0	⅞	⅞	⅞	⅞	⅝	⅞	⅞	⅞	⅞	⅞	⅞	⅞
36,000	0	⅞	⅞	⅞	1⅛	⅞	⅞	⅞	⅞	⅞	⅞	1⅛	1⅛
48,000	0	⅞	1⅛	1⅛	1⅛	⅞	⅞	⅞	1⅛	⅞	1⅛	1⅛	1⅛
60,000	0	1⅛	1⅛	1⅛	1⅜	⅞	1⅛	1⅛	1⅛	1⅛	1⅛	1⅛	1⅜
	33%	1⅛	1⅛	1⅛	1⅜	⅞	1⅛	1⅛	1⅛	1⅛	1⅛	1⅛	1⅜
75,000	0	1⅛	1⅛	1⅛	1⅜	⅞	1⅛	1⅛	1⅛	1⅛	1⅛	1⅜	1⅜
	33%	1⅛	1⅛	1⅛	1⅜	⅞	1⅛	1⅛	1⅛	1⅛	1⅛	1⅜	1⅜
100,000	0	1⅛	1⅜	1⅜	1⅝	1⅛	1⅛	1⅜	1⅜	1⅛	1⅜	1⅜	1⅝
	33% to 50%	1⅛	1⅜	1⅜	1⅝	1⅛	1⅛	1⅜	1⅜	1⅛	1⅜	1⅜	1⅝
150,000	0	1⅜	1⅝	1⅝	2⅛	1⅛	1⅜	1⅜	1⅜	1⅜	1⅜	1⅝	1⅝
	33% to 50%	1⅜	1⅝	1⅝	2⅛*	1⅛	1⅜	1⅜	1⅜	1⅜	1⅜	1⅝	1⅝
	66%	1⅜	1⅝	1⅝	2⅛*	1⅛	1⅜	1⅜	1⅜	1⅜	1⅜	1⅝	1⅝
200,000	0	1⅝	1⅝	2⅛	2⅛	1⅜	1⅜	1⅝	1⅝	1⅜	1⅝	1⅝	2⅛
	33% to 50%	1⅝	1⅝	2⅛	2⅛	1⅜	1⅜	1⅝	1⅝	1⅜	1⅝	1⅝	2⅛
	66%	1⅝	1⅝	2⅛*	2⅛*	1⅜	1⅜	1⅝	1⅝	1⅜	1⅝	1⅝	2⅛*
300,000	0	2⅛	2⅛	2⅛	2⅛	1⅜	1⅝	1⅝	2⅛	1⅝	2⅛	2⅛	2⅛
	33% to 50%	2⅛	2⅛	2⅛	2⅛	1⅜	1⅝	1⅝	2⅛	1⅝	2⅛	2⅛	2⅛
	66%	2⅛	2⅛	2⅛	2⅛	1⅜	1⅝	2⅛*	2⅛*	1⅝	2⅛	2⅛	2⅛
400,000	0	2⅛	2⅛	2⅛	2⅝	1⅝	2⅛	2⅛	2⅛	2⅛	2⅛	2⅛	2⅝
	33% to 66%	2⅛	2⅛	2⅛	2⅝	1⅝	2⅛	2⅛	2⅛	2⅛	2⅛	2⅛	2⅝
500,000	0	2⅝	2⅝	2⅝	2⅝	2⅛	2⅛	2⅛	2⅛	2⅛	2⅛	2⅝	2⅝
	33% to 50%	2⅝	2⅝	2⅝	2⅝	2⅛	2⅛	2⅛	2⅛	2⅛	2⅛	2⅝	2⅝
	66%	2⅝	2⅝	2⅝	2⅝	2⅛	2⅛	2⅛	2⅛	2⅝	2⅝	2⅝	2⅝
600,000	0	2⅝	2⅝	2⅝	3⅛	2⅛	2⅛	2⅛	2⅝	2⅛	2⅝	2⅝	3⅛
	33% to 50%	2⅝	2⅝	2⅝	3⅛*	2⅛	2⅛	2⅛	2⅝	2⅛	2⅝	2⅝	3⅛*
	66%	2⅝	2⅝	3⅛*	3⅛*	2⅛	2⅛	2⅛	2⅝*	2⅛	2⅝	2⅝	3⅛*
750,000	0	3⅛	3⅛	3⅛	3⅛	2⅛	2⅝	2⅝	2⅝	2⅝	2⅝	2⅝	3⅛
	33% to 50%	3⅛	3⅛	3⅛	3⅛	2⅛	2⅝	2⅝	2⅝	2⅝	2⅝	2⅝	3⅛
	66%	3⅛*	3⅛*	3⅛*	3⅛*	2⅛	2⅝	2⅝	2⅝	2⅝	2⅝	2⅝	3⅛*

* Use one line size smaller for vertical riser

Recommended sizes are applicable for applications with evaporating temperatures from -40° F. to 45° F. and condensing temperatures from 80° F. to 130° F.

Figure 8-25 Recommended discharge line sizes. (*Courtesy Copeland Corporation*)

RECOMMENDED SUCTION LINE SIZES

R-12   40° F. Evaporating Temperature

Capacity BTU/hr.	Light Load Capacity Reduction	Equivalent Length, Ft.							
		50		100		150		200	
		H	V	H	V	H	V	H	V
6,000	0	5/8	5/8	5/8	5/8	5/8	5/8	5/8	5/8
12,000	0	7/8	7/8	7/8	7/8	7/8	7/8	7/8	7/8
18,000	0	7/8	7/8	7/8	7/8	1 1/8	7/8	1 1/8	1 1/8
24,000	0	7/8	7/8	1 1/8	1 1/8	1 1/8	1 1/8	1 1/8	1 1/8
36,000	0	1 1/8	1 1/8	1 1/8	1 1/8	1 3/8	1 1/8	1 3/8	1 3/8
48,000	0	1 1/8	1 1/8	1 3/8	1 3/8	1 3/8	1 3/8	1 5/8	1 5/8
60,000	0 to 33%	1 1/8	1 1/8	1 3/8	1 3/8	1 5/8	1 3/8	1 5/8	1 5/8
75,000	0 to 33%	1 3/8	1 3/8	1 5/8	1 3/8	1 5/8	1 3/8	1 5/8	1 5/8
100,000	0 to 50%	1 3/8	1 3/8	1 5/8	1 5/8	2 1/8	1 5/8	2 1/8	1 5/8
150,000	0 to 33%	1 5/8	1 5/8	2 1/8	1 5/8	2 1/8	1 5/8	2 5/8	2 1/8
	50% to 66%	1 5/8	1 5/8	2 1/8	1 5/8	2 1/8	1 5/8	2 1/8	1 5/8
200,000	0	2 1/8	2 1/8	2 1/8	2 1/8	2 5/8	2 1/8	2 5/8	2 5/8
	33% to 50%	2 1/8	2 1/8	2 1/8	2 1/8	2 5/8	2 1/8	2 5/8	2 1/8
	66%	2 1/8	2 1/8	2 1/8	2 1/8	2 1/8	2 1/8	2 1/8	2 1/8
300,000	0 to 50%	2 1/8	2 1/8	2 5/8	2 1/8	2 5/8	2 1/8	3 1/8	2 5/8
	66%	2 1/8	2 1/8	2 5/8	2 1/8	2 5/8	2 1/8	2 5/8	2 1/8
400,000	0 to 50%	2 5/8	2 5/8	3 1/8	2 5/8	3 1/8	2 5/8	3 1/8	3 1/8
	66%	2 5/8	2 5/8	3 1/8	2 5/8	3 1/8	2 5/8	3 1/8	2 5/8
500,000	0 to 50%	2 5/8	2 5/8	3 1/8	2 5/8	3 1/8	2 5/8	3 5/8	3 1/8
	66%	2 5/8	2 5/8	3 1/8	2 5/8	3 1/8	2 5/8	3 5/8	2 5/8
600,000	0 to 66%	3 1/8	2 5/8	3 1/8	3 1/8	3 5/8	3 1/8	3 5/8	3 1/8
750,000	0 to 66%	3 1/8	3 1/8	3 5/8	3 1/8	3 5/8	3 1/8	4 1/8	3 5/8

Recommended sizes are applicable for applications with condensing temperatures from 80° F. to 130° F.

H - Horizontal
V - Vertical

Figure 8-26 Recommended suction line sizes for R-12 at 40°F evaporating temperature. *(Courtesy Copeland Corporation)*

RECOMMENDED SUCTION LINE SIZES

R-12   25° F. Evaporating Temperature

Capacity BTU/hr.	Light Load Capacity Reduction	Equivalent Length, Ft.							
		50		100		150		200	
		H	V	H	V	H	V	H	V
6,000	0	5/8	5/8	7/8	5/8	7/8	5/8	7/8	5/8
12,000	0	7/8	7/8	7/8	7/8	1 1/8	7/8	1 1/8	7/8
18,000	0	1 1/8	1 1/8	1 1/8	1 1/8	1 1/8	1 1/8	1 1/8	1 1/8
24,000	0	1 1/8	1 1/8	1 1/8	1 1/8	1 3/8	1 1/8	1 3/8	1 1/8
36,000	0	1 3/8	1 1/8	1 3/8	1 3/8	1 3/8	1 3/8	1 5/8	1 3/8
48,000	0	1 3/8	1 3/8	1 5/8	1 3/8	1 5/8	1 5/8	1 5/8	1 5/8
60,000	0 to 33%	1 5/8	1 3/8	1 5/8	1 5/8	2 1/8	1 5/8	2 1/8	1 5/8
75,000	0 to 33%	1 5/8	1 5/8	2 1/8	1 5/8	2 1/8	1 5/8	2 1/8	1 5/8
100,000	0 to 33%	1 5/8	1 5/8	2 1/8	1 5/8	2 1/8	2 1/8	2 1/8	2 1/8
	50%	1 5/8	1 5/8	2 1/8	1 5/8	2 1/8	1 5/8	2 1/8	1 5/8
150,000	0 to 33%	2 1/8	2 1/8	2 5/8	2 1/8	2 5/8	2 5/8	2 5/8	2 5/8
	50% to 66%	2 1/8	2 1/8	2 5/8	2 1/8	2 5/8	1 5/8 *2 1/8	2 5/8	1 5/8 *2 1/8
200,000	0 to 50%	2 5/8	2 5/8	2 5/8	2 5/8	3 1/8	2 5/8	3 1/8	2 5/8
	66%	2 5/8	2 1/8	2 5/8	1 5/8 *2 1/8	2 5/8	1 5/8 *2 1/8	2 5/8	1 5/8 *2 1/8
300,000	0 to 50%	2 5/8	2 5/8	3 1/8	2 5/8	3 5/8	2 5/8	3 5/8	2 5/8
	66%	2 5/8	2 5/8	3 1/8	2 5/8	3 1/8	1 5/8 *2 5/8	3 1/8	1 5/8 *2 5/8
400,000	0 to 50%	3 1/8	3 1/8	3 1/8	3 1/8	3 5/8	3 1/8	3 5/8	3 1/8
	66%	3 1/8	2 5/8	3 1/8	1 5/8 *2 5/8	3 1/8	1 5/8 *2 5/8	3 1/8	1 5/8 *2 5/8

Recommended sizes are applicable for applications with condensing temperatures from 80° F. to 130° F.
* Double Riser

H - Horizontal
V - Vertical

Figure 8-27 Recommended suction line sizes for R-12 at 25°F evaporating temperature. *(Courtesy Copeland Corporation)*

RECOMMENDED SUCTION LINE SIZES

R-12   15° F. Evaporating Temperature

Capacity BTU/hr.	Light Load Capacity Reduction	Equivalent Length, Ft.							
		50		100		150		200	
		H	V	H	V	H	V	H	V
6,000	0	7/8	7/8	7/8	7/8	7/8	7/8	7/8	7/8
12,000	0	7/8	7/8	1 1/8	7/8	1 1/8	7/8	1 1/8	7/8
18,000	0	1 1/8	1 1/8	1 1/8	1 1/8	1 3/8	1 1/8	1 3/8	1 1/8
24,000	0	1 1/8	1 1/8	1 3/8	1 1/8	1 3/8	1 1/8	1 3/8	1 3/8
36,000	0	1 3/8	1 3/8	1 3/8	1 3/8	1 5/8	1 3/8	2 1/8	1 5/8
48,000	0	1 3/8	1 3/8	1 5/8	1 3/8	1 5/8	1 3/8	2 1/8	1 5/8
60,000	0 to 33%	1 5/8	1 5/8	1 5/8	1 5/8	2 1/8	1 5/8	2 1/8	1 5/8
75,000	0 to 33%	1 5/8	1 5/8	2 1/8	1 5/8	2 1/8	1 5/8	2 5/8	1 5/8
100,000	0 to 33%	2 1/8	1 5/8	2 1/8	1 5/8	2 5/8	1 5/8	2 5/8	2 1/8
	50%	2 1/8	1 5/8	2 1/8	1 5/8	2 5/8	1 5/8	2 5/8	1 5/8
150,000	0 to 33%	2 1/8	2 1/8	2 5/8	2 1/8	2 5/8	2 1/8	3 1/8	2 1/8
	50% to 66%	2 1/8	2 1/8	2 5/8	2 1/8	2 5/8	1 5/8*2 1/8	2 5/8	1 5/8*2 1/8
200,000	0 to 50%	2 5/8	2 1/8	2 5/8	2 5/8	2 5/8	2 5/8	3 1/8	2 5/8
	66%	2 5/8	2 1/8	3 1/8	2 1/8	2 5/8	1 5/8*2 1/8	2 5/8	1 5/8*2 1/8
300,000	0 to 50%	3 1/8	2 5/8	3 1/8	2 5/8	3 1/8	3 1/8	3 5/8	3 1/8
	66%	3 1/8	2 5/8	3 1/8	2 5/8	3 1/8	1 5/8*2 5/8	3 1/8	1 5/8*2 5/8
400,000	0 to 50%	3 1/8	2 5/8	3 1/8	3 1/8	3 5/8	3 1/8	3 5/8	3 1/8
	66%	3 1/8	2 5/8	3 1/8	1 5/8*2 5/8	3 5/8	1 5/8*2 5/8	3 5/8	1 5/8*2 5/8

Recommended sizes are applicable for applications with condensing temperatures from 80° F. to 130° F.

H - Horizontal
V - Vertical

* Double Riser

Figure 8-28 Recommended suction line sizes for R-12 at 15°F evaporating temperature. *(Courtesy Copeland Corporation)*

RECOMMENDED SUCTION LINE SIZES
R-12   -20° F. Evaporating Temperatures

Capacity BTU/hr.	Light Load Capacity Reduction	Equivalent Length, Ft.							
		50		100		150		200	
		H	V	H	V	H	V	H	V
6,000	0	7/8	7/8	1 1/8	7/8	1 1/8	7/8	1 1/8	7/8
12,000	0	1 1/8	1 1/8	1 3/8	1 1/8	1 3/8	1 1/8	1 3/8	1 1/8
18,000	0	1 3/8	1 3/8	1 3/8	1 3/8	1 5/8	1 3/8	1 5/8	1 3/8
24,000	0	1 3/8	1 3/8	1 5/8	1 5/8	1 5/8	1 5/8	2 1/8	1 5/8
36,000	0	1 5/8	1 5/8	2 1/8	1 5/8	2 1/8	1 5/8	2 1/8	1 5/8
48,000	0	2 1/8	1 5/8	2 1/8	1 5/8	2 1/8	2 1/8	2 5/8	2 1/8
60,000	0 to 33%	2 1/8	2 1/8	2 5/8	2 1/8	2 5/8	2 1/8	2 5/8	2 1/8
75,000	0 to 33%	2 1/8	2 1/8	2 5/8	2 1/8	2 5/8	2 1/8	3 1/8	2 1/8
100,000	0 to 50%	2 5/8	2 1/8	2 5/8	2 1/8	2 5/8	2 1/8	3 1/8	2 1/8
150,000	0 to 50%	2 5/8	2 5/8	3 1/8	2 5/8	3 1/8	2 5/8	3 5/8	2 5/8
	66%	2 5/8	1 5/8 *2 1/8	3 1/8	1 5/8 *2 1/8	3 1/8	1 5/8 *2 1/8	3 5/8	1 5/8 *2 5/8

Recommended sizes are applicable for applications with condensing temperatures from 80° F. to 130° F.

H - Horizontal
V - Vertical
* Double Riser

Figure 8-29 Recommended suction line sizes for R-12 at −20°F evaporating temperature. *(Courtesy Copeland Corporation)*

RECOMMENDED SUCTION LINE SIZES
R-12   -40° F. Evaporating Temperature

Capacity BTU/hr.	Light Load Capacity Reduction	Equivalent Length, Ft.							
		50		100		150		200	
		H	V	H	V	H	V	H	V
6,000	0	1 1/8	1 1/8	1 1/8	1 1/8	1 3/8	1 1/8	1 3/8	1 1/8
12,000	0	1 3/8	1 1/8	1 5/8	1 3/8	1 5/8	1 3/8	1 5/8	1 3/8
18,000	0	1 3/8	1 3/8	1 5/8	1 3/8	1 5/8	1 3/8	2 1/8	1 5/8
24,000	0	1 5/8	1 5/8	2 1/8	1 5/8	2 1/8	1 5/8	2 5/8	1 5/8
36,000	0	2 1/8	2 1/8	2 1/8	2 1/8	2 5/8	2 1/8	2 5/8	2 1/8
48,000	0	2 1/8	2 1/8	2 5/8	2 1/8	2 5/8	2 1/8	2 5/8	2 1/8
60,000	0 to 33%	2 5/8	2 1/8	3 1/8	2 1/8	3 1/8	2 1/8	3 1/8	2 1/8
75,000	0	2 5/8	2 5/8	3 1/8	2 5/8	3 1/8	2 5/8	3 1/8	2 5/8
	33%	3 1/8	2 1/8	3 1/8	2 1/8	3 1/8	2 1/8	3 1/8	2 1/8
100,000	0 to 33%	3 1/8	2 5/8	3 5/8	2 5/8	3 5/8	2 5/8	3 5/8	2 5/8
	50%	3 1/8	1 5/8 *2 5/8	3 1/8	1 5/8 *2 5/8	3 5/8	1 5/8 *2 5/8	3 5/8	1 5/8 *2 5/8

Recommended sizes are applicable for applications with condensing temperatures from 80° F. to 130° F.

H - Horizontal
V - Vertical
* Double Riser

Figure 8-30 Recommended suction line sizes for R-12 at − 40°F evaporating temperature. *(Courtesy Copeland Corporation)*

RECOMMENDED SUCTION LINE SIZES
R-22   40° F. Evaporating Temperature

Capacity BTU/hr.	Light Load Capacity Reduction	50		100		150		200	
		H	V	H	V	H	V	H	V
6,000	0	½	½	½	½	⅝	½	⅝	½
12,000	0	⅝	⅝	⅝	⅝	⅞	⅝	⅞	⅝
18,000	0	⅞	⅞	⅞	⅞	⅞	⅞	⅞	⅞
24,000	0	⅞	⅞	⅞	⅞	⅞	⅞	1⅛	⅞
36,000	0	⅞	⅞	1⅛	⅞	1⅛	⅞	1⅛	1⅛
48,000	0	1⅛	1⅛	1⅛	1⅛	1⅛	1⅛	1⅜	1⅛
60,000	0 to 33%	1⅛	1⅛	1⅛	1⅛	1⅜	1⅛	1⅜	1⅛
75,000	0 to 33%	1⅛	1⅛	1⅜	1⅛	1⅜	1⅛	1⅝	1⅜
100,000	0 to 50%	1⅜	1⅜	1⅜	1⅜	1⅜	1⅜	1⅝	1⅜
150,000	0 to 66%	1⅜	1⅜	1⅝	1⅝	1⅝	1⅝	2⅛	1⅝
200,000	0 to 66%	1⅝	1⅝	2⅛	1⅝	2⅛	1⅝	2⅛	1⅝
300,000	0 to 50%	2⅛	2⅛	2⅛	2⅛	2⅛	2⅛	2⅝	2⅛
	66%	2⅛	2⅛	2⅛	2⅛	2⅛	2⅛	2⅛	2⅛
400,000	0 to 66%	2⅛	2⅛	2⅛	2⅛	2⅝	2⅛	2⅝	2⅛
500,000	0 to 66%	2⅛	2⅛	2⅝	2⅛	2⅝	2⅛	2⅝	2⅝
600,000	0 to 66%	2⅝	2⅝	2⅝	2⅝	2⅝	2⅝	3⅛	2⅝
750,000	0 to 66%	2⅝	2⅝	3⅛	2⅝	3⅛	2⅝	3⅛	2⅝

Equivalent Length, Ft.

Recommended sizes are applicable with condensing temperatures from 80° F. to 130° F.

H - Horizontal
V - Vertical

Figure 8-31 Recommended suction line sizes for R-22 at 40°F evaporating temperature. *(Courtesy Copeland Corporation)*

RECOMMENDED SUCTION LINE SIZES

R-22   25° F. Evaporating Temperature

Capacity BTU/hr.	Light Load Capacity Reduction	Equivalent Length, Ft.							
		50		100		150		200	
		H	V	H	V	H	V	H	V
6,000	0	½	½	⅝	⅝	⅝	⅝	⅝	⅝
12,000	0	⅝	⅝	⅞	⅝	⅞	⅝	⅞	⅞
18,000	0	⅞	⅞	⅞	⅞	⅞	⅞	1⅛	⅞
24,000	0	⅞	⅞	⅞	⅞	1⅛	⅞	1⅛	⅞
36,000	0	1⅛	1⅛	1⅛	1⅛	1⅛	1⅛	1⅜	1⅛
48,000	0	1⅛	1⅛	1⅛	1⅛	1⅜	1⅛	1⅜	1⅛
60,000	0 to 33%	1⅛	1⅛	1⅜	1⅜	1⅜	1⅜	1⅜	1⅜
75,000	0 to 33%	1⅜	1⅜	1⅜	1⅜	1⅝	1⅜	1⅝	1⅜
100,000	0 to 50%	1⅜	1⅜	1⅝	1⅜	1⅝	1⅜	1⅝	1⅜
150,000	0 to 50%	1⅝	1⅝	2⅛	1⅝	2⅛	1⅝	2⅛	1⅝
	66%	1⅝	1⅝	1⅝	1⅝	1⅝	1⅝	1⅝	1⅝
200,000	0 to 50%	2⅛	2⅛	2⅛	2⅛	2⅛	2⅛	2⅛	2⅛
	66%	2⅛	1⅜*1⅝	2⅛	1⅜*1⅝	2⅛	1⅜*1⅝	2⅛	1⅜*1⅝
300,000	0 to 50%	2⅛	2⅛	2⅝	2⅛	2⅝	2⅛	2⅝	2⅝
	66%	2⅛	2⅛	2⅝	2⅛	2⅝	2⅛	2⅝	2⅛
400,000	0 to 50%	2⅝	2⅛	2⅝	2⅛	3⅛	2⅛	3⅛	2⅛
	66%	2⅝	2⅛	2⅝	2⅛	2⅝	1⅝*2⅛	2⅝	1⅝*2⅛
500,000	0 to 66%	2⅝	2⅝	2⅝	2⅝	3⅛	2⅝	3⅛	2⅝
600,000	0 to 66%	2⅝	2⅝	3⅛	2⅝	3⅝	2⅝	3⅝	2⅝
750,000	0 to 66%	3⅛	3⅛	3⅛	3⅛	3⅝	3⅛	3⅝	3⅛

Recommended sizes are applicable with condensing temperatures from 80° F. to 130° F.   H - Horizontal
* Double Riser   V - Vertical

Figure 8-32 Recommended suction line sizes for R-22 at 25°F evaporating temperature. *(Courtesy Copeland Corporation)*

RECOMMENDED SUCTION LINE SIZES

R-22   15° F. Evaporating Temperature

Capacity BTU/hr.	Light Load Capacity Reduction	Equivalent Length, Ft.							
		50		100		150		200	
		H	V	H	V	H	V	H	V
6,000	0	⅝	⅝	⅝	⅝	⅝	⅝	⅝	⅝
12,000	0	⅝	⅝	⅞	⅝	⅞	⅝	⅞	⅞
18,000	0	⅞	⅞	⅞	⅞	1⅛	⅞	1⅛	⅞
24,000	0	⅞	⅞	1⅛	⅞	1⅛	⅞	1⅛	⅞
36,000	0	1⅛	1⅛	1⅛	1⅛	1⅛	1⅛	1⅜	1⅛
48,000	0	1⅛	1⅛	1⅜	1⅛	1⅜	1⅛	1⅜	1⅛
60,000	0 to 33%	1⅜	1⅜	1⅜	1⅜	1⅝	1⅜	1⅝	1⅜
75,000	0 to 33%	1⅜	1⅜	1⅝	1⅜	1⅝	1⅜	1⅝	1⅜
100,000	0 to 50%	1⅜	1⅜	1⅝	1⅝	1⅝	1⅝	1⅝	1⅝
150,000	0 to 50%	1⅝	1⅝	2⅛	1⅝	2⅛	1⅝	2⅛	1⅝
	66%	1⅝	1⅝	1⅝	1⅝	1⅝	1⅝	1⅝	1⅝
200,000	0 to 50%	2⅛	2⅛	2⅛	2⅛	2⅝	2⅛	2⅝	2⅛
	66%	2⅛	1⅜*1⅝	2⅛	1⅜*1⅝	2⅛	1⅜*1⅝	2⅛	1⅜*1⅝
300,000	0 to 50%	2⅛	2⅛	2⅝	2⅛	3⅛	2⅛	3⅛	2⅛
	66%	2⅛	2⅛	2⅝	2⅛	2⅝	2⅛	2⅝	1⅝*2⅛
400,000	0 to 50%	2⅝	2⅝	2⅝	2⅝	3⅛	2⅝	3⅛	2⅝
	66%	2⅝	2⅝	2⅝	2⅝	3⅛	1⅝*2⅛	3⅛	1⅝*2⅛
500,000	0 to 50%	2⅝	2⅝	3⅛	2⅝	3⅛	2⅝	3⅝	2⅝
	66%	2⅝	2⅝	3⅛	2⅝	3⅛	2⅝	3⅛	2⅝
600,000	0 to 66%	3⅛	2⅝	3⅛	3⅛	3⅝	3⅛	3⅝	3⅛
750,000	0 to 66%	3⅛	3⅛	3⅝	3⅛	3⅝	3⅛	3⅝	3⅛

Recommended sizes are applicable with condensing temperatures from 80° F. to 130° F.
* Double Riser

H - Horizontal
V - Vertical

Figure 8-33 Recommended suction line sizes for R-22 at 15°F evaporating temperature. *(Courtesy Copeland Corporation)*

RECOMMENDED SUCTION LINE SIZES
R-22   -20° F. Evaporating Temperature

Capacity BTU/hr.	50 H	50 V	100 H	100 V	150 H	150 V	200 H	200 V
	Equivalent Length, Ft.							
6,000	7/8	7/8	7/8	7/8	7/8	7/8	7/8	7/8
12,000	1 1/8	1 1/8	1 1/8	1 1/8	1 1/8	1 1/8	1 1/8	1 1/8
18,000	1 1/8	1 1/8	1 3/8	1 1/8	1 3/8	1 1/8	1 3/8	1 1/8
24,000	1 3/8	1 3/8	1 3/8	1 3/8	1 5/8	1 3/8	1 5/8	1 3/8
36,000	1 5/8	1 5/8	1 5/8	1 5/8	1 5/8	1 5/8	1 5/8	1 5/8
48,000	1 5/8	1 5/8	2 1/8	1 5/8	2 1/8	1 5/8	2 1/8	1 5/8

Recommended sizes are applicable with condensing temperatures from 80° F. to 130° F.

H - Horizontal
V - Vertical

Figure 8-34 Recommended suction line sizes for R-22 at −20°F evaporating temperature. *(Courtesy Copeland Corporation)*

RECOMMENDED SUCTION LINE SIZES
R-502   25° F. Evaporating Temperature

Capacity BTU/hr.	Light Load Capacity Reduction	50 H	50 V	100 H	100 V	150 H	150 V	200 H	200 V
		Equivalent Length, Ft.							
6,000	0	5/8	5/8	5/8	5/8	5/8	5/8	5/8	5/8
12,000	0	7/8	7/8	7/8	7/8	7/8	7/8	7/8	7/8
18,000	0	7/8	7/8	1 1/8	7/8	1 1/8	7/8	1 1/8	7/8
24,000	0	7/8	7/8	1 1/8	7/8	1 1/8	7/8	1 1/8	1 1/8
36,000	0	1 1/8	1 1/8	1 1/8	1 1/8	1 3/8	1 1/8	1 3/8	1 1/8
48,000	0	1 1/8	1 1/8	1 3/8	1 1/8	1 3/8	1 1/8	1 5/8	1 3/8
60,000	0 to 33%	1 3/8	1 1/8	1 3/8	1 1/8	1 3/8	1 3/8	1 5/8	1 3/8
75,000	0 to 33%	1 3/8	1 3/8	1 5/8	1 3/8	1 5/8	1 3/8	1 5/8	1 5/8
100,000	0 to 33%	1 3/8	1 3/8	1 5/8	1 5/8	1 5/8	1 5/8	2 1/8	1 5/8
	50%	1 3/8	1 3/8	1 5/8	1 5/8	1 5/8	1 5/8	1 5/8	1 5/8
150,000	0 to 50%	1 5/8	1 5/8	2 1/8	1 5/8	2 1/8	1 5/8	2 1/8	1 5/8
	66%	1 5/8	1 5/8	1 5/8	1 5/8	1 5/8	1 5/8	1 5/8	1 5/8
200,000	0 to 50%	2 1/8	2 1/8	2 1/8	2 1/8	2 5/8	2 1/8	2 5/8	2 1/8
	66%	2 1/8	1 3/8 *1 5/8	2 1/8	1 3/8 *1 5/8	2 1/8	1 3/8 *1 5/8	2 1/8	1 3/8 *1 5/8
300,000	0 to 50%	2 1/8	2 1/8	2 5/8	2 5/8	2 5/8	2 5/8	2 5/8	2 5/8
	66%	2 1/8	2 1/8	2 5/8	1 5/8 *2 1/8	2 5/8	1 5/8 *2 1/8	2 5/8	1 5/8 *2 1/8
400,000	0 to 50%	2 5/8	2 5/8	2 5/8	2 5/8	3 1/8	2 5/8	3 1/8	2 5/8
	66%	2 5/8	1 5/8 *2 1/8	2 5/8	1 5/8 *2 1/8	2 5/8	1 5/8 *2 1/8	2 5/8	1 5/8 *2 1/8
500,000	0 to 50%	2 5/8	2 5/8	3 1/8	2 5/8	3 1/8	2 5/8	3 1/8	3 1/8
	66%	2 5/8	2 5/8	3 1/8	1 5/8 *2 1/8	3 1/8	1 5/8 *2 1/8	3 1/8	1 5/8 *2 1/8
600,000	0 to 50%	2 5/8	2 5/8	3 1/8	2 5/8	3 5/8	2 5/8	3 5/8	3 1/8
	66%	2 5/8	2 5/8	3 1/8	2 5/8	3 5/8	2 5/8	3 5/8	1 5/8 *2 5/8
750,000	0 to 50%	3 1/8	3 1/8	3 5/8	3 1/8	3 5/8	3 1/8	4 1/8	3 1/8
	66%	3 1/8	3 1/8	3 5/8	3 1/8	3 5/8	3 1/8	3 5/8	3 1/8

Recommended sizes are applicable with condensing temperatures from 80° F. to 130° F.
* Double Riser

H - Horizontal
V - Vertical

Figure 8-35 Recommended suction line sizes for R-502 at 25°F evaporating temperature. *(Courtesy Copeland Corporation)*

RECOMMENDED SUCTION LINE SIZES

R-502   15° F. Evaporating Temperature

Capacity BTU/hr.	Light Load Capacity Reduction	Equivalent Length, Ft.							
		50		100		150		200	
		H	V	H	V	H	V	H	V
6,000	0	5/8	5/8	5/8	5/8	5/8	5/8	5/8	5/8
12,000	0	7/8	7/8	7/8	7/8	7/8	7/8	7/8	7/8
18,000	0	7/8	7/8	7/8	7/8	1 1/8	7/8	1 1/8	7/8
24,000	0	1 1/8	7/8	1 1/8	7/8	1 1/8	7/8	1 1/8	1 1/8
36,000	0	1 1/8	1 1/8	1 3/8	1 1/8	1 3/8	1 1/8	1 3/8	1 1/8
48,000	0	1 3/8	1 3/8	1 3/8	1 3/8	1 3/8	1 3/8	1 5/8	1 3/8
60,000	0 to 33%	1 3/8	1 3/8	1 5/8	1 3/8	1 5/8	1 3/8	1 5/8	1 3/8
75,000	0 to 33%	1 3/8	1 3/8	1 5/8	1 3/8	1 5/8	1 3/8	1 5/8	1 5/8
100,000	0 to 33%	1 5/8	1 5/8	1 5/8	1 5/8	2 1/8	1 5/8	2 1/8	1 5/8
	50%	1 5/8	1 5/8	1 5/8	1 5/8	1 5/8	1 5/8	1 5/8	1 5/8
150,000	0 to 50%	2 1/8	1 5/8	2 1/8	1 5/8	2 1/8	1 5/8	2 1/8	1 5/8
	66%	1 5/8	1 5/8	1 5/8	1 5/8	1 5/8	1 5/8	1 5/8	1 5/8
200,000	0 to 50%	2 1/8	2 1/8	2 5/8	2 1/8	2 5/8	2 1/8	2 5/8	2 1/8
	66%	2 1/8	1 3/8 *1 5/8	2 1/8	1 3/8 *1 5/8	2 1/8	1 3/8 *1 5/8	2 1/8	1 3/8 *1 5/8
300,000	0 to 50%	2 1/8	2 1/8	2 5/8	2 1/8	3 1/8	2 1/8	3 1/8	2 5/8
	66%	2 1/8	2 1/8	2 5/8	2 1/8	2 5/8	1 5/8 *2 1/8	2 5/8	1 5/8 *2 1/8
400,000	0 to 50%	2 5/8	2 5/8	3 1/8	2 5/8	3 1/8	2 5/8	3 1/8	2 5/8
	66%	2 5/8	2 5/8	3 1/8	2 5/8	3 1/8	2 5/8	2 5/8	1 5/8 *2 1/8
500,000	0 to 50%	2 5/8	2 5/8	3 1/8	2 5/8	3 5/8	2 5/8	3 5/8	3 1/8
	66%	2 5/8	2 5/8	3 1/8	2 5/8	3 1/8	1 5/8 *2 5/8	3 1/8	1 5/8 *2 5/8
600,000	0 to 50%	3 1/8	3 1/8	3 5/8	3 1/8	3 5/8	3 1/8	3 5/8	3 5/8
	66%	3 1/8	3 1/8	3 5/8	3 1/8	3 5/8	3 1/8	3 5/8	2 1/8 *3 1/8
750,000	0 to 50%	3 1/8	3 1/8	3 5/8	3 1/8	3 5/8	3 1/8	4 1/8	3 5/8
	66%	3 1/8	3 1/8	3 5/8	3 1/8	3 5/8	3 1/8	3 5/8	2 1/8 *3 1/8

Recommended sizes are applicable with condensing temperatures from 80° F. to 130° F.
* Double Riser

H - Horizontal
V - Vertical

Figure 8-36   Recommended suction line sizes for R-502 at 15°F evaporating temperature. *(Courtesy Copeland Corporation)*

CHAPTER 9

THERMODYNAMICS FOR REFRIGERATION SYSTEMS

This chapter reviews the operation of the refrigeration system using the Mollier diagram. It is intended for the student who wishes to gain a better understanding of the operation of the refrigeration system. This understanding will allow you to understand how subcooling, suction/liquid heat exchangers, compression ratio, and many other factors affect the capacity and operation of the refrigeration system.

Also included in this chapter are the matching of components. This will enable you to predict the operating changes that will occur when either a compressor, condenser, or evaporator is changed.

The p–h Diagram

In Chapter 1, we calculated how much heat was required in order to bring water to its saturation temperature. Then, by adding its latent heat of vaporization (970 Btu/lb), we were able to calculate the heat content when the sample was completely turned to a saturated vapor. This was all done using numbers that were correct for atmospheric pressure only.

If we recalculated the heat content for the water at a higher pressure, we would need to account for the following differences.

1. At a higher pressure, the boiling point (saturation temperature) is higher. Therefore, more heat is required in order to bring the water up to a saturated liquid condition.

2. At a higher pressure, the latent heat of vaporization is reduced.

If we were to recalculate the heat content for a **saturated liquid** condition at several different pressures and graph them, we would have a graph similar to line A in Figure 9-1. And if we were to add the latent heat of vaporization to the heat of the saturated liquid at each pressure, we would be able to graph a line similar to line B (**p–h diagram**). Line A represents all combinations of pressure and heat content at which a saturated liquid may exist and line B represents all combinations of pressure and heat content at which a **saturated vapor** may exist. Although we developed this graph for water, the same graph can be made for a refrigerant. This **pressure–enthalpy** (enthalpy is heat content) diagram forms the basic building blocks for the **Mollier diagram.**

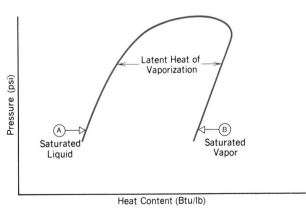

Figure 9-1 Skeleton of a pressure–enthalpy diagram.

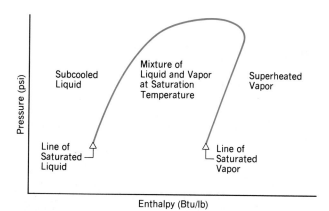

Figure 9-2 Physical condition of refrigerants on the pressure–enthalpy diagram.

Figure 9-2 labels all of the regions on the p–h diagram that do not fall on either the saturated liquid line or the saturated vapor line. All points to the left of the saturated liquid line represent refrigerant that does not contain enough heat to bring it up to the saturated liquid condition. It is subcooled liquid.

Refrigerant that is at a condition between the saturated liquid and saturated vapor lines has absorbed enough heat to vaporize some of the saturated liquid, but not all of it. This region is a mixture of saturated liquid refrigerant and saturated vapor. The percentage of liquid versus the percentage of vapor contained in the sample de-

pends upon whether the total heat content places the point closer to the liquid line or the vapor line. This percentage is called the **quality** of the refrigerant. Zero percent quality is a saturated liquid, 100 percent quality is a saturated vapor, and 50 percent quality is a mixture of half liquid and half vapor (by weight).

The region to the right of the saturated vapor line represents refrigerant that has absorbed more heat than was required to create a saturated vapor. Refrigerant in this area is **super-heated.**

In Figure 9-3 we have added lines of constant temperature. As each constant temperature line cuts through the **subcooled liquid,** saturated liquid, saturated vapor, and **superheated vapor** regions, it shows all the combinations of pressure and enthalpy that will be at the same temperature. From these lines, it is clear how, at 70°F, a low pressure refrigerant is superheated, while at a higher pressure the same refrigerant is at a subcooled condition when it is at 70°F. Some other interesting observations about the constant temperature lines follow.

1. As pressure increases, the temperature of the saturated refrigerant increases.
2. The temperature of the saturated liquid is the same as the temperature of the saturated vapor at any one pressure.

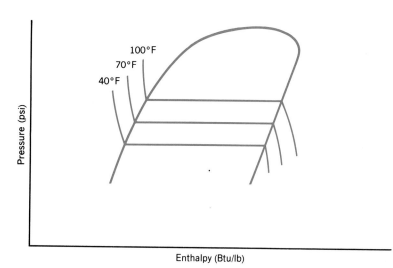

Figure 9-3 Constant temperature lines (isotherms) on a pressure–enthalpy diagram.

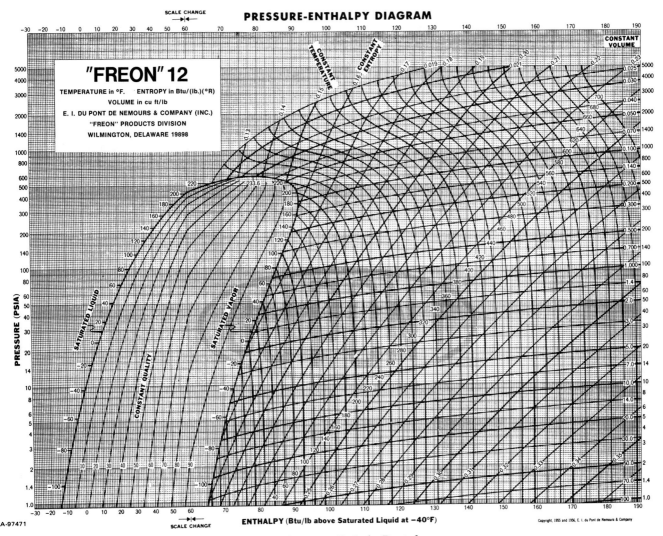

Figure 9-4 Completed Mollier diagram for R-12. (Courtesy E. I. du Pont de Nemours & Co.)

3. When heat is removed from a saturated liquid, its temperature decreases (it becomes sub-cooled). When heat is added, its temperature remains unchanged.

4. When heat is added to a saturated vapor, its temperature increases (it becomes super-heated). When heat is removed, its temperature remains the same.

Figure 9-4 shows a completed Mollier diagram for R-12 refrigerant. The following properties have been added.

1. Constant volume lines, in cubic feet per pound. As pressure is reduced, this specific volume is increased.

2. Constant entropy lines—these are used only by engineers who design systems and components. They are not discussed in this text.

The Refrigeration Cycle

Figure 9-5 shows how a basic refrigeration system would look when traced on the Mollier diagram. It assumes no superheat leaving the evaporator or subcooling leaving the condenser.

At point A, we have a low temperature, low pressure saturated vapor that is leaving the evaporator and entering the compressor. As the refrigerant is compressed, its pressure increases to point B. The heat content also increases because of the work that the compressor performs on the refrigerant. The quantity of heat added to the refrigerant due to the work being done by the compressor is called the **heat of compression.** The discharge temperature from the compressor at point B is quite high. From this point, the refrigerant enters the condenser where it will give up some of its heat to the condensing medium. As the superheated refrigerant enters the condenser, the first Btus of heat that are removed will cause the refrigerant to cool to its saturation temperature (point C1). Then, as the condenser continues to remove heat from the refrigerant, the condition moves horizontally to the left until all of the vapor has been condensed into a saturated liquid (point C2). From here, the refrigerant enters the metering device.

In the metering device, the pressure of the refrigerant is reduced to the low-side pressure. Its heat content remains constant through the metering device. When it leaves the metering device, its quality is between 80 and 90 percent, and its pressure and temperature are low (point D). This is the condition of the refrigerant that enters the evaporator.

In the evaporator, heat is added to the refrigerant. The pressure remains constant, but the remaining liquid is caused to vaporize until we are left with only a saturated vapor leaving the evaporator (point A). From here, the cycle is repeated. The remainder of this chapter deals with

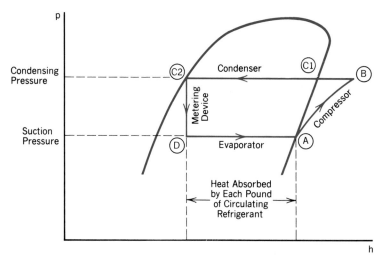

Figure 9-5 A simple refrigeration cycle traced on a pressure–enthalpy diagram.

how the operation and capacity of the system is changed on the Mollier chart as various components are changed.

Subcooling and Superheat

The capacity of the system shown in Figure 9-5 will be the amount of heat (Btu/lb) absorbed by the refrigerant in the evaporator, multiplied by the number of pounds of refrigerant that are being circulated. Figure 9-6 shows the cycle diagram for a system in which the condenser and evaporator sizes have been increased. The condenser is large enough to cool the refrigerant to a temperature below its saturation temperature (point C3). When the subcooled refrigerant passes through the metering device, the quality of the refrigerant leaving (percentage of vapor) is lower. There is more liquid available to absorb heat in the evaporator. The Btu/lb that can be absorbed by each pound of refrigerant is increased. The result increased system capacity.

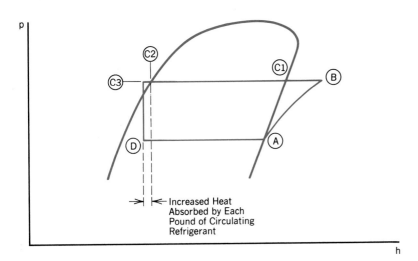

Figure 9-6 Increase in refrigeration capacity available when refrigerant is subcooled in the condenser.

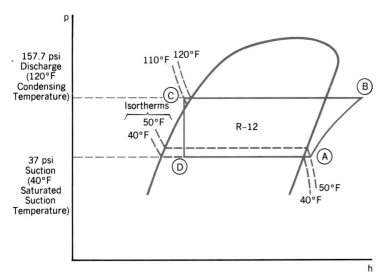

Figure 9-7 Refrigeration cycle with 10°F of subcooling and 10°F of superheat.

In Figure 9-7, we have increased the size of the evaporator once again. The evaporator is now sufficiently large so that the refrigerant entering the compressor is superheated (point A1). There is a slight increase in the Btu/lb that has been absorbed in the evaporator, but it does not require very much heat to superheat the refrigerant compared to the amount of heat it absorbed when it absorbed its latent heat of vaporization. The system capacity increase will therefore be small.

Suction/liquid line heat exchangers that were discussed in Chapter 7 have the same effect upon the refrigeration cycle as shown in Figure 9-7.

Liquid Line Pressure Drop

If there is an obstruction in the liquid line, such as a plugged filter or a restricted liquid solenoid valve, flashing will occur before the metering device as shown in Figure 9-8. If this flashing is allowed to occur, the refrigerant entering the metering device will contain a small percentage of vapor, by weight. This represents, however, a large percentage of the total refrigerant supplied to the metering device, by volume. The evaporator will be starved for refrigerant, the system capacity will be reduced, and the blockage will be

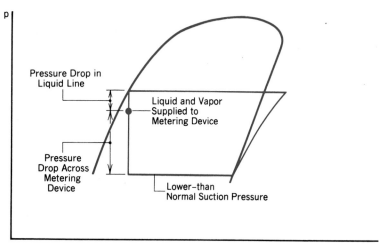

Figure 9-8 Pressure drop in the liquid line.

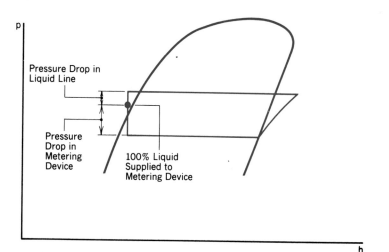

Figure 9-9 Subcooling in the condenser offsets the flashing caused by a liquid line with pressure drop.

easily located because of the cooling that occurs at that point. If the system is equipped with a liquid line sight glass, bubbles will be visible.

Sometimes, **liquid line pressure drop** is unavoidable. This is especially true when the evaporator is located at an elevation much higher than the condenser. Even without friction or a blockage, the pressure at the top of the liquid line will be lower than the pressure at the bottom. In order to prevent flashing in this case, we provide a large enough condenser to supply subcooled refrigerant. On the Mollier chart in Figure 9-9, even though the pressure drops in the liquid line, the refrigerant being supplied to the metering device still remains 100 percent liquid.

paired, the system capacity is reduced. The Btu/lb that can be absorbed in the evaporator is reduced. An impaired condenser, however, also reduces the number of pounds of refrigerant that are being circulated. More of the compressor piston stroke is devoted to creating pressure, with less remaining for pushing the refrigerant out into the system.

As a general guide for compressors and condensers, the head pressure should be kept as low as possible in order to maximize system capacity. The limitation to this guide is that the head pressure must be kept sufficiently high so that the condenser can liquify all of the hot gas from the compressor.

Effect of Head Pressure

Figure 9-10 shows the difference between two refrigeration systems, one with higher head pressure than the other. The system outlined with dashes is operating with a condensing temperature of 120°F, while the solid outlined system is operating at 100°F condensing temperature. When the high pressure saturated liquid in the 120°F system flashes to the low pressure side, there is a smaller percentage of liquid left than with the 100°F system. This partially explains why, when condenser capacity is somehow im-

Low Suction Pressures

We learned in Chapter 3 that low suction pressures mean lower compressor capacity and lower system capacity. Figure 9-11 compares two systems with the same condensing temperature, but one has a saturated suction temperature of 40°F, while the other has a saturated suction temperature of 0°F. Even though the 0°F system is capable of transferring heat at a faster rate, the refrigeration effect is reduced. It cannot absorb as much heat as the 40°F system before all the liquid is vaporized. A further reduction in capac-

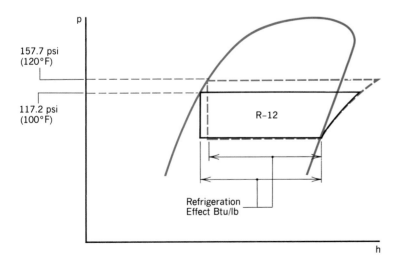

Figure 9-10 A high condensing temperature will cause a reduction in refrigerating effect for each pound of refrigerant being circulated.

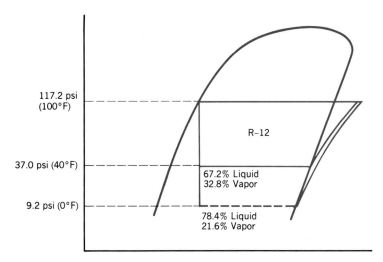

Figure 9-11 Operating a system at a lower suction pressure causes a higher percentage of flash gas entering the evaporator.

ity occurs because the specific volume of the 0°F saturated vapor is far greater than the specific volume of the 40°F saturated vapor. The compressor in the 0°F system will move fewer pounds of refrigerant even though it will pump virtually the same volume of refrigerant as the 40°F system. The Mollier diagram in Figure 9-11 illustrates that the lower the suction temperature becomes, the more exaggerated the loss of system capacity.

The loss in system capacity due to low suction pressures may be due to system design or due to a restricted metering device. If the system design requires such low saturated suction temperatures, there is nothing that can be done. The general guideline should be to try to keep the

saturated suction pressure as high as possible, while keeping it sufficiently low to create the temperature required.

Compound Systems

Figure 9-12 compares a single-stage system with a high compression ratio to a **compound system** with two stages of compression and **interstage cooling.** Note that the single-stage system causes the temperature of the hot gas to become much higher than the compound system. Oil breakdown and compressor damage will be the result unless the compound system is used.

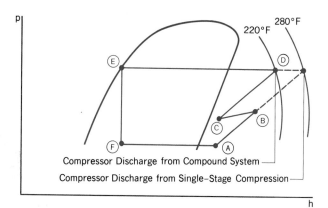

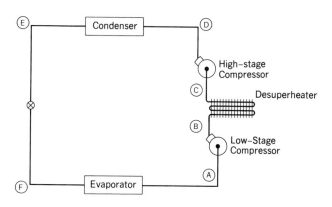

Figure 9-12 Using a compound system with interstage cooling produces lower discharge temperature than a single stage of compression.

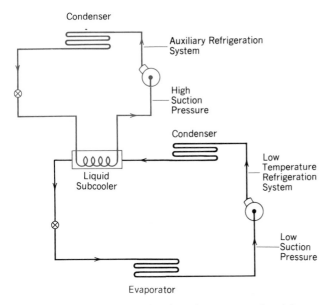

Figure 9-13 The liquid subcooler removes heat from the lower temperature system and shifts it to a more energy-efficient system.

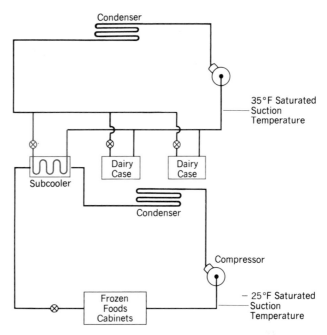

Figure 9-14 A two-temperature supermarket system.

Liquid Subcoolers

For refrigeration systems operating at low suction temperatures, it is necessary that the suction pressure be very low. The result is a system that requires large compressors and consumes a high horsepower per ton of refrigeration. **Liquid subcoolers** can be used to improve the system. Figure 9-13 shows a liquid subcooler installed with an auxiliary refrigeration system, which serves only to subcool the liquid refrig-

erant in the main system, thus increasing its capacity. The advantage to this arrangement is that the load from the low temperature, low pressure system is shifted to the auxiliary system, which operates at more efficient suction pressures. The savings in operating cost can be so dramatic that it will pay off the cost of the auxiliary system within a short period.

Figure 9-14 shows another arrangement of shifting low temperature loads to compressors, which operate at higher suction pressures. In this supermarket system, one compressor is

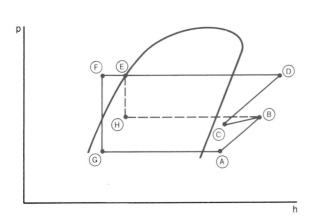

Figure 9-15 Compound system with liquid subcooler. Dotted path on the pressure–enthalpy diagram shows the flow of refrigerant that does not pass through the evaporator.

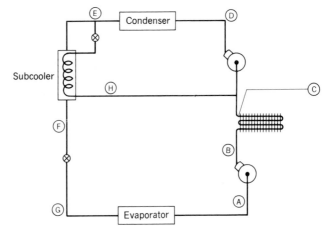

used for low temperature freezer boxes, while a separate system, operating at higher suction pressure, is used for the dairy cases. The subcooler becomes just another one of the loads on the high temperature system.

Figure 9-15 shows a compound system, using a subcooler. A portion of the liquid is passed through a metering device, and then moves to the subcooler where it subcools the remaining liquid. After absorbing heat in the subcooler, the vaporized refrigerant passes directly to the high-stage compressor. This reduces the amount of refrigerant that must be handled by the inefficient low-stage compressor.

Capacities of System Components

From the previous discussions, it is apparent that compressor capacity depends upon both saturated suction temperature and condensing temperature. The evaporator capacity also depends upon the saturated suction temperature. Figure 9-16 shows how the compressor capacity increases as the saturated suction temperature increases, and Figure 9-17 shows how the evaporator capacity decreases as the suction temperature increases. If we put both of these curves on the same graph (Figure 9-18), we see where these two components will match. This type of graph is sometimes called a **marriage curve.** It shows how two different components will work together. Figure 9-19 is another marriage curve, showing another, larger evaporator. Note that if the same compressor is matched with a larger

evaporator, the system saturated suction temperature will be higher, and the system capacity will also be higher. In Figure 9-20, a new, smaller compressor is shown. With two evaporators and two compressors, four different combinations are possible. It happens that for the components shown in Figure 9-20, the same capacities can be attained by a small compressor matched with a large evaporator (point C) or by a large compressor matched with a smaller evaporator, operating at a lower saturated suction temperature (point A).

Figure 9-21 shows the same type of marriage curves for compressors and condensers operat-

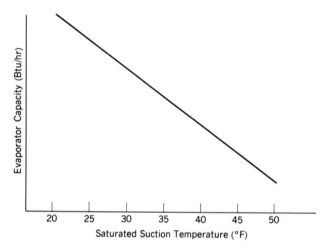

Figure 9-17   A filled evaporator operating at a lower saturated suction temperature will be capable of absorbing more heat.

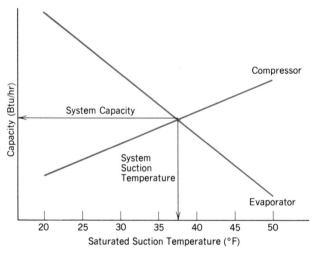

Figure 9-18   A marriage curve showing the balance operating point for a compressor and evaporator.

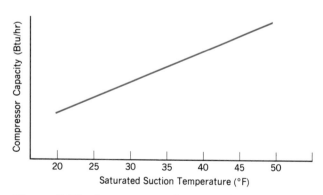

Figure 9-16   A compressor drawing a low saturated suction temperature will have low capacity due to the low suction pressure.

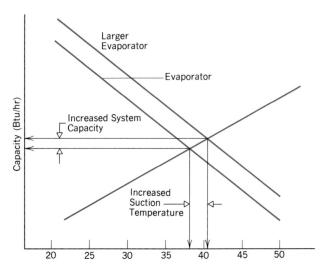

Figure 9-19 By using a larger evaporator, the system capacity will be increased, but the saturated suction temperature will also rise

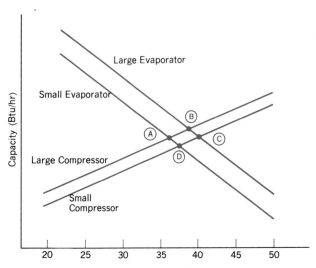

Figure 9-20 A marriage curve showing four different operating conditions.

ing at various condensing temperatures. A large condenser can be matched with a smaller compressor to produce the same effect as a small condenser and large compressor. Years ago, engineers found that additional compressor capacity could be built into the system more cheaply than could condenser capacity. Packaged units tended to operate at high condensing temperatures. In recent years, however, the operating costs have become increasingly important. The more energy efficient units will be physically larger as more condenser capacity is provided. The compressor will operate against a lower head pressure, and will consume less energy.

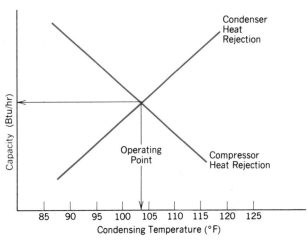

Figure 9-21 Marriage curve for a compressor and a condenser.

KEY TERMS AND CONCEPTS

p–h Diagram	Heat of Compression	Effect of Suction Pressure
Pressure-enthalpy	Refrigeration Cycle	
Mollier Diagram	Refrigeration Effect	Compound System
Quality	Effect of Subcooling and Superheat	Interstage Cooling
Saturated Liquid		Effect of Liquid Subcoolers
Saturated Vapor	Liquid Line Pressure Drop	
Subcooled Liquid		Marriage Curves
Superheated Vapor	Effect of Head Pressure	

QUESTIONS

1. On a p–h diagram, what property of the refrigerant determines the distance between the saturated liquid line and the saturated vapor line?

2. What is meant by the term refrigeration effect?

3. In addition to refrigeration effect, what other factor determines how much useful cooling will be done by a refrigeration system?

4. On the p–h diagram, where will you find a   (a) subcooled liquid? (b) superheated vapor?

5. How much heat will be absorbed by 1 lb of saturated liquid (R-12) at 40°F when it completely vaporizes?

6. How much heat will 1 lb of 40°F saturated R- 12 vapor absorb when it picks up 10°F of superheat?

7. What happens to the refrigeration effect when head pressure is lowered? What happens to system capacity?

8. What is the volume of 1 lb of R-12 that is at 37 psig and is superheated by 10°F?

9. What is the volume of 1 lb of R-12 that is at 0 psig, and is superheated by 10°F?

10. If identical compressors were operating in the systems described in Questions 9 and 10, which compressor would be moving more pounds of refrigerant through the system each minute?

11. What is meant by 10 percent quality on the Mollier chart?

12. What will happen to refrigeration capacity if subcooling leaving the condenser is increased?

13. What is the advantage of using a separate refrigeration system to sub-cool the refrigerant liquid? When is this practical?

14. If you replace an evaporator with a larger evaporator, what will happen to the saturated suction pressure at which the system operates?

15. What will happen to the system capacity in Question 14?

16. If you reduce the rate of airflow across an air-cooled condenser, what will happen to the condensing temperature? What will happen to the flow rate of refrigerant circulating through the system?

17. What will happen to the system capacity in Question 16?

18. What device can you use to cause the condition of refrigerant to move straight down on the Mollier chart?

19. What device can you use to cause the condition of refrigerant to move straight up on the Mollier chart?

20. What device can you use to cause the condition of refrigerant to move horizontally to the left on the Mollier chart?

CHAPTER 10

REFRIGERATION APPLICATIONS

With the basic study of the refrigeration cycle now completed, this chapter deals with its many different uses and applications.

Simple Household Refrigerator

The simplest arrangement for a **household refrigerator** is shown in Figure 10-1. The compressor is located under the food storage compartment, and the static condenser is mounted on the back of the box. The metering device is a capillary tube, carrying the refrigerant to the plate-type evaporator located at the top of the food compartment. A separate door is provided on the evaporator, inside the box, and the section inside the evaporator box is the freezer. Air that is cooled at the top of the box is allowed to fall to the rest of the box. The freezer section is main-

tained at 0°F to −10°F, while the refrigerator section is kept at 35°F to 45°F. A thermostat may be located to sense temperature either in the refrigerator or freezer section. The compressor is cycled on and off to control the temperature in that section. The temperature in the other section is controlled by adjustment of a damper, which adjusts the quantity of air that is being supplied to the refrigerator section from the freezer section. Refrigerators that have completely separate frozen food sections and refrigerated food sections sometimes use two evaporators. The refrigerant flows first through the freezer evaporator, but there is sufficient liquid refrigerant remaining to then provide cooling to the evaporator in the refrigerator section.

The most common operating pressures for the domestic refrigerator are 0–2 psi on the low side, and 100–125 psi on the high side, although some units operate outside these ranges. The refrigerant that returns to the compressor picks up heat from the capillary tube and returns to the compressor at a temperature sufficiently high to prevent the suction line from sweating on the outside surface. Some refrigerators use insulated suction lines to assure that they will not sweat.

Some compressors used in refrigerators are equipped with two additional connections. These are merely the opposite ends of a tube that runs through the oil reservoir at the bottom of the compressor. They are **oil cooler** connections, and are piped as shown in Figure 10-2. The hot gas from the compressor discharge flows through an auxiliary condenser, which is sometimes called an oil cooler. From the outlet of the auxiliary condenser, the refrigerant flows through the loop in the bottom of the compressor where it picks up heat from the oil. It then circulates to the main condenser and through the rest of the cycle in the conventional fashion.

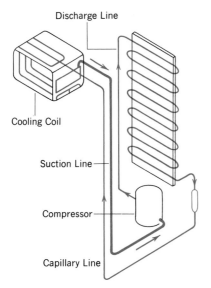

Discharge Line

Cooling Coil

Suction Line

Compressor

Capillary Line

Figure 10-1 Refrigeration system for a simple household refrigerator/freezer.

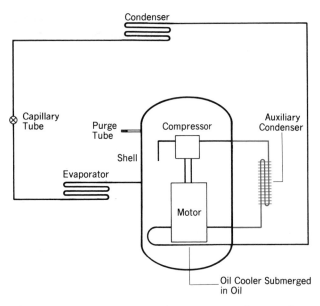

Figure 10-2 Installation of a compressor having five tubing connections.

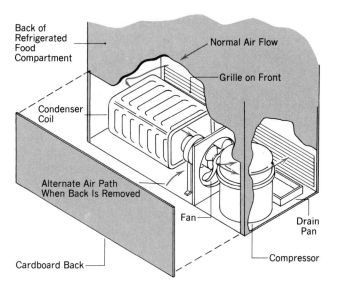

Figure 10-3 Condenser air (from the room) is drawn into the front grille of the refrigerator, passes over the condenser, through the condenser fan, and over the compressor and drain pan.

The door of the refrigerator has a soft rubberized seal that should be airtight when the door is closed. If it is not, the compressor will have to run longer than normal to remove the heat due to the air leakage. Frost will build up rapidly on the evaporator due to the humidity that leaks into the box. To test the door seal, see if it will hold a scrap of paper tightly between the door seal and the box. Try this test at several positions around the seal. Another door feature is a switch on the box frame that is wired into a circuit that turns on a light inside the box whenever the refrigerator door is opened.

Household Refrigerator Features

Several additional features have been incorporated into many refrigerator models. These features are usually designed either to create more food storage space within the same outside dimensions of the box or to provide more convenience to the user.

In order to save the space required for the condenser on the back of the box, the condenser is moved under the food compartment, next to the compressor. In order to make everything fit, the condenser is made smaller, but is provided with a condenser fan to increase the airflow and the rate of heat transfer. It is important that the cardboard back covering the compressor-condenser compartment remain in place. If it is removed, the air moved by the condenser fan will not be drawn over the condenser coil (Figure 10-3).

Evaporators may also be made smaller by using a finned tube element and blowing the refrigerator air across the coil through plastic or Styrofoam ducts inside the box. The space between the fins is small, so these units are provided with **automatic defrost.** Figure 10-4 shows a finned evaporator with electric heating elements mechanically attached. A timer activates the heaters and deactivates the compressor, condenser fan, and evaporator fan motors. The condensate melts off the evaporator, and flows through a tube to a pan under the box. The heat from the condenser and compressor reevaporate this water back into the kitchen when the normal cycle is resumed. When there appears to be excessive humidity on the inside of the refrigerator, one possible cause may be that the passage for removing the condensate has become plugged.

An alternate method used for defrosting is to energize a hot gas solenoid valve that supplies hot gas to the evaporator.

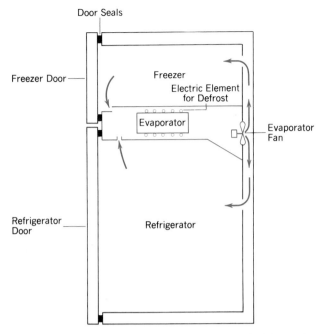

Figure 10-4 A household refrigerator with a finned tube evaporator and electric defrost. The airflow into the freezer and evaporator can be adjusted for different refrigerator and freezer temperatures.

Another feature incorporated into refrigerators to increase storage space is the use of thinner insulation in the walls. The insulating value is sufficient to keep the exterior surfaces from sweating, except near the door openings. Resistance heaters called **mullion heaters, stile heaters,** or **case heaters** are provided to heat these areas, which would otherwise sweat. Obviously, this increases the operating cost, both to operate the heaters and to operate the compressor to remove the additional heat that is added to the food compartment. Some models provide an **energy saver switch.** When the switch is On, it actually turns the heaters off. Many boxes operate satisfactorily in this mode until the room humidity rises. If the cabinet sweats, the energy saver must be turned Off (turning the heaters back on).

Drinking Fountain

The drinking fountain (Figure 10-5) is sometimes referred to as an **electric water cooler.** Its function is to cool the city water from an entering temperature of between 60°F and 80°F to a more desirable temperature of 50°F. The refrigeration system for the electric water cooler is quite simple, as shown in Figure 10-6. The sides of the unit are removable, making the hermetic R-12 compressor, condenser, and condenser fan visible. The metering device is a capillary tube. The evaporator consists of a water holding tank that has a bare tube evaporator coil in direct contact with the outside surface. This whole evaporator assembly is located just below the top of the unit, and is filled with insulation. It is not easily accessible. A thermostat senses the temperature of the water inside the tank, and starts the compressor and condenser fan whenever the water temperature is above the thermostat set point.

The storage of water in a tank makes it possible to satisfy the cooling demand with a smaller refrigeration system. When water is used, it is replenished with warm city water. The tank warms and turns on the compressor, usually before the person has finished getting a drink. When the dispensing is complete, the water tank may be several degrees higher than the set point. The compressor will continue to run, spreading out the cooling load over a longer period of time. If a drinking fountain were subjected to continuous use, it would not be able to maintain the 50°F water temperature.

One additional feature is included in the electric water cooler to reduce the size of the required refrigeration system. The entering city water line is wrapped around the outside of the drain line. The only time water enters the tank is when there is also cold water going down the drain. This crude heat exchanger recovers some of the cooling that has already been done and would otherwise, literally, go down the drain.

Ice Makers

Refrigeration units to make ice come in capacities ranging from a few pounds of ice per day to building-size systems used by the food industry for packing and shipping. These systems produce ice in either **flake** or **cube** form (Figure 10-7). The unique feature of ice makers is how

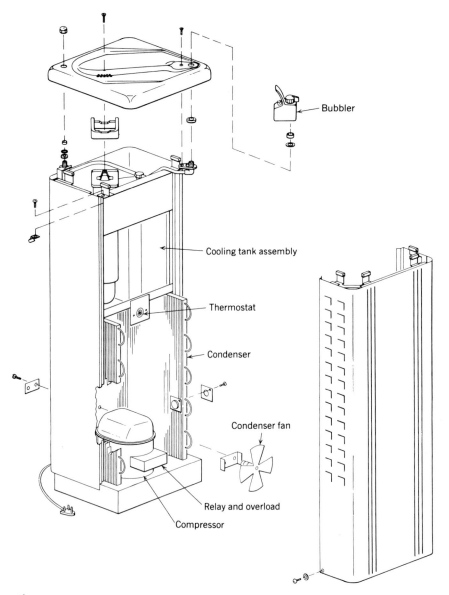

Bubbler

Cooling tank assembly

Thermostat

Condenser

Condenser fan

Relay and overload

Compressor

Figure 10-5 Drinking fountain. *(Courtesy Ebco Manufacturing Company)*

the ice is formed and handled. The refrigeration systems are quite similar to other common systems.

Figure 10-8 shows the operation of an ice flaker. The evaporator is shaped like a cylinder and operates at a saturated suction temperature of 0°F. The area around the cylinder is filled with water that enters through a float chamber. The water freezes at the evaporator surface. An auger, driven by an external motor, rotates around the evaporator cylinder. It scrapes the ice from the

surface of the evaporator, into a chute, where it is pushed into a storage bin. The ice flakes are made continuously until a level sensor in the storage bin shuts the system down.

There are two methods of making ice cubes in commercial ice makers. One method is to have an evaporator in the form of an ice cube tray, which forms the cubes individually. The second method is to make a slab of ice, and then cut it into cubes with a hot wire grid.

Figure 10-9 shows a system with an evap-

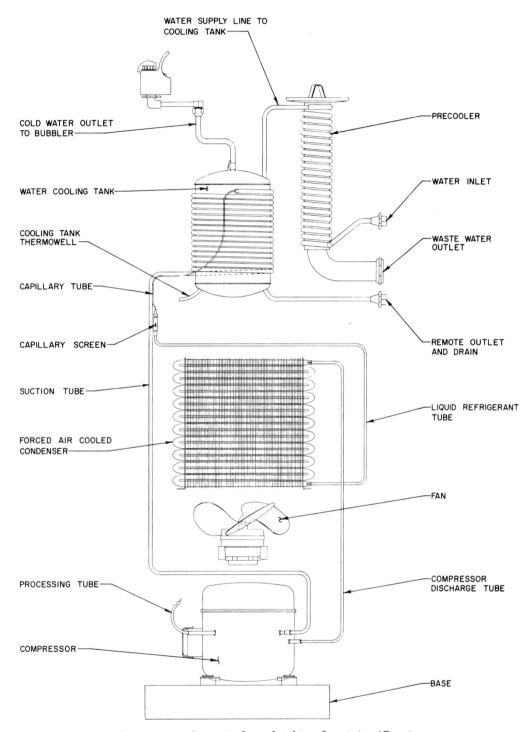

WATER SUPPLY LINE TO COOLING TANK

COLD WATER OUTLET TO BUBBLER

WATER COOLING TANK

COOLING TANK THERMOWELL

CAPILLARY TUBE

CAPILLARY SCREEN

SUCTION TUBE

FORCED AIR COOLED CONDENSER

PROCESSING TUBE

COMPRESSOR

PRECOOLER

WATER INLET

WASTE WATER OUTLET

REMOTE OUTLET AND DRAIN

LIQUID REFRIGERANT TUBE

FAN

COMPRESSOR DISCHARGE TUBE

BASE

Figure 10-6 Refrigeration schematic for a drinking fountain. *(Courtesy Ebco Manufacturing Company)*

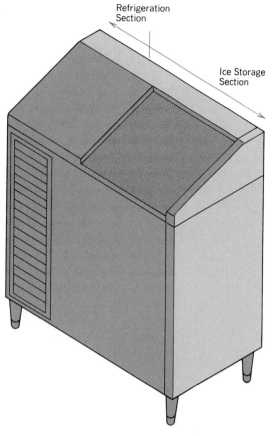

Figure 10-7 Ice maker for either cubed ice or flaked ice.

orator that is an inverted ice cube mold. The recirculation pump continuously sprays water up onto the evaporator. The ice builds up until cubes have been formed. The system then goes into a hot gas defrost. The hot gas from the compressor enters the evaporator, melting the cubes where they contact the evaporator. The cubes fall out into a storage bin. During the defrost cycle, the compressor is on and the condenser fan is off. Liquid that forms as the hot gas gives up heat in the evaporator is trapped in the suction line accumulator, and is reevaporated when the system returns to the normal cooling mode.

Figure 10-10 shows the second type of ice cuber. The recirculating pump causes water to flow over an inclined evaporator plate. With time, the thickness of the ice formed on the plate builds. A thickness switch senses when the ice has grown to three quarters of an inch, when it puts the system into the defrost cycle. The slab of ice slides off the evaporator and onto a wire grid that is electrically heated. The weight of the ice pushes it down onto the grid until the ice slab has been cut into cubes that fall into the storage bin.

One of the most troublesome parts about mak-

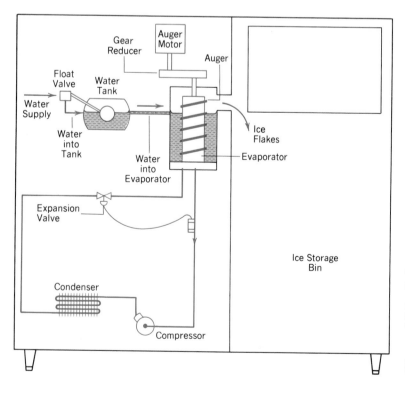

Figure 10-8 Refrigeration schematic for an ice flaker.

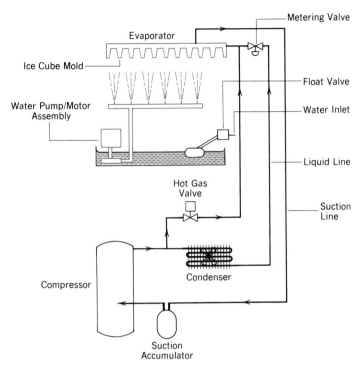

Figure 10-9 Ice cube maker. The evaporator is an inverted ice cube mold.

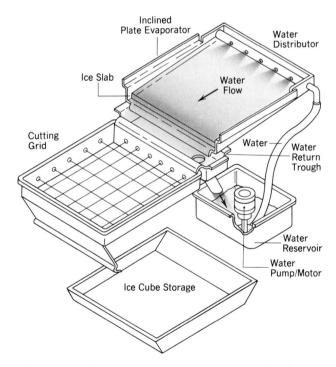

Figure 10-10 Ice cube maker using an inclined plate evaporator and an electric cutting grid.

ing ice cubes is that owners frequently complain about ice cubes that are not clear. Cloudy ice cubes can be attributed to impurities in the water. There are many filtering systems available for the water that enters the ice maker. These must be maintained and changed upon the recommended intervals. But even the finest filters cannot filter out 100 percent of the impurities in the water. Therefore, each time the ice maker goes through a harvest cycle, the water that has been used and recirculated during the freezing process, and which does not eventually become part of the ice cubes, is discarded. There are many systems in use to assure that all this unused water is drained after each cycle. If recirculated water is allowed to remain in the system from cycle to cycle, the concentration of impurities will continue to build, and cloudy ice cubes will result.

There is a belief by some that hot water will make clearer ice cubes than cold water. The origin of this myth is that when hot water is allowed to remain in a storage tank, the impurities tend to settle out. But without the settling process, hot and cold water will have the same impurities and will make comparable ice cubes.

Most ice making machines will use either R-12 or R-502 as the refrigerant. The material of construction for the parts in contact with the ice is usually stainless steel to avoid contamination.

Ice makers in household refrigerator/freezers are merely mechanical devices that handle the water and ice (Figure 10-11). Although there are many physical arrangements available, they all follow the same general sequence of operation:

1. A solenoid valve is energized, admitting a predetermined quantity of water into a mold.

2. Some time later, a thermostat set at 25°F senses that the ice is ready for harvest. It energizes a motor and a mold-release heater.

3. As the heater releases the ice from the mold, the motor pushes the ice out. The motor also operates a cam and an ice-sensing feeler arm.

4. If the feeler arm senses a full bin, the sequence stops. Otherwise, the cam operates electrical switches that energize the solenoid valve, starting the process over.

Note that the rate of ice production is dependent upon the freezer temperature. Low ice production can be a symptom of a too-warm freezer.

Dehumidifiers

Dehumidifiers (Figure 10-12) are a simple arrangement of a refrigeration system to condense water vapor out of the air. They are commonly used to dry out basements or other areas where high humidity encourages the growth of mold.

Figure 10-13 shows a typical arrangement of a compressor, condenser, and evaporator. The unique feature of the dehumidifier is that the air being cooled across the evaporator is the same air that is heated as it, in turn, cools the condenser. As the air is cooled across the evaporator, the moisture drops out and drains into a tank that must periodically be emptied. This cooled air is then passed over the condenser. The temperature of the air as it leaves the condenser is warmer than that of the air that entered the evaporator, but it contains less moisture.

Dehumidification requirements range from 10 pt (light duty) to 14 pt (severe duty) per 24 hr for a 500-sq-ft area. A 2000-sq-ft area would require 22 to 32 pt per 24 hr of moisture removal.

Figure 10-12 Packaged dehumidifier contains a complete refrigeration system. Room air is drawn over the evaporator coil and discharged over the condenser coil back to the room. (*Courtesy Dayton Electric Mfg. Co.*)

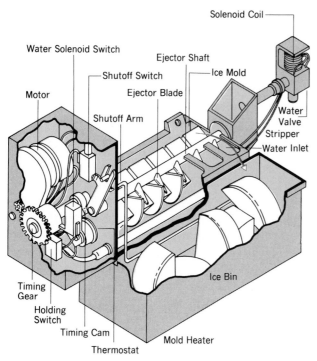

Figure 10-11 Whirlpool ice maker for installation in a domestic refrigerator/freezer.

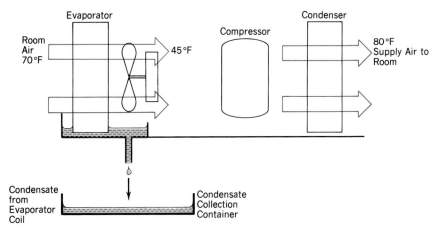

Figure 10-13 The dehumidifier cools the room air to dehumidify it, and then uses the cooled air to remove heat from the refrigerant in the condenser.

Compressed Air Drier

The system shown in Figure 10-14 is used to cool compressed air to lower its moisture content. Compressed air is used in control systems, paint spraying, and other processes. When the air is compressed its moisture-holding capability is reduced. The **compressed air drier** prevents water droplets from forming that could ruin air operated controls, paint finishes, pneumatic tools, and other water related problems.

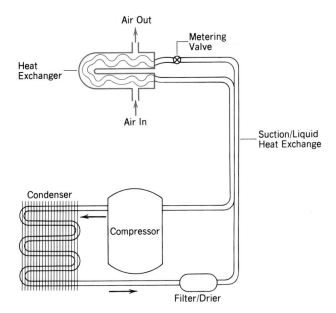

Figure 10-14 Compressed air drier.

Window Air Conditioner

The **window air conditioner** provides the lowest cost air conditioning for many applications. Figure 10-15 shows the basic arrangement of components. There are normally no refrigerant system accessories other than the filter dryer on the window air conditioner. These units use R-22 for the maximum cooling effect for the smallest size compressor. A single double-shafted motor is used to turn both the evaporator and condenser fans. A thermostat cycles the compressor to satisfy the room temperature demand. High, Medium, and Low Cool selections do not affect the operation of the compressor. It only sets the speed of the evaporator/condenser fan motor. Some units have a ventilation switch that oper-

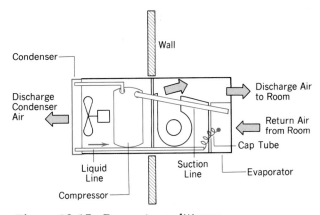

Figure 10-15 Room air conditioner.

ates a cable to open a small door between the outdoor and indoor section. A small quantity of outside air may then be bled into the room air for odor dilution.

As an aid to consumers purchasing air conditioners, the federal government has mandated that each manufacturer provide an energy-efficiency rating for each air conditioner, called the **Seasonal Energy Efficiency Ratio (SEER).** High SEER ratings indicate more energy-efficient units and lower operating costs. Figure 10-16 shows how to translate differences in SEER into differences in operating cost.

The major advantage of room air conditioners is low cost. But there are three major drawbacks compared to central air conditioning.

a. The noise of the compressor and fans can be objectionable within the conditioned space.

b. The cooled air is distributed from one discharge outlet. This may cause objectionable uneven room air temperature distribution.

c. Each room must be provided with sufficient capacity to meet the peak demand of that room. With a central unit, rooms with east

exposures will not require peak cooling at the same time as the rooms with west exposures, due to the position of the sun. Therefore, the total installed capacity for window units may be higher than for central units.

Heat Pumps

The **heat pump** is an air conditioning system that operates as a heating system. It is sometimes called a reverse cycle system. It is a very efficient system to use when electrical energy must be used for heating. It can make operating costs comparable to those obtained when gas or fuel oil are used.

In order to understand why a heat pump is an attractive system, we must compare it to the alternative—electric **resistance heat.** Figure 10-17 shows an electric resistance heater. Regardless of the type of heater, the voltage, the size, or any other factor, this heater will convert the energy in 1 kilowatt (kW) of electricity into 3414 Btu/hr. Those people who sell electric resistance heating equipment may be quick to point out that it is 100 percent efficient, which is true, in that 100

The chart below makes comparing air conditioners with different S.E.E.R. ratings very simple.
1. Assume your cooling bill is $100 per month.
2. Determine the S.E.E.R. ratings of both air conditioners.
3. Locate the higher of the S.E.E.R. ratings on the top of the chart, and the lower S.E.E.R. rating on the side of the chart.
4. Where the two columns intersect, you'll find how many dollars you'll save every month you use your air conditioner, if you buy the unit with the higher S.E.E.R. rating.

One other thing to remember! As electricity costs go up, the amount you save will be even larger.

		Higher S.E.E.R. Rating					
		6	7	8	9	10	11
Lower Rating SEER	6	0	$14.30	$25.00	$33.00	$40.00	$45.50
	7	NA	0	$12.50	$22.20	$30.00	$36.40
	8	NA	NA	0	$11.10	$20.00	$27.30
	9	NA	NA	NA	0	$10.00	$18.20
	10	NA	NA	NA	NA	0	$ 9.10
	11	NA	NA	NA	NA	NA	0

Figure 10-16 Chart for comparing air conditioner monthly operating costs. *(Courtesy Coleman Company).*

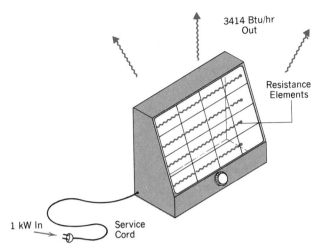

Figure 10-17 The electric resistance heater converts 1 kw of electricity into 3414 Btu/hr of heat.

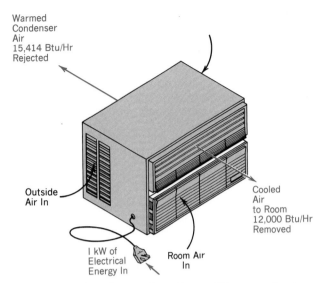

Figure 10-18 The window air conditioner or heat pump moves much more heat than the heat energy equivalent of the power consumed by the compressor.

percent of the electricity purchased is converted into heat. The rest of the story, however, is that heat energy in the form of electricity is two to four times as expensive per Btu as heat purchased in the form of natural gas or fuel oil.

Let's see how a heat pump can cut the cost of electric heating. Consider a 1-ton window air conditioner as shown in Figure 10-18. It moves 12,000 Btu/hr from the room into the refrigerant. The compressor input is 1 kW, so another 3414 Btu/hr is added to the refrigerant. The total of 15,414 Btu/hr is rejected in the condenser to the outside. If were were to turn the air conditioner around in the window, the outside air would be cooled, and 15,414 Btu/hr would be rejected to the room! This represents more than a

fivefold increase in the heating available from the same 1-kW input.

In practice, we do not rearrange the refrigeration components each time we want to operate the system as a heater. Instead, we use a **four-way reversing valve.** For normal air conditioning, the hot gas from the compressor discharge is routed to the outdoor coil, then to the metering device, indoor coil, and compressor suction via the four-way valve. On the heating cycle, the four-way valve changes its position (Figure 10-19). The hot gas from the compressor discharge is

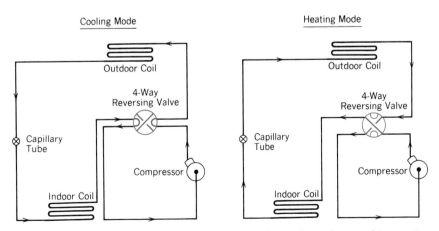

Figure 10-19 The four-way reversing valve allows the indoor coil to receive the hot gas instead of cold refrigerant during the heating mode.

routed backwards through the indoor coil, metering device, outdoor coil (where it picks up heat from the outdoor air), and back to the compressor suction via the four-way valve. A capillary tube is a handy metering device to use with this type of system, as it permits flow in either direction. Where thermostatic valves are used as the metering devices, one is provided for each coil, and a bypass with a check valve is provided around each TXV (see Figure 10-20).

The details of the four-way valve operation is shown in Figure 10-21. The solenoid valve operates to release pressure from one end of the barrel or the other. With the unbalanced pressures, the cylinder inside the valve body will move toward the end that has had the pressure bled off. The normal setup is for the thermostat to energize the solenoid valve to call for heating. In the deenergized position, the valve will operate in the cooling position. Figure 10-22 shows the outdoor portion of a split-system heat pump.

As outside temperatures get lower, the heating output of the heat pump also drops. Heat pump systems are provided with auxiliary electric heaters that turn on only when the heat pump cannot meet the demand for heat.

The system just described is called an air-to-air heat pump. The same principle may also be used to have heat pumps transfer heat from air to water, water to air, and water to water.

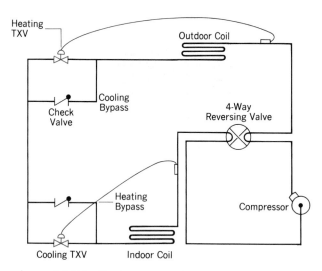

Figure 10-20   Heat pump system using two TXVs for metering devices.

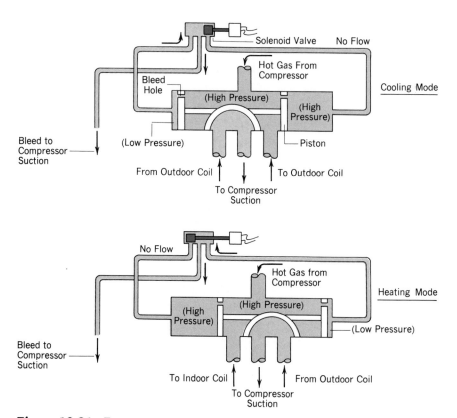

Figure 10-21   Four-way reversing valve.

Outdoor Coil
Discharge

Liquid

Service
Ports

Suction

4-Way
Reversing
Valve

Outdoor Coil

Figure 10-22 Outdoor heat pump, 1 to 5 tons. *(Courtesy Coleman Co., Inc.)*

Residential Central Air Conditioning

A popular way of providing **residential central air conditioning** is shown in Figure 10-23. The system shown utilizes a forced-air heating system. The compressor and condenser (condensing unit) are located outdoors. The condensing unit looks almost identical to the heat pump unit in Figure 10-22, except for the four-way reversing valve. The preferred location is on the north side of the house to shelter it from the midday sun, and not under a bedroom window where the noise will be objectionable.

The condensing unit provides a supply of saturated liquid refrigerant at temperatures approximately 20°F – 30°F warmer than the outside air. It is routed to the evaporator coil, which is mounted on top of the furnace. The metering device is mounted at the inlet to the evaporator coil. The furnace fan runs, blowing the room air through the evaporator coil. The suction line is insulated, and returns the superheated vapor to the outdoor condensing unit.

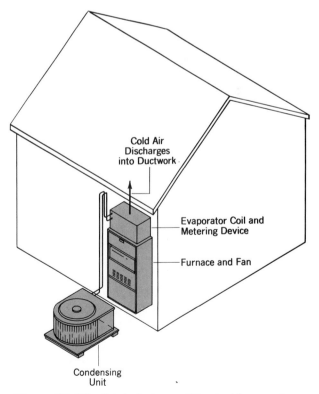

Cold Air
Discharges
into Ductwork

Evaporator Coil and
Metering Device

Furnace and Fan

Condensing
Unit

Figure 10-23 Central air conditioning for a residence.

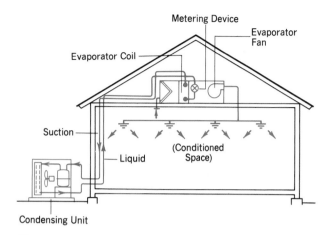

Figure 10-24 Split-system air conditioner.

Where the heating system is not of the forced-air type, a separate low side is provided. It may be located in the attic, as in Figure 10-24, and contain the evaporator coil, blower and motor, metering device, and air filters.

Other Split Systems

The term **split system** refers to a certain physical arrangement of components, such as those shown in Figure 10-25 and 10-26. A split system is one in which the compressor, condenser, receiver (if the system has one), and controls are located outdoors in a unit called a condensing unit (note that a condensing unit is different from just a condenser). The condensing unit is sometimes referred to as the high side of the system. The refrigerant coil, metering device, and blower for the conditioned air are located indoors (sometimes called the low side). This indoor unit may also contain heating coils and filters. This indoor unit may be called an air-handling unit (for air conditioning systems), a unit cooler (for refrigerated walk-in boxes), or various other terms for different applications. The high side and the low side are connected together by the liquid line and an insulated suction line. The condensing unit may be as small as one quarter of a ton, to supply evaporator coils in a small display case, or as large as several hundred tons, to supply a large, built-up air conditioning or process cooling system. The compressors may be reciprocating, rotary, centrifugal, or

Figure 10-25 Commercial outdoor condensing unit contains a condenser, condenser fan, semihermetic compressor, and electric controls. *(Courtesy Russell Coil Company)*

Figure 10-26 Indoor coil and fan assembly used in walk-in refrigerators or freezers. The evaporator coil is located behind the fans. *(Courtesy Russell Coil Company)*

screw type. The condensers may be water-cooled or air-cooled, with the air-cooled normally being of the forced-draft type.

Self-contained Air Conditioners

The term **self-contained** refers to any system in which all of the refrigeration components have been preassembled at the factory. Self-contained units are frequently installed on rooftops of

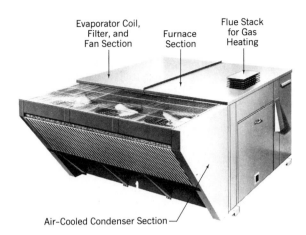

Figure 10-27 labels: Evaporator Coil, Filter, and Fan Section — Furnace Section — Flue Stack for Gas Heating — Air-Cooled Condenser Section

Figure 10-27 Packaged rooftop heating/cooling unit. *(Courtesy The Trane Company)*

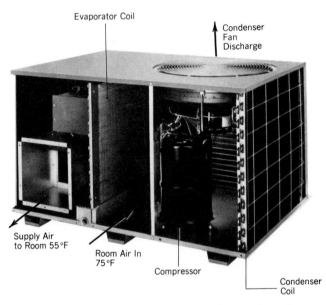

Figure 10-28 labels: Evaporator Coil — Condenser Fan Discharge — Supply Air to Room 55°F — Room Air In 75°F — Compressor — Condenser Coil

Figure 10-28 Packaged rooftop cooling-only unit. Filters for the evaporator coil would be located in the ductwork leading from the room to the air conditioner. *(Courtesy Coleman Co., Inc.)*

buildings, with cool air being supplied down into the conditioned area by a duct, and the room air returning to the rooftop unit through another duct (Figures 10-27 and 10-28). The advantage of this system is that it requires no floor space in the building. The disadvantage is that all service work must be done outdoors. When the self-contained unit also contains the heating system, winter service can be a real challenge to the service technician.

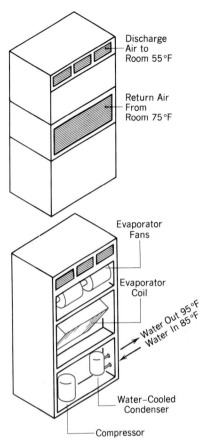

Figure 10-29 labels: Discharge Air to Room 55°F — Return Air From Room 75°F — Evaporator Fans — Evaporator Coil — Water Out 95°F — Water In 85°F — Water-Cooled Condenser — Compressor

Figure 10-29 Self-contained water-cooled air conditioner designed to be installed within the air conditioned space.

Figure 10-29 shows another type of self-contained air conditioner. This unit is located in the space that is to be cooled, so its cabinet and grilles must be of suitable appearance. You will find this type of unit in small commercial applications. An enhanced version of this unit is also used in computer rooms, and referred to as a computer air conditioning unit. There is usually no ductwork associated with this unit, with cool air being supplied to the room directly from the unit, and room air returning freely to the return-air side of the unit. This unit may be water-cooled or air-cooled. The water-cooled variety will usually use either a coil-within- a-coil condenser, or a shell-and-coil condenser. An external source of condenser water is required, either from city water or from a cooling tower. The air-cooled variety is supplied with a centrifugal condenser fan, so that the condenser air may be ducted to the air conditioner from outdoors.

Central Water Chiller

Figure 10-30 shows a large central **water chiller** that supplies a large quantity of chilled water between 40°F and 46°F to air conditioning units in a building. In concept, the water chiller is no more complex than the simple drinking fountain. In sizes over 100 tons, the compressors are centrifugal or screw-type instead of reciprocating (Figure 10-31). Many of the larger units use R-11, which is a low pressure, high volume refrigerant.

Figure 10-30   Large (50 tons) reciprocating water chiller supplies 44°F water to the building air-handler units. *(Courtesy The Trane Company)*

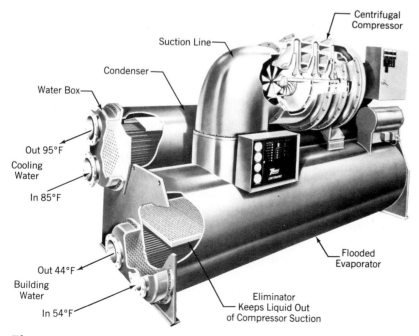

Figure 10-31   Large (300 tons) centrifugal water-cooled water chiller. *(Courtesy The Trane Company)*

On these systems, the low-side pressure will be between 15 and 20 in. Hg (for saturated suction temperatures between 24°F and 42°F). The discharge pressures will be between 3 psi and 21 psi (at condensing temperatures between 85°F and 125°F). A unique feature of systems that operate normally at a relatively deep vacuum is that any leakage on the low side of the system will result in air getting into the system rather than refrigerant leaking out. These systems are provided with purge systems that separate the air from the refrigerant, and operate periodically to expel this air.

Display Cabinets

Figure 10-32 shows a type of refrigerated cabinet that displays merchandise as well as cools it. **Display cabinets** are designed to be attractive and provide lighting for the merchandise inside. They are also called self-service cases and **reach-in boxes.** They are commonly used for beverages, frozen foods, ice cream, bakeries, delicatessens, and cheese storage. Each cabinet contains a complete refrigeration system.

Automobile Air Conditioning

Figure 10-33 shows the conventional arrangement of components for an **automobile air conditioner.** There are a number of unique features to the automobile system.

The R-12, open-drive compressor that is used is driven by the rotation of the engine, which means the speed of the compressor (and its capacity) will vary widely. At idling speed, the compressor may rotate at only 600 rpm, while at maximum speeds it may reach 5000 rpm. Also, the condenser, which is placed in front of the car radiator, will receive significantly more cooling effect when the car is traveling at 50 mph than when it is stopped and idling in traffic. This adds to the wide spread in capacity of the automobile system.

The operation of the air conditioner is not always desired when the engine is running, so a means must be provided to stop the rotation of the compressor, even when the car engine is running. This is done by the use of a **magnetic clutch** on the compressor shaft (Figure 10-34). When 12 V are applied to the magnetic coil in the clutch assembly, the rotating pulley engages the compressor shaft, and the compressor then rotates according to the speed of the engine. When the clutch is not energized, the pulley simply rotates as an idler pulley. In this mode of operation, the compressor is not running, and there is virtually no additional load on the car engine. During the winter months, some people disconnect

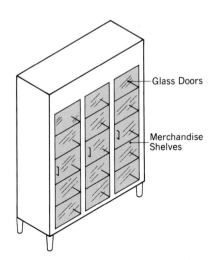

Figure 10-32 Food and beverage refrigerated cabinet.

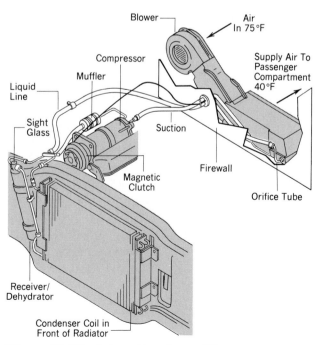

Figure 10-33 Automotive air conditioner.

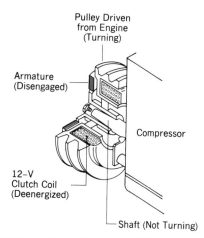

Figure 10-34 Magnetic clutch. When 12 V is applied to the clutch coil, the armature is pulled in to engage the compressor shaft.

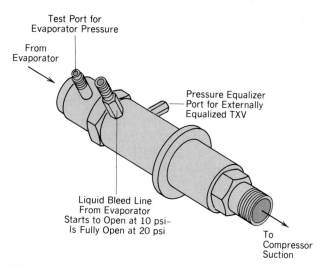

Figure 10-35 The POA valve maintains the evaporator pressure at higher than 30 psi.

the wire that energizes the compressor, mistakenly believing that this will save money.

The metering device used in an automobile system is either an expansion valve (most common), or an orifice tube. The expansion valves are usually nonadjustable, and are provided with a filter screen in the inlet. If this screen becomes plugged, frost forms at the inlet to the TXV, cooling is reduced, and the low-side pressure drops too low. Plugging of the inlet screen often indicates that the flexible, rubberized hoses that are used to connect the components are deteriorating. In this case, simply cleaning or replacing the TXV will only solve the problem for a short while.

Control of the suction pressure in automotive systems is a unique problem due to the wildly varying system capacity and cooling load. The normal suction operating pressure is between 28 and 32 psi. At these temperatures (30°F to 34°F), ice will not form on the evaporator surface, and the air to the passenger compartment will be cooled to 40°F–44°F. When the compressor is operating at high speed, however, it is easily capable of pulling lower suction pressures. This cannot be allowed to happen, as the automobile system has no means of defrosting the evaporator. Three methods used to control the evaporator pressure are (1) a **CCOT system,** (2) an **EPR or STV valve,** and (3) a **POA valve.**

CCOT stands for cycling clutch orifice tube system. In the CCOT system either a pressure switch senses low-side pressure, or a thermostat attached to the evaporator surface senses when the evaporator pressure is low enough to begin

forming ice. The contacts from this switch are wired in the clutch circuit to stop the compressor whenever the evaporator is too cold, and restart it when the danger of frosting has passed.

The EPR (evaporator pressure regulator), STV (suction throttling valve), and POA (pressure-operated altitude) valves are all valves that are installed between the evaporator outlet and the compressor suction (Figure 10-35). When the evaporator pressure begins to drop below 30 psi, the valve begins to throttle closed. The evaporator pressure is thus maintained above 28 psi, regardless of how low the compressor suction pressure may drop.

A continuing source of service work on the automobile system is refrigerant leakage. As there is a compressor shaft seal, seal leakage is always a potential problem. When a seal is leaking at a rate that requires system recharging once each year or two, many owners prefer this option to the expense of changing the seal. There is a popular myth that all air conditioning systems should be run once a month (although the reason usually does not get carried along with the myth). The source of this myth is that it is true on automotive systems, since periodically running it will keep the shaft seal lubricated, and prevent it from drying and leaking. Obviously, this reason does not hold for the majority of non-automotive systems that have hermetic compressors.

There are two major designs of compressors used on automobile systems. One, a reciprocating compressor, is similar to most other air conditioning compressors. The other, a swash plate compressor (Chapter 3), is a unique design commonly used on General Motors systems. In this compressor rotation of the shaft causes the swash plate to move back and forth in the same direction as the shaft. The cylinders, located on both sides of the swash plate, are oriented in the same direction as the shaft. This swash plate produces a double action, as it is always compressing the gas in a cylinder on one side or the other, regardless of its instantaneous direction of travel.

The high-side operating pressure of the automotive system may be as low as 150 psi when operating at low ambient temperatures, or upwards of 300 psi when ambient temperatures exceed the 100°F mark. Charging to the sight glass or using a charging cylinder are the best methods to use when charging these systems. Not all systems will have sight glasses, however. Discharge air temperatures will range from 33°F to 48°F at 70°F outdoor air temperature, depending upon the automobile make and model. When outdoor air temperature is 100°F, the discharge air temperature will range from 38°F to 56°F.

Another unique feature of automotive systems is that they are the only systems in which untrained people recharge refrigerant. Refrigerant charging kits may be purchased at any automobile parts store. The 1 lb cans of refrigerant have a maximum allowable pressure of 170 psi, and many cans have been exploded by untrained people who mistakenly expose the can to the high-side pressure in the system. In order to protect untrained consumers, some manufacturers are putting a nonstandard size fitting on the high-side access port. The service technician can obtain an adaptor from this nonstandard size to the normal one-quarter-in. flare connection.

Compound Systems

Systems designed to produce very low temperatures create some unique problems. Consider, for example, an R-12 system that is to keep a box at −40°F. Air for condensing is available at 95°F.

In order to cool the box down to −40°F, the saturated suction temperature would have to be no higher than −50°F. The suction pressure

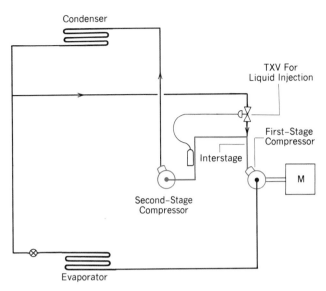

Figure 10-36 Compound system with interstage cooling.

would be 15.4 in. Hg, or 8.3 psia, the condensing temperature no lower than 115°F, and the condensing pressure 147 psig, or 162 psia. The compression ratio required of a compressor to do this duty would be

$$\text{compression ratio} = 162/8.3$$
$$= 19.5$$

The chapter on compressors (Chapter 3) explained that this is too high for the normal materials and oils commonly available. Therefore, this job would need to be done in two steps.

Either two separate compressors or a single **compound compressor** may be used. The compound compressor shown in Figure 10-36 is actually two compressors mounted on a common shaft and driven by a single motor. The low-stage compressor discharges into the high-stage compressor suction. In order to avoid overheating of the high stage, however, the interstage refrigerant must be cooled. This is accomplished by running a small line from the liquid line, through an expansion valve, and into the interstage. The liquid refrigerant flashes, thus providing sufficient cooling at the inlet to the high-stage compressor.

Cascade Systems

While compound systems are suitable for producing temperatures in the −40°F to −50°F range, lower temperature applications make two- and even three-stage compound systems unsuitable. The extremely low pressures required at the suction to the low-stage compressor reduce the refrigerant to an extremely low density. This means that the compressor would have to pump an extremely large volume of refrigerant in order to move the required weight flow through the system. **Cascade systems,** however, can easily produce box temperatures of −100°F. These boxes are used primarily for testing of electronic components. This thermal shock testing involves subjecting electronic circuits to cycles between very cold temperatures and very hot temperatures. The largest markets for such testing are the military and aerospace industries.

The key to the operation of a cascade system is the use of a very low temperature refrigerant called R-13. R-13 has a pressure temperature relationship as shown in the following list.

Temperature (°F)	Pressure (psi)
− 100	7.5
− 60	43.5
− 20	111.7
+ 20	225.7
+ 60	402.3
+ 80	520.8

Several observations can be made from these data. R-13 is a wonderful refrigerant to use at saturated suction temperatures of down to −100°F, where it can maintain a reasonable pressure. The drawback to using R-13 is that if you want to condense it, even at a relatively low temperature of 80°F, your system would need to be capable of producing a high-side pressure of 521 psig. This would be quite impractical.

Instead, we operate the R-13 system at a condensing temperature of around 20°F, which gives

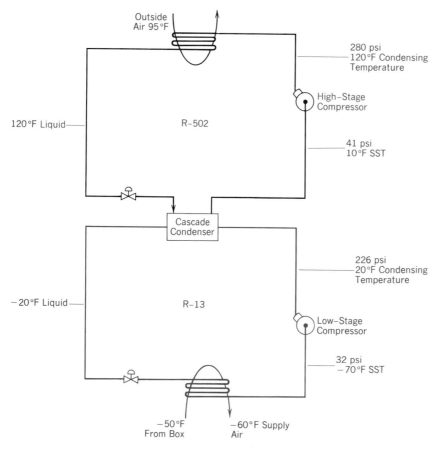

Figure 10-37 A cascade system using R-502 in the high stage and R-13 in the low stage.

us a reasonable high-side pressure of 226 psi. So far, so good—except that in order to maintain a condensing temperature of 20°F, we need to have a condensing medium available at 10°F! The answer to this problem lies in the introduction of a second refrigeration system that uses one of the more conventional refrigerants. Figure 10-37 shows how these two refrigerant systems work together. R-502 is used in the high stage, and R-13 is used in the low stage. The two stages come together at a heat exchanger called the cascade condenser. This heat exchanger actually serves as the condenser for the R-13 system, while the other side of the condenser serves as the evaporator for the R-502 system. In this configuration, extremely low temperatures can be produced by using commonly available components and materials.

Figure 10-38 shows a cascade system with the major operating controls. The function of each of these is explained in the following.

1. The liquid solenoid valves (LSV) and hot gas bypass valves (HGBP) in the low stage operate together to control the system capacity. When very close temperature control is required, these valves can cycle many times each minute in order to precisely match the required load.

2. The expansion tank on the suction side of the R-13 system serves two functions. First, it provides additional volume in the R-13 system, so that when the system is shut down and the temperature in the R-13 system rises, the pressure is maintained at reasonable levels of approximately 150 psi. Second, when the system pulls down, the 150 psi R-13 in the expansion tank bleeds back into the suction side. This maintains the suction pressure higher

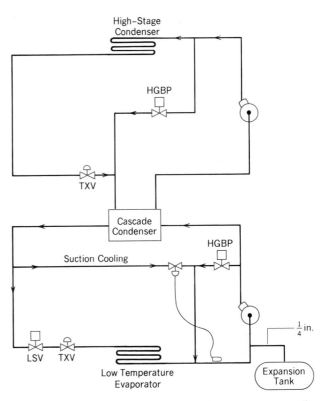

Figure 10-38 Cascade system with capacity control and suction cooling.

than it would be without the tank, and it makes the capacity of the low-stage compressor higher during the pull-down period.

3. The suction cooling line provides cooling for the compressor suction gas during periods when the temperature controller has the system in bypass. If the suction gas gets too warm to provide adequate cooling to the compressor motor, the suction cooling line provides liquid injection similar to the interstage cooling in the compound system.

KEY TERMS AND CONCEPTS

Household Refrigerator

Oil Cooler

Door Seal

Automatic Defrost

Stile Heaters

Energy Saver Switch

Electric Water Cooler

Flaked Ice

Cubed Ice

Dehumidifier

Compressed Air Dryer

Window Air Conditioner

SEER

Heat Pump

Resistance Heat

Four-Way Reversing
Valve

Residential Central Air
Conditioning

Split Systems

Self-Contained Systems

Water Chiller

Display Cabinets

Reach-In Box

Automobile Air
Conditioning

Magnetic Clutch

CCOT System

EPR Valve

POA Valve

STU Valve

Compound
Compressor

Cascade Systems

QUESTIONS

1. Name three types of units that use R-12 as the refrigerant.

2. Name three types of units that use R-22 as the refrigerant.

3. What types of system commonly uses R-502?

4. What are the normal operating pressures for the domestic refrigerator?

5. What are two ways in which the domestic refrigerator may be automatically defrosted?

6. How can a door seal be tested?

7. When a refrigerator automatically defrosts, where does the water go?

8. A dehumidifier causes room temperature to (increase, decrease, neither). Why?

9. How are the compressors arranged in a compound system?

10. Name three differences between a compound system and a cascade system.

11. What type of system would use R-13?

12. What is the major cause of cloudy ice cubes?

13. What types of applications are ideal candidates for heat pumps?

14. A split system is connected by two refrigerant lines. What is each line called? Which line is smaller? Which line is insulated? Which line is cold? Which line carries refrigerant from the indoor coil to the outdoor condensing unit?

15. What type of ice maker uses a thickness switch and a wire grid cutter?

16. Which of the major components (compressor, outdoor coil cap tube, indoor coil) in a heat pump has refrigerant flow in the same direction during both the heating and the cooling modes of operation?

17. Why do some large water chillers require air purging systems?

18. What device is used to allow the compressor on an automobile system to stop running?

19. What types of metering devices are found on automotive systems?

20. Why do automotive systems use an EPR valve?

CHAPTER 11

REFRIGERATION TOOLS

The purpose of this chapter is to describe some of the tools and meters that are unique to the refrigeration and air conditioning trades. It is presumed that you have some experience with the use of basic hand tools.

Manifold Gauges

The **manifold gauges** are to the service technician what the stethoscope is to a physician. It is undoubtedly the most universally used refrigeration tool. A common set of manifold gauges is shown in Figure 11-1. The service technician uses the manifold gauges to read system operating pressures, to evacuate a system, to add refrigerant to a system, and a multitude of other service operations.

A cutaway of the manifold gauge is shown in Figures 11-2 and 11-3. The pressure gauge on the left always reads the pressure in the left chamber, while the one on the right always reads the pressure in the right chamber. Depending on the position of the valves on the ends of the manifold, the center chamber may be connected to either or both of the end chambers.

The gauge on the left is called a **compound gauge.** That is, it can read either pressure or vacuum. Its range is 0 to 30 in. Hg vacuum, and 0 to 120 psig pressure. This gauge is usually blue in color. The hose on the left-hand chamber is normally connected to the low pressure side of an operating system.

The gauge on the right is used to read high-side pressure. Its range is 0–500 psig, and it is most commonly red in color.

There are a number of other scales on both the high- and low-side gauges. They are temperature scales for several different refrigerants. It is im-portant to remember, however, that these are pressure gauges and not temperature gauges. The temperature scales are only handy reproductions of the pressure–temperature charts. They tell the service technician, for any pressure, what the saturation temperature happens to be for several refrigerants. If the refrigerant happens to be either subcooled or superheated, however, the gauges will tell nothing about the actual refrigerant temperature.

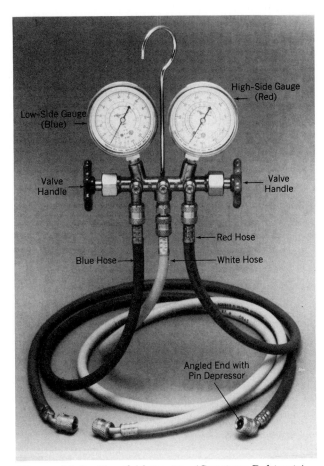

Figure 11-1 Manifold gauges. *(Courtesy Robinair)*

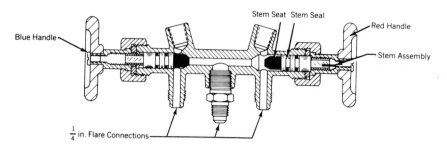

Figure 11-2 Cutaway of manifold set. (*Courtesy Robinair*)

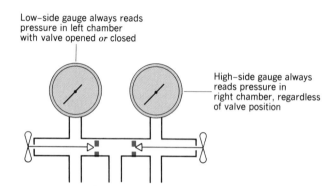

Figure 11-3 Schematic of manifold gauge set.

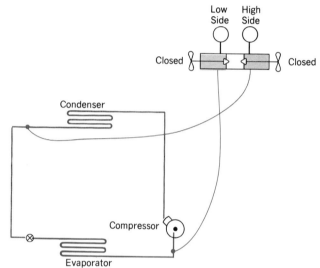

Figure 11-4 Manifold gauge attached to an operating system to read high- and low-side operating pressures.

The gauges are initially used by the service technician to read system operating pressures (Figure 11-4). The low-side hose is connected to any fitting between the metering device and the compressor suction, and the high-side hose is connected to any point between the compressor discharge and the metering device (connection downstream of the condenser is preferable to sensing pressure in the hot gas line, as pressure pulsations from the compressor will be significantly dampened). Both valves will be in the closed position, screwed all the way clockwise. The center hose is not used, so it is screwed onto a keeper that keeps the hose out of the way and also prevents dirt or other contaminants from entering the open end of the hose.

If the technician interprets the pressure readings as indicating that refrigerant must be added, the gauges are set up as shown in Figure 11-5. The refrigerant bottle is connected to the center hose, the low-side valve is open, and the high-side valve is closed. The refrigerant flows from the can, through the manifold, and into the low side of the refrigeration system.

The manifold gauges may also be used for evacuating a system, as shown in Figure 11-6. Here, the center hose is connected to a vacuum pump, and both valves are open to allow vacuum to be drawn from both the high and low sides of the system.

The hoses used on the manifold gauges are available in 36- to 72 in.-lengths, and are color coded to match the red and blue gauges, with the center hose being either white, yellow, or black. You will notice in Figure 11-1 that the two ends of each hose are different. One end is straight, and is designed to fit onto the manifold gauge. The other end is offset, and has a pin in the center. This is the end that will be connected to the system. The pin is required for use with a Shrader-type access fitting. With some hose sets,

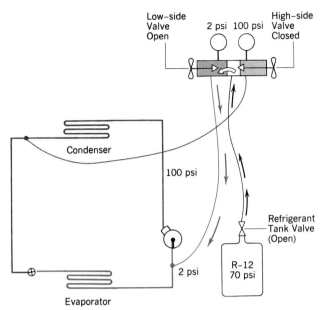

Figure 11-5 Manifold gauges used to add refrigerant to an operating system.

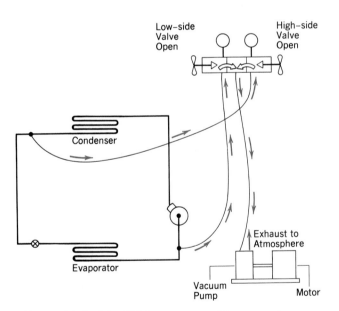

Figure 11-6 Manifold gauges used to evacuate a system that is not running.

Figure 11-7 Heat pump manifold gauges. Two high pressure gauges plus one compound gauge. *(Courtesy Robinair)*

the length of the pin depressor may be adjusted to match the length of the pin in the Shrader fitting. Each of the hose ends contains a tubular gasket. This allows a leaktight connection to be made with only moderate finger tightening. Never overtighten these connections, as the gasket is easily ruined. The gaskets may be changed,

but it is a tedious, time-consuming process that, with care, can usually be avoided.

There are many variations on the basic manifold gauges (Figure 11-7) that provide added conveniences. With proper use, however, the basic manifold set may be used for all common service operations.

Thermometers

In the field of heating, air conditioning, and refrigeration, the measurement of temperature is obviously an important task. A common type of thermometer (illustrated in Figure 11-8) is frequently found in a service technician's shirt pocket. This type of thermometer is referred to as a **dial-stem thermometer.** The thermometer dial is quite easy to read compared to a mercury-filled glass thermometer. It is also quite rugged. The thin ($\frac{1}{8}$-in. diameter) stem makes for easy insertion into ducts or suppy diffusers. Figure 11-9 shows a number of ways the dial-stem thermometer may be used, including measurement of room temperatures (Figure 11-9a), pipe temperatures (Figure 11-9b), and duct temperatures (Figure 11-9c). The dial-stem thermometer is probably not quite as accurate as some others, but its versatility, compactness, and rugged con-

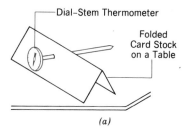

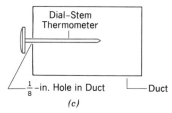

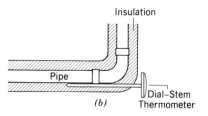

Figure 11-9 Common uses for a dial-stem thermometer. (a) Measuring room temperature. (b) Measuring fluid temperature in a pipe. (c) Measuring air temperature in a duct.

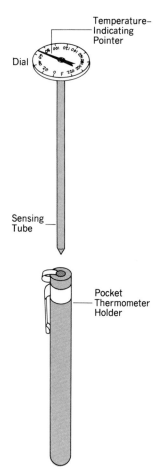

Figure 11-8 Dial-stem thermometer.

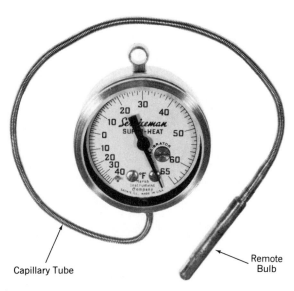

Figure 11-10 Remote bulb thermometer used for measuring superheat. Bulb is placed under the insulation on the suction line. (*Courtesy Robinair*)

struction make it the favorite for field technicians.

Sometimes it is necessary to read the temperature in a location in which it is not practical to place the thermometer. For example, it may be required to have a temperature readout indicating the temperature in an overhead duct, without using a ladder each time the reading is

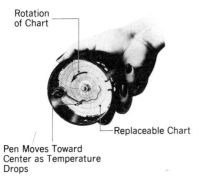

Rotation
of Chart

Pen Moves Toward
Center as Temperature
Drops

Replaceable Chart

Figure 11-11 Windup recording thermometer.
(Courtesy Airserco Manufacturing Company)

required. Or, it may be desirable to be able to read the temperature inside a refrigerated box without first opening the door. For applications requiring reading temperature in a remote location, a thermometer with a **remote-type sensing bulb** is used (Figure 11-10). The bulb is filled with a volatile fluid and placed in the area where the temperature is to be sensed. It is connected to the thermometer dial by a capillary tube.

Recording thermometers are used where it is necessary to monitor temperature changes over a period of time. The unit shown in Figure 11-11 is quite useful in monitoring the temperature of air conditioned rooms. The sensor causes movement of a pen in the door section of the unit. As temperature decreases, the pen moves toward the center. The rear portion of the unit contains a clock mechanism that causes a recording chart to rotate once every 24 hr. The pen touches the recording chart, and a permanent record of the temperature variations over time is created.

Figure 11-12 shows a chart that might be typ-

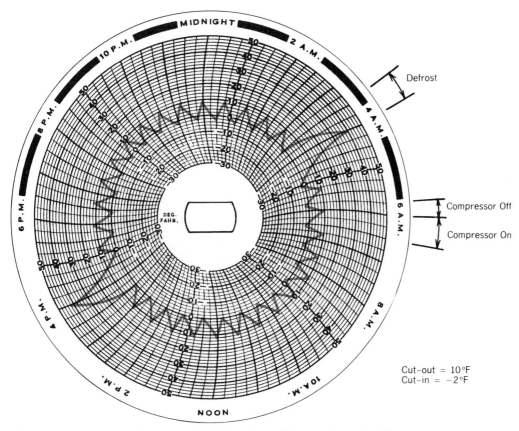

Figure 11-12 Typical temperature recording of temperatures inside a refrigerator.

ical for a domestic refrigerator. We can easily determine the cut-in and cut-out temperature settings for the thermostat, and we can see the unit going through a normal defrost cycle twice during the 24-hr period.

There are many variations on the recording thermometer. Units with remote bulb sensing are often used (Figure 11-13). Units with two pens and two remote bulbs are available to record two temperatures simultaneously, such as refrigerator and freezer temperature. Various temperature ranges are available, and clock mechanisms are available for either 24 hr/revolution or 7 days/revolution.

Figure 11-14 shows a closeup view of the pen end. A special ink is placed in the V-shape using a glass dropper. The ink drains through a small-diameter tube, where it is deposited onto the recording chart. Occasionally, the tube will become clogged, at which point it will be necessary

to pass a thin-diameter wire through the tube. A wire is provided on the end of the dropper for this purpose. Keen eyes and a steady hand are real assets when inserting the wire through the tube.

With the advent of microelectronics and microprocessors, a new generation of temperature-sensing equipment has become popular. **Electronic thermometers** (Figure 11-15) are extremely easy to use, and they provide a digital temperature readout, either to the nearest degree or the nearest tenth of a degree. Various probes are available for measuring air temperatures, liquid temperatures, or surface temperatures. Electronic thermometers are available in either 120-V or battery-operated models.

Leak Detectors

A common failure of refrigeration systems is the loss of refrigerant over a long period of time due to a small leak in the system. Before adding more refrigerant, the technician must locate and repair the leak. Otherwise, a callback on the job is a virtual certainty. There are a variety of detectors available to help the technician locate the source of a leak.

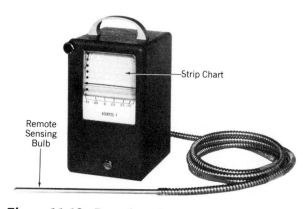

Figure 11-13 Recording thermometer with remote temperature-sensing bulb. *(Courtesy Airserco Manufacturing Company)*

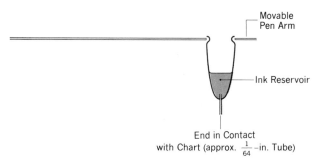

Figure 11-14 Ink reservoir and pen assembly used in recording thermometers (enlarged).

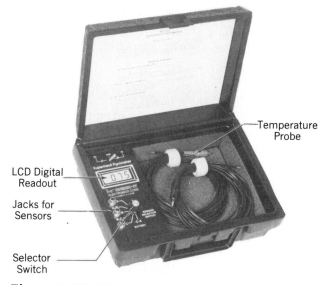

Figure 11-15 Digital superheat thermometer may be used to read up to three different temperatures, or the difference between any two pairs. *(Courtesy Check-It Electronics Corp.)*

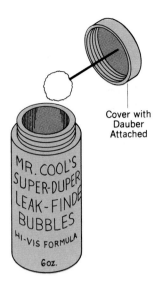

Figure 11-16 Soap bubbles are a quick, easy, and inexpensive method of checking for leaks.

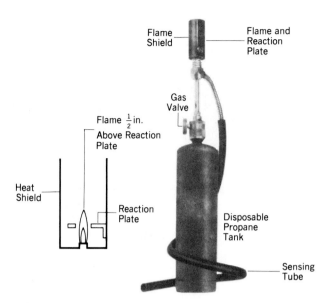

Figure 11-17 Propane leak detector. *(Courtesy Robinair)*

The least complex type of leak detector is essentially a solution of soap and water (Figure 11-16). The solution is formulated to have a high viscosity, and it may even have fluorescent coloring for easy visibility. The solution is applied to the suspected leak with a dauber. A leak will announce itself by the appearance of bubbles. In order to detect very small leaks, the dauber may be held against the joint being tested to form a small puddle of solution between the dauber and the joint. This type of detection is quite useful where there are only a limited number of potential leaking joints that are easily accessible. The liquid leak detector would be used after a system has been opened at a few connections, and then reconnected.

A **propane leak detector** or halide leak detector (Figure 11-17) is a relatively inexpensive tool that is used to find out if there is a leak in the vicinity without having to directly touch the potential leak. This leak detector works on the principle that the blue propane flame is allowed to heat a copper **reaction plate.** The velocity of the propane gas draws an air sample through the sniffer (sensing) tube. If any refrigerant vapor is drawn into the sample tube, it will cause the flame to turn green in color (bright blue for a large leak). You may encounter some difficulty in lighting the torch unless you first plug up the end of the sniffer tube with your finger.

Several safety precautions need to be observed in using the propane leak detector. First, there is an open flame, and obvious safety procedures should be followed. Second, when the refrigerant is burned, small quantities of **phosgene** gas are produced. Phosgene gas is very poisonous, and therefore the propane leak detector may be used only in well-ventilated areas. It is perhaps because of these limitations that the propane leak detector is not as widely used as some of the others.

The **electronic leak detector** is available in both a 120-V model (Figure 11-18) and a battery-operated model. They both work on the same principles. A pump draws an air sample into the detector through the probe. The sample is passed over a hot wire within the detector circuitry. Any refrigerant contained within the air sample will chemically react with the wire, causing a change of resistance. The detector will emit a loud squeal when a leak has been detected. In order to allow for use in noisy areas, some detectors also use a flashing light in the end of the probe to indicate a leak.

Some electronic leak detectors have sensitivity adjustments for small, medium, and large leaks. Generally, the setting for a small leak (most sensitive setting) is used to determine the general area of a leak. The sensitivity setting is then increased to allow pinpointing the exact location.

There is a small cotton filter located in the end

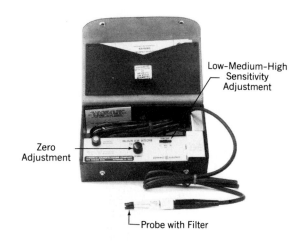

Figure 11-18 Electronic leak detector. *(Courtesy Airserco Manufacturing Company)*

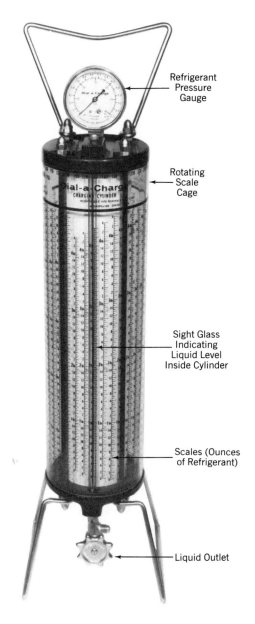

Figure 11-19 A charging cylinder used to measure a predetermined weight of refrigerant into a system. *(Courtesy Robinair)*

of the probe. It must be in place to protect the pump and electronic parts inside the detector. It must also be clean to prevent causing a restriction to the airflow sample. This instrument is the most sensitive of the leak detectors, and also the most delicate. It should not be exposed to concentrated doses of refrigerant, as it can cause burn-out of the detector.

Charging Cylinder

Small systems using 10 lb or less of refrigerant may be charged using a **charging cylinder** (Figure 11-19). Most systems will state the required number of pounds and ounces of refrigerant that the system should contain. Note that we are talking about a required weight of refrigerant in pounds, and not about operating pressures, which are expressed as pounds per square inch (psi).

The charging cylinder is nothing more than a calibrated refrigerant storage tank. Some are furnished with an electric heater in order to be able to add heat (and pressure) to the stored refrigerant. The graduations on the side of the cylinder indicate the number of ounces of refrigerant contained in the cylinder. There are scales for several different refrigerants. Each refrigerant in turn has several scales, corresponding to the pressure in the cylinder at the time the weight is being read. When preparing to add charge to a unit from the charging cylinder, the outside barrel of the cylinder is rotated so that the appropriate scale lines up with the liquid level in the cylinder. The initial quantity of liquid is noted and recorded to the nearest quarter ounce. The refrigerant may then be dispensed either as a vapor from the top valve, or as a liquid from the bottom valve.

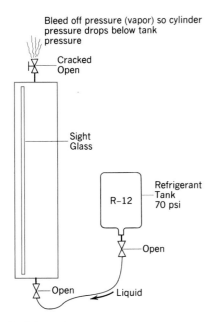

Figure 11-20 Filling the charging cylinder.

When dispensing vapor, you will be able to observe boiling occurring in the cylinder as the pressure is reduced. By closing the dispensing valve and reading the number of ounces of refrigerant remaining, you may easily figure how much has been dispensed. Simply subtract the remaining ounces of refrigerant from the original quantity of refrigerant.

In order to recharge refrigerant into the charging cylinder, connect a refrigerant bottle to the cylinder as shown in Figure 11-20. First, crack open the top valve on the charging cylinder so that a small quantity of refrigerant vapor bleeds out, thus reducing the cylinder pressure. Then, with the tank valve open, liquid refrigerant will flow into the cylinder. With the refrigerant tank elevated, the flow of refrigerant will be aided by gravity.

Pinch-Off Tool

The **pinch-off tool** shown in Figure 11-21 is used to squeeze a soft copper line together so that no pressure can pass from one side of the pinched tube to the other. Figure 11-22 shows the appearance of a tube that has been pinched. The seal is complete, as long as the pinch-off tool is in place. This tool is useful for closing off process

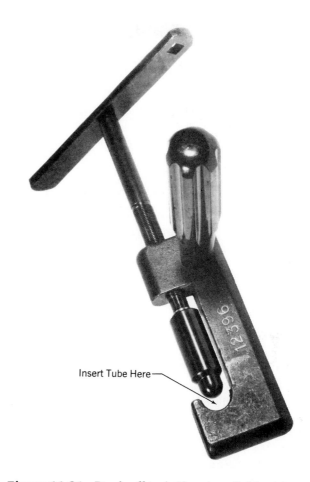

Figure 11-21 Pinch-off tool. *(Courtesy Robinair)*

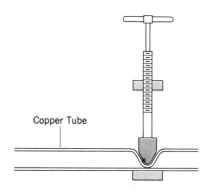

Figure 11-22 Pinch-off tool used to close off a copper tube.

tubes after recharging a unit, and for placing on a compressor suction line in order to check compressor pumping capacity.

Figure 11-23 shows a different type of pinch-off tool. It is actually a combination of a pinch-off

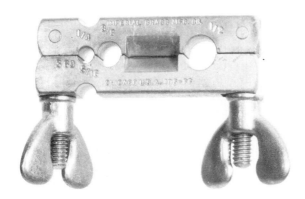

Figure 11-23 Pinch-off tool with rerounding holes. *(Courtesy Robinair)*

Figure 11-24 Pinch-off tool. *(Courtesy Robinair)*

tool and a **rerounding tool.** The rerounding tool may be used to return the tubing to its approximate original shape by working the smooth-bore cylinders along the tube where it was pinched. This tool should not be confused with a flare block, even though they are similar in appearance. The holes in the flare block are grooved to get a bite on the tubing, and they are also chamfered on one side for flaring and swaging operations, which are described in the following section. Another hand-held pinch-off tool is shown in Figure 11-24.

Flaring and Swaging Tools

Figure 11-25 shows a **flaring tool** and a **flare block.** Soft copper tubing is inserted and held in the flare block. The yoke of the flaring tool is then slipped on to the block, and the spinner cone is forced into the tube end, creating a flare as shown in Figure 11-26.

A different set of cones are available for the flaring tool to form double-thickness flares. Fig-

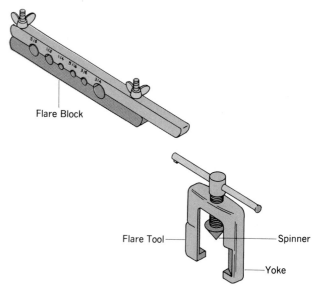

Figure 11-25 Flare tool.

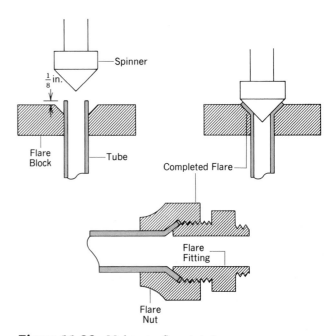

Figure 11-26 Making a flare joint.

ure 11-27 shows the two different cone shapes and the sequence in which they are used to form a double flare. Double flares are capable of withstanding more pressures than single-thickness flares. For most refrigeration and air conditioning applications, however, the single-thickness flare is totally satisfactory.

Figure 11-28 shows a punch-type **swage.** This is used in conjunction with a swage block (same as a flare block) in order to enlarge the end of a tube as shown in Figure 11-29. A hammer is

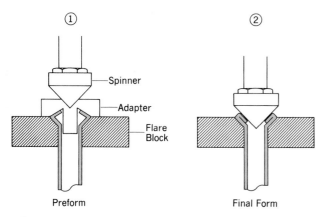

Figure 11-27 Two-step operation to form a double flare.

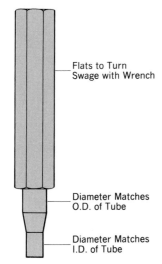

Figure 11-28 Punch swage.

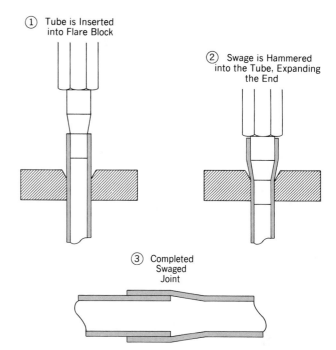

Figure 11-29 Making a swage joint.

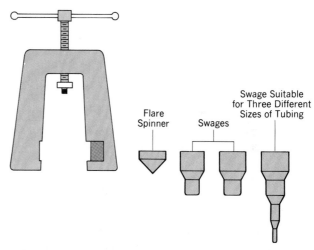

Figure 11-30 A combination flaring/swaging tool. By eliminating the hammer required for a punch swage, this tool can be used in close quarters.

used to force the swage into the tube end. The tube must extend far enough above the swage block to avoid interference between the swage and the block; otherwise, the swage can be easily broken. The punch-type is relatively inexpensive, but is of limited value in close quarters.

Figure 11-30 shows a more expensive but very versatile combination flaring and swaging set. The end of the flaring tool may be fitted with either a flaring cone or one of several different size swages.

Vacuum Pump

The **vacuum pump** is used each time a refrigeration system has been opened. Before recharging with refrigerant, it is necessary to remove all air and moisture that has entered the system. Vacuum pumps have two important ratings, cfm capacity and vacuum microns. Cubic feet per minute, or cfm, refers to how fast the pump can move air when not pumping against any pressure differential, while vacuum microns refers to how deep a vacuum can be created by the pump when pulling against a closed container. A well-made large vacuum pump will not draw any deeper vacuum than a small vacuum pump of similar quality. It will, however, reach that deep vacuum level in a shorter period of running time.

Figure 11-31 shows a standard single-stage vacuum pump, with a free air capacity of 1.2 cfm, that is capable of attaining a vacuum of 200 microns (a **micron** is $\frac{1}{1000}$ of a millimeter, and there are 25,400 microns in 1 in.). This vacuum pump is suitable for use in all air conditioning and appliance service work.

There are more sophisticated vacuum pumps that, although similar in appearance to the pump described previously, have more capability. There are two- and even three-stage pumps that are capable of higher flow rates and deeper vacuums (Figure 11-32). They may be either direct-drive or belt-drive (Figure 11-33). In a two-stage pump, one pump draws a vacuum from the system. The discharge from that pump is internally routed into the suction side of the second stage. With this arrangement, vacuums of 10 microns are attainable, and under laboratory-controlled conditions, even vacuums of $\frac{1}{10}$ micron. The larger flow-rate pumps are used on physically larger systems to save time. The two-stage deep vacuum pumps are used in low temperature applications when the removal of air and water vapor is more critical.

One note of caution is in order about vacuum pumps. Although some single-stage vacuum pumps may look identical to compressors found in small refrigeration systems, they are quite different in capabilities. Do not attempt to get by using a refrigeration compressor as a vacuum pump. This type of pump cannot reach a suffi-

Figure 11-32 Two-stage direct-drive vacuum pump, 4.5-cfm free air displacement, 20-micron vacuum rating. *(Courtesy Robinair)*

Figure 11-31 Single-stage vacuum pump is direct drive, 1.2-cfm free air displacement, and capable of 200 microns of vacuum. *(Courtesy Robinair)*

Motor →   → Drive Pulley

→ Pump

Figure 11-33 Belt-drive vacuum pump. *(Courtesy Airserco Manufacturing Company)*

ciently deep vacuum, and is unsuitable for this purpose.

Micron Gauge

The micron gauge shown in Figure 11-34 is an excellent tool for determining whether the system has been properly evacuated and prepared for recharging. Its scale reads from 50 microns to over 20,000 microns. A vacuum that may have looked adequate on a set of gauges may show up as a lesser vacuum than expected. In essence, the micron gauge takes the portion of the manifold gauge scale between 29 in. Hg and 30 in. Hg, and expands it into a full scale.

Once evacuation has been completed, the micron gauge can be used to show any loss in vacuum. This would indicate that either there is a leak, or that there is moisture in the system, which is boiling and creating water vapor.

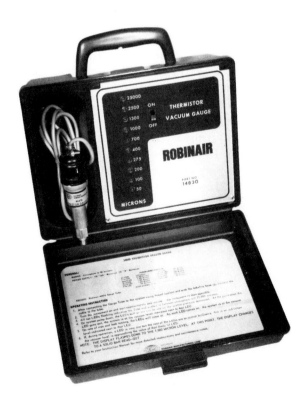

Figure 11-34 The micron gauge expands the scale between 29 in. Hg and 30 in. Hg. (*Courtesy Robinair*)

Cap Tube Chaser

The **cap tube chaser** shown in Figure 11-35 is a hydraulic pump. It is manually pumped like an automobile jack to force oil from the reservoir through a cap tube. Any buildup of wax or scale in the capillary tube will thus be washed out. For cap tubes that are totally clogged, the hydraulic unit will build up to a pressure of 5000 psi. Some units have a gauge that indicates the pressure being imposed on the system. When the restriction has been cleared, it will be indicated by the reduction of pressure shown on the gauge.

For partial restrictions that cannot be washed out with oil, you can use a kit that contains several different diameters of lead wire, each just slightly smaller than the inside diameter of a popular size of cap tube. A $\frac{1}{4}$-in. length of lead wire is inserted into the cap tube prior to attaching the cap tube hydraulic unit. The pump is then used to push the wire through the cap tube until it lodges against the partial blockage. Because the wire virtually fills the cross section of the cap tube, the full 5000 psi will be available to force the blockage loose.

Once the lead wire has been forced through the cap tube, it winds up in the evaporator. Because it is relatively heavy wire, it does not move any farther through the system.

In the case of restrictions due to mechanical damage to the cap tube, such as kinking or crushing, no amount of coaxing from these tools will correct the problem. In that case, replacement of the cap tube would be required.

Pump —⟶   Cap Tube —⟶

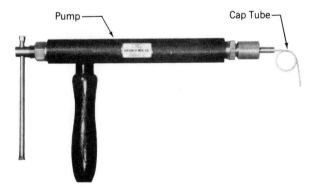

Figure 11-35 The cap tube chaser is used to pump oil through the cap tube to remove an obstruction or wax deposit. (*Courtesy Airserco Manufacturing Company*)

Tube Benders

Whenever refrigeration systems are modified, field piped, or have components replaced, the technician may be required to bend tubing. From the standpoint of system reliability, bends are preferred over soldered fittings. Small-diameter (¼-in.) soft copper is easily bent by hand, using no special tools. Larger size tubing, however, requires some help to prevent it from flattening or kinking.

The simplest form of tube bender is the **spring tube bender** shown in Figure 11-36. It comes in a variety of sizes to match either inside or outside diameters of standard-size tubing. It is placed on (or inside) the tube. The bends are then made by hand, while the spring prevents flattening of the tube. Once the bend is made, the spring may be difficult to remove. With the bender on the outside of the tube, one end of the spring can be twisted in the direction shown. The spring can be slowly worked off the tube end as it is twisted. Removal will be easier for wide radius bends. Do not attempt to bend tubing to a radius of less than five times the tubing diameter. Bending springs are available in sizes up to $\frac{5}{8}$ in., but they are more difficult to use in the larger sizes.

For professional looking bends, a **mechanical tube bender** is required. One type of mechanical bender is illustrated in Figure 11-37. These benders use either a lever action or a gear-assisted action to bend tubing up to $1\frac{1}{8}$ in. diameter. Each size of tubing requires a bender for that specific size. Some mechanical benders have replaceable forming wheels and blocks that allow them to be used on a variety of tubing sizes.

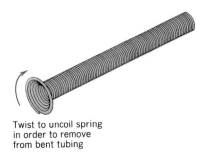

Twist to uncoil spring
in order to remove
from bent tubing

Figure 11-36 Spring tube bender.

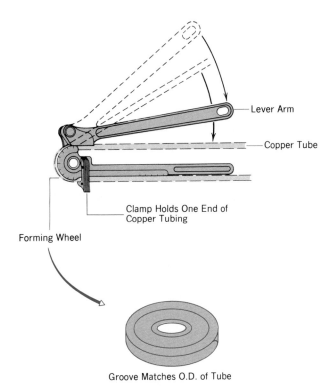

Lever Arm

Copper Tube

Clamp Holds One End of
Copper Tubing

Forming Wheel

Groove Matches O.D. of Tube

Figure 11-37 Mechanical tube bender.

Tube Cutters

The most widely used method of cutting tubing is the **tube cutter** shown in Figure 11-38. Simply rotate the cutter around the tube, tightening the knob at the end a half-turn each revolution, and a quick, square cut will result. This type of cutter is suitable for tubing up to $2\frac{5}{8}$ in. in diameter. Some models have grooves in the rollers, which are handy when you have flared the end of a piece of tubing that has already been cut to length, and the flare turns out damaged. The flare may then be fitted into the groove, as shown in Figure 11-39, and cut off with only a minimal loss of tubing length.

Sometimes there is insufficient room to rotate

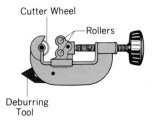

Cutter Wheel

Rollers

Deburring
Tool

Figure 11-38 Tubing cutter.

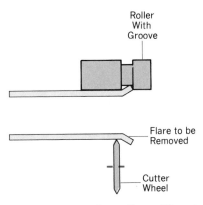

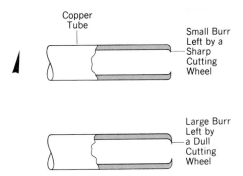

Figure 11-39 Using a tube cutter with a grooved roller to remove a flare.

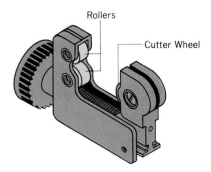

Figure 11-40 Mini tube cutter.

Figure 11-41 Cross section of tubing that has been cut with a tubing cutter.

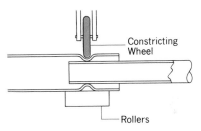

Figure 11-42 Tube cutter with constricting wheel.

the cutter around the tube. In that case, a mini tube cutter, such as the one in Figure 11-40, can be invaluable. This cutter requires only $1\frac{1}{4}$ in. swing clearance, but it is limited to cutting tubing up to only $\frac{1}{2}$ in. outside diameter.

The cut made by a tube cutter is actually not as clean as it may appear. Figure 11-41 shows what the actual cross section of the tube looks like after cutting. It is deformed in that its inside diameter has been reduced to less than the original dimension. After cutting, the tube end must still be cleaned up with either a reamer or a file. The standard-size tube cutters come with a built-in reamer blade. For safe operation, the reamer blade should always be retracted during cutting operations.

The cutter wheels can get dull, especially if the cutter is used on anything besides soft copper tube. It should be replaced whenever the cutting efficiency deteriorates. There is a special wheel called a **constricting roller,** which can be used in place of the cutting wheel. Figure 11-42 shows how the constricting wheel may be used to seat a larger tube onto a smaller tube.

Access Fittings

On small systems, such as refrigerators and window air conditioners, manufacturers frequently do not provide any connection points to attach manifold gauges. There are several types of fittings available to pierce a line containing refrigerant without losing any of the refrigerant charge. Figure 11-43 is typical of this type of line tap available in sizes from $\frac{3}{16}$ to $\frac{5}{8}$ in. outside tube diameter. The body halves of the **piercing valve** are screwed together around the tube to be pierced. The top half of the valve will provide the seal around the hole with a rubber gasket, while the bottom half may be used with or without the adapters, depending on the size of the tube being pierced. The valve inside the upper half of the valve is similar to a tire valve, holding pressure inside the tube until the pin is depressed. A special pin extension is inserted into the top of the valve pin. The valve cap is then screwed on. The pin extension causes the valve pin to be depressed further than normal, and a point on the bottom of the valve is forced into the tubing. When the cap is removed, the pin extension is removed, and a fitting is now available for attaching manifold gauge hoses.

Figure 11-44 shows a slightly different arrangement for an access fitting. The hole is pierced in a similar fashion to the one described above, but there is no automatic closing valve. The valve in this case is opened and closed manually, either with a hand wheel or an allen wrench. The pressure is then read from the side port of the valve.

Access fittings represent an additional potential source for refrigerant leakage. For the most reliable system, the access valve should be removed after it has served its intended use and the hole permanently brazed shut. Some service technicians have had success in letting the access valve remain installed on the low side of R-12 systems.

Fin Comb

The fins on the evaporators and condensers are quite thin and are easily bent. If too many fins become bent, there will be a reduction in airflow through the coil. A **fin comb** (Figure 11-45) is used to restore the fins to their original shape. There are several types of fin combs available. The ones that work best are molded from nylon, and have a different head for each different fin spacing one might encounter. The appropriate fin comb teeth are inserted into the coil at a point where the fins are still undamaged. They are then worked up or down through the damaged fin area, pulling the bent fins forward.

Epoxies

Epoxy is a very strong, hard, and chemically non-reactive glue. Its chief use in air conditioning and refrigeration work is as a substitute for soldering. Although there are claims of epoxies that can seal holes in high pressure lines, epoxy use is best restricted to low pressure systems. For example, they are quite useful for sealing a small hole in the evaporator of an R-12 system.

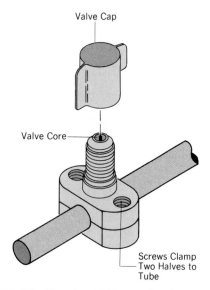

Figure 11-43 Piercing fitting with valve core similar to that used in a tire valve.

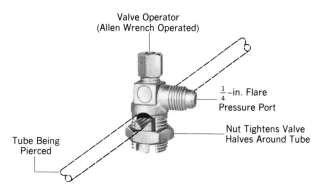

Figure 11-44 Piercing fitting with manual valve operator. *(Courtesy Robinair)*

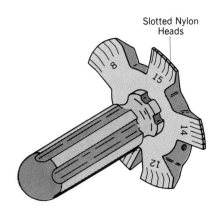

Figure 11-45 Fin comb. Numbers on each slotted head indicate the number of fins per inch on the coil.

Figure 11-46 is a typical two-part epoxy kit. After all the pressure inside the system has been released, the joint is first cleaned down to bare metal with sand cloth. An MEK (methyl ethyl ketone) solution is then swabbed on to remove any traces of oil that would interfere with the epoxy bond. The two parts of the epoxy are then dispensed next to each other on a scrap of paper or cardboard and thoroughly mixed together. The glue may be applied to the hole being patched by a tool or by your finger (CAUTION: BE SURE TO CHECK ANY SAFETY WARNINGS ON THE EPOXY PACKAGE). The epoxy will harden within an hour. A faster cure may be accomplished by applying heat. A small wattage light bulb placed near the curing epoxy can speed the process to 10 min or less.

Although epoxies are normally two-part mixtures, there is another epoxy-type glue available in stick form, as shown in Figure 11-47. The part to be sealed is warmed with a heat gun or propane torch while the epoxy stick is applied (be extremely careful not to melt aluminum evaporators while using any type of torch). The melted epoxy seals quite well and requires no curing time.

Can Tapping Valve

The assembly in Figure 11-48 attaches to a small (1- or 2-lb) refrigerant container, which is supplied without a valve. The **can tapping valve** is screwed onto the top of the refrigerant can to form a seal. The valve is then screwed through the seal, piercing the top of the can. The seal made by this valve will not hold the refrigerant pressure indefinitely. Professional service technicians do not frequently use small cans of refrigerant, since it is more convenient to use 30- or 50-lb refrigerant containers. The automobile air conditioning field is a notable exception to this rule.

Service Valve Wrench

The **service valve wrench** (Figure 11-49) is used to operate the stem on all types of service valves.

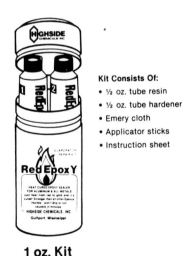

Kit Consists Of:
- ½ oz. tube resin
- ½ oz. tube hardener
- Emery cloth
- Applicator sticks
- Instruction sheet

1 oz. Kit

Figure 11-46 Two-part epoxy kit. *(Courtesy Highside Chemicals Inc.)*

Figure 11-47 Heat-activated epoxy stick.

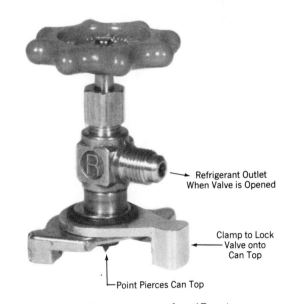

Refrigerant Outlet
When Valve is Opened

Clamp to Lock
Valve onto
Can Top

Point Pierces Can Top

Figure 11-48 Can tapping valve. *(Courtesy Robinair)*

Figure 11-49   Service valve wrench. *(Courtesy Robinair)*

The stems are easily deformed, so it is important to use the correct size square service wrench. The ratchet feature also makes opening and closing of the service valve quicker when operating in cramped quarters.

Inspection Mirror

The **inspection mirror** (Figure 11-50) is used whenever direct line of sight is not available. The most valuable use for the inspection mirror is inspecting the back side of tubing joints when searching for leaks.

If there is insufficient light to see the area being viewed, you can point a flashlight along your line of sight into the mirror. This will provide reflected light onto the part being inspected.

Manometer

Manometers, in general, are used to measure a difference in pressure. The manometer in Figure 11-51 is unique in that one end is closed and sealed. When a vacuum is applied to the open end, the level in the closed end will drop. This type of manometer is used in testing the vacuum-producing capability of vacuum pumps.

Oil Charging Pump

If inspection of a compressor sight glass indicates a shortage of oil, refrigeration oil must be

Figure 11-50   Inspection mirror. *(Courtesy Robinair)*

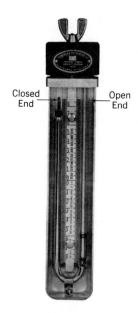

Figure 11-51   Mercury-filled manometer. *(Courtesy Robinair)*

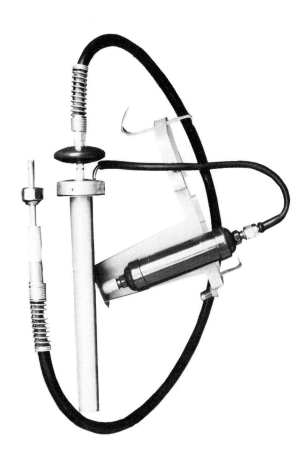

Figure 11-52 Oil charging pump. (*Courtesy Airserco Manufacturing Company*)

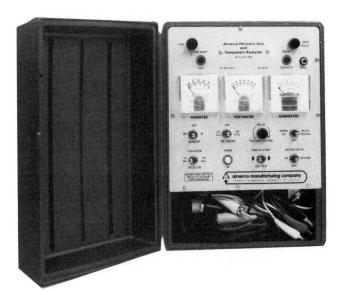

Figure 11-53 Hermetic analyzer. (*Courtesy Airserco Manufacturing Company*)

added to the system. The hand-operated pump in Figure 11-52 is able to draw oil out of a can and discharge it into the system through an access fitting or service valve. When adding oil to a system, purge the air from the pump lines by pumping some oil prior to attaching the discharge line to the refrigerant system.

Hermetic Analyzer

The **hermetic analyzer** (Figures 11-53 and 11-54) is a convenient packaging of components used to start single-phase compressors. It contains meters to allow you to first determine whether the motor windings are open or shorted to the casing. If the windings are good, the relay is removed from the compressor and the her-

Figure 11-54 Compressor analyzer stand. (*Courtesy Airserco Manufacturing Company*)

metic analyzer wires are attached. When the unit is turned on, voltage is applied to the run winding. Then, when the start button is depressed, voltage is applied to the start winding. A capacitor is automatically wired into the circuit and a reverse switch is also provided. The reverse switch reconfigures the internal wiring of the capacitor in the hermetic analyzer so that the compressor will tend to start backwards. This is useful for compressors that have become mechanically stuck.

Core Removal Tool

The **core tool** (Figure 11-55) makes it possible for the service technician to remove the core from a Shrader valve without opening the system to the atmosphere. The primary use for this tool is removing the valve core while evacuating the system. By providing an increased-size opening, evacuation can be accomplished more quickly than if the core remained in place.

$\frac{1}{4}$ in. Flare Connection

Figure 11-55   Core removal tool. *(Courtesy Robinair)*

KEY TERMS AND CONCEPTS

Manifold Gauges

Compound Gauge

Dial-Stem
    Thermometer

Remote Bulb
    Thermometer

Recording
    Thermometer

Electronic
    Thermometer

Propane Leak Detector

Reaction Plate

Phosgene

Electronic Leak
    Detector

Charging Cylinder

Pinch-Off Tool

Rerounding Tool

Flare Tool

Flare Block

Swage

Vacuum Pump

Micron

Cap Tube Chaser

Spring Tube Bender

Mechanical Tube
    Bender

Tube Cutter

Constricting Roller

Piercing Valve

Fin Comb

Epoxy

Can Tapping Valve

Service Valve Wrench

Inspection Mirror

Manometer

Oil Charging Pump

Hermetic Analyzer

Core Tool

QUESTIONS

1. When reading operating pressures, the manifold gauge valve should be (opened, closed, partially opened)?

2. What is a compound gauge?

3. What do the temperature numbers on the manifold gauges tell you?

4. How are the two ends of the manifold gauge hoses different?

5. When adding refrigerant to an operating system, the manifold low-side valve is (open, closed) and the high-side valve is (open, closed).

6. How tightly should the manifold gauge hoses be screwed onto the access fittings?

7. What is a remote sensing thermometer?

8. Name three different types of leak detectors.

9. What flame color on the propane leak detector indicates the presence of refrigerant?

10. How do you get the refrigerant into the charging cylinder?

11. On what types of systems is the charging cylinder most commonly used?

12. How are the holes on a flare block different from the holes on a rerounding tool?

13. Why is a vacuum pump used on a system that has been opened?

14. Which is a deeper vacuum, 1000 microns or 100 microns?

15. Under a vacuum test, the system vacuum is gradually lost, but there is no leak. How do you explain this?

16. What does a cap tube chaser do?

17. What is the major disadvantage to using spring benders for bending copper tubing?

18. How would you be able to tell from the quality of the cut that your tube cutter wheel is getting dull?

19. What does a fin comb allow you to do?

20. How does heat affect the curing time of an epoxy?

CHAPTER 12

SOLDERING, BRAZING, AND TUBE WORKING

Copper Tubing

Copper tubing is the most common material used to connect the various parts of the refrigeration system together. The tubing used in the air conditioning and refrigeration industry is similar to that commonly used in plumbing. The major differences are as follows.

1. Air conditioning and refrigeration **(AC&R) tubing** is guaranteed to be clean inside. When it leaves the factory, it has been dehydrated (dried), cleaned, and the ends capped or crimped. Sometimes it is shipped with a holding charge of nitrogen gas inside in order to keep air out.

2. The AC&R industry refers to the tubing size differently from the plumbing industry. Plumbers refer to tubing size by the nominal inside diameter, but AC&R technicians refer to tubing size by its exact actual outside diameter. For instance, Figure 12-1 shows a tube that is approximately $\frac{1}{2}$-in. inside diameter, and exactly $\frac{5}{8}$-in. outside diameter. To the plumber, this is $\frac{1}{2}$-in. tubing, but the AC&R technician would refer to it as $\frac{5}{8}$-in tubing.

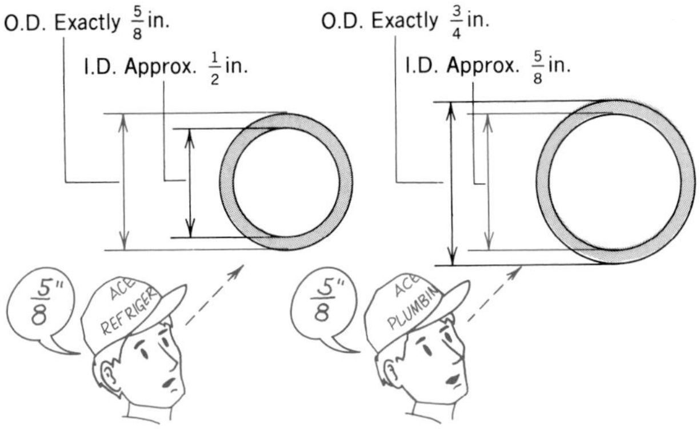

Copper tubing is manufactured in two different forms, **soft** and **rigid.** The only difference between the manufacture of these two types of tubing is the heat treatment process. Copper becomes soft when it has been **annealed.** Annealing is a heat treatment process in which copper is heated until it reaches a dull red color, and is then allowed to cool slowly. Copper becomes hard and rigid when it has been **work hardened.** Work hardening involves the repeated bending or stretching of the tube to change the molecular crystal structure. Hard copper may be made softer by annealing it with a torch. Soft copper may be work hardened by unrolling it and stretching it with a come-along. Stretching it to increase its length by 5 to 10 percent will stiffen it considerably. Its diameter and wall thickness will also both be reduced, but not appreciably.

Unrolling soft copper can present a challenge to the beginner. If it is merely unrolled by hand, a very crooked length of copper will result. A better method is shown in Figure 12-2. Unrolling the

O.D. Exactly $\frac{5}{8}$ in.
I.D. Approx. $\frac{1}{2}$ in.

O.D. Exactly $\frac{3}{4}$ in.
I.D. Approx. $\frac{5}{8}$ in.

Figure 12-1 The AC&R ⅝-in. tube is smaller than the tube called ⅝-in. by the plumber.

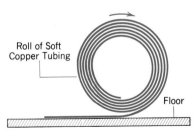

Roll of Soft Copper Tubing

Floor

Figure 12-2 Unrolling a roll of soft copper tubing to produce a straight length.

197

copper against a flat surface such as a floor will result in a relatively straight piece.

Soft copper is used primarily in the smaller sizes (less than 1-in. outside diameter). In sizes smaller than ¾ in., it can be easily bent to accommodate the required routing. Hard drawn copper is used only in straight lengths. Turns or bends are made by using fittings such as those shown in Figure 12-3. The tubing and fittings are soldered or brazed together.

The advantages to using soft tubing are as follows.

1. Installation cost is low, as long lengths can be assembled quickly.

2. Long lengths (up to 100 ft) can be installed with no joints or connections. Besides the quicker installation time, there is less likelihood of a leak developing at a joint.

Following are the advantages to using hard drawn copper.

1. A very neat, workmanlike installation can be made.

2. Tubing can be run over much greater unsupported lengths, especially in the larger sizes.

This can be important when tubing is to be run overhead and supported from roof trusses that are 10 to 15 ft. apart.

3. Where tubing is exposed, hard drawn copper is more resistant to leaks or restrictions caused by mechanical damage.

Hard and soft copper are manufactured in the same diameters and most of the same wall thicknesses. These standard dimensions are shown in Figure 12-4. Copper is manufactured in three different wall thicknesses, **type K, type L, and type M.** Type K is the thickest, and is only used where there is potential of corrosion and for underground service; type L is the most commonly used for normal applications; type M has the thinnest wall, is available only in the hard temper, and is used for drainage and other nonpressure applications.

Soft copper tubing should not be used for any refrigerant piping in sizes larger than 1⅝-in. outside diameter.

Other Refrigerant Piping Materials

In some applications, noncopper refrigerant piping is used. Systems using ammonia as the refrigerant cannot tolerate any copper, brass, or aluminum in the system. Other large applications may require iron or steel pipe to be used to

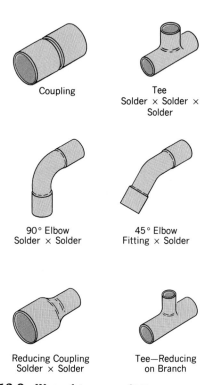

Figure 12-3 Wrought copper fittings.

Coupling

Tee
Solder × Solder × Solder

90° Elbow
Solder × Solder

45° Elbow
Fitting × Solder

Reducing Coupling
Solder × Solder

Tee—Reducing
on Branch

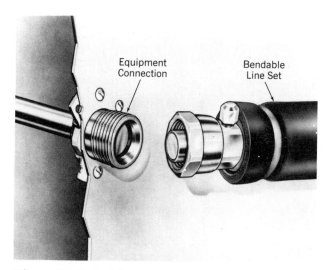

Figure 12-5 Precharged refrigerant tubing. *(Courtesy Aeroquip Corporation)*

Equipment Connection

Bendable Line Set

DIMENSIONS AND PROPERTIES OF COPPER TUBE

(Based on ASTM B-88)

Line Size O.D.	Type	Diameter		Wall Thickness In.	Surface Area Sq. Ft./Lin. Ft.		Inside Cross-section Area, Sq. In.	Lineal Feet Containing 1 Cu. Ft.	Weight Lb/Lin. Ft.	Working Pressure Psia
		OD In.	ID In.		OD	ID				
⅜	K	0.375	0.305	0.035	0.0982	0.0798	0.0730	1973.0	0.145	918
	L	0.375	0.315	0.030	0.0982	0.0825	0.0779	1848.0	0.126	764
½	K	0.500	0.402	0.049	0.131	0.105	0.127	1135.0	0.269	988
	L	0.500	0.430	0.035	0.131	0.113	0.145	1001.0	0.198	677
⅝	K	0.625	0.527	0.049	0.164	0.138	0.218	660.5	0.344	779
	L	0.625	0.545	0.040	0.164	0.143	0.233	621.0	0.285	625
¾	K	0.750	0.652	0.049	0.193	0.171	0.334	432.5	0.418	643
	L	0.750	0.666	0.042	0.193	0.174	0.348	422.0	0.362	547
⅞	K	0.875	0.745	0.065	0.229	0.195	0.436	331.0	0.641	747
	L	0.875	0.785	0.045	0.229	0.206	0.484	299.0	0.455	497
1⅛	K	1.125	0.995	0.065	0.295	0.260	0.778	186.0	0.839	574
	L	1.125	1.025	0.050	0.295	0.268	0.825	174.7	0.655	432
1⅜	K	1.375	1.245	0.065	0.360	0.326	1.22	118.9	1.04	466
	L	1.375	1.265	0.055	0.360	0.331	1.26	115.0	0.884	387
1⅝	K	1.625	1.481	0.072	0.425	0.388	1.72	83.5	1.36	421
	L	1.625	1.505	0.060	0.425	0.394	1.78	81.4	1.14	359
2⅛	K	2.125	1.959	0.083	0.556	0.513	3.01	48.0	2.06	376
	L	2.125	1.985	0.070	0.556	0.520	3.10	46.6	1.75	316
2⅝	K	2.625	2.435	0.095	0.687	0.638	4.66	31.2	2.93	352
	L	2.625	2.465	0.080	0.687	0.645	4.77	30.2	2.48	295
3⅛	K	3.125	2.907	0.109	0.818	0.761	6.64	21.8	4.00	343
	L	3.125	2.945	0.090	0.818	0.771	6.81	21.1	3.33	278
3⅝	K	3.625	3.385	0.120	0.949	0.886	9.00	16.1	5.12	324
	L	3.625	3.425	0.100	0.949	0.897	9.21	15.6	4.29	268
4⅛	K	4.125	3.857	0.134	1.08	1.01	11.7	12.4	6.51	315
	L	4.125	3.905	0.110	1.08	1.02	12.0	12.1	5.38	256

Figure 12-4 Dimensions of AC&R tubing. *(Courtesy Copeland Corporation)*

carry refrigerant. Where low first cost is a primary consideration in manufactured systems, aluminum or steel tubing may be used.

In automotive and other transport applications, flexible hose is used. Figure 12-5 shows a bendable line set, which is often used in residential applications to connect the outdoor condensing unit to the indoor evaporator. These lines may be precharged with refrigerant. When connected to precharged components, it is possible to field assemble a system without ever exposing the interior of the system to the atmosphere.

Tubing Connections

Tubing connections fall into the following two categories: (1) **soldered** or **brazed connections** and (2) **mechanical connections.**

Soldering and brazing involve using a molten metal between two closely fitting surfaces to bond them together. If the melting temperature of this filler metal is lower than 800°F, the process is called soldering, but if the filler metal melting point is higher than 800°F, the process is called brazing. Neither of these processes is considered to be welding. Welding involves the use of a filler material that is the same as the base metal. Both the base metal and the filler metal are melted and allowed to reform. With soldering and brazing, only the filler metal melts.

Mechanical connections can include flared, screwed, compression, or other specialized devices. Flaring techniques are explained in Chapter 11, on tools.

Heat Sources for Soldering/Brazing

Figure 12-6 shows a **propane torch** that is commonly found in the home workshop. It may be used with soft solders that have melting points up to 450°F. The torch mixes propane gas and a small quantity of air to create a flame sufficiently hot to solder tubing up to 1-in. diameter.

Figure 12-7 shows a different torch that mixes a greater quantity of air with the fuel and creates a high degree of turbulence that aids in transferring heat to the work. With a propane fuel, this type of torch can be used with solders that have melting points that approach 800°F. For higher temperature applications, acetylene gas or artificial manufactured gas such as Mapp (trademark of Uniweld products) gas may be used. The fuel-air torch may also be used on a refillable fuel cylinder with a torch shown in Figure 12-8. The regulator with a pressure gauge is attached to the cylinder and connected to the torch handle with a single hose.

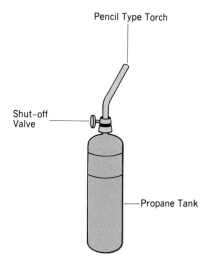

Figure 12-6 Propane torch.

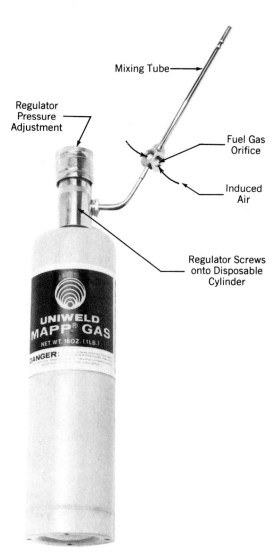

Figure 12-7 High turbulence torch and regulator. *(Courtesy Uniweld Products, Inc.)*

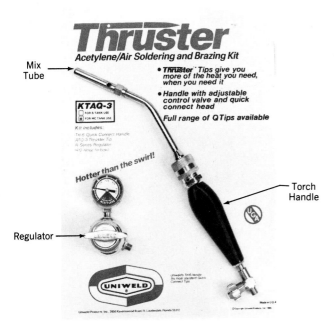

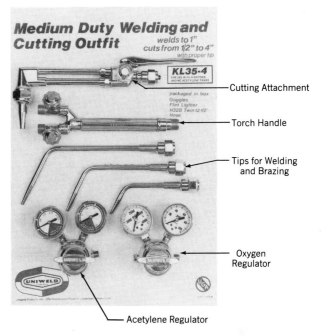

Mix Tube

Torch Handle

Regulator

Figure 12-8 Fuel–air torch used on a refillable fuel cylinder. *(Courtesy Uniweld Products, Inc.)*

Cutting Attachment

Torch Handle

Tips for Welding and Brazing

Oxygen Regulator

Acetylene Regulator

Figure 12-10 Welding and brazing kit including cutting attachment. *(Courtesy Uniweld Products, Inc.)*

When the highest flame temperatures are desired, as in silver brazing, the acetylene is mixed with oxygen in a torch such as that shown in Figure 12-9. This arrangement produces a flame that can be made hot enough to weld and cut through steel. Figure 12-10 shows a kit that has attachments for cutting, welding, and brazing.

The Soldered Joint

Figure 12-11 shows a cutaway view of a soldered joint. The clearance between the original pieces is only a few thousandths of an inch. The joint is assembled, and the molten filler material is allowed to flow into the crack by means of **capillary action.** Capillary action can be demonstrated by holding two pieces of glass in a tray of water as shown in Figure 12-12. When the panes of glass are held together so that the clearance between them is quite small, the water will actually climb up between them, apparently defying the laws of gravity. It is this capillary action that causes the filler metal to be drawn into the soldered joint, even if this means that the solder will need to flow uphill. When the base metals have been well

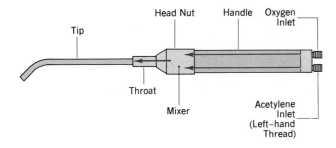

Tip, Head Nut, Throat, Mixer, Handle, Oxygen Inlet, Acetylene Inlet (Left–hand Thread)

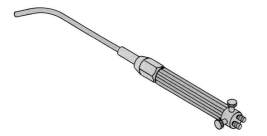

Figure 12-9 Oxy-acetylene torch.

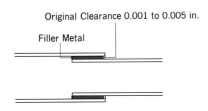

Original Clearance 0.001 to 0.005 in.

Filler Metal

Figure 12-11 Cutaway of a soldered or brazed joint.

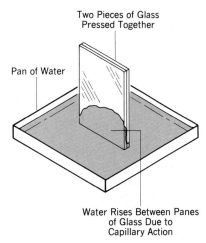

Figure 12-12 Capillary action causes the water to rise into the space between the glass plates.

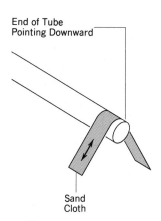

Figure 12-13 Cleaning the tubing end that is to be soldered or brazed.

cleaned, the molten solder will make a molecular bond with the base metal. Many students, while trying unsuccessfully to solder an improperly cleaned joint, will complain that the solder doesn't stick.

Cleaning the Joint

Immediately following the manufacture of copper tubing, an oxide coating begins to form where the copper is exposed to air. Old pennies look darker than new pennies because of the advanced stage of this oxidation. Before soldering or brazing can be done, this oxide coating must be removed from the surfaces that are to be joined.

The outside of the male surface is cleaned by using a strip of sand cloth, as shown in Figure 12-13. Note that in cleaning the tube, the end is held downward so that pieces of sand from the cloth do not fall into the tube and remain inside the system when it is assembled. The inside of the fitting may be cleaned with sand cloth or with a tube cleaning brush. Sand cloth is difficult to use on sizes less than ¾ in. For these smaller sizes, the brush shown in Figure 12-14 is used. Tube cleaning brushes come in a variety of sizes, so only the correct size brush that matches the tubing size should be used. The brush is inserted into the fitting and turned clockwise a few times. The brush is removed by pulling it out while

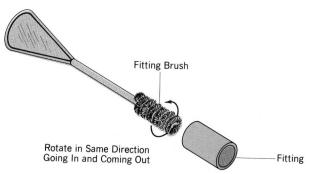

Figure 12-14 The inside of small fittings required the use of a fitting brush to remove oxidation.

continuing to turn it clockwise. Using too large a brush, or turning the brush counterclockwise will immediately ruin the brush.

The tubing should be cleaned just prior to soldering or brazing. As soon as the cleaning is completed, the oxide once again begins to form.

When you are finished cleaning, do not use your hand to brush away any loose sand or oxide. Your hand has natural oils on it that will be deposited onto the freshly cleaned surfaces.

Filler Materials

A solder that is commonly used in plumbing work is 50/50 solder. The numbers denote that the solder is made of 50 percent tin and 50 percent lead. Other common compositions for tin/lead solders are 60/40 and 40/60. Where a some-

Figure 12-15 Soft solder has a melting point between 350°F and 450°F, depending on its composition.

what stronger joint is required, 95/5 solder is used (95 percent tin and 5 percent antimony). Some municipalities have disallowed the use of 50/50 and all other lead-bearing solders for potable (drinking) water systems, as there is evidence that the lead content may contaminate the water.

These **soft solders** are supplied in 1-lb. rolls of $\frac{1}{8}$-in. wire (Figure 12-15). Soft solders are suitable for all services at room temperature and for low pressure steam. It may also be used for other services at moderate pressures where temperatures will not exceed 250°F. The melting point of these soft solders is in the range of 350°F to 450°F. The base metal can be easily heated to this temperature with a propane torch for soldering tube sizes smaller than 1 in.

On soldered joints that are to be used to carry refrigerant, soft solders are not recommended due to the possibility of loosening from vibration. They have been successfully used on the low side

by some technicians; however, the joints are not as reliable as the higher temperature solders. Solders containing silver, phosphorus, and copper as the major ingredients are recommended for refrigerant service. There are two different types of refrigerant-type solders (actually brazing materials). The first is comprised of mostly copper, and contains about 5 percent phosphorus, plus anywhere from 6 to 15 percent silver. This solder is usually purchased in stick form, as shown in Figure 12-16. Many technicians call this type of material **Sil-Fos,** although this is actually a brand name of one manufacturer (Handy & Harman) and for one formulation. With the rise in silver prices during the early 1980s, formulations with reduced silver content (actually down to 0 percent silver) were introduced. They found some acceptance into the marketplace, and are still being used.

Silver–phosphorus brazing materials have melting points between 1200°F and 1500°F. They require temperatures close to those that can be produced by oxy-acetylene torches for effective application. These solders are suitable for joining nonferrous material, such as copper, brass, monel, or any combination. Some have used this type of filler with steel tubing, but this is not recommended because of the brittle joint that results. It may be leaktight when the joint is made, but it will be susceptible to failure due to cracking.

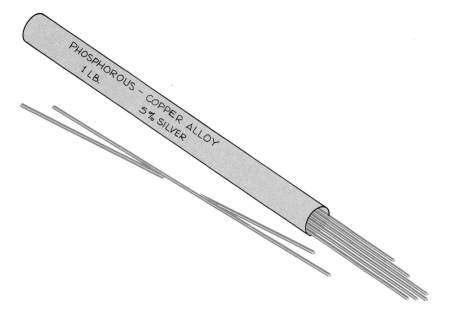

Figure 12-16 Silver–phosphorous–copper brazing alloy in stick form.

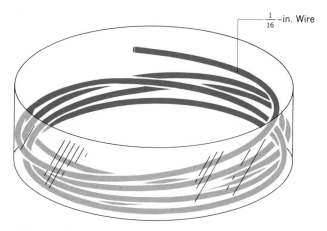

$\frac{1}{16}$-in. Wire

Figure 12-17   High-silver-bearing brazing alloy.

The third category of solders, sometimes called either **silver solder** or **Easy-Flo,** contains a much higher silver content. As with Sil-Fos, Easy-Flo is a Handy & Harman brand name that has become generic over the years. For many years, the primary materials in these compounds were silver (between 35 percent and 55 percent), copper, zinc, and cadmium. After many years in the marketplace, however, it was discovered that the fumes associated with melting the cadmium were hazardous to breathe. Silver solders in all of the silver percentages, but with no cadmium, were then introduced. From a strength and reliability standpoint, there is no reason to risk using anything other than the cadmium-free compounds. The silver solders melt in the 1100°F to 1200°F range, and are suitable for joining copper, brass, and monel, as well as any of these

alloys to steel. The most common form for silver solder is the $\frac{1}{16}$-in. wire shown in Figure 12-17. These solders can be quite expensive because of the high silver content. As a matter of fact, they are sold by weight according to the same troy system of weights used to weigh precious metals.

Because so many filler materials are commonly called by their brand names, a comparison chart of many of the popular fillers is presented in Figure 12-18.

Fluxing the Joint

Flux is a material that is sometimes applied to the cleaned surfaces prior to soldering or brazing. The function of the flux is to protect the cleaned surface from reoxidizing when heat is applied. It also promotes the easy flow of the molten filler material. It is not the function of flux to clean the joint: that is your job.

The type of flux to be used depends on the melting temperature of the filler metal being used. The flux should have a melting point that is only slightly lower than the filler melting temperature. When the joint is being heated, the technician should watch to see when the flux melts. That is the clue that tells you when the metal is almost hot enough.

Flux is usually in a paste form, and is applied to the joint with a brush (remember, your fingers have natural oil on them). With joints that are to be used on plumbing or drainage systems, ap-

AWS Number	J.W. Harris	Hardy & Harman	All State	Engelhard	Airco	United Wire	Silver	Copper	Zinc	Cadmium	Nickel	Phosphorus	Tin	Melting Range (°F)
BCuP 2	Stay-Silv 0	Fos-Flo 7	21 + Silflo 0		Phos-Copper	Phoson 0		92.9				7.1		1310–1475
BCuP 3	Stay-Silv 5	Sil-Fos 5	Silflo 5	Silvaloy 5	Aircosil 5	Phoson 5	5.0	89.0				6.0		1190–1500
BCuP 5	Stay-Silv 15	Sil-Fos	Silflo 15	Silvaloy 15	Aircosil 15	Phoson 15	15.0	80.0				5.0		1190–1480
BAg 2	Stay-Silv 35	Easy-Flo 35	S-135	Silvaloy 35	Aircosil 35	Sil-Bond 35	35.0	26.0	21.0	18.0				1125–1295
BAg 1	Stay-Silv 45	Easy-Flo 45	S-145	Silvaloy 45	Aircosil 45	Sil-Bond 45	45.0	15.0	16.0	24.0				1125–1145
BAg 1a	Stay-Silv 50	Easy-Flo	S-150	Silvaloy 50	Aircosil 50	Sil-Bond 50	50.0	15.5	16.5	18.0				1160–1175
BAg 7	Safety-Silv 1200	Braze 560	155	Silvaloy 355	Aircosil J	Sil-56 T	56.0	22.0	17.0				5.0	1145–1205

Figure 12-18   Comparison of filler metals.

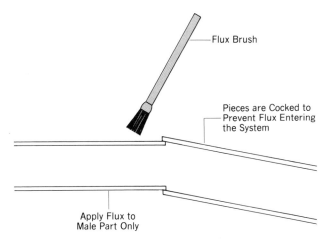

Figure 12-19 Fluxing the cleaned joint.

WRONG

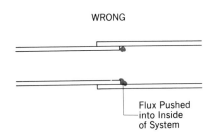

Figure 12-20 An improperly fluxed joint. Flux was applied to the inside of the female part of the joint.

plication of the flux is not critical. For refrigeration systems, however, it is imperative that the flux not be allowed to get on the inside of the tube. In order to assure no flux on the inside of the tubing, the flux is applied as in Figure 12-19. The male piece is inserted only $\frac{1}{8}$ in. into the fitting, and cocked to hold it in place. Then the flux is applied to the male piece only, after which the joint may be assembled. Never apply flux to the inside of the fitting, as it will be pushed into the tube, as shown in Figure 12-20, when the joint is assembled.

Flux should be used in all cases except one. The phosphorus in the silver–phosphorous materials have a unique property. When they are used on copper, the phosphorus actually attacks the copper oxide coating. It is therefore unnecessary to use flux when using a silver–phosphorous filler material on a copper-to-copper joint.

Making the Soldered Joint

After the joint has been cleaned, fluxed, and assembled, you are ready to make the soldered joint. The objective is to heat the metal to a high enough temperature that when the filler metal is touched to the joint, it melts and is drawn into the space between the pieces being joined. The heat from the torch is not used to melt the filler. The base metal melts it. Figure 12-21 shows the application of heat to the joint while the solder is applied at the same time. You should use a length of solder approximately equal to the value shown in Figure 12-22. If you use too much solder, it will wind up on the inside of the tube and will cause a flow restriction.

The entire joint should be heated. When the flux turns to liquid, keep testing the joint for the proper temperature by momentarily touching the solder to the tube. Once the solder begins melting, concentrate the heat on the fitting, as the liquid solder will tend to flow toward the heat. On small joints (diameters of $\frac{1}{2}$-in. or less), the joint may sometimes be made by applying heat to one side of the joint and the solder to the other. The melted solder will then encircle the tube on its way toward the heat source (Figure 12-23).

It is important to bring the base metal up to temperature as quickly as possible, and complete the soldering or brazing process without unnecessary heating. When the heat is applied to the

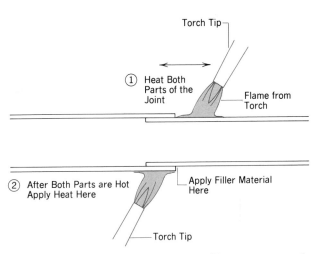

Figure 12-21 Proper application of heat causes molten filler alloy to flow into the joint.

tubing, it accelerates the formation of copper oxide. The flux protects the joint, but the inside of the tube is not fluxed. Working quickly minimizes the amount of oxide formed inside the tube. Excess copper oxide inside the tube can break loose after the system is fully assembled and cause restrictions of small openings. In manufacturing refrigeration systems, and in some limited field applications, this oxidation is prevented by allowing a small amount of nitrogen to flow through the inside of the tubing while the

Normal Tubing Size	3/64" Wire	1/16" Wire	3/32" Wire	1/8 x .050 Rod
1/4"	1 1/4"	3/4"		
3/8"	1 1/2"	1"		
1/2"	2"	1 1/2"	3/4"	7/8"
3/4"	3"	2"	1"	1 1/8"
1"		3"	1 1/2"	1 5/8"
1 1/4"		4"	2"	2 1/2"
1 1/2"			2 1/2"	2 3/4"
2"			3 3/4"	4 1/2"
2 1/2"			6"	7 1/2"
3"			10"	11 1/2"
3 1/2"			12"	13 3/4"
4"			14"	16"
6"			21"	23 3/4"

Figure 12-22   Estimated amounts of brazing alloy required per joint. *(Courtesy J. W. Harris Co., Inc.)*

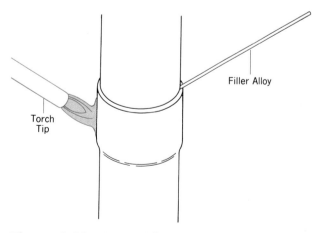

Figure 12-23   Heat and filler alloy should be applied at different points on the circumference to prevent melting the filler with the torch.

brazed joint is being made. This excludes the air (and oxygen), thus preventing the scale formation.

Overheating should also to be avoided to ensure a leaktight joint. Too much heat will cause the flux to vaporize and the joint will no longer be protected from oxidation.

When the joint has been completed, use a wet rag to remove excess flux. The flux is mildly corrosive and can cause leaks to occur long after you have completed your work.

When joining a **ferrous** tube (steel, iron) to a nonferrous tube (copper, brass, bronze), most of the heat should be directed to the nonferrous tube. The nonferrous alloys conduct heat more rapidly, so applying the heat to the nonferrous tube will result in more uniform heating of the joint.

Oxy-Acetylene

Oxy-acetylene torches and equipment deserve a specific discussion because of their general popularity within the AC&R industry. A typical oxy-acetylene setup is shown in Figure 12-24. It consists of the following components.

1. Oxygen tank with tank valve.
2. Acetylene tank with tank valve.
3. Gas pressure regulator on each tank.
4. Twin hose.
5. Torch butt and tip.
6. Dolly or frame to support tanks.

The oxygen cylinder is available in sizes from less than 20 cu ft to more than 300 cu ft. When it is filled with oxygen, its pressure will be 2200 psi. The **tank valve** is operated with a handwheel. When the gas is in use, the oxygen tank valve should be fully opened. With a partially opened oxygen valve there will be a tendency for the oxygen to leak out around the valve packing. Every oxygen tank valve is also equipped with a safety disc. If, for any reason, the cylinder's internal pressure reaches an unsafe level, the safety disc will burst, releasing the contents of the cylinder.

The acetylene cylinder is more complex than the oxygen cylinder. Acetylene is a manufactured

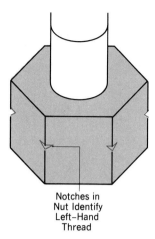

Notches in
Nut Identify
Left-Hand
Thread

Figure 12-25 Acetylene fitting with left-hand thread.

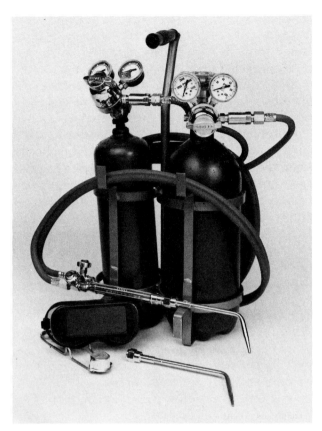

Figure 12-24 Oxy-acetylene setup. *(Courtesy Uniweld Products, Inc.)*

fuel gas that, when combined with oxygen and burned, will produce the hottest commonly available flame for soldering, brazing, welding, or even cutting. One of the peculiar and potentially dangerous characteristics of acetylene is that it is unstable at pressures above 15 psi. That means that if it is subjected to a severe shock, it could explode without any other source of ignition. Therefore, safety codes forbid the use of free acetylene at pressures above 15 psi. In the acetylene tank, the acetylene is mixed with another chemical called **acetone** (a common solvent) and absorbed into a rigid, porous material. In this form, the acetylene–acetone mixture is stable even up to the maximum fill pressure, which is 250 psi. In order to protect the acetylene cylinder, a **fusible metal plug,** which melts at 212°F, is provided either on the tank or on the valve. The plug is designed so that if an acetylene tank is exposed to a fire, the gas will be released slowly through the melted plug. This avoids the possibility of a

ruptured tank and a release of the contents all at once.

An acetylene tank that has reached 25 psi should be considered empty, because if all the pressure is released, it becomes more difficult for your supplier to refill the tank.

The acetylene tank valve may be equipped with a handwheel or it may be operated with a tank valve wrench. When the valve is opened, it should be turned no more than $\frac{1}{4}$ to $\frac{1}{2}$ turn. This is to assure that in an emergency, you will be able to quickly close the acetylene tank valve. It is also for this reason that on valves operated by a wrench, the wrench must be left in place while the valve is in the open position. Acetylene tanks may hold, in dissolved form, 390 to 10 cu ft.

In order to reduce the tank pressure down to a usable level, a regulator is screwed into the tank valve on each of the cylinders. These regulators cannot be inadvertently placed on the wrong tanks, since the oxygen regulator has a right-hand female inlet, while the acetylene regulator has a left-hand male thread on its inlet. **Left-hand threads** can usually be identified by the notches on the nut, as shown in Figure 12-25. Left-hand threads are common on other fuel gasses as well.

A simplified cutaway of a **regulator** is shown in Figure 12-26. There are two pressure gauges. One shows the pressure coming to the regulator from the tank and is called tank pressure. The

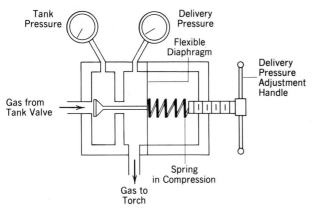

Figure 12-26 Gas regulator.

other shows the pressure that is available for the torch and is called delivery pressure. The available delivery pressure is determined by the spring pressure on the flexible diaphragm, which is set with the pressure-adjusting screw. When the gas is being used, the pressure in the delivery chamber tends to be reduced, which causes the diaphragm to move towards the left. The valve then opens, preventing the delivery pressure from falling any further.

The gas is delivered from the oxygen and acetylene regulators to the torches through two color-coded hoses, red for the acetylene (also with left-hand thread) and green for the oxygen.

For brazing small-diameter tubing the acety-

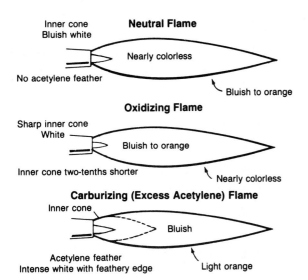

Figure 12-27 Various mixtures of oxygen and acetylene will produce different flames. (*Courtesy J. W. Harris Co., Inc.*)

lene and oxygen pressures are set at about 7 psi. For larger jobs, try 10 psi. Once you become proficient, you may try other pressures, but these are good starting points.

Once the gas pressures are set on the regulators, the acetylene valve on the torch is opened. The acetylene is then lighted with a friction lighter. The acetylene flow is reduced at the torch valve, until the yellow flame just begins to show some smoke. Then oxygen is added by gradually opening the oxygen torch valve. As the flow of oxygen is increased, the size of the flame becomes shorter. When the outer cone is larger than the inner cone (Figure 12-27), we have a **carburizing flame** (excess acetylene). Adding more oxygen causes the outer cone to match the inner cone (**neutral flame).** As still more oxygen is added, the flame becomes very short and hard. This is an **oxidizing flame** (excess oxygen). For silver brazing, use a carburizing flame, with the outer cone twice as long as the inner cone. The neutral and the oxidizing flames are used for welding and cutting, respectively.

When shutting down the oxyacetylene torches, follow the following sequence.

1. Shut off the oxygen torch valve first, and then the acetylene torch valve. Some service technicians prefer the reverse order, causing a loud pop when the acetylene is shut. This practice is less safe than shutting the oxygen first.

2. Shut down the acetylene and oxygen tank valves.

3. Open the acetylene torch valve, bleeding all the acetylene pressure from the regulator and hoses. Repeat for the oxygen.

4. Screw each of the regulator screws counterclockwise until you feel all the spring tension released. If this is not done, the spring pressure on the diaphragm during long periods of nonuse will cause premature failure of the diaphragm.

5. Rerack the hoses onto the dolly to protect them from damage.

Troubleshooting

If the filler metal melts but doesn't flow into the joint, it may be due to one of the following causes.

1. The outside of the joint may be hot, but not the inside. Heat the tube first, and then the fitting (or swaged end).
2. The joint may be overheated, causing the flux to vaporize. Try using a softer flame (less oxygen), and apply the filler as soon as the base metal is hot enough.

If the filler metal melts but doesn't wet the surface of the base metal, check for the following.

1. Make sure that the filler metal is not being heated by the torch. Apply heat to the base metal, and let the base metal melt the filler.
2. Base metals may not have been sufficiently clean.

If the filler metal cracks after it solidifies, it could be for one of these four reasons.

1. If the joint is copper-to-steel, cracking may result from different rates of expansion. Make the assembly so that the steel tube is inside the copper tube.
2. Brazing a steel joint with any alloy containing phosphorus can lead to cracking.
3. Excessive joint clearance will cause cracking under vibration.
4. Cooling the joint too quickly can cause cracking. Let the joint cool naturally in air before washing off the flux residue.

Safety Precautions for Oxy-Acetylene

1. Use no oil or grease around the oxy-acetylene equipment. It can burn violently in the presence of oxygen. The regulators require no lubrication.
2. Do not use oxygen as a substitute for compressed air.
3. Never use acetylene at a pressure higher than 15 psig. It becomes unstable.
4. Never use equipment that is in need of repair. Need for repair is indicated by a regulator that will not hold a set pressure, a torch valve that will not shut off tight, or a connection or hose that leaks.
5. Wear goggles when using the torch.
6. Before starting work, check the area for potential fire hazards that could be ignited by the torch flame.
7. Work only in areas with adequate ventilation.
8. Protect cylinders from damage by keeping them chained to a solid structure or to a stable cart.
9. Open tank valves slowly, never standing directly in front of the regulator.
10. Always leave the wrench in place on the acetylene tank, and do not open the acetylene tank valve more than one-quarter of a turn.

KEY TERMS AND CONCEPTS

AC&R Tubing	Oxyacetylene Torch	Ferrous Materials
Soft Tube	Capillary Action	Tank Valves
Rigid Tube	Removing Oxidation	Acetone
Annealing	Soft Solder	Fusible Metal Plug
Work Hardening	Sil-Fos	Left-hand Threads
Unrolling Copper	Silver Solder	Regulator
Type K, L, M Copper	Easy-Flo	Carburizing Flame
Soldered Connections	Fluxing	Neutral Flame
Brazed Connections	Nonfluxed Joints	Oxidizing Flame
Mechanical Connections	Making the Soldered Joint	Oxy-Acetylene Safety
Propane Torch	Internal Oxidation	

QUESTIONS

1. You want to buy a piece of tubing that is ¾- in. outside diameter. What size tubing would you ask for in a refrigeration supply house? What size would you ask for in a hardware or building supply store?

2. What are the properties of annealed copper?

3. What is work hardening?

4. How can a roll of copper tubing be unrolled in a workmanlike manner?

5. What physical characteristic of copper is described by type K, L, and M?

6. What is the difference between soldering and brazing?

7. What is the difference between brazing and welding?

8. What is a common melting temperature for soft solders, such as those used in plumbing or drain lines?

9. What must you do to a joint to which the solder won't "stick"?

10. What is capillary action?

11. What happens to cleaned copper when you touch it with your hand?

12. Name two common compositions for soft solder.

13. What are the major ingredients in Sil-Fos?

14. Which of these ingredients are not found in Easy-Flo?

15. When can you make a soldered joint without using any flux?

16. What is a common melting temperature for Sil-Fos?

17. What type of filler material should be used to join a copper hot gas line to a steel condenser?

18. What type of filler material should be used to join a copper refrigerant liquid line to a brass liquid solenoid valve? (Cheapest acceptable method.)

19. What is the function of flux?

20. How does the application of heat affect where the molten filler material will flow?

21. Why is it necessary to make brazed joints as quickly as possible?

22. There are two gauges on an oxygen regulator. What does each indicate?

23. What is the maximum allowable (red line) pressure for acetylene?

24. How far should the acetylene tank valve be opened?

25. How far should the oxygen tank valve be opened?

26. What is the pressure in a full acetylene tank?

27. What is the pressure in a full oxygen tank?

28. What color is the oxygen hose? the acetylene hose? Which has left-hand threads?

CHAPTER 13

SERVICE PROCEDURES

This chapter describes many of the specific service operations that must be performed routinely on systems. They are applicable to both air conditioning and refrigeration systems, regardless of the specific application.

Check Compressor Windings

When a compressor does not run, it is most likely because of one of the following problems.

1. The compressor motor is burned out.
2. The compressor is mechanically stuck.

3. There is no voltage to the compressor.

In order to check the electrical integrity of the compressor motor windings, disconnect all wiring from the three compressor terminals. Using an ohmmeter set to the $R \times 1$ scale, measure the resistance between each two pairs (Figure 13-1). If the power source is three-phase, all three readings on the three compressor windings should be equal. If the compressor operates on single-phase power (115 or 230V), there should be one resistance reading that equals the sum of the other two. Most resistance readings will fall within the range of 1 to 20 ohms(Ω). A compressor with a failed motor will often have one or more of the

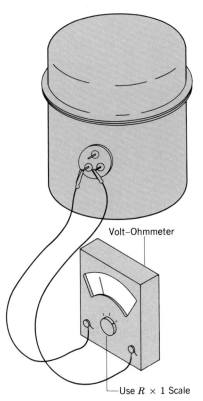

Figure 13-1 Reading the resistance of compressor motor windings.

Volt–Ohmmeter

Use $R \times 1$ Scale

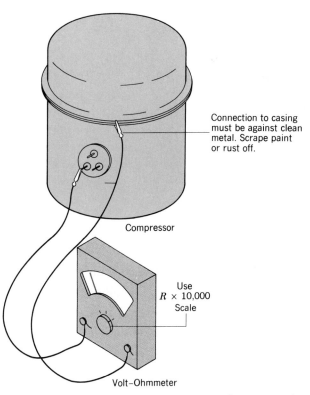

Connection to casing must be against clean metal. Scrape paint or rust off.

Compressor

Use $R \times 10,000$ Scale

Volt–Ohmmeter

Figure 13-2 Checking the compressor for a ground to casing. If the ohmmeter needle moves at all, the compressor is grounded.

readings equal to zero (winding is **shorted**) or infinite resistance (winding is **open**).

The compressor winding can also be **grounded** to the casing (see Figure 13-2). For this test, set the ohmmeter on the $R \times 10,000$ scale (or the highest resistance scale available on your ohmmeter), and check the resistance between each terminal and the casing. To make sure the probe on the casing is touching bare metal, you may need to scrape away some paint. The resistance reading should be infinity. If there is any movement at all on the ohmmeter, there is some continuity to ground, and the compressor motor should be considered unserviceable.

If the compressor motor windings are not shorted, not open, and not grounded, then electrically the motor is all right.

Identifying Compressor Pins

Single-phase compressors have two motor windings, the start winding and the run winding. These two windings are wired together, in series, inside the compressor shell. Three wires are attached to these windings, one at each end, and one at the common junction of the two windings. These three wires are routed through the compressor shell, usually terminating at three pins. When any sort of wiring is done, you will need to be able to identify which of these pins is attached to the free end of the start winding, which is attached to the free end of the run winding, and which is attached to the common junction of the two windings. The pins are called the **start, run, and common pins.** The key fact you need to know is that the start winding always has a higher resistance than the run winding. Using an ohmmeter set on the $R \times 1$ scale, there are three different resistance readings you can measure (Figure 13-3). They will all be different, and the lowest two should add together to match the value of the third reading. The following procedure will enable you to determine which pin is which.

a. Find the pair of pins that give you the highest resistance reading. The pin that is not being used for this measurement is the common pin.

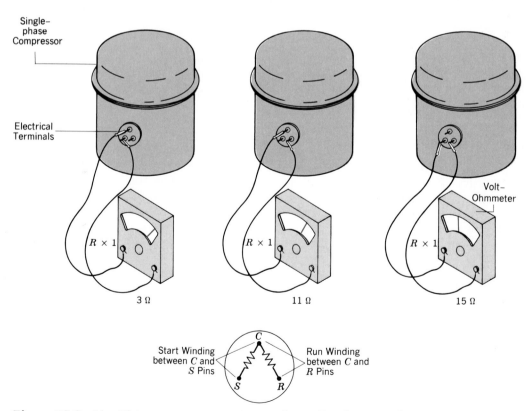

Figure 13-3 Identifying compressor motor windings. (Readings are for a single-phase compressor.)

b. Find the pair of pins that give you the lowest resistance reading. The one that is not the common pin is the run pin.

c. The remaining pin is the start pin.

Example

You have measured the following resistances between pins W, X, and Y. W to X is 7 Ω; X to Y is 9 Ω; and W to Y is 2 Ω. Which of these pins is common, which is start, and which is run?

Solution

The highest reading of 9 ohms was obtained between pins X and Y, so W is the common pin. The lowest reading was obtained between pins W and Y. Since we have already determined that W is the common pin, the run pin must be Y. The remaining pin, X, is the start pin.

There is another clue that will often give you a hint of which pin will be common, which start, and which run. The most commonly used color scheme for the wiring attached to the compressor is black for common, red for run, and white for start.

Evacuating the System

Whenever a system has been opened to atmosphere, air and water vapor have been allowed to enter. A vacuum pump is used to remove these contaminants. With the compressor not running, the pump is attached to the system as shown in Figure 13-4. If there are pressure taps on both the high and low side, they should both be used. If the system has been pressurized for leak testing prior to **evacuation,** all pressure must first be bled off. If the vacuum pump is attached to a pressurized system, the oil in the pump may be blown out.

Turn on the vacuum pump. You should see an immediate drop on the low-side gauge. As the pump continues to run, the pressure should continue to drop. For small systems such as refrigerators, the pressure should drop to 25 in. Hg within a minute or two, and to 29 or 30 in. Hg within 5 min. For very large systems, it may take an hour or more tō reach a deep vacuum. The

pump is allowed to continue to run, even after 30 in. Hg has been attained. For small systems, evacuate for at least 15 min, and for a residential split system, 40 min. For a large water chiller (several hundred tons), it is not uncommon to leave the vacuum pump operating overnight.

When evacuation is complete, close the manifold gauge valves before turning off the vacuum pump. Otherwise, the vacuum inside the system will draw air in through the pump.

If the pump cannot pull the system down to 29–30 in. Hg, either the system has a leak or the pump is not operating properly. To check for proper pump operation, attach the pump directly to the manifold as shown in Figure 13-5. When

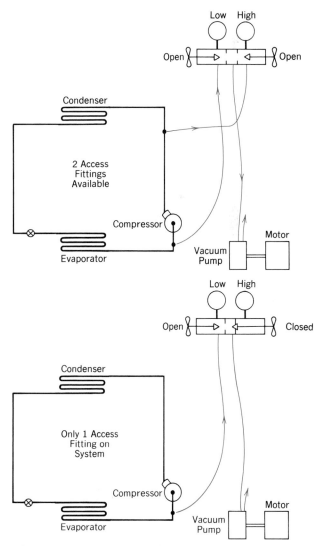

Figure 13-4 Evacuation of the system. The compressor is off.

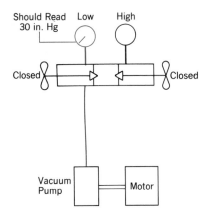

Figure 13-5 When the pump is turned on, the low-side gauge should immediately drop to 30 in. Hg.

the pump is turned on, the gauge should immediately go to 29–30 in. Hg (assuming correct calibration of the gauge). If it does not, change the oil in the vacuum pump using vacuum pump oil specifically designed for this service.

The evacuation process may be speeded by applying heat to the system in the form of an electric heater or a heat lamp. Do not use a torch for this purpose, because it may cause decomposition of the refrigerant oil. In the absense of an external source of heat, the compressor may be run for a short period (1 min. or less). This will provide the heat from the motor to the system. JUST BE CAREFUL NOT TO RUN THE COMPRESSOR TOO LONG AS IT WILL BE OPERATING WITHOUT THE BENEFIT OF COOLING FROM THE REFRIGERANT.

Checking for Leaks after Reassembly

Whenever a system has been opened and reassembled, the work that you did in reassembling the unit must be leak tested. Before you evacuate and recharge the system, pressurize the system with the compressor not running. Ten to fifteen psi of pressure should be sufficient. Use the refrigerant that you will use when charging the system. If the system is large, you may want to use nitrogen or compressed air in order to save the expense of the refrigerant that will be bled off to atmosphere.

Using a soap and water solution, carefully check each connection. An effective way to test

the joint, is to hold the dauber on the pipe to form a little puddle, then move the puddle slowly around the joint. Work slowly, carefully, and thoroughly. A couple of minutes saved in haste can cost a callback and a complete new refrigerant charge. If there are places that you cannot see, use a mirror placed behind the connection.

Leaking mechanical connections (flare or compression fittings) may be tightened with the pressure in the system. Leaking brazed connections require that the pressure be bled from the system and the joint repaired. Following the repair, repeat the leak test until you have found that each of the joints you have made is leaktight. If there had been no indication of refrigerant leakage prior to your servicing the unit, the rest of the connections in the system are presumed to be all right.

Checking for Leaks: Pressure Method

The leak-check method described previously will always work, assuming that the leaking joint is accessible and that every suspicious point in the system is checked. Unfortunately, this is not always possible. In order to prove that the entire system is leaktight without checking every inch of it, we can use either a pressure test or a vacuum test.

In order to run the pressure test, the entire system is pressurized to 20 to 30 psi with refrigerant, air, or nitrogen. The system is allowed to sit undisturbed for 30 min to 1 hour, depending on the physical size of the system, with the manifold gauges attached. There should be zero loss in pressure noted on the manifold gauges. Even the slightest drop in pressure indicates a leak that will, over a period of time, render the system inoperative.

There is one trap to avoid in running the pressure test. If there is only a low-side pressure fitting, then only the low side may be pressurized from the refrigerant can. Even if there are no leaks in the system, it may appear that you have a leak. You must therefore allow the pressures to equalize before you can be sure of leaks. The pressure in the low side will leak into the high side through the compressor valves and the metering device. If it is a cap tube system, this may not take very long. If it is a TXV system, it may take an hour or more until the high- and low-side pressures have equalized.

If there are both high- and low-side access fittings, turn off the valve on the refrigerant can after the system is pressurized, and open both valves on the manifold gauges for a few seconds. This will allow any difference between the high- and low-side pressures to equalize through the manifold before starting the test. Then, any drop in pressure will indicate a leak to atmosphere, and not an internal leakage from one section of the system to the other.

The pressure test is useful to prove that your system is leaktight, but it is of no value in helping you pinpoint the location of any leaks.

Checking for Leaks: Vacuum Method

The vacuum method of checking for leaks is similar to the pressure method. Instead of pressurizing the system to higher than atmospheric pressure, we pull a vacuum of 30 in. Hg. Any leakage of air into the system will cause a loss in vacuum. The maximum pressure difference we can create with this method is 14.7 psi between atmosphere and the system. With the vacuum test, however, we have a tool available to us that is far more sensitive than the manifold gauges in sensing a change in pressure (vacuum). Figure 13-6 shows the hookup for a **micron gauge.** A micron is a unit of length equal to a millionth of a meter. The micron gauge spreads the vacuum scale between 29 in. Hg and 30 in. Hg into 25,400 microns, so it can easily detect even the smallest change in vacuum because of leakage.

With the micron test, you can be fooled into thinking there is a leak if there is moisture in the system. Under the deep vacuum of the test, moisture that may be being held in a filter dryer or dissolved in the oil will vaporize, causing the same loss of vacuum as would be caused by a leak. Therefore, allow the micron gauge test to continue for 10 or 15 min to determine if the pressure increase stops (moisture), or if it continues its rise (indicating a leak).

As with the pressure method, this test can only tell you if the system is tight. If there is a

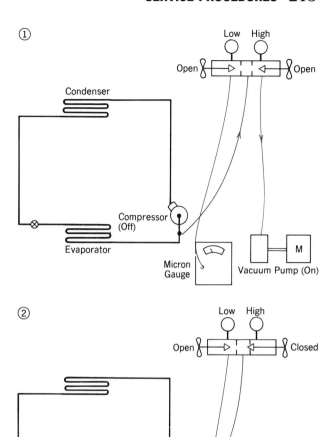

Figure 13-6 Method for evacuating the system and testing for leaks with a micron gauge. Note that the high-side valve must be closed before turning off the vacuum pump.

leak, other tests must be used to locate the source of the leak.

Finding Leaks in Operating Systems

When systems have operated for some time and lost enough refrigerant charge to cause a loss in cooling, the location of the leak must be pinpointed and repaired. The quickest and easiest method of finding the leak involves no tools at all. Simply look for an oily, dirty deposit near each of the connections. When refrigerant leaks out of a system, a small quantity of oil also leaks out. The refrigerant vaporizes into the atmosphere, but the oil remains behind, deposited on the outside of the tubing. Dust and dirt in the atmosphere stick to oil, and the telltale oily, dirty deposit will result. There are **refrigerant dyes** available, usually red in color, that may be added to a system. The dye is soluble in the refrigeration oil, and system leakage will then produce an oily, dirty, red deposit. Some service technicians believe that the dye makes the leak somewhat easier to spot.

When leaks cannot be easily located visually, either the **propane (halide) leak detector** or the **electronic leak detector** may be used. In either case, the system must be pressurized with refrigerant in order for the leak detector to sense leakage. After turning off the unit, search the system methodically, inch by inch, connection by connection. The refrigerant is heavier than air, so search underneath each component. For evaporators located inside closed boxes, allow the door to remain closed to allow any leaking refrigerant to collect in the box. Then open the door and insert the leak detector probe into the bottom of the box. The most common source of the leakage will be at the joints, soldered, brazed, flared, or other. Other possibilities are aluminum condensers or evaporators that have been subjected to mechanical damage or a corrosive environment. Fertilizers and the salt air near the ocean can be especially damaging to aluminum condensers. Cast brass fittings have been known to leak because of a casting imperfection. Occasionally, a leak is detected where the electrical connections pass through the shell of a hermetic compressor. Copper tubing is normally pretty reliable and leak free, except where it has been routed so that two pieces touch each other. Then vibration, created when the system runs, can cause a hole to wear into one or both tubes.

If no leaks are found, do not overlook the possibility that the caps on the access fittings to which your manifold gauges are now attached were not tight.

Sometimes, you just cannot find the leak in a reasonable period of time. This leaves you with several choices that should be discussed with the customer. If the leak is very slight, it may be cost effective to simply let it go, and add a slight amount of refrigerant each time the unit is regularly serviced. (This is a good opportunity to recommend the need for regular preventative

maintenance, aside from the issue of the refrigerant leak.) A second choice is to advise the customer of the status of the system and to get authorization to spend additional time searching. If you continue your search, there are additional measures available to you to make the leak easier to find. They involve increasing the pressure inside the unit in order to increase the rate of leakage.

a. Search the high side of the system again, but this time with the compressor in operation.

b. If step (a) yields no results, a high head pressure condition may be produced by partially disabling the condenser. Airflow through the condenser may be partially blocked, a panel may be opened to cause condenser air to bypass the coil (Figure 13-7), or the unit may be overcharged. USE EXTREME CAUTION TO MAKE SURE THAT YOU DO NOT GET WITHIN 10 PERCENT OF THE MAXIMUM ALLOWABLE PRESSURE RATING OF THE COMPONENTS. You, or a helper, must moni-

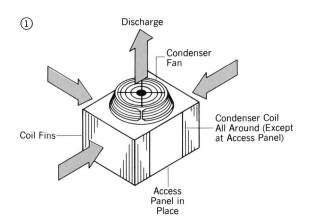

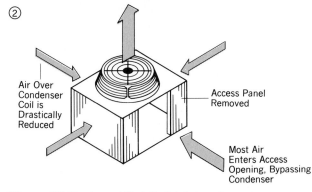

Figure 13-7   An artificially high head pressure can be created by allowing condenser air to bypass the condenser coil through an open access panel.

tor the pressures closely, and be ready to pull the disconnect switch should the pressure approach to within 10 percent of the maximum allowable working pressure.

c. If the actual leak is in the low side, neither step (a) nor (b) will produce results. With the system off, the low side must be pressurized with nitrogen to not higher than 10 percent below its maximum allowable working pressure. Following this test, the entire system must be evacuated and recharged.

Occasionally, these pressure methods will result in causing leakage in components that were already corroded and had thin walls. For all practical purposes, these components were failed before your test, and should be replaced.

Finding Leaks in Ammonia Systems

The pressure and vacuum methods described previously will work for all systems, regardless of the refrigerant being used. The electronic and halide leak detectors, however, depend on the chemical composition of the fluorocarbon refrigerants. Therefore, they will not work with systems using **ammonia** as the refrigerant. But there are two other leak-detection methods available for use on ammonia systems. The sulfur

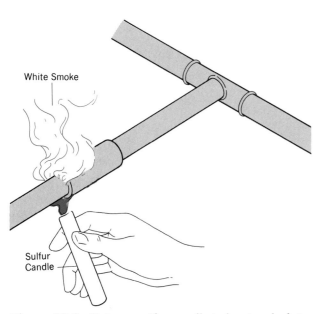

Figure 13-8 Using a sulfur candle to locate a leak in an ammonia system.

candle (Figure 13-8) is used around suspected leak areas. The presence of ammonia will react with the smoke from the candle, creating a white smoke, thus making the leak visible. Also a specially treated paper called litmus paper may be used to detect an ammonia leak. In the presence of ammonia, the color of the litmus paper will change.

Pumping the System Down

Pumpdown is a term that should not be confused with system evacuation. Pumpdown is a service technique that causes all the refrigerant to be stored in the receiver and/or condenser (Figure 13-9). This allows components in the low side of the system (actually between the receiver and the

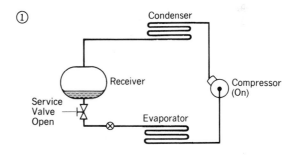
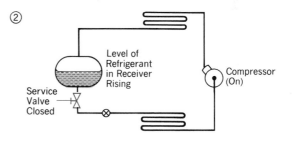
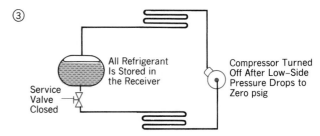

Figure 13-9 Sequence used to store the system refrigerant in the receiver (manual pumpdown).

compressor discharge valve) to be removed and replaced without losing the entire refrigerant charge.

In order to pump down the refrigeration system, there must be a service valve located in the liquid line leaving the receiver. If there is no receiver, a valve at the outlet of the condenser can be used.

With your gauges attached to both the high and low sides and the system running, close the liquid line service valve. If there is no receiver, watch the high-side pressure to make sure it remains below the maximum allowable working pressure. As the system runs, the low-side pressure will drop. Any liquid in the low side will flash and be pumped into the high side where it will be condensed. When the pressure in the low side drops to zero psig, pull the disconnect switch to stop the compressor discharge. The refrigerant will now be trapped in the high side. It will not flow backwards through the compressor, as the discharge valves on each cylinder will act as check valves. If there is a service valve on the compressor discharge, it may also be closed (front seated) to eliminate leakage through the compressor.

When the work has been completed on the low side, it should be evacuated to remove all air and moisture. Then the liquid line service valve may be reopened, and the system returned to normal operation.

If the low side has not been left opened for a long period, it may be suitable to simply **purge** the air out using some of the refrigerant that is stored in the receiver. The liquid line service valve is cracked open in order to pressurize the low side. Then, with the liquid line closed, an access fitting in the low side may be opened, bleeding out a mixture of air and refrigerant. Repeat this process for a total of **three evacuations,** and the system should be sufficiently purged.

This pumpdown sequence is done automatically **(automatic pumpdown)** upon each shutdown by some systems. First, the thermostat closes a solenoid valve in the liquid line. Then a pressure controller senses low-side pressure, and turns off the compressor automatically when the low-side pressure drops to zero psig.

Charging the System with Refrigerant

Refrigerant may be charged into an empty system that has just been assembled or reassembled, or it may just be necessary to add some refrigerant to an operating system. In either case, the conventional method is to attach manifold gauges to the high- and low-side fittings and connect the refrigerant cylinder to the center hose (Figures 13-10 and 13-11). Starting with all valves closed, the following sequence is used to charge a system that has **Shrader-type fittings** on both the high and low side.

a. Open the valve on the refrigerant cylinder. This will pressurize only the center hose.

b. Loosen the center hose at the manifold gauges. Allow refrigerant from the cylinder to escape for a couple of seconds. This will push all the air out of the center hose. Retighten the center hose.

c. If the system is empty, it was evacuated prior to charging, and the low-side hose is still under a vacuum. If the system has refrigerant, loosen the low-side hose at the manifold gauge and allow refrigerant from the system to escape for a couple of seconds. This will push all the air out of the low-side hose.

d. If the system contains some refrigerant, start the compressor. The pressure in the low side will drop below the pressure of the refrigerant

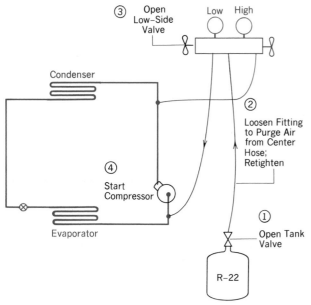

Figure 13-10 Sequence for adding refrigerant to a system that has been evacuated.

in the cylinder. If the system is empty, open the low-side valve on the manifold prior to starting the compressor.

e. Open the low-side valve on the manifold gauge, allowing refrigerant vapor to be drawn from the cylinder into the system. Use the low-side valve to control the rate of flow of refrigerant into the system. While charging, you should not allow the suction pressure to exceed the normal operating pressure. To do so may overload the compressor motor.

If the system has service valves instead of Shrader fittings, the service valves should be cracked off backseat after step (b). If the system does not have any access fitting on the high side, then only the low side is used.

As refrigerant is charged from the cylinder into the system, you will note that the cylinder gets cold. The pressure in the can is being bled off, and that reduction in pressure causes some of the liquid to flash, lowering the can temperature. The problem is that after some time, the pressure in the can becomes so low that it approaches the suction pressure in the system. At that point, the charging process becomes very slow. In order to speed up the process, heat may be added to the can in order to increase the can pressure. Electric heaters work well, or the heat from the condenser may be quite convenient to use.

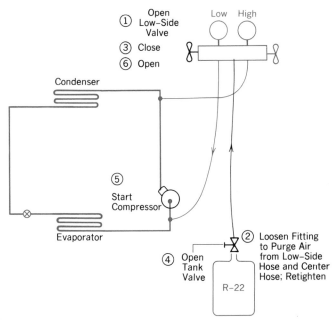

Figure 13-11 Sequence for adding refrigerant to a system that is low on charge.

To speed up the charging process, some technicians turn the refrigerant cylinder upside down so that it delivers liquid instead of vapor refrigerant. This can be a timesaver, or it can ruin the compressor, depending on your technique. If the liquid is fed into the system too fast, it may become entrained into the flow that enters the compressor cylinder. As the liquid is not compressible, this can cause tremendous pressures to build up at the top of the piston up stroke. The result can be a broken cylinder head, piston, piston rod, piston pin, or crankshaft.

In order to prevent ruining the compressor, the liquid must be charged into the system *very slowly*. There are fittings available that may be attached to the low side of your manifolds that provide a small orifice designed to prevent charging liquid at too fast a rate. Charging of liquid into the low side should not be attempted by the novice. The experienced technician who has developed a sense of the correct charging rate should also charge vapor first, until the cylinder pressure has been significantly reduced.

Determining the Correct Charge

Knowing how to add refrigerant to a system is not enough. You must know how much refrigerant to add. Not enough refrigerant will cause the system to produce lower than its rated capacity, may cause an evaporator that normally operates at a temperature warmer than 32°F to freeze up (on the outside surface), and could provide insufficient cooling to the motor of hermetic and semihermetic compressors. Too much refrigerant charge can also cause a loss in system capacity, or it could cause liquid slugging of the compressor in cap tube systems, high head pressure, and high operating cost. There are several ways of determining the correct charge. Sometimes you will use two or more methods at the same time in order to check yourself.

Charging by Weight

The most accurate method of charging, **charging by weight,** is used on small capillary tube systems, such as household refrigerators. Figure

13-12 shows a charging cylinder hooked up to the manifold gauges. The charging cylinder is calibrated in ounces of refrigerant. There are different scales, depending on the type of refrigerant being used and the pressure of the refrigerant. The unit being charged must state the correct number of ounces required in order to use the charging cylinder. Starting with the refrigeration system evacuated, you can dispense the exact weight of refrigerant required. This method has the advantage of being accurate and fast. You do not need to spend time allowing the system to get down to temperature in order to confirm that the charge is correct.

For larger units, the charging cylinder may be replaced by a scale that weighs the standard refrigeration tank. Simply note the weight of refrigerant and the cylinder before and after charging. The difference will be the weight of refrigerant charged into the system.

The weight charging method may be used only when there is no initial charge of refrigerant in the unit. For small systems, it is practical to release the total charge of refrigerant so that the charging cylinder can be used. In the case of larger systems, releasing a significant quantity of refrigerant is not cost effective, and a different charging method should be used.

Charging to Pressures

When a system does not have sufficient charge, both the suction pressure and the head pressure will be lower than normal. As refrigerant is added, both pressures will increase until they are correct. If the service technician knows the correct operating pressures, it will be easy to charge until those pressures are reached.

Figure 13-13 shows charging information that may be provided with packaged air conditioning systems. Note that the correct operating pressures depend upon the outdoor air temperature for air-cooled units. The same information may also be provided in tabular form. This information is furnished by manufacturers either in the installation instructions or, conveniently for the service technician, on the unit inside one of the removable service panels.

The specific pressures required for each unit

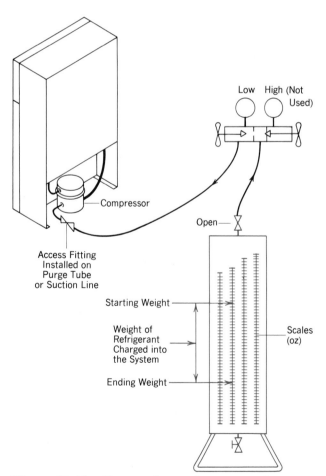

Figure 13-12 Charging by weight, using a charging cylinder.

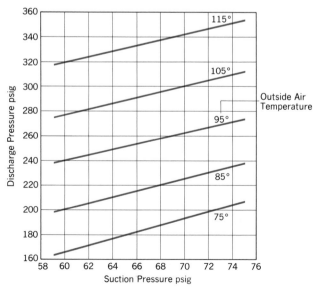

Figure 13-13 Sample operating pressures for charging a specific R-22 air conditioner.

may not be readily available, however. Figure 13-14 gives some typical operating pressure ranges for various types of equipment. Using these pressures as guides will get you reasonably close to the correct charge. They should be used in conjunction with other indicators of the correct charge whenever possible.

Charging to Superheat or Frostline

When the system is low on refrigerant, the liquid that enters the evaporator becomes completely vaporized too soon. There is insufficient liquid available to last to the end of the evaporator, so only the beginning portion of the evaporator gets to the saturation temperature. Once the liquid refrigerant is completely vaporized, the remaining portion of the evaporator simply superheats the refrigerant.

As the system nears the correct charge, the superheat will begin to drop rapidly. An accurate way to charge a system is to take a pressure and actual temperature reading at the compressor suction, and continue to add refrigerant until the superheat has dropped to the correct value. Many air conditioning and refrigeration applications will be quite happy with a superheat of between

7°F and 14°F. You can charge air conditioning systems to an approximately correct level simply by holding your hand on the suction line. With a suction pressure of, say, 68 psi (40°F), when the correct charge is reached the suction line temperature will be between 47°F and 55°F, which will feel cold to the touch. In most cases, **sweating** of the suction line will also occur at this temperature unless the surrounding air is quite dry.

When charging by holding your hand on the suction line, the temperature will not change much when the charge goes from zero to 90 percent of the correct charge. As the last 10 percent of the charge is added, the suction line temperature will change rapidly.

Household refrigerators are a notable exception. The suction line for this application should not be cold enough to sweat because we do not want to cause a puddle of water on the kitchen floor. The refrigerator may be charged until the suction line begins to get cold. Then a slight amount of refrigerant is bled off so that the suction line does not sweat.

When charging to a suction line temperature or superheat value, it is important that the manifold gauge pressures be checked. Otherwise, if there happened to be a problem in the system, you might overcharge and overpressurize the sys-

Type of Unit	Refrigerant	Air Temperature Entering Condenser	Suction Pressure	Discharge Pressure (psi)
Air conditioner	22	70°F	60–65 psi	170–200
		100°F	70–75 psi	260–300
Automobile	12	70°F	25–31 psi	150–250
		100°F	25–36 psi	230–330
Household refrigerator	12	70°F	2 in. Hg–2 psi	95–125
		100°F	0–4 psi	145–155
Ice maker	502	70°F	10–20 psi	180–215
		100°F	15–25 psi	280–320
35°F refrigerated box	12	70°F	15–20 psi	100–125
		100°F	20–25 psi	140–160

Figure 13-14 Estimates of correct operating pressures for some air-cooled applications where the manufacturer's data are not available.

tem while fruitlessly waiting for a cool suction line.

In systems where the evaporator is easily visible, service technicians watch the evaporator filling by noting the formation of frost on the evaporator surface. It is easy to see at what point the evaporator is beginning to get warm by simply noting at what point the frost is no longer being formed. This point is called the **frostline.** When the frostline has moved to the end of the evaporator, the correct charge has been attained.

Charging to the Sight Glass

When the system you are charging has been furnished with a sight glass, you are in luck. The sight glass provides an excellent guide to the correct amount of charge.

When the unit is undercharged, the high-side pressure is too low. The corresponding condensing temperature will also be low, and the condenser will not have sufficient capacity to condense all the hot gas to a liquid at this low temperature. The outlet from the condenser will then be a mixture of liquid and vapor. The sight glass provides a window into the liquid line, where the mixture of liquid and vapor will appear white and frothy. As the correct charge is approached, the appearance will change to mostly liquid, with a few vapor bubbles. When the correct charge has been attained, the sight glass will be absolutely clear. No movement of the refrigerant will be apparent.

Again, watch your pressures while charging to the sight glass. A system that contains air (noncondensible) may never show a clear sight glass even when the correct charge has been reached and passed.

Setting the Low Pressure Cut-out

Many refrigeration applications use a low pressure cutout as the primary operating control to cycle the compressor. The primary advantage in omitting the thermostat is eliminating the problem of location. For example, in order to control the temperature inside a bottle cooler, there is no

place where the thermostat can be easily located where it would not be subject to mechanical damage. A second advantage is that the expense of the thermostat is eliminated. The only disadvantage is that service technicians who set the pressures on the LPC incorrectly will be called back again and again until it is correct.

The LPC has one set of contacts that open or close to control the operation of the compressor. It has two pressure settings. When the pressure sensed by the LPC rises to the higher setting (called the cut-in pressure), the contacts will close and the compressor will start. When the pressure sensed by the LPC drops to the lower setting (called the cut-out pressure), the contacts open and the compressor will stop. The arithmetic difference between these two pressures is called the differential. Figure 13-15 gives the approximate correct pressure settings for many applications. These are not to be used as settings that you can set and then forget, but are starting points that you can use to arrive at the correct settings. The procedure to do this is described as follows.

a. Set the cut-in pressure higher than what you think will be the correct pressure. Set the cut-out at a very low set point. These settings should cause the compressor to run.

b. When the box has cooled to lower than the required temperature, raise the cut-out set point until the compressor turns off. Return the cut-out to a low set point.

c. With the compressor off, the box temperature (and evaporator pressure) will rise. When the box temperature rises to the maximum desired temperature, lower the cut-in set point slowly until the compressor starts.

d. After the compressor has run for 5–10 min, or after the box temperature falls to the minimum desired temperature (whichever happens sooner), raise the cut-in slowly until the compressor stops.

There are two different LPC arrangements available (Figure 13-16). One type provides one adjustment for the cut-in pressure, and a separate adjustment for the cut-out pressure. The second type provides two scales, with a label that says CUT OUT IS **CUT IN** MINUS **DIFFEREN-TIAL.** One scale is the cut-in pressure, and the other is the differential. The cut-out pressure

	R-12		R-22		R-502	
	Cut-in	Cut-out	Cut-in	Cut-out	Cut-in	Cut-out
Walk-in refrigerator	34	14	64	32	75	40
Reach-in refrigerator	36	19	68	40	78	40
Flower display case	42	28	77	55	89	65
Open dairy case	35	10	66	26	77	33
Closed freezer	8	1	22	11	29	16

Figure 13-15 Approximate LPC settings (psi).

must be calculated, and when you change either the cut-in pressure or the differential pressure, you will also be changing the cut-out pressure.

The graph in Figure 13-17 shows how the LPC will operate when it is properly set. When the compressor is off, the pressure in the low side of the system corresponds to the temperature of the box in which the evaporator is located. In our example, when the box temperature reaches 40°F, the pressure sensed by the LPC rises to the cut-in pressure of 37 psi. The compressor turns on, and the low-side pressure is drawn down to 22 psi within a few seconds. As the compressor runs, the box cools down slowly. As it cools, the load on the evaporator is reduced, and the low-side pressure gradually drops. After 10 min of operation, the box temperature drops to 37°F, and the low-side pressure drops to 19 psi. When the LPC senses that the low-side pressure has dropped to the cut-out pressure of 19 psi, its

Figure 13-16 Low pressure cut-outs.

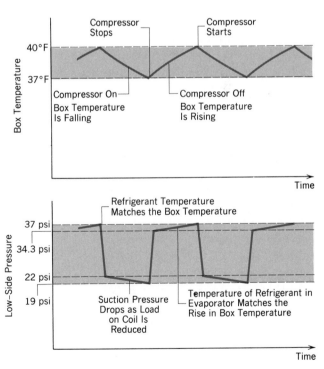

Figure 13-17 Variation in temperature and pressure for a refrigeration system operated by a low pressure cut-out.

contacts open and the compressor stops. Within seconds, the low-side pressure rises to 34.3 psi, which is the pressure corresponding to the box temperature of 37°F. The box slowly warms, and when it reaches 40°F, the cycle will be repeated again.

If the low pressure cut-out is not set properly, any of the following problems can result.

a. If the cut-in setting is too high, the box will be too warm.

b. If the cut-in setting is too low, the box will be too cold.

c. If the differential is too high (cut-out too low), the unit will experience a long on cycle, and then a long off cycle, resulting in possibly objectionable temperature swings in the box.

d. If the differential is too low (cut-out too high), the unit will cycle excessively, and may cause premature failure of the compressor. A cycle of 5 to 10 min on and an equal time off is a good compromise between temperature variation and minimizing the number of starts per hour by the compressor.

Checking Safety Pressure Controls

There are two levels of checking the operation of low and high pressure cut-outs that are used as safety controls (as opposed to an LPC used as an operating control). One method merely checks the operation of the switch contacts to confirm that when they open, the compressor will, in fact, shut off. The second method also checks to make sure that the switch will operate at the correct pressure.

The pressure settings of safety controls are outside the range of the normally encountered system pressures. That's why they're called safety controls. To make the switch operate, we can either change the pressure setting of the switch to a normal operating pressure, or we can cause the system pressure to become abnormal.

In the case of a high pressure cut-out, if the normal maximum system operating pressure is 260 psi, the HPC might be set for 290 psi. In no case should the setting of the HPC be higher than the maximum allowable working pressure

stamped on the unit nameplate. In order to check the switch contacts, the system should be operating. The set point of the HPC is then lowered until it reaches the actual operating pressure, at which point the switch should open, shutting down the compressor. The set point of the HPC is then returned to its normal setting. A low pressure cut-out switch would be checked in a similar fashion, except that the set point would have to be raised until it reached the normal operating low pressure.

In order to check that the pressure control actually operates at the intended pressure, there are more options to consider. Lets look at the HPC first. If the sensing location from the HPC is on the head of the compressor, and if the compressor has a discharge service valve, our job is easy. With the manifold gauges attached to the pressure ports on the service valves (Figure 13-18), the discharge valve is slowly turned toward the front seat position. Not much will happen until the front seat position is approached. Then, the compressor discharge pressure will begin to climb very rapidly. Watching your gauges carefully, note the pressure at which the HPC causes the compressor to cut out. Also, **BE PREPARED TO QUICKLY PULL THE DISCONNECT IF THE HPC FAILS TO OPERATE AT THE ANTICIPATED PRESSURE.**

The low pressure cut-out can be similarly tested at the compressor suction line service valve (Figure 13-19). If the switch fails to operate at the desired low pressure, however, it is no emergency. When the desired cut-out pressure is reached, simply increase the set point of the LPC until it does cause the compressor to shut down.

Sometimes the pressure switch sensing lines are not located so that an artificially high or low pressure can be created by simply turning a valve. In that case the high or low pressure condition can be simulated by other methods. On the high pressure side, high pressures can be created by doing one of the following.

a. Shut down one or more fans on an air-cooled condenser.

b. Block the airflow through an air-cooled condenser.

c. Open an access panel on an air-cooled condenser, allowing air to bypass the condenser coil.

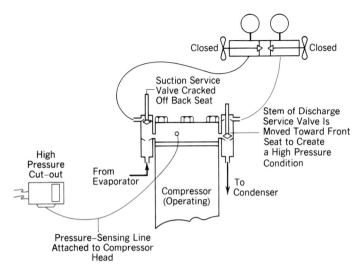

Figure 13-18 Checking the pressure set point of the high pressure cut-out.

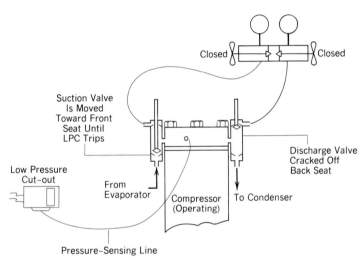

Figure 13-19 Checking the pressure set point of the low pressure safety cut-out.

d. Close down a water valve on a water-cooled condenser.

e. Readjust the condenser water regulating valve.

On the low pressure side, low pressure can be created by the following.

a. Shut down the evaporator fan motor.

b. Open an access panel, allowing the air into the evaporator fan to bypass the evaporator coil.

c. Restrict the flow across the evaporator coil by blocking the airflow across the filters.

d. In the case of water chillers, the flow of the water being cooled may be restricted. This re-quires extreme caution to assure that sufficient water flow is maintained to avoid freezing the water inside the evaporator, because that would cause major damage to the evaporator.

Adjusting the Thermostatic Expansion Valve

The thermostatic expansion valve (TXV) is relatively reliable, but it is blamed for problems far more often than is deserved. Before making any

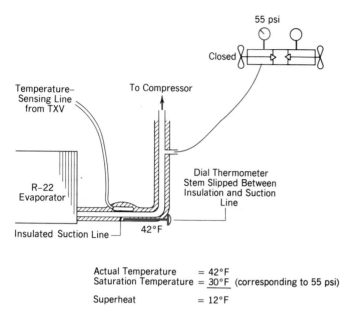

Actual Temperature   = 42°F
Saturation Temperature = 30°F (corresponding to 55 psi)

Superheat   = 12°F

Figure 13-20   Measuring the superheat entering the compressor.

adjustment to the expansion valve setting, use your gauges and a thermometer as in Figure 13-20 to determine the actual superheat being maintained. For example, if you expect a suction pressure of 65 psi on an air conditioner, and you measure 55 psi, you may suspect that the thermostatic expansion valve is not allowing a sufficient flow of refrigerant to pass. Using your thermometer, you measure a suction temperature of 42°F, which is 12°F superheat. You can now forget about the TXV. Its only job is to maintain a reasonable superheat, and it is doing that job just fine. You will have to look for the problem elsewhere.

If you measure a superheat that is too low, you may safely conclude that the TXV is overfeeding the evaporator. It may be badly adjusted, or the sensing bulb may have worked loose from the suction line. If the superheat is too high, the problem may be with the expansion valve, but there are lots of other potential culprits also. The actual superheat adjustment, once made during the initial set up of the system, rarely will get out of adjustment, so look elsewhere for mechanical failures first.

When setting an expansion valve, if the initial superheat reading is too high, the valve must be adjusted for increased flow (turn the adjustment screw counterclockwise). Make adjustments of no more than one-half of one turn. Then give the system 5 min to reach a new, stable operating condition before taking a new superheat reading. In the absence of other specifications, shoot for a superheat of 10°F to 15°F. A good setting will be one that causes the least fluctuation of suction pressure while the TXV is in control.

Adding Oil

A properly operating system will never require the addition or changing of oil. The only time oil must be added to a system is to replenish any that has been lost through leakage, replacement of components, or a catastrophic failure in which oil was blown out of the system. With semihermetic and open compressors, a sight glass is provided in the crankcase, which allows you to see the oil level (Figure 13-21). When the unit is operating, the oil level should be visible in the sight glass, preferably at the halfway level in the glass. Automotive compressors may use a dipstick to check the level against a specified number of inches. For other systems, there is no good way to read the oil level. The system may be completely drained of oil, and then refilled to the correct number of ounces as specified by the manufacturer.

When hermetic compressors are changed, the

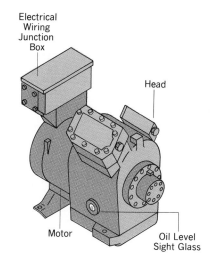

Figure 13-21 Sight glass on a compressor.

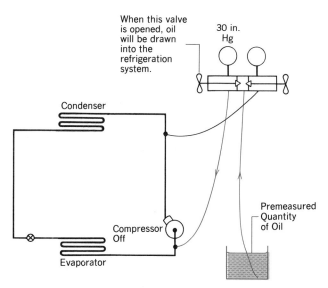

Figure 13-22 Adding oil to the refrigeration system, using the vacuum method.

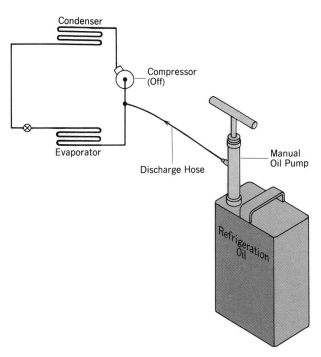

Figure 13-23 Using a hand pump to add oil to the refrigeration system.

new compressor is charged with oil. Simply changing the compressor will automatically change the oil, and provide the correct quantity.

For units that do need to have lost oil replaced, you may either use a vacuum in the system to pull in the oil, or use a pump to push oil into the system. Figure 13-22 shows a setup to use the vacuum method. The system has been evacuated, and the valves on the manifold gauges closed. A premeasured quantity of oil is placed in a container. The center hose from the manifold is then inserted into the container. As the suction valve on the manifold is opened, the system vac-

uum will draw the oil into the system. The system may then be reevacuated and charged with refrigerant.

Figure 13-23 shows the pressure method of adding oil. A hand-operated oil pump dips oil out of the container and delivers it into the suction line or the crankcase of the compressor, overcoming the pressure of any refrigerant in the system.

Starting a Stuck Compressor

Compressors may become mechanically **stuck** (sometimes called frozen, although it has nothing to do with temperature or heat), so that the electric motor on the compressor is unable to supply enough torque to get the compressor running. There are three methods that may be used to attempt to unfreeze a compressor.

a. First, 115-V single-phase compressors may be hooked up to a 230-V power source for a few seconds only.

b. A starting capacitor may be wired into the circuit. If the compressor already uses a start capacitor, it may be temporarily replaced with a larger capacitor.

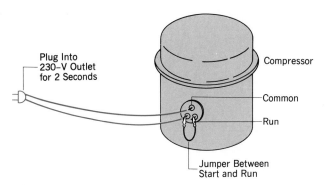

Figure 13-24 Starting a stuck 115-V compressor using 230 V.

c. A capacitor may be wired into the circuit of a single-phase compressor in a way to make the compressor run in the reverse direction.

The wiring schemes for these three methods are shown in Figures 13-24 through 13-26. A compressor analyzer may also be used to accomplish methods (b) and (c).

If the compressor has been stuck because of close manufacturing tolerances on a new compressor or a lubrication failure, the compressor will usually be all right after it is unstuck and the problem corrected. If it is an old compressor that has become stuck, it is probably getting ready to seize up and will in all likelihood become stuck again.

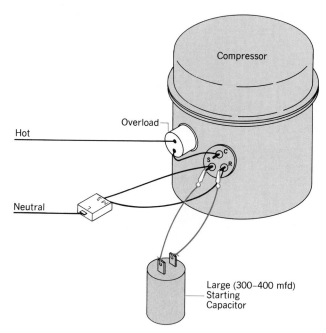

Figure 13-25 A larger capacitor wired between start and run. Energize the circuit for no longer than 2 seconds. This may break a stuck compressor loose.

There are **hard-start kits** available for air conditioning compressors that may be permanently wired into the circuit. These place a capacitor in the circuit, along with a switch that takes the capacitor out of the circuit a few seconds after the compressor starts.

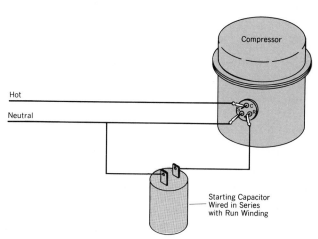

Figure 13-26 Breaking loose a stuck compressor by making it turn backwards. Energize the circuit for no longer than two seconds.

KEY TERMS AND CONCEPTS

Open Windings

Shorted Windings

Grounded Windings

Identifying
Compressor Pins

Common, Start, Run
Pins

Evacuation

Leak Checking

Micron Gauge

Refrigerant Dye

Electronic Leak
Detection

Ammonia Leaks

Triple Evacuation

Pumpdown

Purging

Automatic Pumpdown

Refrigerant Charging
by Weight

Refrigerant Charging
by Pressure

Refrigerant Charging
by Superheat

Refrigerant Charging
to Frostline

Shrader Fittings

Liquid Slugging

Sweating

Refrigerant Charging
to Sight Glass

LPC Setting and
Testing

HPC Setting and
Testing

Setting the TXV

Adding Refrigerant Oil

Stuck Compressors

Hard Start Kit

QUESTIONS

1. Which resistance scale on the ohmmeter should be used when checking compressor windings for continuity?

2. Which resistance scale should be used when identifying compressor pins?

3. Which resistance scale should be used when checking for grounded windings?

4. Three compressor pins (left, center, right) yield the following resistance readings: left to center, 11 Ω, left to right, 4 Ω, center to right, 15 ohms. Which pin is common? Which start? Which run?

5. You have found a system to be low on charge. You located and tightened a leaking flare fitting in the suction line. Under what circumstances would you consider evacuating the entire system prior to recharging?

6. In checking the tightened flare nut in Question 5, what method would you use to leak check? Would you do it with the compressor on or with it off? Why?

7. When reading normal system operating pressures, the manifold gauge valves are (open, closed, half open)?

8. When reading system operating pressures, the service valves are (front seated, back seated, cracked off front seat, cracked off back seat)?

9. Name five different indicators that you might look for while charging a system in order to conclude that you are approaching the correct amount of refrigerant charge.

10. While evacuating a small system, the pressure drops quickly to 26 in. Hg, and no further. What are two potential problems.

11. How can you detect leaks without the aid of any tools?

12. List all the steps you would perform to pump all the refrigerant from a system into the condenser.

13. On which types of systems is it most practical to charge using a charging cylinder?

14. In Figure 13-13, you have charged a unit to 65 psi on the low side, and 240 psi on the high side. The outside air temperature is 95°F. Is the unit still undercharged, overcharged, or just right?

15. How much superheat is there in the refrigerant entering the compressor of a refrigerator?

16. A low pressure cut-out is set for a cut-in of 40 psi and a differential 25 psi. What is the cut-out pressure?

17. The unit controlled by the LPC in Question 16 holds the box temperature correctly, but the unit cycles on for 1 min, and then off for 1 min. What new settings would you try in order to reduce cycling while maintaining the same maximum box temperature?

18. If the system in Questions 16 and 17 uses R-12 refrigerant, what would be the range of temperatures seen by a thermometer inside the refrigerated space?

19. In order to check the HPC switch, you would (raise, lower) the set point until the compressor shut down?

20. In order to check the pressure setting of the HPC, you would move the service valve toward the (back seat, front seat, midseat) position?

CHAPTER 14

MECHANICAL TROUBLESHOOTING

The purpose of this chapter is to enable you to recognize which mechanical malfunctions may be responsible for abnormal system operation. Troubleshooting of electrical devices and circuits is covered in Chapter 33.

What Is the Problem?

The best advice you can get on troubleshooting is to listen to your customer. Their description of the system behavior will very often lead the trained listener to the exact problem. Listen for how the problem arose. Was it a gradual deterioration or a sudden event? Did anything make a noise? Did it stop working while some new construction work was in progress? Has somebody else been working on the unit? If you are able to translate a layperson's description of the problem, you will increase your effectiveness in locating troubles.

The most common service complaint will be either insufficient or no cooling. With the system operating, you must take your operating pressures and compare them to what you would expect to be normal pressures. There are four basic combinations that you may see.

1. Suction and discharge pressures are both too low.
2. Suction and discharge pressures are both too high.
3. Suction is too low and discharge is too high.
4. Suction is too high and discharge is too low.

For the purpose of this chapter, we will assume that you are troubleshooting a system that, in the recent past, was doing the same job it cannot now handle. We will not be dealing with design problems or with undersized units that have never operated properly.

Suction and Discharge too Low

The most common culprit will be that the refrigeration system is **low on charge.** Check the sight glass. If there are bubbles, locate and repair the leak, and then recharge the system. If there is no sight glass in the system, add some refrigerant to the low side. After adding just a small quantity of refrigerant, you should notice that both the suction and discharge pressures will begin to come up to their normal values. If they do not, you will need to look for a different cause.

The second potential cause of low suction and discharge pressures is that there is insufficient load on the evaporator. The load on the evaporator may be lower than normal for one of several reasons.

1. The evaporator fan is slipping on the shaft. This can sometimes be heard. Tighten the set screw that holds the fan to the shaft.

2. The belt-drive evaporator fan is not rotating because the belt is broken. Replace the belt.

3. The belt-drive evaporator fan is rotating too slowly because the belt is slipping. Inspect the belt. If it appears to be in good condition, re-tension the belt. Usually a slipping belt will become glazed (shiny) on the sides and should be replaced.

4. The evaporator fan is not rotating because the motor is inoperative.

5. The airflow is restricted. There are lots of potential causes of this condition. For example, the air filters may be quite dirty. If the unit has been operated for any period of time without air filters in place, it is likely that the cooling coil is dirty, and the air passages between the fins are restricted. A sheet-metal panel that is not in place on the air conditioner can allow the evaporator fan air to

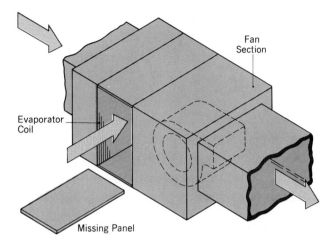

Figure 14-1 A missing sheet-metal panel can cause low airflow through the evaporator coil, causing low suction pressure.

bypass the evaporator coil (Figure 14-1). Occupants in a building who, at one time may have felt too cold may have adjusted dampers in ductwork or grilles to reduce airflow. In the realm of less common occurrences, ducts with fiberglass insulation on the inside may have become blocked when the insulation came loose. And, of course, there is the possibility of duct collapse on the return duct, or mechanical damage that has crushed the ductwork.

Suction too High, Discharge too High

The most common cause here is a condenser that is not rejecting sufficient heat. Possible malfunctions are as follows.

1. The condenser fan is not rotating properly because of slippage on the shaft, a slipping or broken belt, or an inoperative motor.
2. The path for air through an air-cooled condenser is blocked. Look for debris on the fins of the air inlet side of the coil, corrosion of the aluminum fins, or external barriers such as walls.
3. The water supply to a water-cooled condenser is inadequate. The temperature may be too high, or the flow rate may be insufficient. If the entering water temperature is all right, check the condenser water regulating valve.

4. The tubes on the water side of the water-cooled condenser are fouled. This would be indicated by a large temperature difference between the condensing temperature and the leaving water temperature, and not much temperature difference between the entering and leaving condenser water temperature.
5. Air in the system. Air is referred to as a **noncondensible** gas. That means that when it is cooled in the condenser, it remains a vapor. It takes up room in the condenser that should be available for refrigerant and reduces the condenser capacity. Air can only get into a system if there is a leak in the low side and the suction pressure drops lower than atmospheric. Or improper service techniques may have allowed air into the system. Find out if anyone else has serviced the unit before you conclude that air is the problem.

Another possible cause for the **high head pressure** condition is an overcharge of refrigerant or the wrong type of refrigerant. These are unlikely unless an unqualified service technician has preceded you.

Suction too Low, Discharge too High

There is a **restriction** to the refrigerant flow in the system—anywhere in the system. Restrictions may be caused by the following.

1. **Scale** that was formed inside the refrigerant tubing during brazing operations breaks loose and collects in the smallest refrigerant passages. This would normally be either at the filter dryer or at the metering device itself. Capillary tubes are especially prone to restrictions.
2. **Moisture** circulating in the system can freeze at the metering device, and cause an annoying restriction that comes and goes as the ice forms, blocks the system, melts, and then reforms. The moisture can be removed by evacuating the system and recharging. The tip-off to moisture in the system is the customer who says, "Sometimes it cools all right, but other times it runs but there is no cooling."
3. The oil in the system is breaking down because of high temperatures. A sludge forms

and can cause a restriction. An oil sample should be taken. If it is dark in color, the oil should be replaced.

4. Wax is precipitating out of the oil, and restricting the metering device. If this is because of extremely low suction temperatures, there is little that can be done without some design changes. The problem can be minimized by using a refrigerant oil with a low floc point.

5. Mechanical damage. Look for crushed or kinked refrigerant lines that are exposed to potential mechanical abuse.

6. The thermostatic expansion valve is closed. This can be because of a power element that has lost its charge. Even though the temperature leaving the evaporator is high, the bulb cannot exert a force on the diaphragm to open the valve.

Restrictions that occur in the liquid line (such as at the filter dryer) are easy to spot. The reduction in refrigerant pressure because of the restriction will cause the liquid refrigerant to flash. A temperature drop at the point of the restriction will become evident.

Systems that have a receiver will exhibit low suction pressure, but the head pressure may still appear to be normal.

Technicians who service household refrigerators like to troubleshoot using the low-side pressure only. A low suction pressure (below zero psig or the specified pressure for that system) can mean either a restriction or a low refrigerant charge. By adding a small quantity of R-12, you can determine which is the cause. If the suction pressure remains low after adding refrigerant, it is a restriction. If it comes up, it was a low refrigerant charge, and you will have to locate the leak.

Suction too High, Discharge too Low

There is an insufficient pressure differential being maintained between the high side and the low side. The high side is too low to allow all the hot gas to condense, and the low side is too high and too warm to do any effective cooling. There are two possible causes.

1. The metering device is malfunctioning, and it is open too far.

2. The compressor is losing its pumping capacity, and it cannot create the required pressure differential.

The metering device may be open too far because of a mechanical problem, or the sensing bulb may have come loose from the suction line, and is sensing a too-warm temperature.

The compressor may have lost its pumping ability because of wear in the pistons and cylinder walls or piston rings. The valves may also be seating incorrectly.

In order to determine which is the culprit, close a valve on the suction line and watch the compressor suction pressure. The compressor should be capable of pulling a vacuum pressure of 20 to 25 in. Hg. If it cannot pull that much vacuum, it has lost its ability to pump properly. A word of caution here before you condemn a perfectly good compressor. While most compressors are capable of pulling the vacuum stated, you may want to obtain specific performance from a manufacturer to confirm this performance for a suspected malfunctioning compressor.

Too Much Moisture in the Refrigerated Space

When air is cooled, moisture is removed. In closed boxes such as household refrigerators, the moisture that condenses on the coil is melted off by periodic defrost, and allowed to drain out of the box. There are two malfunctions that can result in too much moisture in the box, thus causing the contents to become wet.

1. The **drain line** from the evaporator to the pan below the box is plugged, and the condensate cannot drain out. The blockage may be from food, or it may be caused by a defective drain heater.

2. The **door seal** is not sealing sufficiently because it is either deteriorated or misaligned. The moisture flows into the box even when the door is closed, and the ice formation on the evaporator occurs more quickly than designed.

Noise

Noise is a common complaint. Sometimes the noise is normal, sometimes it is an indication of impending trouble, and sometimes it is merely an annoyance. One time-tested technique used as part of preventive maintenance programs in large plants with lots of mechanical equipment is to listen to each piece of equipment each day. If it sounds different than it did the day before, something is about to go wrong. Noises are not as easy to describe as, say, operating pressures. Therefore, we will describe some of the more common sources of noise.

1. Mechanical vibration is a common source of the noise. Often, the noise can be traced to a misplaced drain pan, a misaligned shelf, vibrating items in the box, or vibration of refrigerant tubing against sheet metal. Use your eyes and ears to get to the root of these problems.

   Figure 14-2 shows a common vibration source. Manufacturers name plates, which are mounted in the center of fan guards, frequently come loose and vibrate. Needle-nose pliers may be used to pierce the plate in several places, and turn additional tabs around the wire guard.

2. Airflow can cause noise. Air flowing through a loose damper, grille, or diffuser can cause a rattling noise. Air that flows through a too small opening attains a high velocity, and may cause a whistling type of noise. A strange rattling sort of noise can also be caused by a loose piece of paper that has found its way into the airflow and is flapping in the breeze. The source of the paper may even be operating instructions that have been left by the installer, or a schematic wiring diagram that has become unglued from an inside panel.

3. Fan noise can be caused by a fan that is loose on a shaft, or by fan bearings failing, or motor bearings failing. Check for noise with the fan not running (power disconnected) by rotating it by hand. This will allow you to check for fan blades hitting a shroud, a wire, or any other object.

4. Compressor noise has two common causes. One is serious, the other is not. Hermetic com-

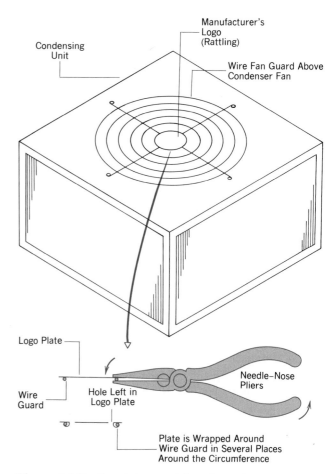

Figure 14-2   Repairing a rattling logo plate.

pressors have been known to start and/or stop with a loud clunk. This sounds much worse than it is. The usual problem is that one of the mounting springs that supports (and isolates) the compressor from the shell has weakened or become broken. When the compressor starts or stops, the sudden change in torque causes the compressor to actually bang into the inside of the shell. Compressors may do this for their entire service life. The only remedy is to change the compressor.

5. The second common compressor noise is a valve noise. Noisy valves indicate impending compressor failure. Sometimes the noise will be apparent for only a few moments after start-up. The valves need to be changed, except for welded hermetic compressors, which must be replaced.

6. **Rain shields** are shrouds fastened to the shaft of a motor to exclude rain (see Figure 14-3).

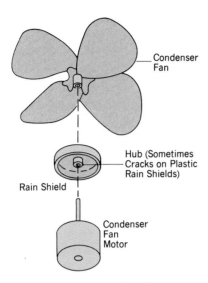

Figure 14-3 A cracked rain shield can cause a "chirping" noise.

When they are manufactured from plastic, they will frequently develop cracks in the hub. The noise from a rain shield is similar to a bird chirping. Fix it by either replacing the rain shield or by using a hose clamp around the hub to help it hold on to the shaft tightly.

For complex equipment such as large water chillers with a large number of rotating parts, noise can be used by engineers to pinpoint sources of trouble. Noise analyzing equipment is available that determines the loudness of each frequency. The peaks can then be analyzed to determine which part in the system is rotating at that frequency. This, however, would be beyond the normally used troubleshooting techniques available to the average service technician.

Ice on the Evaporator

There are two classifications of problems involving **iced-up evaporator coils.** One is for units that normally operate at refrigerant temperatures above 32°F, and the other is for units that normally operate with refrigerant temperatures below freezing.

For units such as air conditioners, the refrigerant temperature should never get so low that it causes icing. If you find ice on an air conditioner evaporator, the unit is most probably low on refrigerant. Being low on refrigerant, when the unit starts up it runs with a too low suction pressure (and temperature). The unit produces less than rated capacity, but the inlet end of the evaporator is below freezing. Ice forms and blocks the airflow across the beginning of the coil. This insulates the refrigerant from the warm air, and then allows the liquid refrigerant to last longer through the coil. The frost will gradually move along the coil, until the entire evaporator becomes encased in a block of ice. Turn the unit off for several hours, allowing the ice to melt before attempting to service the system.

Units that normally form ice on the evaporator are provided with defrost systems. If an accumulation of ice develops, there are two possible causes.

1. The defrost system is malfunctioning.

2. The amount of ice that needs to be melted is more than the defrost system can handle.

Inspect the box for all possible sources of moisture entry. This could be poor door seals, higher than normal frequency of door openings, or high latent loads from stored product. High humidity can also be caused by improper draining of the water that forms during the defrost cycle.

Check the airflow across the coil. A clogged coil, air filter, or a malfunctioning evaporator fan can cause ice to build up too rapidly.

If the moisture load appears normal, and if the airflow is proper, put the system into defrost. For electric defrost systems, you should be able to hear a sizzling sound as the heater gets hot and boils some of the ice away.

Units that use off-cycle defrost can become encased in ice if the unit is not cycling off. This can be caused by any problem that has reduced the overall system capacity.

Unit Runs Continuously

A unit that runs continuously is usually not producing its rated capacity. Check suction and discharge pressures in order to start troubleshooting.

If operating pressures appear normal, or only

slightly high, the problem is that the unit is simply too small to handle the load. Before reaching that conclusion, however, you should find out whether the unit was ever capable of handling the same load under similar outside conditions.

Short Cycling

A unit that **short cycles** (runs for a short time before turning off, and then keeps repeating the cycle) has an electrical switch that is opening. Use a voltmeter across each of the switches in order to determine which one is operating. (Any switch that shows a voltage across its terminals is the one that is open.) Determine if the fault is within the switch, or if the system is operating outside normally expected pressures, and the switch is merely doing its job of protecting the system.

High Operating Cost

High operating cost may very likely be caused by high head pressure. It increases the horsepower required per ton of capacity, and it reduces the capacity of the system, causing the compressor to run longer. Other causes may be low evaporator airflow, high load (unusually severe weather conditions or product loads), or the customer may simply be seeing the effects of a recently enacted price increase from the local utility company.

KEY TERMS AND CONCEPTS

Defining the Problem

Troubleshooting from Pressures

Low Charge

Low Load

Low Airflow

High Head Pressure

Low Condenser Water Flow

Noncondensibles

Scaled Tubes

Moisture in the Refrigerant

Restrictions

Plugged Evaporator Drain

Door Seal

Noise

Rain Shield

Iced-up Evaporator Coil

Short Cycling

QUESTIONS

1. A 30-ton water-cooled water chiller is shutting down on the high pressure cut-out. What is the first potential problem you would suspect? How would you check to determine if this is the cause?

2. Suppose the water chiller in Question 1 is air-cooled. What would you suspect, and how would you check it?

3. The evaporator coil in a walk-in refrigerator box is frozen into a solid block of ice. Name three potential causes.

4. Suppose the evaporator coil in Question 3 is in a walk-in freezer. What additional problem might you suspect?

5. The evaporator of a window air conditioner is covered by a block of ice. What are two potential causes? How would you determine which of these two is the culprit?

6. A split system air conditioner is not doing sufficient cooling. The condensing unit runs continuously, and the condenser discharge air does not feel warm. What is the most likely cause?

7. An automobile air conditioner cools properly for a while, and then doesn't. What is the problem?

8. A system is providing insufficient cooling. The suction line is warm, the discharge line is hot, and the liquid line is cold. What is the problem?

9. What two problems could be indicated by bubbles in the sight glass?

10. An R-22 split system air conditioner is doing insufficient cooling. The outside air temperature is 95°F. The suction and discharge pressures are 55 psi and 230 psi, respectively. The suction line is quite cold. Name three potential causes.

11. Same problem as Question 10, except that the suction line is not at all cold. What is the problem?

12. A side-by-side refrigerator has a forced-draft evaporator fan in the freezer section that supplies air to the freezer, and through a damper to the refrigerator. The thermostat is located in the refrigerator section. Refrigerator and freezer sections are both too cold, and the compressor runs continuously. What is the problem?

13. In Question 12 the freezer is not sufficiently cold, but the refrigerator is all right. The compressor cycles on for 10 min out of every 20 min. What is the problem?

14. In Question 12 everything in the freezer section seems to be covered with frost. What problem would you suspect? Name two potential causes.

15. A compressor makes a clunk noise each time it starts. What is the problem? What is the solution?

16. A walk-in R-12 freezer has a suction pressure of 5 in. Hg and a discharge pressure of 155 psi. What is the most likely problem? What are two potential causes?

17. An R-22 air conditioner is providing insufficient cooling. The suction pressure is 80 psi and the discharge pressure is 200 psi. It is 90°F outside. What is the problem?

CHAPTER 15

SPECIAL SYSTEMS

Except for this chapter, this entire text pertains to cooling and refrigeration systems that use a compressor and a refrigerant to produce a cooling effect. This is called a **vapor-compression** cycle. There are, however, other types of systems that produce cooling in a very different fashion. This chapter describes those specialized cooling systems.

Absorption Cooling

Absorption cooling is similar to the vapor-compression cycle in that a refrigerant is produced at a high temperature and pressure, condensed into a liquid, allowed to flash to a low pressure, and absorb heat from a material to be cooled. One major difference between the absorption system and the vapor-compression system is the method that is used to transport the refrigerant from the low pressure side of the cycle to the high pressure side. Instead of using a compressor to transform a low temperature, low pressure vapor into a high pressure vapor, the absorption system uses a chemical process. A second major difference between the absorption system and the vapor compression system is the refrigerant that is used. The refrigerant used in an absorption system is not a halogenated fluorocarbon of any sort. It is either water or ammonia.

In order to describe the operation of the absorption cycle, let's start with a tank of saturated water in a container that has been evacuated to a pressure of one-tenth of an atmosphere (Figure 15-1). The temperature of the refrigerant water at this low pressure is 110°F. The 110°F saturated liquid is allowed to flow through an orifice to the low pressure side, which is at a pressure of 0.12 psia (almost a perfect vacuum). The saturated liquid flashes down to 40°F, which is the saturation temperature that corresponds to 0.12 psia for water. This 40°F water is then used in the

evaporator section to cool a flow of chilled water that circulates through the building. The cool water vaporizes as it absorbs heat from the building chilled water flow. The recovery of this vapor is where the absorption process is unique.

A second fluid, called the **absorbent,** circulates in the **absorber** section. In large absorption systems **lithium bromide** is used as the absorbent. Its most important property is that it absorbs large quantities of water vapor. As it does, it becomes diluted. The diluted lithium bromide is then pumped up to a section called the **generator** or the **concentrator** (Figure 15-2). This section is at the high-side pressure of 1.3 psia. Heat is applied in the generator by steam, causing the water to boil out of the dilute solution. The water vapor thus formed migrates to the condenser section where an external source of cooling water causes the refrigerant water to condense at 110°F, ready to repeat the cycle. Meanwhile, the lithium bromide solution that has been concen-

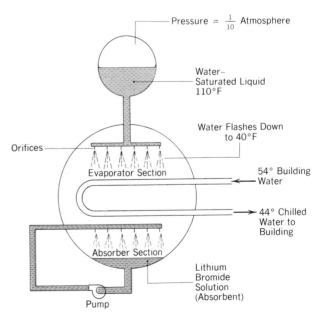

Figure 15-1 Low pressure side of the absorption cycle.

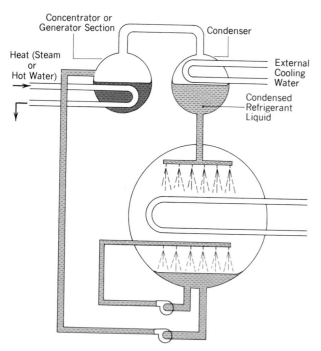

Figure 15-2 Recovery of the refrigerant water for reuse.

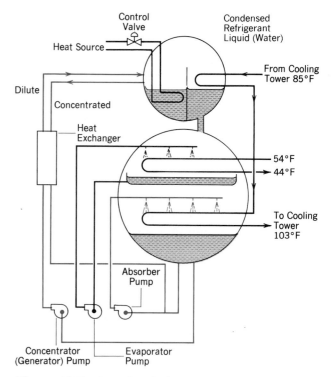

Figure 15-3 Completed absorption cycle.

trated in the generator section is allowed to flow back to the absorber section to absorb additional water vapor.

The preceding description is an oversimplification. Actual operating systems require additional plumbing, which has been added in Figure 15-3.

1. A pump has been added to the evaporator section to continually recirculate refrigerant water that does not happen to vaporize on its first pass through the chilled water tube bundle.

2. A source of cooling water has been added in the absorber section. This is required because

Figure 15-4 Large-capacity (100–1500 tons) absorption water chiller. *(Courtesy The Trane Company, LaCrosse, Wis.)*

the chemical reaction that occurs when the lithium bromide absorbs the water vapor is **exothermic.** That is, it gives off heat. This heat must be removed.

3. A heat exchanger has been added to transfer heat from the hot, concentrated lithium bromide on its way to the absorber to the dilute lithium bromide on its way to the generator. This reduces the amount of steam required in the generator to bring the solution to its boiling point. It also reduces the temperature and the need for cooling water for the lithium bromide in the absorber section.

4. A steam control valve has been added. The

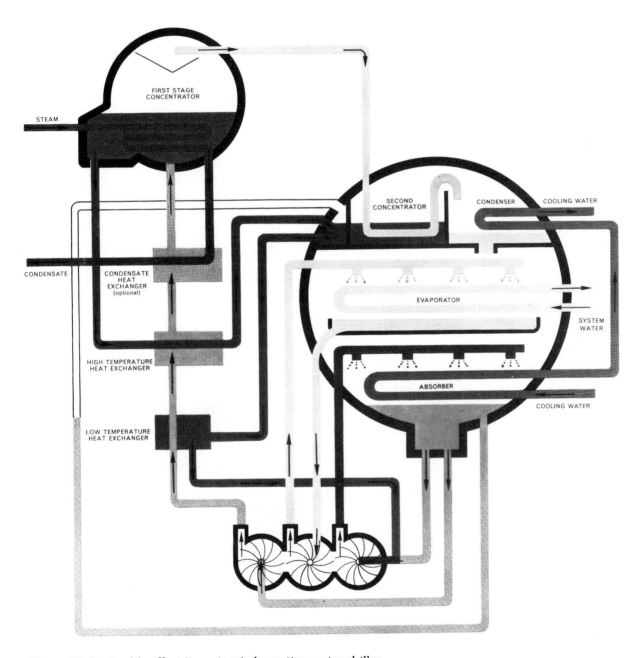

Figure 15-5 Double-effect (two-stage) absorption water chiller.
(Courtesy The Trane Company, LaCrosse, Wis.)

Figure 15-6 Double-effect absorption water chiller. *(Courtesy The Trane Company, LaCrosse, Wis.)*

control scheme senses the temperature of the chilled water leaving the evaporator section. The flow rate of the steam is then modulated to match the refrigeration effect to the load.

The primary cost of operating an absorption system is in the cost of producing the heat used in the generator section. The electrical requirements are quite minor relative to the heating costs. Absorption systems are used when the cost of producing heat is low compared to the cost of buying electricity to run a vapor-compression system. An electric vapor-compression water chiller will require roughly 0.75 kilowatts (kW) of electricity for each ton of compressor capacity. An absorption water chiller will require 18 to 20 lb per hour of steam at 12 psi for each ton of refrigeration being produced. In most cases, the cost of operating the electric system is lower. Even if the absorption system is slightly lower in cost, the electric system may be chosen because of its lower initial cost. Practical solar heated absorption systems have not yet been popularized. The high temperatures required in the generator would make for an inefficient solar system, and would add dramatically to the system's first cost. A photograph of a large-capacity absorption water chiller is shown in Figure 15-4. Figures 15-5 and 15-6 show an adaptation called **2-stage** or **double-effect** absorption. Where a high temperature source of heat is available, the refrigerant generated in the first-stage concentrator is sufficiently hot to be used as the heat

source in the second-stage concentrator. By using this refrigerant twice, the overall steam rate is reduced by 30 to 40 percent per ton of refrigeration.

Other popular options available for absorption systems are as follows.

1. Instead of using lithium bromide as the absorbent and water as the refrigerant, small systems may use water as the absorbent, and ammonia as the refrigerant.
2. Other sources of heat commonly used in the generator are hot water (minimum temperature is 200°F), or a gas flame (called a **direct-fired absorption** system).
3. Capacity may be controlled by a solution control valve rather than by modulating the heat input.

As might be expected, the absorption system has a unique set of service problems. The first is the extremely deep vacuums that must be maintained in order for the water to boil at 40°F. The slightest leakage will quickly reduce the system capacity. A purge system is provided to expel air. A second problem is one of corrosion. Lithium bromide is a type of salt water; if air is allowed to enter the system, the lithium bromide will quickly corrode the steel parts, rendering the system useless. The third problem is one of **crystallization.** The absorption system is a temperamental one; quick changes in load or in the temperature of the cooling water can cause the lithium bromide to solidify into a form similar to rock candy. Some technicians call this a freeze-up. In fact, it is not. It is a change in the crystalline structure that causes a solid instead of a liquid.

Absorption systems have been used in residential and small commercial applications in the past when gas companies subsidized their installations. Many of these systems are now being changed over to electric systems, as the days of inexpensive gas have long passed. Most new absorption systems from 25 to 1600 tons in a single unit, are now used in large bulding air conditioning systems. Small systems are still available for special applications, however (Figures 15-7 and 15-8). Extremely small absorption systems are sometimes used to cool bottled water in a home or office dispenser. A small electric heating ele-

Figure 15-7 Direct-fired small absorption chiller (3–5 tons). *(Courtesy Servel Gas Air Conditioning Company)*

ment is used as the heat source. In these applications, the absorption system is cheaper to manufacture than a conventional compressor system. Absorption systems may also be found in refrigerators for use in campers. They can oper-ate on propane gas only, and require no external source of electricity.

Steam-Jet Refrigeration

Steam-jet refrigeration has some similarities to absorption refrigeration. It relies on the principle of exposing water to a low pressure, thereby caus-ing it to flash down to a cooler temperature. In an absorption system, the deep vacuum is main-tained by spraying an absorbent that effectively absorbs all the water vapor. In the steam-jet sys-tem, the vacuum is maintained by a steam-jet **ejector** as in Figure 15-9. High pressure steam (above 200 psi) is used in the steam ejector in order to draw a vacuum on the evaporator tank (sometimes referred to as a flash tank). The warm water that is returning to the evaporator flashes and exists as chilled water. The vapor formed by the flashing is drawn off through the steam ejec-tor. The steam that flows into the ejector is con-densed by an external source of cooling water and discarded. Commercially built steam-jet systems are available in the range from 10 to 1000 tons, and are capable of producing chilled water at temperatures down to 32°F. The primary draw-backs to the steam-jet refrigeration system are

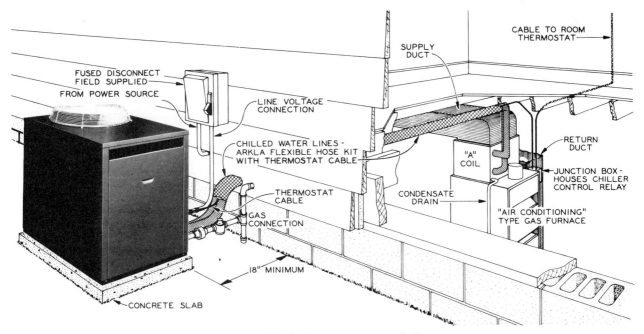

Figure 15-8 Installation of a residential gas-fired absorption water chiller. *(Courtesy Servel Gas Air Conditioning Company)*

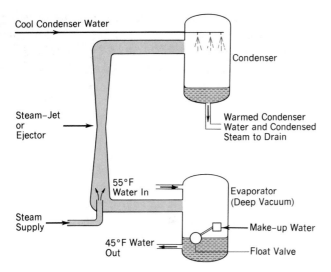

Figure 15-9 Steam-jet cooling.

very high steam consumption and high noise levels from the steam-jet ejector. A steam-jet system producing 40°F chilled water temperature with a condensing temperature of 100°F will consume in excess of 25 lb of steam per hour for each ton of capacity.

Turbine-Absorption Systems

Figure 15-10 shows an interesting arrangement of refrigeration components that have already been discussed. It is of interest because it is a

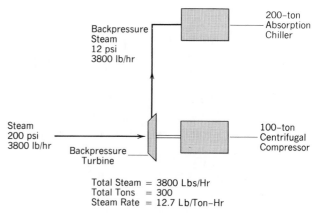

Total Steam = 3800 Lbs/Hr
Total Tons = 300
Steam Rate = 12.7 Lb/Ton–Hr

Figure 15-10 Turbine-absorption system.

steam system that is far more energy efficient than an absorption system. High pressure steam is supplied to a steam turbine, which drives an open centrifugal compressor. For each ton of capacity in the centrifugal water chiller, 38 lb/hr of steam are consumed, which does not seem that impressive. But the 38 lb/hr of steam leaves the turbine at 12 psi, which is a perfect match for the steam requirement for 2 tons of absorption cooling. We wind up with 3 tons of cooling capacity for the 38 lb/hr of steam, or 12.7 lb/hr/ton.

The **turbine-absorption** system is only suited to very large chilled water plants (1000 tons and up). For smaller size plants to achieve this low **steam rate,** two-stage absorption systems can be used.

KEY TERMS AND CONCEPTS

Vapor Compression	Exothermic	Steam-jet
Absorption	Two-stage Absorption	Ejector
Absorbent	Direct-fired Absorption	Turbine-absorption
Lithium Bromide	Crystallization	Steam Rate
Generator		

QUESTIONS

1. Name the three distinctly different water circuits used in a large absorption water chiller.

2. Why is the vacuum so critical in systems that use water as a refrigerant?

3. What property of water makes it such an excellent refrigerant?

4. Why isn't water used as a refrigerant in vapor-compression systems?

5. Steam is available at a cost of $0.50 per thousand pounds. Electricity is available at a cost of $0.10 per kilowatt hour. How much would it cost to run an absorption system for 1000 hours? How much would it cost to run a centrifugal system for 1000 hours? How much would it cost to run a steam-jet system for 1000 hours?

6. What device is used to create the vacuum in the evaporator of a steam-jet system?

7. What is crystallization?

8. Name three potential causes of crystallization.

9. Which would require a larger cooling tower, a 100-ton absorption system, or a 100-ton vapor compression system?

10. In a lithium bromide–water absorption system, which fluid is the refrigerant?

11. In an ammonia–water absorption system, which fluid is the refrigerant?

12. What is meant by a direct-fired absorption system?

13. What does the heat exchanger do in the absorption system?

14. How is the capacity of a large absorption system modulated?

15. Name three effects of air leaks in an absorption system.

16. What are the disadvantages associated with a steam-jet refrigeration system?

PART TWO

Drill Size	Natural Gas ① Capacity (000 Btu/hr)	Propane ② Capacity (Btu/hr)	Butane ② Capacity (Btu/hr)
20	74.0		
22	69.0		
24	63.0		
26	57.0		
28	52.0		
30	47.0		
32	36.5	106	117
34	35.2	97.0	107
36	32.4	89.2	98.8
38	29.4	81.0	89.6
40	27.4	75.4	83.5
42	25.0	68.7	76.2
44	21.1	58.0	64.4
46	18.7	51.5	57.0
48	16.5	45.5	50.3
50	14.0	38.5	42.8
52	11.5	31.7	35.1
54	8.6	23.9	26.3
56	6.2	17.0	18.8
58	5.0	13.8	15.3
60	3.6	12.6	13.8

① Based on 1050 Btu/ft³ 3.5 in. w.c. pressure.
② Based on 11 in. w.c. pressure.

Figure 16-9 Heating capacities of gas orifices.

The convention for identifying the size of the orifice is that it is the size of the drill bit used to make the orifice hole. Figure 16-9 shows the orifice size required for natural gas or propane at normal delivery pressures.

Example

A 100,000 Btu/hr furnace with four burners is to be converted from natural gas to propane. The orifice size must therefore be changed to maintain the same heat input. What size orifices are required for use on propane?

Solution

Each of the burners is rated at 100,000/4 = 25,000 Btu/hr. From Figure 16-9, the orifice size for natural gas should be a number 42. If propane is used, the same Btu rating will be obtained using a number 54 orifice.

If the actual heat content of the natural gas is other than 1050 Btu/ft³, the actual heat rate of the fuel used will be different than the table value. The actual heat rate may be calculated as

$$\text{Actual heat rate} = \left[\frac{\text{actual heat value}}{1050}\right] \times [\text{tabulated heat rate}]$$

The same applies for other heat values for propane.

The actual gas supply pressure may also differ from the 3.5 in. w.c. for natural gas or the 11 in. w.c. for propane used in the table. In order to adjust the table value for supply pressures other than 3.5 in. w.c. for natural gas

$$\text{Actual heat rate} = \left[\sqrt{\frac{\text{actual gas pressure}}{3.5}}\right] \times [\text{tabulated heat rate}]$$

For propane pressures other than 11 in. w.c.

$$\text{Actual heat rate} = \left[\sqrt{\frac{\text{actual gas pressure}}{11.0}}\right] \times [\text{tabulated heat rate}]$$

For fuel gases other than natural gas or propane, proper orifice sizes may be obtained from your local utility or LPG supplier.

The orifice size is critically matched to the burner. To adjust an atmospheric burner, the orifice size or gas pressure behind the orifice is selected so that the heat input rate matches the burner rating. The primary air shutter is then adjusted so that there is a blue flame.

If the primary air shutter is adjusted badly, the appearance of the flame will be changed. A slight shortage of primary air will cause yellow tips to form on the flame. A severe shortage of primary air will cause the entire flame to turn yellow, except for the small blue inner cone. A shortage of primary air has the following effects:

a. Incomplete combustion, allowing the formation of carbon monoxide.

b. Formation of **soot,** which is unburned carbon. This soot collects on the furnace surfaces and can eventually interfere with the proper venting of the products of combustion.

c. Poor combustion efficiency, resulting in high fuel costs.

If the combustion air contains dust particles, they will cause orange streaks in the flame. These should not be confused with the yellow tips caused by insufficient primary air. Unless the dust is causing other problems, such as plugged burner ports, it is not a cause for concern.

The primary air shutters may also be badly adjusted so that too much air is being admitted into the mixing tube. In that case, the symptoms that might be observed are a small, hard blue flame that may be lifting off the burner head. The flame may be noisy and may also cause resonance of the furnace. Resonance is a condition where the flame noise causes the furnace sheet-metal sides to begin vibrating. The resulting noise is similar to that of a bass guitar, and can be quite loud. In some cases of too much primary air, flame instability results. The flame may blow itself out along with the pilot light (the pilot light will be discussed in a later section).

5, 4, 3, 2, 1 . . . Light Off

The gas burners described in the previous sections have no capability for reduced capacity. They are either full on or full off. Capacity control is accomplished by a thermostat that switches a gas valve to the fully open position when heat is required. When the space is warm enough, the thermostat causes the gas valve to fully close.

One popular way to ignite the main gas is by the use of a **standing pilot** flame. That is, a very small gas flame is allowed to burn all the time in a **pilot burner.** When the main gas valve opens, the main gas is ignited by the standing pilot (Figure 16-10). When more than one main burner is to be ignited by a single pilot flame, a **crossover slot** or carryover wing is used. This is a slot that allows main gas to cross over to all the burners, and be ignited from the common pilot. If the crossover

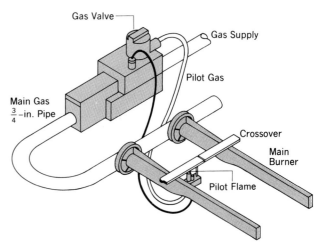

Figure 16-10 A standing pilot flame ignites the gas from the main burners whenever the main gas valve opens.

slot becomes blocked or misaligned, it will result in **delayed ignition.** Delayed ignition occurs when the main gas is not ignited immediately. Gas continues to enter the furnace until a large gas cloud is ignited by the pilot flame. Depending upon the length of the delay, delayed ignition can cause noisy light-off or a minor explosion, which remains inside the furnace. A dangerous explosion may be caused if sufficient gas is introduced to the furnace prior to ignition.

A blocked crossover may be cleaned with a tool fashioned from a scrap of sheet metal. With the burner off, the sheet metal is worked through the slot to dislodge dust or rust.

On light-off, another combustion problem called **flashback** may occur. This happens when the velocity of the gas/primary air is too low to keep the flame outside the burner head. The flame flashes back into the mixing chamber and burns at the gas orifice. This causes an objectionable noise, as well as incomplete combustion, sooting, and overheating of the orifice. The most common cause for flashback is a low pressure in the main gas supply line. Less common causes are an obstructed or nicked burner orifice, and a damaged or obstructed main burner.

The Pilot Burner and Safety

The safety implications of a reliable pilot burner cannot be overemphasized. Before we can allow

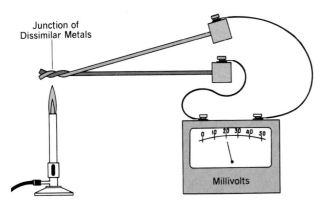

Figure 16-11 A thermocouple heated by a flame, producing 20 mV.

the main gas valve to open, we must prove that a pilot flame is available. Most furnaces with a pilot flame use a thermoelectric generator, commonly called a **thermocouple,** to prove the pilot flame.

The thermocouple operates on the principle that when two dissimilar metals are joined at one end and heated, a small electrical voltage is produced (Figure 16-11). The voltage produced depends on the temperature reached at the hot junction. The higher the temperature, the higher the voltage that will be produced. The voltage produced by a common thermocouple is 15 to 30 millivolts (mV) (1000 mV equals 1 V). This millivoltage is used to energize an electromagnet that is used in the gas flow control circuit.

Figure 16-12 shows a cutaway of a thermocouple of the type used in gas furnaces. Figures 16-13 and 16-14 show a pilot safety valve that uses the voltage generated in order to remain

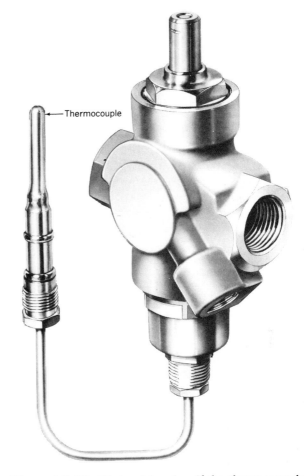

Figure 16-13 Pilot safety valve. If the thermocouple is allowed to cool, the valve will close. *(Courtesy Johnson Controls, Inc.)*

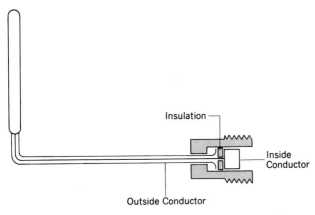

Figure 16-12 Thermocouple cutaway. To avoid crushing the insulation, connection to the safety valve must not be overtightened.

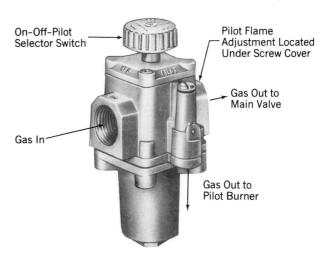

Figure 16-14 Manual gas safety valve and cock used on gas heaters with a single-function main valve (no safety monitoring of pilot). Safety valve is located in series with the main gas valve. *(Courtesy Dayton Electric Mfg. Co.)*

open. Figures 16-15 and 16-16 show the pilot burner. It accommodates the thermocouple in a position that will accurately sense the heat from the pilot flame. The orifice inside the pilot burner maintains a very small pilot flame. When removing a pilot burner, care must be taken not to lose the orifice.

Figure 16-17 shows how a pilot safety valve is used in conjunction with the main gas valve to assure a safe light off. A transformer provides 24 V to energize a coil in the main gas valve whenever the switch in the room thermostat is closed. In order for the fuel gas to reach the burners,

however, the pilot safety valve must also be open. During normal operation, this valve will always be open as long as there is a pilot flame. If the pilot flame goes out for any reason, the thermocouple voltage output will drop to zero, and the pilot safety valve will close.

Figure 16-18 shows a main gas valve. It opens whenever 24 V are applied to the coil. Another type of pilot safety valve is illustrated in Figure 16-19. Instead of a separate pilot valve in series with the main gas valve, this device places an electrical switch in the circuit of the main gas valve. If the pilot flame goes out, the switch opens and the main gas valve coil cannot be energized.

The arrangement shown in Figure 16-17 will shut off only the main gas in the event of a pilot flame failure. The pilot gas continues to be supplied through the pilot burner. This small quantity of unburned gas is allowed to rise and escape

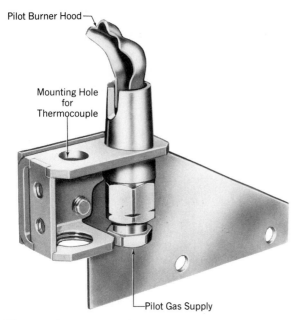

Figure 16-15 Pilot burner. *(Courtesy Johnson Controls, Inc.)*

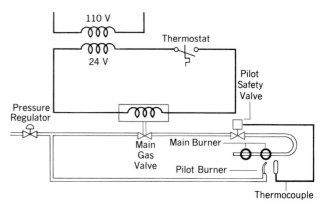

Figure 16-17 Main gas valve in series with a pilot-operated safety valve. If the pilot flame goes out, the pilot safety valve will close.

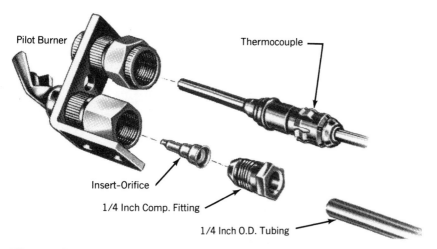

Figure 16-16 Pilot burner with thermocouple. *(Courtesy Honeywell)*

through the flue gas vent to the outside. There is no safety hazard involved, as long as it is a natural gas system.

For propane systems, this sort of system is not permitted. Propane is heavier than air, meaning the unburned pilot gas would sink, creating a dangerous collection of unburned gas. For all propane systems and most natural gas systems, a **100 percent shutoff** is used. When the pilot flame is not proved, not only is the main gas shut down, but the pilot gas is also automatically shut off. Once this happens, the pilot flame must be manually relighted.

For the thermocouple system to work reliably, the following installation notes are in order.

a. The thermocouple must be screwed into the valve only finger tight, plus one quarter of a turn. If the thermocouple is overtightened, the insulation separating the two thermocouple conductors will be crushed.

b. The end $\frac{1}{4}$ in. to $\frac{1}{2}$ in. of the thermocouple should be in the pilot flame. Figure 16-20

shows the correct and incorrect mounting positions.

c. The pilot flame must be large enough to supply sufficient heat to the thermocouple. Where pilot gas is supplied from a pilot safety valve,

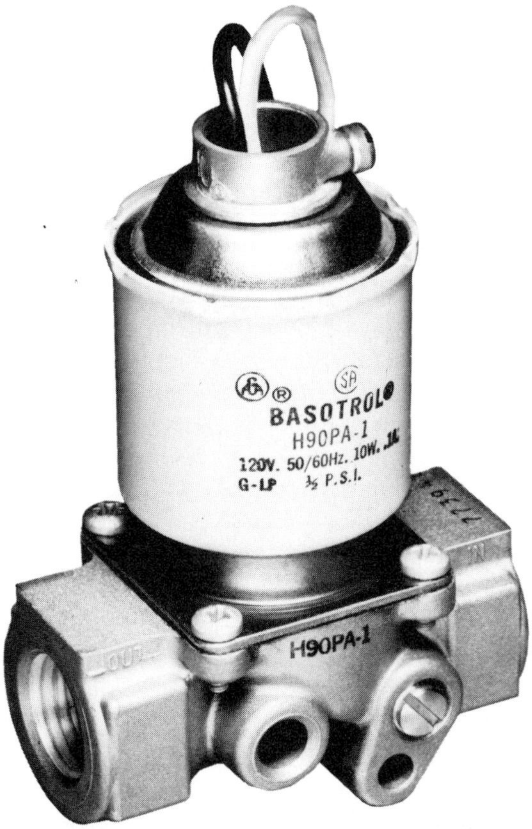

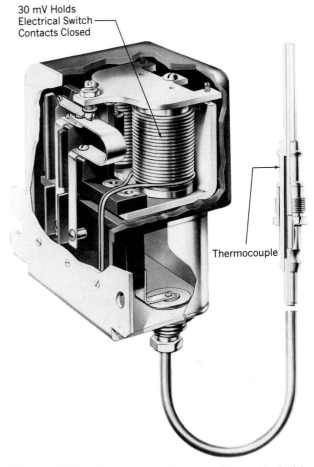

Figure 16-19 Thermocouple voltage is used to hold an electrical switch closed. (*Courtesy Johnson Controls, Inc.*)

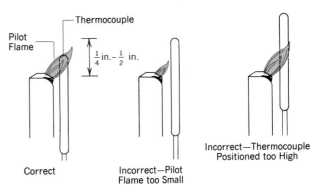

Figure 16-20 Correct positioning of the thermocouple relative to the pilot flame.

Figure 16-18 Single-function main gas valve. (*Courtesy Johnson Controls, Inc.*)

there is usually an adjustment screw on the valve for the size of the pilot flame. Sometimes the small orifice in the pilot burner can become obstructed, and must be cleaned by disassembling the pilot burner. The pilot burner gas supply line is normally $\frac{1}{4}$-in. aluminum. Older systems using $\frac{1}{4}$-in. copper should be changed over to aluminum in order to minimize the potential of orifice blockage due to internal scaling of the gas line.

Combination Gas Valve

In the previous sections, separate functions were performed by a pressure regulator, a pilot safety valve, and a main gas valve. A **combination gas valve** (Figure 16-21) performs all of these functions. The pressure regulator inside the valve body is adjustable from 2 in. to 5 in. w.c. gas supply pressure to the manifold. Inside the valve there are actually two valves in series. One can only be open when the thermocouple proves a pilot flame. The other opens when the thermostat completes a control voltage circuit.

But if that were all that were included, it would be impossible to light off the system. The valve will not supply gas to the pilot burner until there is a proven pilot flame. And there can be no pilot flame until there is a supply of gas to the pilot

burner! In order to solve this dilemma, a manual override for pilot gas is supplied as part of the valve. Pushing down on the red button (or other manual override device) will allow gas to be supplied to the pilot burner for as long as the button remains depressed. The pilot flame must be lit while holding down this red button. After one minute, the pilot flame will have produced sufficient heat for the thermocouple to provide 18 mV or more. An electromagnet in the gas valve builds sufficient magnetic field strength to hold open the valve, and the red button may then be released.

Valves of different design may use a red button, pushing down on the selector switch, or any other method of manually holding the valve open.

The selector switch has three positions, On, Off, and Pilot. While lighting the pilot flame, the selector switch is turned to Pilot. Once the pilot flame has been established, the selector switch is turned to On. This then allows main gas to be supplied to the burner whenever the thermostat calls for heating.

Checking the Thermocouple

The output voltage of a thermocouple may be checked with a millivoltmeter. A minimum output of 18 mV is required to reliably hold the magnetic coil that holds the pilot valve open. A more reliable check can be made using a thermocouple tester, which is merely a duplicate of the coil and spring loaded button in a gas valve. The thermocouple must be capable of holding in the test button against the spring pressure, or the thermocouple must be replaced.

Bimetal-type Pilot Safety

Figure 16-22 shows another type of device used to prove a pilot flame. The pilot flame is allowed to impinge directly onto a bimetal element. When heated sufficiently, the bimetal element bends, causing a switch to close. Otherwise, the switch remains open. When this type of device is used, it is wired in series with a simple 24-V single-function automatic gas valve.

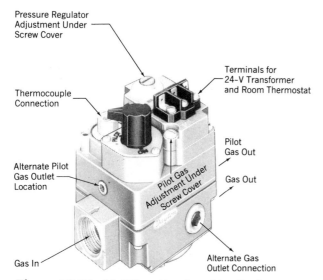

Figure 16-21 Multifunction (combination) gas valve. *(Courtesy Dayton Electric Mfg. Co.)*

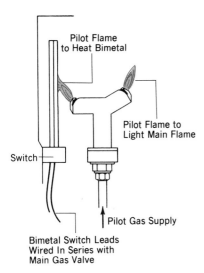

Figure 16-22 Bimetal pilot safety switch is wired in series with the main gas valve.

Automatic Reignition

The simple two-wire bimetal device described in the last section served only one function. If the pilot flame goes out, the circuit to the main gas valve opens, and the switch remains open until somebody physically relights the pilot.

Figure 16-23 shows a slightly more sophisticated bimetal safety switch that incorporates a third wire and a glow coil. This type of arrangement is frequently found on rooftop furnaces where the pilot flame is more susceptible to being blown out by the wind. Whenever the pilot flame goes out, a 2.4-V transformer is energized,

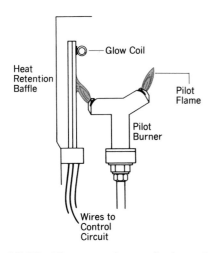

Figure 16-23 Three-wire bimetal pilot safety switch.

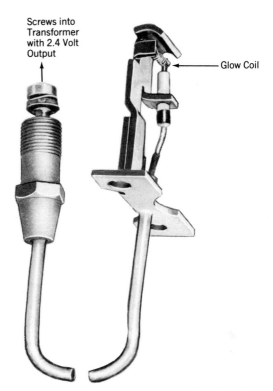

Figure 16-24 Glow coil to reignite the pilot flame. *(Courtesy Johnson Controls, Inc.)*

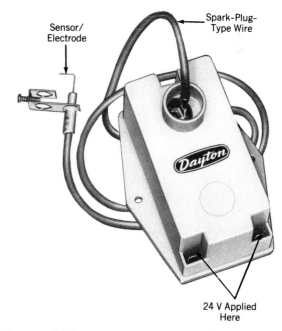

Figure 16-25 Automatic gas pilot lighter. The electrode will begin to spark whenever a flame is not sensed, and 24 V is applied to the sparker module. *(Courtesy Dayton Electric Mfg. Co.)*

providing power to the **glow coil** (Figure 16-24). thus relighting the pilot flame. When the flame is reestablished, the glow coil is deenergized and the gas valve circuit through the pilot safety is once again completed. Note that this system cannot be used with a 100 percent shut-off system. The flow of pilot gas must be maintained even during a pilot flame outage.

A second method can be used to relight the pilot flame. Figure 16-25 shows a sparking module that senses the pilot flame. Whenever the flame goes out, the sensor/electrode produces a high voltage spark to the pilot burner hood, thus relighting the pilot flame.

Low Gas Pressure Cut Out

Figure 16-26 shows a device that senses gas pressure, and opens a switch should the gas pressure fall below the set point. This device is used in conjunction with the automatic reignition system. If the pilot has gone out because the gas supply to the heater has been turned off, the reignitor glow coil would continue to attempt to relight the pilot until it burned out. The low gas pressure switch is wired in series with the glow coil circuit. Automatic reignition is then not even attempted unless there is sufficient gas pressure available.

Heat Exchanger

To get the heat from combustion out of the furnace and into the room, a **heat exchanger** is used between the heat from combustion and the room air (Figure 16-27). The heat exchanger may be constructed from cast iron, stamped steel, or ceramic coated steel. Various shapes are used to make the path of travel for the flue gas longer, so it has more time to allow its heat to transfer to the room air. The design of the heat exchanger is the key to furnace efficiency. The more heat-transfer surface area provided, the less heat will be wasted up the stack. Most residential and commercial furnaces are provided with a heat exchanger that will transfer 80 percent of the heat input into the room air.

The heat exchanger is the heart of the furnace, and the most expensive component. It is rarely repaired or replaced. When it fails, it is usually time to replace the entire furnace.

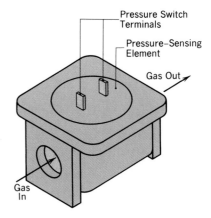

Figure 16-26　Low gas pressure switch.

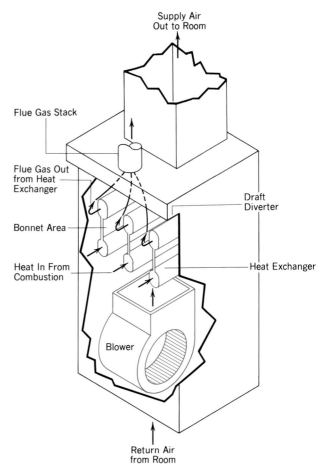

Figure 16-27　Major parts of a residential forced-air furnace.

Furnace Fan

The **furnace fan** shown in Figure 16-27 draws air from the heated space at room temperature. As the air moves over the heat exchanger, it experiences a temperature rise of between 45°F and 100°F. Many fans have two- or three-speed motors. At the higher fan speeds, the temperature rise of the air is lower and the furnace efficiency is improved. If the room air flow is too high, however, the low discharge air temperature may be perceived as a draft by the room occupants. Also, the increased fan operating cost can offset the decreased gas consumption.

The relationship of the fan to the heat exchanger may be as shown in Figure 16-28a. This is called a **vertical upflow** furnace, and is the most common arrangement for residential applications. Figure 16-28b shows a **vertical downflow** furnace, and Figure 28c, a **horizontal** furnace. The downflow furnaces discharge into under the floor ductwork. The horizontal furnace is commonly found in attic installations and on rooftop furnaces. The furnaces may also be **blow-through** or **draw-through** arrangements. That is, the fan may blow through the heat exchanger, or it may draw the room air through the heat exchanger. From a safety standpoint, blow-through units are preferable. If a hole develops in the heat exchanger because of corrosion or any other reason, a draw-through unit will be more likely to

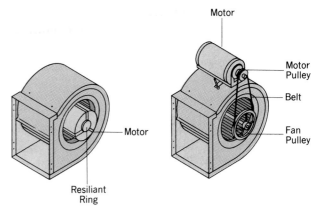

Figure 16-29 Direct-drive and belt-drive fans.

draw the products of combustion into the room air.

The fan (or furnace blower as it is sometimes called) may be either belt-drive or direct-drive. Both types are shown in Figure 16-29. The direct-drive blower and motor operate at 1075 rpm, the belt-drive blower motor runs at 1750 rpm, and the fan runs at 1075 rpm by using a large blower pulley and a smaller motor pulley. Many blower motors, especially those that are also used for air conditioning, will have three or four different speeds available. On multispeed motors, the operating speed may be changed by choosing which winding in the motor is connected into the circuit. On heating plus cooling systems, the fan will be automatically switched to the higher speed when the thermostat calls for cooling.

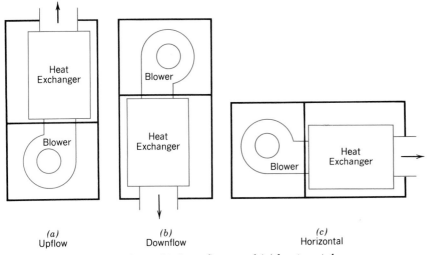

Figure 16-28 (a) Upflow, (b) downflow, and (c) horizontal furnace arrangements.

Fans are used on most furnaces, except some small room space heaters and very old central heating furnaces. These are referred to as **gravity-type furnaces,** because they rely on the effects of natural convection to move the air around the occupied space.

Fan Switch

In the normal furnace operating sequence, the gas valve is opened by the room thermostat. Then, only after the heat exchanger has been sufficiently heated, does the fan turn on. When the room is heated sufficiently, the thermostat causes the gas valve to close. Then, after another time delay, the fan is allowed to turn off. These delays in fan operation are accomplished by the use of a **fan switch.** The delay in turning the fan on is to prevent blowing uncomfortably cool air onto the occupants on each furnace start-up. The delay in turning the fan off is to increase furnace efficiency. Once the gas has been burned to heat up the heat exchanger, it makes sense to run the fan for an extra minute or two to capture that heat.

Several kinds of fan switches are shown in Figures 16-30 to 16-32; each of them senses the temperature of the air as it passes through the heat exchanger. There are two set points on the switch. Typically, the switch will close, turning on the fan when the bonnet temperature reaches 120°F or 130°F. The switch will open, turning off the fan when the **bonnet** (area around the heat exchanger) temperature reaches 100°F–110°F. If these set points seem backwards to you, take a look at Figure 16-33. It shows bonnet temperature versus time for a complete cycle of operation. With a cut-in temperature of 130°F and a cut-out temperature of 100°F, the

Figure 16-31 High limit control with fixed 25°F differential. The switch will close when the temperature drops to 25°F lower then the setpoint. *(Courtesy Dayton Electric Mfg. Co.)*

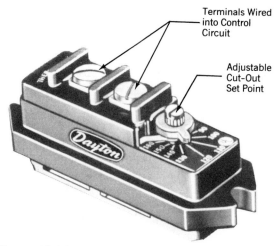

Figure 16-32 Flush-mount fan switch or limit switch. *(Courtesy Dayton Electric Mfg. Co.)*

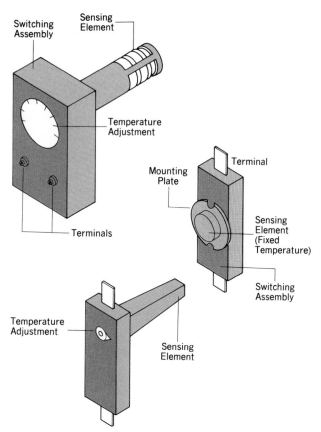

Figure 16-30 Fan switches turn the fan on after the heat exchanger has warmed to the Fan On setting.

fan will operate at any temperature of 130°F or higher, and the fan will be off at any temperature of 100°F or lower. Between 100°F and 130°F, the fan may be on or off, depending on whether the furnace is heating up or cooling down.

For downflow furnaces, there is a peculiar problem in finding a suitable location for the temperature-sensing fan switch. If the switch were to be located downstream from the heat exchanger, it would never turn the fan on. This is because before the fan comes on, the heat from the heat exchanger rises and would not be sensed by the fan switch. If the fan switch were to be located before the heat exchanger, it would turn the fan on. But as soon as it did, it would sense the cool room air returning to the furnace, and promptly turn the fan back off. In order to solve this dilemma, downflow furnaces use the type of **time-delay fan switch** shown in Figure 16-34. Instead of trying to sense temperature to control the fan, this switch uses a fixed time delay. A schematic cutaway of the inside of the time-delay fan switch is shown in Figure 16-35. It consists of a heater element and a bimetal switch. The heater element is nothing more than a 24-V resistor. It is energized whenever the room thermostat calls for heat. While the gas valve opens immediately, it takes about 1 min for the heater

element to build up heat. When sufficient heat from the resistor reaches the bimetal element, it warps, closing the 120-V fan switch. On shutdown, when the thermostat opens the heater element is deenergized. After 2 min, it cools sufficiently for the bimetal switch to return to its original position, turning the fan off. Time-delay

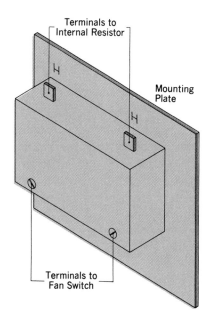

Figure 16-34 Time-delay fan switch.

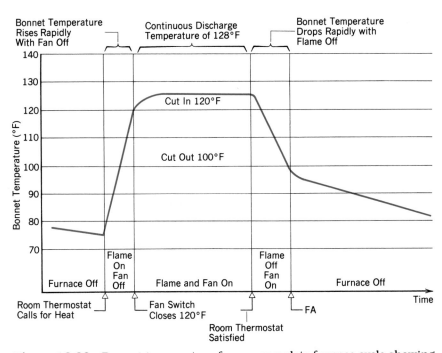

Figure 16-33 Bonnet temperature for one complete furnace cycle showing the fan switch settings.

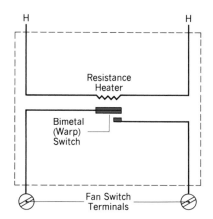

Figure 16-35 Internal schematic of the time-delay fan switch.

relay switches are used in most of the downflow and horizontal furnaces, as well as in some upflow furnaces.

Limit Switches

The normal discharge temperature of the air from the furnace is below 140°F. There are several abnormal situations that can cause this temperature to be higher. Therefore, a temperature switch is provided to sense bonnet temperature and shut off the gas valve if the bonnet temperature reaches 190°F. These switches, called **limit switches,** may have either a fixed set point, which will be printed on the switch, or an adjustable cut-out temperature (Figure 16-36). All limit switches have a fixed differential. That is, the switch recloses at a temperature of 15°F–30°F lower than the temperature at which it opened. The limit switch can look exactly like a fan switch, except that it opens on a rise in temperature instead of closing.

Downflow furnaces use two limit switches. One, located at the top of the furnace, will sense if the fan does not turn on. The second, located downstream from the heat exchanger, will sense overfiring or a low air flow condition.

Because the limit switch and the fan switch both sense the same temperature, they are sometimes combined into a single unit, as in Figures 16-37 to 16-39. This is called a **combination fan/ limit switch.** There is only one temperature-sensing element, but there are two separate switches

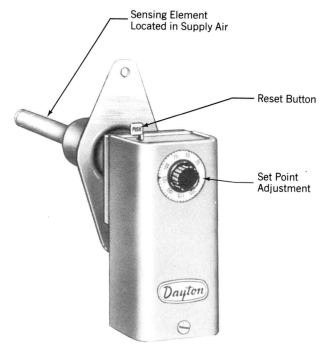

Figure 16-36 Manual reset limit control with adjustable set point. *(Courtesy Dayton Electric Mfg. Co.)*

inside. There are also three different set points on the combination fan/limit switch (Figure 16-40). The highest set point is the cut-out temperature for the limit. The other two temperatures control the operation of the fan.

Under normal circumstances, the limit switch will not open during the entire life of the furnace. If the safety cutout of the limit switch is required, it will probably be caused either by loss of airflow or overfiring of the burner. Loss of airflow can be because of any of the following reasons.

1. The fan switch is defective and has not closed its switch.
2. The blower motor has failed.
3. The air filters are extremely dirty, blocking the airflow.
4. A duct has collapsed or otherwise become blocked.
5. On a direct-drive blower, the set screw attaching the blower to the motor shaft has loosened.
6. On a belt-drive blower, the belt is slipping or has broken.

The second potential cause for overtem-

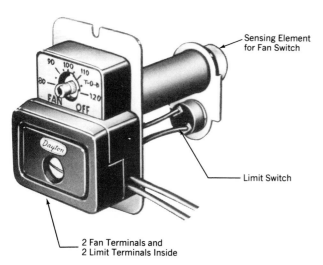

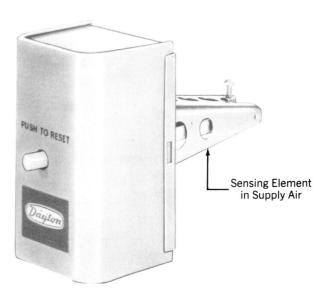

Figure 16-37 Combination fan/limit switch. The Fan On temperature is adjustable from 80°F to 120°F, with a fixed differential. The limit switch is not adjustable.

Figure 16-39 Manual reset high limit control. Once the set point is reached, the reset button must be pushed before the contacts will reclose. *(Courtesy Dayton Electric Mfg. Co.)*

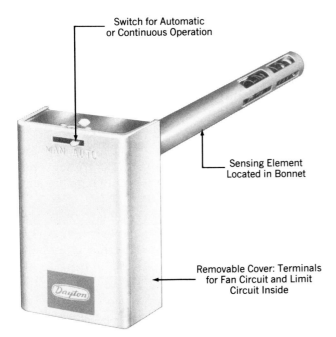

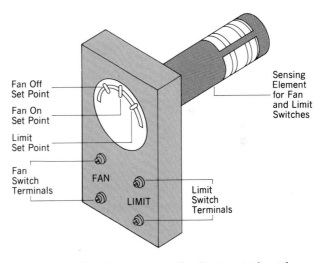

Figure 16-40 Combination fan/limit switch with three set points.

Figure 16-38 Combination fan/limit switch with a single temperature-sensing element. *(Courtesy Dayton Electric Mfg. Co.)*

2. The fuel supply pressure is too high.

3. The orifice has become enlarged or is missing.

Figure 16-41 shows a thermocouple with a terminal block. When used, the terminals are wired to a limit switch. If the limit switch opens, the system will shut down as if the pilot flame went out and the thermocouple output voltage was lost.

perature, overfiring, if it occurs, would most likely be because of one of the following reasons.

1. The wrong orifices have been installed for the fuel gas being used (most likely on new installations).

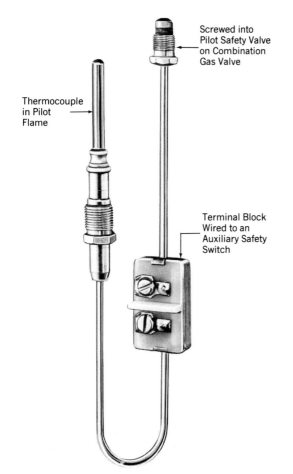

Figure 16-41   Thermocouple with a terminal block.

Millivoltage Systems

On gravity systems (heaters using no fan), the control system is different from the 24-V systems discussed to this point. The heart of the **millivolt system** is the **pilot generator** or **powerpile.** Figure 16-42 shows schematically what a pilot generator does. The end of the pilot generator is positioned in the pilot flame as if it were a thermocouple. But it is actually a number of thermocouples, wired together in series to provide a much higher voltage than a single thermocouple. Where a standard thermocouple might normally produce 18 to 25 mV, pilot generators will produce between 250 and 1000 mV (remember, 1000 mV is still only 1 V). This millivoltage is the only electrical power source available, and is used to provide the power through the thermostat, limit switch, and gas valve.

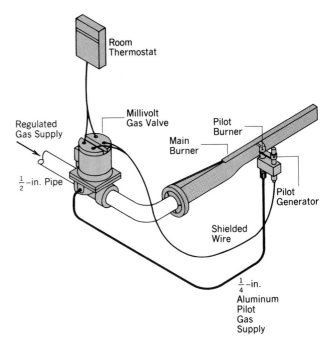

Figure 16-42   The pilot generator provides 250–1000 mV to operate the gas valve.

The pilot generator shown in Figure 16-42 is different in appearance from the thermocouple in several ways (Figure 16-43). The end of the pilot generator is physically larger than the end of the thermocouple. The voltage is transmitted over two wires, protected by metal shielding. The end of the pilot generator has normal wire terminations rather than the screw-in arrangement familiar to the thermocouple.

The thermostat used in the millivoltage system is also different from the thermostat used in a standard 24-V system. A 24-V thermostat usually contains a small resistor called a heat anticipator. Its function is discussed in Part Four, on electricity. A millivoltage thermostat cannot allow this added electrical resistance, as there is already precious little voltage available to open the gas valve.

And, of course, the millivoltage gas valve is going to be very different from the 24-V gas valve. The millivoltage valve, although similar in appearance to a 24-V valve (Figure 16-44) will be a **diaphragm-type valve,** shown schematically in Figure 16-45. When the valve is closed, there is inlet gas pressure available on both the top and the bottom of the diaphragm. In fact, the area of

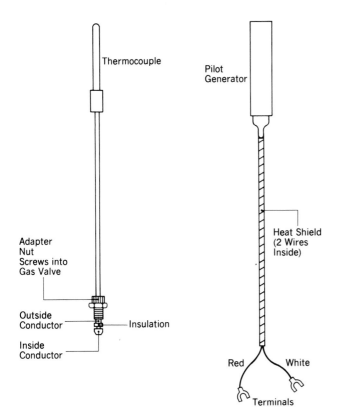

Figure 16-43 Thermocouple compared to a thermopile (pilot generator).

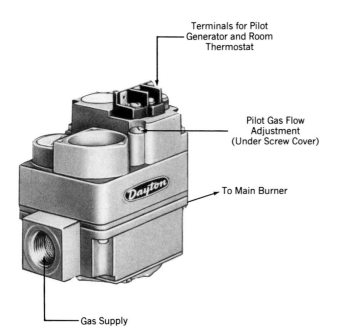

Figure 16-44 Millivoltage gas valve. Single function valve is suitable for natural gas only. (*Courtesy Dayton Electric Mfg. Co.*)

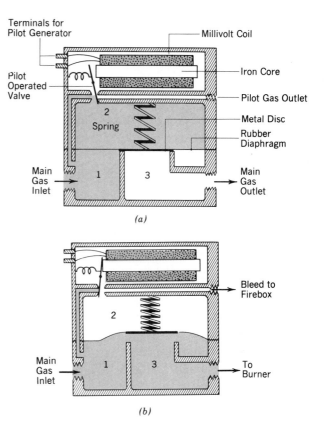

Figure 16-45 With the millivolt coil not energized, the pilot valve allows the pressure in chamber 2 to equalize with the inlet pressure in chamber 1. The pressure in chamber 2 plus spring pressure hold the valve closed. (*b*) When the millivolt coil is energized by the pilot generator, the pilot valve bleeds off the pressure in chamber 2. The pressure in chamber 1 overcomes the spring pressure, lifting the metal disc off its seat.

the top of the diaphragm that "sees" the gas pressure is larger than the area on the bottom. Therefore, the gas pressure actually tends to help the small spring hold the valve closed. When the room thermostat closes, the millivoltage from the pilot generator is supplied to the coil, creating a magnetic field. The top of the small pilot valve is pulled to the coil. The passage from the gas inlet is shut off, while the trapped gas pressure above the diaphragm is allowed to bleed off through the vent tube (where it is routed to be burned off by the pilot flame). This type of diaphragm valve is sometimes used in 24-V systems, but it is always used in automatic millivoltage systems.

Automatic Flue Damper

The **automatic flue damper** shown in Figure 16-46 is an energy-savings device popularized during the shortages of the 1970s. It is a damper that is installed in the flue gas stack. When the room thermostat calls for heating, the damper motor is energized, and the damper is moved to a fully open position. A micro-switch senses when the damper has in fact been fully opened. It deenergizes the damper motor, and then allows the ignition sequence to begin. When the room thermostat has been satisfied, it shuts down the gas valve in the normal fashion. It also energizes the damper motor once again, moving the damper inside the flue stack to the closed position. The full travel of the damper requires 15 sec. During this closing time, any residual flue gas remaining in the heat exchanger section is vented.

Where local codes require outside air ventilation of the closet where the furnace is installed, no energy savings will result from the closing of the vent damper. The same is true for furnaces installed in unheated garages, attics, or outdoors. Vent dampers may only be used on furnaces that are certified by the American Gas Association (AGA) as suitable for use with vent dampers.

Electronic Ignition Systems

During the energy shortages of the 1970s, electronic ignition systems were popularized. They save energy by eliminating the burning of gas to support the continuously burning of the pilot flame. Electronic ignition systems fall into two categories

1. Intermittent pilot ignition.

2. Direct-spark ignition.

The intermittent pilot system goes into action when the thermostat calls for heat. Gas is supplied to the pilot burner, and at the same time a spark is provided near the pilot gas. When the pilot has been lit and sensed, the main gas is then allowed to open and the sparking stops. When the room is heated sufficiently, the ther-

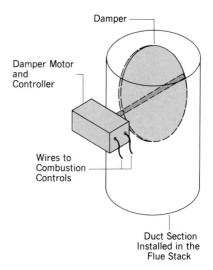

Figure 16-46   Automatic vent damper.

mostat stops both the main and pilot gas flows, leaving them ready for the next heating cycle.

The direct-spark ignition (DSI) system is different in that, instead of using a pilot to light the main burner, the spark is used directly. With this system, the gas valve has two steps of opening. On trial for ignition, the valve allows a reduced quantity of gas to flow. When ignition of the main flame has been proved, the main valve is allowed to fully open. If the flame is not proved within the trial for ignition time, the system will go into lockout. That is, the gas valve will close fully and the sparking will stop. The system will not try to relight again until after it is reset. Resetting the system may be accomplished by pushing a reset button on the control module, if one is provided. Otherwise, the system is reset by turning the thermostat down to minimum for 10 sec, and then returning it to a call for heat.

Direct ignition is not suitable for use with propane systems. Intermittent pilot systems may be used with either natural gas or propane. On natural gas systems, trial for ignition may go on indefinitely, but propane systems must be supplied with 100 percent lockout if pilot flame is not proved.

There are three popular methods in use to prove the existence of a flame in electronic ignition systems. In place of the thermocouple, these systems may use any of the following.

1. **Flame rectification.** With this system, the same wire that carries the spark senses the existence of the flame (Figure 16-47). This is

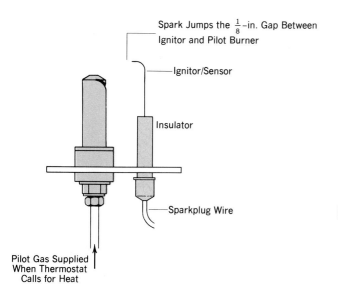

Spark Jumps the $\frac{1}{8}$-in. Gap Between Ignitor and Pilot Burner

Ignitor/Sensor

Insulator

Sparkplug Wire

Pilot Gas Supplied When Thermostat Calls for Heat

Figure 16-47 Electronic ignition lighting of pilot flame.

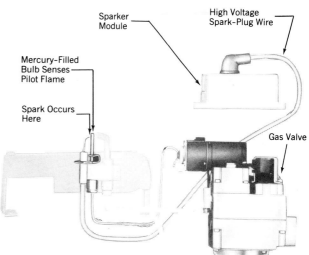

Sparker Module

High Voltage Spark-Plug Wire

Mercury-Filled Bulb Senses Pilot Flame

Spark Occurs Here

Gas Valve

Figure 16-48 Retrofit gas pilot ignition system. *(Courtesy Dayton Electric Mfg. Co.)*

done through some sophisticated electronics that sense the ionized gas molecules produced in a flame. The automatic pilot relighter shown in Figure 16-25 works on this same principle. Some designs use one wire for ignition, and a separate wire for pilot sensing through flame rectification.

2. **Bimetal pilot safety switch.** This is identical to the three-wire bimetal safety used in conventional furnaces with automatic pilot relighting. It incorporates a normally open and a normally closed switch. The pilot gas valve is energized through the normally closed contact. When the heat of the pilot flame moves the switch, the main gas valve is energized through the normally open set of contacts.

3. **Bulb-type sensor.** A liquid- or gas-filled bulb is positioned close to the pilot flame. The presence of a flame causes the pressure to build and a switch on top of the valve changes position. This type of sensing is not as quick to respond as the flame rectification type of sensor.

Retrofitting to Electronic Ignition

There are **retrofit** kits available, as shown in Figure 16-48. They allow a furnace that has a standing pilot for ignition to be changed over to an electronic ignition system. The retrofit kit consists of a new gas valve, a control module, and

a spark-ignition/flame-sensing module. They are easy to install because the component sizes are engineered to match the components being replaced. The spark/sensor is especially convenient, fitting into the same brackets that were used to hold the pilot burner and thermocouple.

The only advantage to the customer in changing over to electronic ignition is the reduction in

Figure 16-49 Electronic ignition gas valve—an early design. *(Courtesy Johnson Controls, Inc.)*

fuel required to keep the pilot burning. Normally, the cost of the retrofit cannot be justified on a furnace that is working properly. When a furnace has been diagnosed as needing a new gas valve, however, the retrofit kit becomes an attractive option at very little additional cost. The electronic ignition gas valve (Figure 16-49) is not interchangeable with a conventional valve without a complete retrofit.

Condensing Furnaces

Conventional furnaces described to this point produce heating efficiencies of between 75 and 80 percent. This means that for every 100 Btu of fuel burned, 75 or 80 Btu are converted into useful heat and delivered into the heated space. This also means that 20 or 25 Btu go up the stack with the flue gas. **Condensing furnaces** (Figure 16-50) effectively reduce this loss. An auxiliary heat exchanger is provided in which the 300°F flue gas that would normally be vented is allowed to be cooled to 100°F by the cool air returning to the furnace from the room. Also, a small auxiliary fan is provided to mechanically remove the flue gas. The major amount of heat recovery is attributable to the condensing of the water vapor in the flue gas, which occurs when it is cooled to 100°F. The volume of the flue gas is reduced dramatically. The products of combustion and the condensed water vapor are removed from the furnace through a small (2 in) PVC drain line.

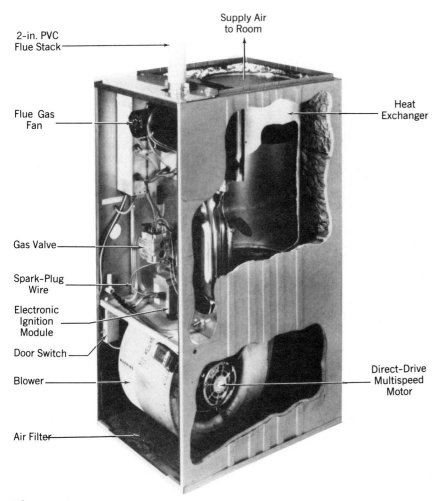

Figure 16-50 Condensing furnaces produce efficiencies in the 80 to 90 percent range and incorporate automatic ignition. *(Courtesy Coleman Co., Inc.)*

Other Furnace Applications

Figures 16-51 through 16-54 show some other furnace applications that use the same heating systems as described for furnaces. Unit heaters are usually mounted in an exposed location, within the heated space. For commercial and industrial applications, unit heaters provide a low-cost installation.

The rooftop gas-fired furnace is usually a horizontal furnace that is designed for exposure to weather on the building roof. They do not provide the ideal working environment for the service technician who needs to perform repairs. They do, however, provide a simple installation in commercial applications without occupying rentable indoor space. Technicians commonly refer to the rooftop gas-fired furnace as a "gas pack."

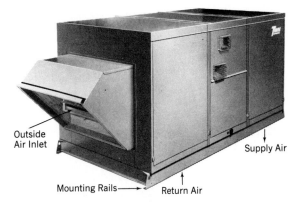

Figure 16-52 Rooftop furnace has supply- and return-air ductwork connections from below. *(Courtesy The Trane Company, LaCrosse, Wis.)*

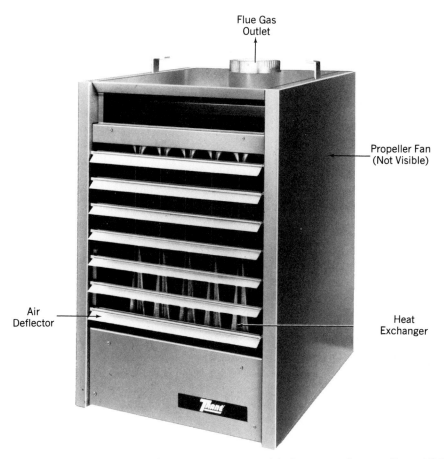

Figure 16-51 Propeller fan unit heater suitable for exposed mounting within the heated space. *(Courtesy The Trane Company, LaCrosse, Wis.)*

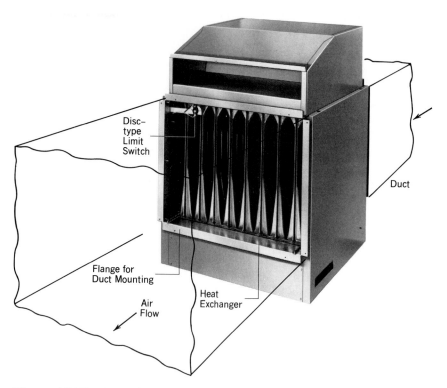

Figure 16-53   Centrifugal fan unit heater may be installed with or without distribution ductwork. *(Courtesy The Trane Company, LaCrosse, Wis.)*

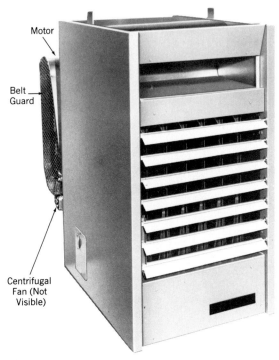

Figure 16-54   Indoor duct furnace unit heater. Airflow is provided by a separate fan. *(Photo courtesy The Trane Company, LaCrosse, Wis.)*

KEY TERMS AND CONCEPTS

Natural Gas

Hydrocarbons

Methane

Ethane

Propane

Butane

Mercaptain

Specific Gravity

Heating Value

LPG

Excess Air

Flue Gas

Bunsen Burner

Primary Air

Secondary Air

Venturi

Crossover Slot

Overfiring

Incomplete
  Combustion

Carbon Monoxide

Carbon Dioxide

Water Vapor

Nitrogen

Atmospheric Burner

Gas Manifold

Orifice

Flame Colors

Soot

Standing Pilot

Delayed Ignition

Flashback

Thermocouple

Pilot Burner Orifice

100 Percent Shutoff

Combination Gas
  Valve

Glow Coil

Heat Exchanger

Furnace Fan

Bonnet

Upflow, Downflow,
  Horizontal

Blow Through

Draw Through

Gravity-Type Furnace

Fan Switch

Limit Switch

Combination Fan/limit
  Switch

Millivolt System

Pilot Generator

Diaphragm Valve

Automatic Flue
  Damper

Intermittent Pilot
  Ignition

Direct Spark

Flame Rectification

Bimetal Pilot Safety
  Switch

Retrofit

Condensing Furnace

QUESTIONS

1. Name two important differences between the properties of natural gas and propane.

2. How many cubic feet of natural gas will be burned each hour of operation for a furnace that has a 125,000-Btu/hr input rating?

3. What delivery pressure is used for natural gas burners?

4. What delivery pressure is used for propane burners?

5. What is excess air?

6. What is the effect of too much excess air? Not enough excess air?

7. What is primary air? Secondary air?

8. What is an atmospheric gas burner?

9. What size orifice would you use to convert a four-burner 120,000-But/hr input propane furnace to burn natural gas?

10. A flame appears to be longer than normal and yellow in color. What problem would you suspect?

11. What causes the formation of soot on the heat exchanger?

12. The flame appears to have intermittent orange streaks. What is the probable cause? What would you change?

13. What is a standing pilot?

14. What is a crossover slot?

15. What is flashback? What are the most likely causes?

16. Name three devices that are used to sense the standing pilot flame.

17. What is the problem with copper piping to the pilot burner?

18. Name all the functions performed by the combination gas valve.

19. What is a timed start fan switch? In what applications must it be used?

20. Which systems use two limit switches?

21. Describe the sequence of operation for an electronic ignition furnace.

22. What is a condensing furnace? What is its major advantage?

CHAPTER 17

RESIDENTIAL AND COMMERCIAL OIL-FIRED FURNACES

This chapter describes the differences between the gas-fired furnaces in Chapter 16 and oil-fired furnaces. In addition to the objectives stated at the beginning of Chapter 16, this chapter adds the objective of being able to troubleshoot problems with the oil-fired furnace.

The Oil-Fired System

The sequence of operation for the oil-fired furnace is similar to that for the gas-fired furnace.

1. Thermostat calls for heat.
2. Burner fires up.
3. Fan turns on.

The difference between the oil-fired burners and the gas-fired burners are as follows.

1. The physical configuration of the burner used for burning oil is different from the gas burner.
2. The safety checks for safe ignition are different.

Oil as a Fuel

Fuel oil is a product of the refining of crude oil, the same process that produces gasoline, kerosene, jet fuel, gases, lubricating oil, wax, coke, and asphalt. All of these products consist principally of hydrogen and carbon. The refining process separates the crude oil into lighter products and heavier products. There are several different grades of fuel oils that can be produced within the refining process. They are classified by the numbers 1 through 6, with the low numbers signifying **light oil** and the higher numbers indicating **heavier oil.**

Number 1 oil is highly refined, and the most expensive of all the fuel oils. It contains only trace quantities of water or sediment, and cannot contain more than 0.5 percent sulfur, by weight.

Number 2 oil is the general-purpose household heating oil. It may contain 0.10 percent maximum water and sediment, and 1.0 percent sulfur. It has a higher viscosity than number 1 oil, but it still flows easily at normal temperatures.

Number 4 oil may contain up to 0.50 percent water and sediment, and there is no limit to the percent of sulfur it may contain. High sulfur contents can make fuel oils difficult to use where there are regulations governing the emission of sulfur compounds from the combustion process. The viscosity of number 4 fuel oil is quite high.

Number 5 oil contains up to 1.0 percent water and sediment and, like number 4 oil, may contain any percentage of sulfur. It has an even higher viscosity, and requires the use of a preheating system in order to reduce the viscosity sufficiently for it to be pumped.

Number 6 is the most difficult of all the fuel oils to burn. It is only used because it is by far the cheapest of the oils. It may contain up to 2.0 percent water and sediment, and there is no specification as to the amount of sulfur or ash that it may contain. It may only be used in burners equipped with a preheater because of its high viscosity.

Contrary to what some may believe, fuel oil is not very flammable. If you were to apply a propane torch flame to a bucket of fuel oil at room temperature, it would not ignite (Figure 17-1). It is the job of oil burners to **atomize** the oil supply into a mist in order to provide an extreme amount of surface area relative to the weight of oil. Without these measures, the oil will not burn.

The physical properties of fuel oils can vary depending upon the source and the refining process. The **heating value** will generally fall in the range of 18,000 to 20,000 Btu/lb. Its density can vary from 7.4 to 8.3 lb/gal (it is lighter than water). Heating value per gallon is relatively con-

sistent between different fuel oils, providing approximately 140,000 usable Btu/gal (slightly higher for number 5 and number 6 oil).

Fuel Oil Burners

Figures 17-2 and 17-3 show an **oil burner** that is similar to many that are used on a variety of

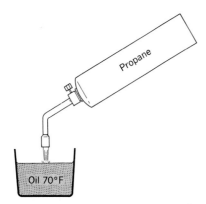

Figure 17-1 Oil at room temperature will not ignite. It must be either heated above its flash point or dispersed into a fine spray.

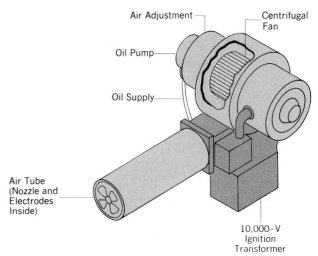

Figure 17-2 Oil burner for residential use.

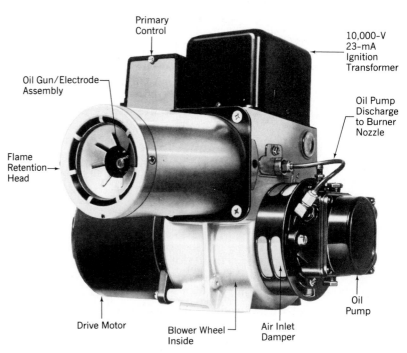

Figure 17-3 Flame retention oil burner available in capacities from 0.40 to 3.00 gal/hr of oil. *(Courtesy R. W. Beckett Corporation)*

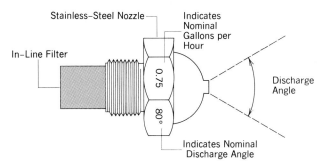

Figure 17-4 Oil burner nozzle and filter.

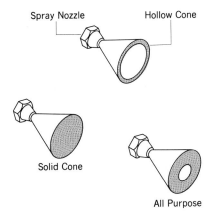

Figure 17-5 Oil spray patterns.

furnaces or boilers. The motor (1725 or 3450 rpm) drives a fan and a fuel pump. The fan draws in the combustion air that is mixed with the atomized fuel oil. The pump draws the fuel to the burner and supplies sufficient pressure so that the oil can be pushed through an **atomizing nozzle.** There is a filter in line with the nozzle (Figure 17-4).

Once the fuel oil is atomized, it may be easily ignited. Figure 17-5 shows an oil nozzle producing a fine oil spray. A pair of **electrodes** are used to ignite the oil (Figure 17-6). These electrodes are positioned so that there is a $\frac{1}{8}$-in. air space between the ends. The control system applies a high voltage of 10,000 V or more, creating a spark between the electrodes. The position of the electrodes relative to the cone formed by the oil spray is important. It must not be within the oil spray to prevent the oil residue from bridging between the electrodes. But it must be close enough so that the primary air stream blows the arc into the spray. The actual location for the electrodes is given by the manufacturer of the burner. It depends upon the shape of the oil spray cone that, in turn, depends upon the nozzle design and the pressure behind the nozzle. Where manufacturers' data are not available, for 80-degree hollow-cone nozzles, set the

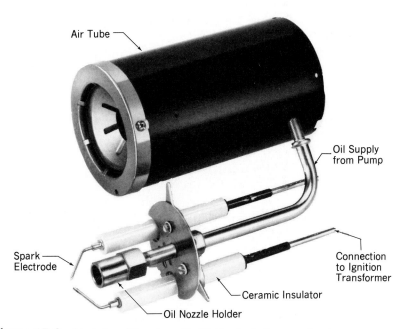

Figure 17-6 Air tube. *(Courtesy R. W. Beckett Corporation)*

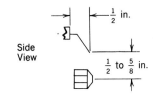

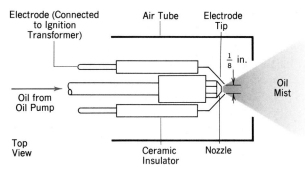

Figure 17-7 Atomization and ignition of fuel oil.

ends of the electrodes $\frac{1}{2}$ in. to $\frac{5}{8}$ in. above the centerline of the nozzle, and $\frac{1}{2}$ in. in front of the nozzle face (Figure 17-7). For cones of less than 80 degrees, the electrodes should be placed within $\frac{3}{8}$ in. of the face of the nozzle. It is important that no part of the electrode be less than $\frac{1}{4}$ in. from any metal part. Otherwise, sparking might take place at that point instead of between the electrodes.

The ends of the electrodes should have a pointed shape, but not to the point of being sharp. They will gradually disappear because of the high voltage spark, but they can be reshaped with a file and bent to adjust the gap to $\frac{1}{8}$ in. The nozzle and pair of electrodes, insulated with ceramic, are called the **oil gun assembly.**

Once the oil flame has been established, the ignition spark may remain on or it may turn off,

depending on the manufacturer. Those that remain on are called **continuous ignition.** Where **intermittent ignition** is used, the flame itself provides sufficient heat to burn the entering fuel and keep the chain reaction going.

The 10,000-V supply to the electrodes is provided by the **ignition transformer.** It is a step-up transformer (increases the input voltage), using 115 V as the input voltage. The connection from the transformer to the electrodes may be either through a buss-bar connection (Figure 17-8) or snap-on terminals.

Burner Motor

The burner motor may be either split-phase or capacitor start. Residential burners will most commonly use either $\frac{1}{6}$ or $\frac{1}{8}$ hp split-phase. Manual overload protection is required, with the reset button provided on the motor housing. The combustion air blower is fastened directly to the motor shaft. The **fuel oil pump** (Figure 17-9) may be either directly coupled to the opposite end of the motor shaft, or it may be belt driven. The belt-drive arrangement for the pump is commonly used on the 3450-rpm motors to allow the pump to run at 1725 rpm.

Burner Maintenance

Annual maintenance of the oil burner is required to prevent oil residue from causing problems.

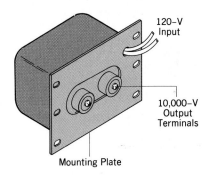

Figure 17-8 Ignition transformer.

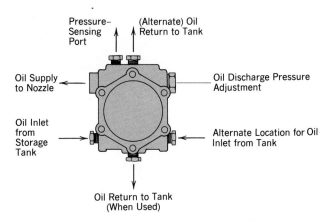

Figure 17-9 Fuel oil pump.

Following are steps that should be performed for routine maintenance.

1. Close the fuel oil shut-off valve from the fuel storage tank. Replace the filter cartridge in the fuel supply line and clean out the filter bowl.

2. Remove the nozzle assembly, and clean with a brush and solvent.

3. Place a pressure gauge in the pressure port connection to measure pump discharge pressure. With the fuel supply line opened, start the burner. Adjust the pressure output setting to the manufacturers' recommendation. In the absence of manufacturer's data, set the pump discharge pressure to 100 psi.

Figure 17-10 Stack-mounted combination flame detector and primary control. *(Courtesy Honeywell)*

Primary Control

All the control for the oil burner is centered in a "black box" called a **primary control** (Figure 17-10) It may also be referred to as a **stack control.** The function of the primary control is:

1. On a call for heat from the thermostat, start the burner motor and energize the ignition transformer.

2. When the oil flame has been established, turn off the ignition transformer (intermittent ignition models only).

3. If the oil flame has not been established within

a specified time (usually around 1 min), shut down the burner motor. In this case, the system will not automatically recycle. The reset button on the primary control must be reset before a new trial for ignition may proceed.

4. When the thermostat is satisfied, the primary control turns off the burner motor and the

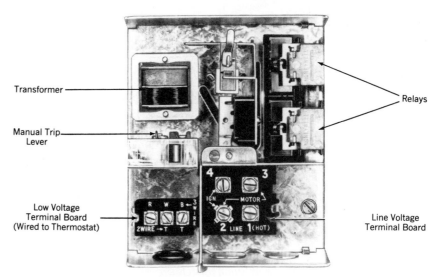

Figure 17-11 Internals of stack-mounted combination flame detector and primary control. *(Courtesy Honeywell)*

ignition transformer (continuous ignition models).

5. In the event of a power failure or a flame failure for any other reason, some models will allow the fan to run for 1 min (scavenging period) to remove fumes in the firebox, and then make one attempt to restart.

There are several ways in which the primary control determines whether the trial for ignition has been successful. The primary controls shown in Figures 17-10 and 17-11 have a heat-sensing element that senses flue gas temperature. This controller is mounted directly on the flue stack, and for that reason it is sometimes called a stack controller. Stack controllers are also available in two pieces. One piece is mounted on the stack (Figure 17-12), and the rest of the control is mounted elsewhere. The stack sensing switch may actually be two switches. They are called the hot switch and the cold switch.

Non-stack-mounted controllers use a remote flame-sensing device. Most common of these for residential use is the cadmium sulfide cell, or **cad cell** (Figures 17-13 through 17-15).

Correct installation of the element that senses stack temperature is important if nuisance trip-outs are to be avoided. The element must be installed so that it senses stack temperature only. If

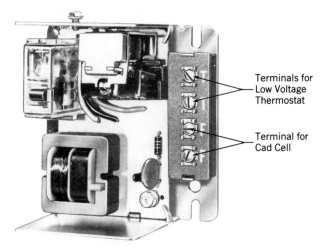

Figure 17-13 Primary control using a cad cell flame detector. *(Courtesy Honeywell)*

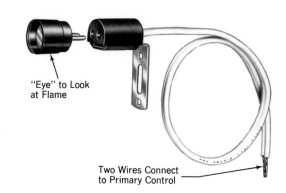

Figure 17-14 Cadmium sulfide cell (cad cell). *(Courtesy Honeywell)*

Figure 17-12 Twenty-four-volt stack switch senses stack temperature, signaling the presence of a flame to the primary control. *(Courtesy Honeywell)*

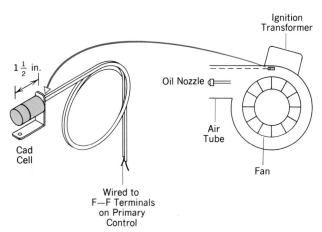

Figure 17-15 Mounting of cad cell used to prove the flame to the primary control.

the flue gas piping leaks near or around the temperature-sensing element, the temperature around the element will be reduced. The sensor will mistakenly determine that the flame has gone out, and will shut down the system.

.

Cadmium Cell Flame Detectors

The cad cell is a light-sensitive device that is mounted where it can see the oil flame. A cad cell exhibits a very high resistance when it senses darkness, but its resistance becomes very low when it sees light. It therefore behaves as a switch in the primary control circuit. But the cadmium cell will not react to just any light. For example, while it will react to an oil flame, it will not react to the wavelength of the light that is emitted from a well-adjusted gas flame.

A burner that is properly adjusted will cause the cad cell to have a resistance of between 300 and 1000 Ω. Higher resistance could mean that the burner is poorly adjusted, or that the cad cell needs to be cleaned.

The cad cell is located so that it is cooled by the combustion air. The cad cell must be kept at a temperature below 140°F. The two wires from the cad cell will be connected to terminals F-F on the primary control (flame). On some controls, the terminals may be labeled S-S (sensor).

Stack Detector

The **stack detector** is located where it will be able to sense heat when the oil flame is established. Usually, this is on the flue stack, but it may also be found on the front of the furnace above the combustion chamber. In any event, it must be mounted ahead of any draft regulator (which would dilute the heat). If it is mounted in an elbow in the flue stack, it should be on the outside of the elbow. The stack detector may have either two or three wires. The two-wire model has a single, normally closed switch. When the sensor detects heat, the switch opens.

The three-wire detector contains two switches (Figure 17-16). In the cold starting position, the switch between R and B is closed. On a temperature rise, R to B will open and R to W will

Figure 17-16 Switching arrangement inside the stack detector.

close. While the sensor element is heating, you may find both switches open or both switches closed simultaneously.

Stepping the Stack Control

One popular stack controller uses a friction clutch that is mounted on a rod connected directly to the heat-actuated element. It is this clutch that operates the hot and cold contacts. If the clutch has slipped, the cold contacts will not be closed. Without the closed cold contacts, the burner cannot start. For Honeywell controllers, the cold contact can be placed in the closed position by pulling a lever on the drive shaft outward about one-quarter inch, and then releasing it. On Penn stack controls, the same function is accomplished by turning the clutch knob clockwise.

Checkout Procedure

In order to determine that a new or existing system is operating properly, the function of the controller must be checked. With the burner operating, a flame failure can be produced by shutting off the fuel oil supply hand valve. Recycle models will shut down the burner, and then attempt one restart before locking out. Other models maintain ignition until the burner locks out on safety in the safety switch timing.

Checking Combustion

The combustion in gas-fired furnaces can be checked visually. Oil flames can only be checked properly by the use of instruments. The measurements that can be taken are as follows.

1. Draft reading in the flue.

2. Draft reading in the combustion chamber.

3. Smoke in the flue gas.

4. CO_2 in the flue gas.

5. Stack temperature.

For accurate readings, the above measurements should be taken only after the burner has operated for at least 10 min.

Figure 17-17 shows a **draft gauge** that can be used to read pressures only a few hundredths of an inch below atmospheric pressure. The probe of the draft gauge is inserted into the flue stack near the furnace. It should read between 0.03 and 0.04 in. of draft. If it is different, the **baro-**

Figure 17-17 Combustion analysis kit. *(Courtesy Bacharach Instrument Co.)*

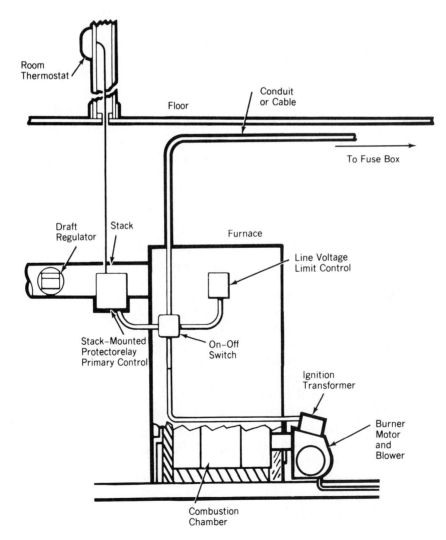

Figure 17-18 Complete oil furnace system with stack control, room thermostat, and oil burner. *(Courtesy Honeywell)*

metric draft regulator (Figure 17-18) may be adjusted to obtain the correct draft. If there is not enough draft, and the barometric draft regulator is closed, there may be a restriction in the path of the flue gas.

Insert the probe of the draft gauge through the inspection door in the firebox. It should be no more than 0.02 in. higher (closer to atmospheric pressure) than the reading obtained in the flue stack. A difference greater than this would indicate that the heat exchanger is restricted and should be cleaned.

Figure 17-19 shows an instrument used to sample an air stream for smoke. A pump is manually operated to pull a sample of air through a piece of filter paper. The sampling tube of the tester should be inserted into the flue stack at the furnace outlet. The pump handle is then stroked the number of times designated by the instrument manufacturer (ten strokes for the tester manufactured by Bacharach Co.). The filter paper is then removed, and the smoke spot compared to a series of spots provided on a card supplied with the pump. The smoke reading should be less than 1 on the standard scale. If it is higher, increase the combustion air setting. It is assumed that the draft has already been checked, and is correct.

Carbon dioxide (CO_2) in the flue gas is measured in order to determine the combustion efficiency. The instrument shown in Figure 17-20 is used to measure the percentage of CO_2 in the flue gas. A sampling tube is placed in the flue gas just outside the furnace (ahead of the barometric damper). The aspirating bulb is then pumped 18 times, pulling a measured sample of flue gas into the chamber. The aspirating assembly is then removed, closing the valve on the top of the container. Turn the container upside down, and

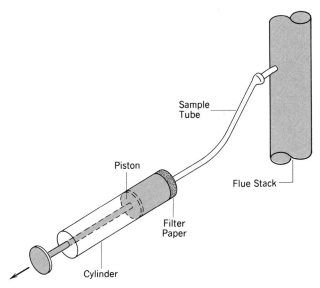

Figure 17-19   Smoke sampler.

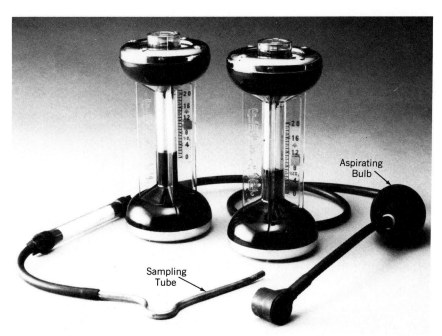

Figure 17-20   Apparatus for sampling oxygen and carbon dioxide. *(Courtesy Bacharach Instrument Co.)*

then right side up. Repeat. This allows the chemical inside the container to absorb the carbon dioxide in the flue gas sample. The vacuum thus created in the upper portion of the chamber will cause the liquid to be pulled up into the tube. The percentage of CO_2 is then read. It should be between 8 and 10 percent. If the CO_2 reading is less than 8 percent, it means that the flue gas is being overdiluted by too much combustion air. The combustion air should be decreased, and the smoke test repeated. If the combustion air is at its minimum setting according to the smoke reading, and the CO_2 is still less than 8 percent, then there is dilution air coming from somewhere else. Check all potential sources of air leakage, especially around the duct-to-furnace connection and around the access door to the firebox. If no leakage is detected, replace the nozzle. It may be defective, causing incomplete combustion.

A thermometer is then inserted into the flue stack at the same point where the CO_2 reading was taken. A chart that is supplied with the sampling equipment can then be used to figure the combustion efficiency.

When using the CO_2 detector, be sure to read all the safety precautions associated with its use. The fluid inside is very corrosive, and should not be allowed to escape from the container.

A microprocessor type of sampler (Figure 17-21) is able to measure many properties of the flue gas and provide a digital readout of the flue gas components and combustion efficiency.

Oil Storage Tanks

Oil storage tanks of between 200 and 1000 gal are provided for most residential and commercial installations. The tank may be located indoors, outdoors, or buried outdoors. Regardless of the location, if the elevation of the tank is above the burner (no higher than 25 ft), fuel will be supplied to the burner by gravity flow (Figure 17-22). If the tank is located indoors, local codes should be checked to make sure that it is not too close to the oil burner. The oil from the tank is routed to the burner through a $\frac{3}{8}$-in. or $\frac{1}{2}$-in. copper tube. On installations where the tank is more than 200 ft away from the burner, use $\frac{1}{2}$-in tubing. Installation of the tank must make provisions for the accumulation of dirt and water at a location where neither will be picked up by the oil supply line. The tank is vented to prevent pressure or vacuum from forming inside the tank.

The tank may be located at an elevation below the burner and the oil pump in the burner will lift the oil from the tank. The tank may not be located lower than 15 ft below the burner with this method. The installation of a check valve at the outlet from the tank will assure that there is always oil available at the burner upon a call for heat. This installation uses a supply and a return oil line to prevent air from entering the fuel line.

Where the oil storage must be located more than 15 ft below the burner (as with suspended heaters or rooftop heaters), an oil booster pump must be used.

Figure 17-21 Multiuse stack gas analyzer. *(Courtesy Bacharach Instrument Co.)*

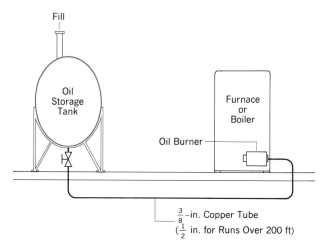

Figure 17-22 Oil storage tank and piping.

KEY TERMS AND CONCEPTS

Oil Numbering System	Intermittent Ignition	Stack Detector
Atomization	Continuous Ignition	Draft Gauge
Heating Value	Ignition Transformer	Barometric Damper
Oil Burner	Fuel Oil Pump	Flue Gas Sampling
Electrodes	Primary Control	Oil Storage Tanks
Nozzle	Stack Controller	
Oil Gun Assembly	Cad Cell	

QUESTIONS

1. Which fuel oil would most likely be used in a residential oil burner?
2. Why do some oils require preheaters?
3. What devices are driven by the motor on an oil burner?
4. How is atomized oil ignited?
5. What is intermittent ignition for an oil burner?
6. What function is served by the ignition transformer?
7. What is an average oil pump discharge pressure?
8. How long will the primary control allow for attempting to ignite the oil?
9. What will happen if the oil in Question 8 fails to ignite?
10. Name two methods that are used to sense that the oil flame has been established.
11. What problem will be encountered if the stack temperature-sensing element is not properly sealed to prevent leakage?
12. Where do the wires from a cad cell go?
13. How much draft should you be able to measure in the flue stack?
14. How is flue gas tested for smoke?
15. What does a high percentage of CO_2 indicate about the combustion efficiency?
16. When is a booster pump used on an oil storage tank?
17. What is a barometric damper?
18. What will happen if electrodes are positioned too close to the oil gun assembly?

CHAPTER 18

BOILERS

The purpose of this chapter is to describe the features of boilers and boiler combustion systems that are different from gas- and oil-fired furnaces.

Boiler Uses

The function of a boiler is to burn a fuel in order to heat water. The output of the boiler may be hot water or it may be steam, depending upon the boiler design. Hot water boilers are used for space heating. Steam boilers may be used for space heating, but the steam output may also be used for a variety of other purposes. Most notably, the steam from a boiler may be used to drive a turbine (Figure 18-1) which in turn, may drive a generator, a pump, or any other type of mechanical equipment. These boilers are referred to as **power boilers.**

The smallest boilers (steam and hot water) are used in residential heating systems. The largest boilers may be as large as a multistory building, and generate sufficient quantities of steam to drive turbines that power the electric generators for an entire region.

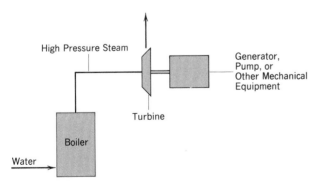

Figure 18-1 A boiler that is used to drive a steam turbine that in turn drives a generator is called a power boiler.

Boiler Codes

A boiler has the capacity to cause significant loss of life and property in the event of equipment failure. Large quantities of water stored in the boiler, under pressure, at temperature above 212°F will flash with explosive force if the pressure is too great for the boiler to contain. Therefore, codes have been written covering the manufacture, operation, maintenance, and testing of boilers. The nationally accepted code governing low pressure boilers is the ASME Boiler Code for Heating Boilers. The code for high pressure boilers is the ASME Code for Power Boilers and Fired Pressure Vessels. **ASME** stands for American Society of Mechanical Engineers. While the ASME codes do not have legal governing authority in and of themselves, state and local governments either make the ASME code a part of the local codes, or write their own codes, which are patterned after the ASME codes.

The use and design of boilers is heavily influenced by the large insurance companies who carry the liability insurance for boiler users. Factory Mutual and Industrial Risk Insurers (FM and IRI) are the most widely involved. These insurance companies can influence the design of boilers and components by offering more attractive premium rates to the users who have boilers designed to meet the company requirements.

Boiler Types

One way to classify boilers is either **low pressure** or **high pressure** (Figure 18-2). For steam boilers, if the maximum operating pressure is 15 psi or less, it is considered to be a low pressure boiler. High pressure steam boilers are commonly designed for 150 psi steam pressure.

	Low Pressure	High Pressure
Steam boilers	below 15 psi	15 psi or higher
Hot water boilers	below 160 psi and below 250°F	above 160 psi or above 250°F

Figure 18-2 Low pressure and high pressure boiler classifications.

In order for a hot water boiler to be classified as a low pressure boiler, it must be designed for a maximum pressure of 160 psi and a maximum temperature of 250°F. Low pressure hot water boilers are commonly operated at 30 psig. Note that the pressure inside a hot water boiler is determined by the design of the water pump and piping system, and it has little to do with the temperature of the water being supplied. The water temperature is usually far lower than the saturation temperature that corresponds to the operating pressure.

Another way of classifying boilers is by the type of construction. There are **watertube** boilers and **firetube** boilers. Watertube boilers have, as the name implies, water inside tubes that is heated from a flame on the outside of the tubes. Firetube boilers (Figures 18-3 and 18-4) have tubes that pass through a water tank. The flame is directed through the inside of the tubes, transferring heat from the inside to the water on the outside of the tube.

Boilers may also be classified as **packaged** or **field erected boilers.** In the smaller sizes, boilers are packaged at the factory and shipped to the jobsite, ready for installation. It includes the boiler, burner, controls, and all auxiliary equipment assembled in a single unit. Larger boilers are erected in the field.

Figure 18-5 shows a commonly used arrangement of components for industrial applications. It is called a **D-tube boiler** because of the cross-sectional shape formed by the tubes, the **steam drum,** and the **mud drum.** The heating effect from the flame causes a natural circulation of water within the boiler. In the steam drum, eliminators separate the steam from the boiling water so that dry steam may be supplied. The mud drum is aptly named due to the collection of solids that tends to collect in this section. Boilers of

Figure 18-3 Firetube boiler. *(Courtesy Cleaver-Brooks, Div. of Aqua-Chem, Inc.)*

Figure 18-4 Cutaway of a firetube boiler showing four passes for the products of combustion. *(Courtesy Cleaver-Brooks, Div. of Aqua-Chem, Inc.)*

this type will use either a manual or an automatic **blowdown** to prevent accumulation of solids.

Solids Control

When water in a boiler is evaporated (turned to steam) the solids that may have been dissolved in that water do not evaporate. If this is allowed to

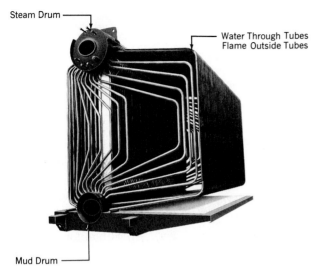

Figure 18-5 D-tube watertube boiler. *(Courtesy Cleaver-Brooks, Div. of Aqua-Chem, Inc.)*

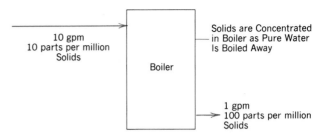

Figure 18-7 The blowdown rate will be much lower than the make-up water rate, but will remove an equal amount of solids.

continue over a period of time, the solids will form a coating on the boiler heat transfer surfaces and reduce boiler efficiency. If solids are allowed to accumulate on the tubes, the heat transfer will be impaired sufficiently to weaken the tubes because of overheating (Figure 18-6). Three methods are used to prevent these dissolved solids from causing trouble.

1. The **make-up water** (new water) supplied to the boiler can be treated through water softeners. This removes the incoming solids to a low level.
2. Chemicals can be added to the boiler water to keep the solids in solution.

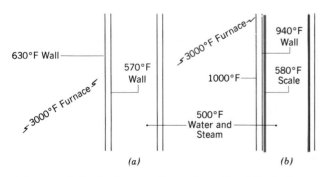

Figure 18-6 Scaling on the water side of the boiler tubes can cause the tube temperature to rise sufficiently to cause damage to the tube. (*a*) Clean boiler tube (tube = 570°F–630°F). (*b*) Scaled boiler tube (tube = 940°F–1000°F).

3. A portion of the water in the boiler can be periodically or continuously removed. This blowdown will remove solids at the same rate as they are being supplied by the make-up water (Figure 18-7). The rate of blowdown required is determined by measuring the total dissolved solids (**TDS**) in the boiler water (Figure 18-8). Figure 18-9 shows continuous blowdown from the surface of the water in the steam drum (called **surface blowdown**). Figure 18-10 shows how the different methods of solids control are used in combination with each other.

Capacity Ratings

Ratings of boilers can be somewhat confusing due to the many different methods of describing capacity. The simplest description of boiler capacity is Btu/hr of fuel input and output (the amount of heat actually transferred into the water). A second simple method of rating boiler capacity is the number of pounds per hour of steam that can be produced per hour at a given pressure. Or, similarly for hot water boilers, the flow rate, entering temperature, and leaving water temperature present a complete description of capacity.

There are also methods of rating boilers that attempt to describe the amount of work or the amount of radiation units that may be supplied from the boiler. Rating methods have been developed by the Hydronics Institute (formerly the Institute of Boiler and Radiator Manufacturers), the American Gas Association (AGA), and other

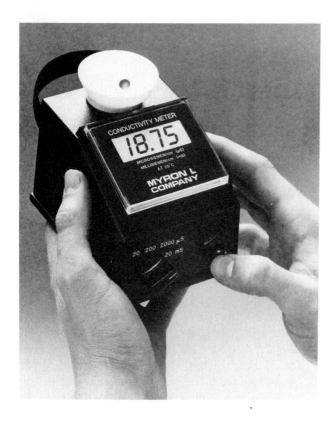

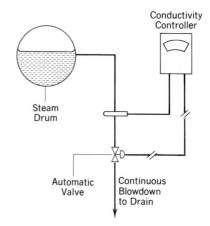

Figure 18-9   Surface blowdown with automatic controller.

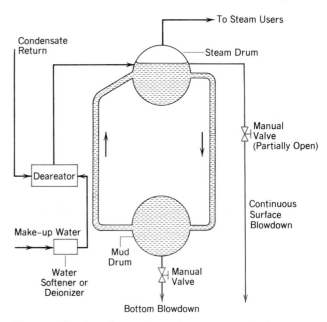

Figure 18-10   Elements used to control the formation of sludge and scale inside the boiler include surface blow, bottom blow, treatment of the make-up water, and deareating the condensate return.

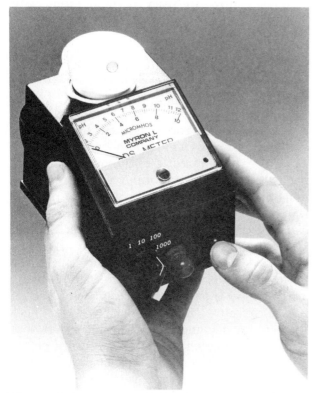

Figure 18-8   Meters used for sampling the condition of boiler water. *(Courtesy Myron L Company)*

industry organizations. A single boiler may be rated at least four different ways. Consider, for example, a boiler rated according to AGA rating methods. It has an **input** of 1700 **MBH** (thousands of Btus per hour), and an **output** rating of 1275 MBH. According to the I-B-R ratings methods, this very same boiler might be rated as 4000 EDR, or 960 MBH output. **EDR** stands for **equivalent direct radiation,** and is an attempt to state how many square feet of radiation surface may be served by this boiler. The difference in MBH output ratings may be explained by the attempt of the I-B-R ratings to present useful output figures, after allowances for piping and other losses have been considered.

And as if that were not enough, the same boiler might also be rated as a 38-hp boiler. Rating boilers in **horsepower** is a carryover from the days when boilers were rated according to how much power could be created from a turbine that used the full output capacity of the boiler. A **boiler horsepower** is equivalent to the capacity to vaporize 34.5 lb/hr of water at 212°F, or 33,475 Btu/hr.

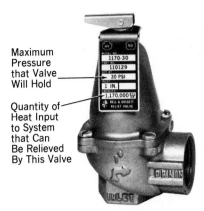

Maximum Pressure that Valve Will Hold

Quantity of Heat Input to System that Can Be Relieved By This Valve

Figure 18-11 Relief valve to protect boiler from over-pressure. *(Courtesy ITT Fluid Handling Division)*

Operational Hazards

There are three major potential hazards that can cause a sudden, devastating loss of a boiler.

1. **Overpressure,** caused by overfiring of the burner.
2. **Overheating,** caused by the water level in the boiler being insufficient to remove the heat from the flame.
3. **Explosion,** caused by the ignition of an accumulation of fuel (gas or oil).

The protection against overpressure on a boiler is the **relief valve** or **safety valve** (Figure 18-11). The relief valve or valves must have sufficient capacity to relieve steam pressure so that even if the burner is firing continuously, the relief valves will release all the steam produced without allowing the boiler pressure to exceed its maximum rating.

Low water level is protected against by the use of a **low-water cut-out switch** (Figure 18-12). If the water level in the boiler drops below a safe level, the low-water cut-out switch will open, causing the burner to turn off.

A piping arrangement called the **Hartford loop** was invented by the Hartford Insurance Company after the loss of several boilers that failed by losing the boiler water through a damaged condensate system. The Hartford loop is an arrangement whereby the feedwater is introduced into a

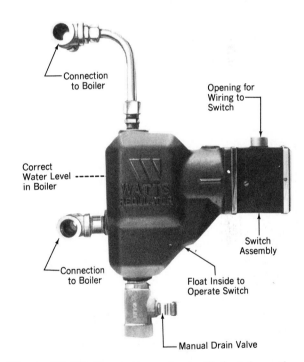

Connection to Boiler

Opening for Wiring to Switch

Correct Water Level in Boiler

Switch Assembly

Connection to Boiler

Float Inside to Operate Switch

Manual Drain Valve

Figure 18-12 Low-water cutout will shut down the burner if the boiler water level drops too low. *(Courtesy Watts Regulator Co.)*

loop above the level of the normal operating water level in the boiler. If the condensate system becomes open, the water in the boiler is prevented from siphoning out.

In larger boilers, the light-off of the boiler is monitored by sophisticated electronic equipment. Generally, the sequence for light-off on these boilers is first to purge the firebox with fresh air to remove any possible accumulation of fuel. Then a pilot flame is ignited and proved. Once the pilot is proved, the main fuel valves will be allowed to open, to be ignited from the pilot. If, during the trial for ignition on either the pilot flame or the main flame, the flame is not proved within a preprogrammed amount of time, the trial for ignition will end. The next attempt at light off will not take place until the boiler firebox is once again purged. Once the main flame is established, any loss in the detection of the main flame will cause the electronic control scheme to return to the original starting point. This is referred to as a **nonrecycling** system.

Flame Detection Methods

Residential and commercial gas-fired units proved the existence of a pilot flame by the use of thermocouples, pilot generators, bimetal switches, liquid filled bulbs, or electronic flame rectification units. Oil-fired units used a heat-sensitive stack switch or a light-sensitive cadmium sulfide cell. Boilers use some of these same methods, as well as others.

All types of flames, while appearing to be of different colors, all emit similar types of **ultraviolet** (UV) and **infrared** (IR) light waves. An oil burner will produce a yellow flame, and a gas burner will produce a blue flame, but both flames emit similar UV and IR characteristics. The naked eye cannot detect these wavelengths, but ultraviolet radiation may be detected by the use of a photocell. Infrared radiation may be detected by the use of a lead sulfide detector. The UV photocell and the lead sulfide infrared detector both respond to varying levels of UV and IR radiation by changing in resistance. When a flame is not seen, the resistance increases causing the detector to act as an open switch.

The **flame rod** is another method of proving the existence of a flame. It is similar in operation to the flame rectification principles used in residential gas-fired furnaces. The flame rod is constructed of a high temperature alloy to withstand temperatures up to 2400°F. Flame rods are used only on gas flames, as they would build up a coating and be rendered useless on oil burners. Sometimes a flame rod is used in an oil-fired burner in order to prove the existence of a gas pilot flame. In this type of application, the flame rod is used in conjunction with another infrared or ultraviolet flame detector, which senses the main flame.

Power Burners

The gas burners described in Chapter 16, on residential and commercial furnaces, were described as atmospheric burners. The combustion air was supplied to the flame at atmospheric pressure.

Large boilers must burn large quantities of fuel, and the mixing of sufficient quantities of gas and air becomes a problem. A combustion air fan is used to supply combustion air to large

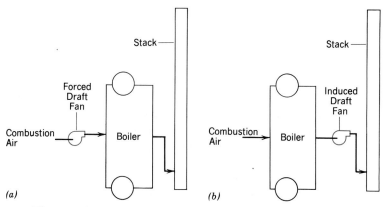

(a)   *(b)*

Figure 18-13 (*a*) Forced- and (*b*) induced-draft boilers.

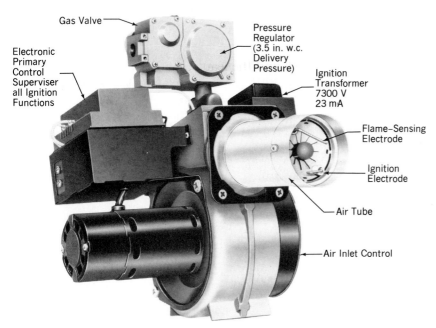

Figure 18-14 labels:
- Gas Valve
- Electronic Primary Control Superviser all Ignition Functions
- Pressure Regulator (3.5 in. w.c. Delivery Pressure)
- Ignition Transformer 7300 V 23 mA
- Flame–Sensing Electrode
- Ignition Electrode
- Air Tube
- Air Inlet Control

Figure 18-14 Power gas burner uses electronic primary controls (flame rectification), supervised ignition, and can be used with natural gas at 5 to 13in.w.c. or LPG at 11 to 13in.w.c. *(Courtesy R. W. Beckett Corporation)*

Figure 18-15 Oil-fired burner. *(Courtesy Weil-McLain Co.)*

burners, so that mixing sections may be used to create a uniform gas–air mixture. These may be either **induced-draft** or **forced-draft** systems (Figure 18-13). Forced draft is more common because the combustion air fan does not need to be capable of handling the hot products of combustion.

A second method of obtaining the energy required to premix the air–fuel mixture is to use higher gas pressures than with atmospheric burners. The gas may be supplied at a pressure of between 1 and 10 psi, compared with 3.5 in. of water column for an atmospheric burner.

Oil Burners for Boilers

Small boilers use the same types of burners that are used on the residential and commercial units consisting of a pump to produce high pressure (100 psi) atomization of a light oil (number 1 or 2) (Figures 18-14 through 18-17). Industrial-type boilers can use this same type of burner with heavier oils, but the oils may have to be preheated to as high as 200 to 300°F in order to reduce the viscosity. Also, the pumps for the heavier oils may be designed to produce higher pressures for atomization of the oil. A pump discharge pressure of 250 psi would not be uncommon.

A second method for providing atomization of the heavier oils is to mix high pressure steam with the oil prior to discharging the mixture through a nozzle (Figure 18-18) **(steam atomiza-**

Figure 18-16 Oil-fired burner. *(Courtesy Weil-McLain Co.)*

Figure 18-17 Oil burner. *(Courtesy Weil-McLain Co.)*

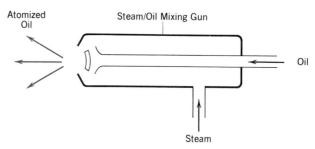

Figure 18-18 Steam atomization of oil.

A fourth method of burning oil in boilers is by the use of a rotary burner. Rotary burners are used for the burning of number 1 through 4 or 5 without the need for preheating the oil. Preheating is necessary with a rotary burner only with number 6 and some number 5 oils.

Combination Burners

Many burners for boilers are capable of burning both gas or oil. In many locations, it is preferred to burn gas rather than oil due to its cleaner burning characteristics and lower operating cost. Many utilities, however, offer **interruptible rates** for the purchase of gas. This means that the consumers will pay a lower than normal rate for the gas used. During periods of extreme de-

tion). The steam reduces the viscosity of the oil producing smaller size oil droplets. The steam used is in the range of 100 to 200 psi.

The third method of atomizing heavy oils is to use compressed air instead of steam. Although compressed air is usually more expensive to produce than steam, it is required on many steam systems for starting up the system. The steam atomization system obtains its steam from the boiler being fired. But if the boiler being fired cannot be fired without steam to atomize the oil, then compressed air is substituted for the steam.

Figure 18-19 Cast iron boiler. The sections of the heat exchanger are bolted together to match the size requirements of the installation. *(Courtesy Weil-McLain Co.)*

Cast Iron Boilers

Cast iron boilers may be constructed in sections (Figure 18-19). The manufacturer may offer capacity ratings to match the job requirements exactly by merely using more or fewer sections. The sections are bolted together as required with draw rods. Sealing rope is used between the sections to provide a gas-tight seal for the products of combustion. The sectional approach also has the feature that if the boiler is large, it can be moved through small access areas for installation or removal. The cast iron heating sections are housed in a sheet-metal enclosure, insulated to prevent heat loss to the mechanical space.

Tankless Water Heater

Small boilers used in residential and small commercial applications (Figures 18-20 and 18-21) may have, as an accessory, a **tankless water heater.** It is merely a heat exchanger that pro-

mand, however, the gas utility may ask the owner of large gas users to stop burning gas. In order to be able to take advantage of the interruptible rates for gas, the dual fuel capability is provided for periods of gas curtailment.

Flue Gas Fan

Pressure Switch

Temperature Controller

Transformer

Spark Ignition

Gas Valve

Burners

Figure 18-20 Gas-fired residential boiler. *(Courtesy Weil-McLain Co.)*

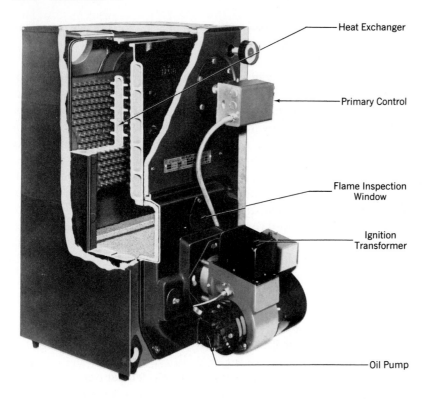

Figure 18-21 Oil-fired residential boiler. *(Courtesy Weil-McLain Co.)*

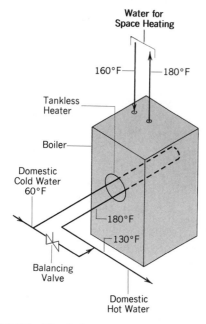

Figure 18-22 The balancing valve can be adjusted to blend cold water with the outlet from the tankless water heater.

vides a path for household hot water through the steam or hot water being stored inside the boiler. The heated domestic hot water may then be used for kitchen, laundry, and baths. The advantage of the tankless hot water heater is it eliminates the requirement for a separate hot water heater.

One disadvantage is that the boiler must remain on line year round in order to satisfy the continuing requirement for domestic hot water. A second disadvantage is the lack of a way to control temperature. The space heating requirements will determine the temperature required inside the boiler. This may, however, produce hot water temperatures that are too high according to the user's taste. In order to overcome this difficulty, the bypass piping shown in Figure 18-22 may be used in order to bleed unheated water into the water stream leaving the tankless water heater.

Compression Tank

When a hot water system is totally filled with liquid, there is no room available for expansion of

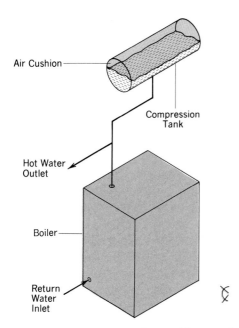

Figure 18-23 Compression tank provides an air cushion to allow space for water expansion.

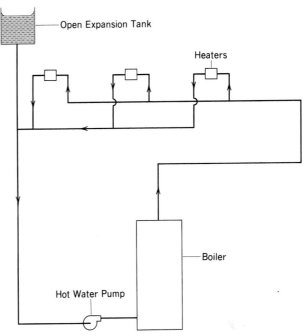

Figure 18-24 Open expansion tank provides room for expansion of water when it is heated.

the water when it heats up to operating temperature. While the amount of expansion of the water will be small, unless there is some place for the water to go, the hydrostatic pressures that may be built up can cause failure of pipes or fittings. The **compression tank** shown in Figure 18-23 has only a single inlet. The tank is filled with roughly half water and half air. When the water expands, it compresses the air cushion, and the air pressure increases very slightly.

Compression tanks are sometimes called **expansion tanks,** although this is really a misnomer. Expansion tanks were used on early systems (Figure 18-24) where an open tank was located above the highest point in the system. It served the same purpose as a compression tank, but in a different manner.

Capacity Control

For residential and small commercial applications, boilers are commonly provided with simple On/Off control. In the simplest system, a thermostat provides a single zone of control, and turns the boiler on or off according to the demands of the space temperature. More commonly, the boiler temperature is maintained

constant at 180°F to 200°F and the room thermostat controls the operation of a hot water pump.

Figure 18-25 shows a system in which a hot water boiler is maintained at a constant water temperature by a thermostat on the boiler itself. **Zone valves** are provided on the boiler outlet so that several thermostats can each operate one specific zone valve. The zone valves are operated by a 24-V motor to open the valve, and is spring-loaded to close when the valve is deenergized. The zone valve is also provided with a set of auxiliary contacts, which are open when there is no call for heat. The auxiliary contacts from each of the zones are wired in parallel with the hot water pump. If any one or more of the zone thermostats calls for heat, the auxiliary contacts close, and the hot water pump is energized.

Figure 18-26 shows another method of obtaining zones of control from a small hot water boiler. Separate booster pumps are provided, one for each zone. Each zone thermostat, when calling for heat, will energize a relay that energizes one of the pumps.

Larger boilers can be equipped with modulating-type control that will vary the firing rate according to demand. Modulating control is ac-

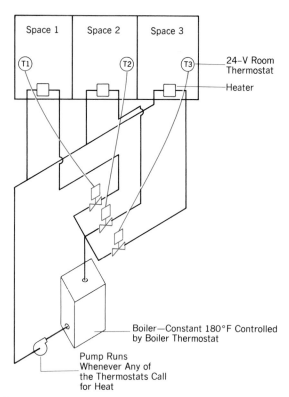

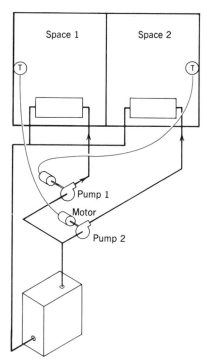

Figure 18-25 Three-zone residential hot water system using zone valves. *(Photo courtesy Watts Regulator Co.)*

Figure 18-26 A two-zone hot water system in which each thermostat operates a pump. The thermostat can be either line voltage or can operate the pump through a relay.

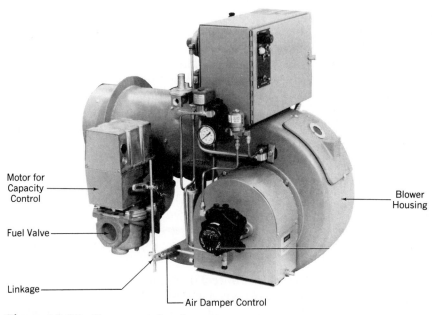

Figure 18-27 Burner used to fire light oil, gas, or a combination of the two. Controls are available for prepurge and postpurge programming, high- and low-fire control with low-fire start, modulating flame control, and automatic fuel changeover. *(Courtesy Weil-McLain Co.)*

complished by using a modulating control valve on the fuel, and dampers on the combustion air that operate with the control valve. The operation of the dampers and the control valve together is usually accomplished by the use of complex mechanical linkages (Figure 18-27). Calibration of the boiler involves a complex series of trial and error adjustments to these linkages in order to optimize the air–fuel mixture at all load conditions. The boiler is usually calibrated so that the minimum quantity of excess air is used at the highest firing rates. As the air and fuel mixture is modulated to a part load condition, the percentage of excess air is increased, and boiler efficiency is reduced.

There is a limitation to how low the capacity of the boiler can be modulated without creating an unstable flame or incomplete combustion. The minimum allowable load depends on the burner and boiler design, and is called **turndown ratio.** A boiler with a 5:1 turndown ratio can operate down to 20 percent of its full load rating.

Combustion Safety Valves

With large boilers it is important to assure that the main fuel valve does, in fact, shut off the flow of fuel when required by the system controls. There are several levels of sophistication that are used to prove proper operation of these valves. The insuring agency will have guidelines about which methods are required for each size boiler. The simplest method of shutting off fuel flow involves only a single valve. Proof of closure may be required. That is, when the valve closes, it mechanically operates a microswitch that is part of the control system.

The second level of sophistication uses two shut-off valves in series. If one fails to shut tightly, the other will still prevent fuel from entering the boiler. Figure 18-28 shows the most sophisticated and most reliable approach to positive shut-off control. It is called a **double block and bleed** arrangement. When the control system shuts down the boiler the two fuel valves close and the bleed valve opens. This way, if the first fuel valve leaks, there can be no fuel pressure buildup between the valves. Therefore, there

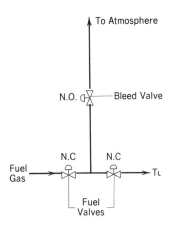

Figure 18-28 The double block and bleed arrangement prevents the accumulation of fuel gas in the boiler because of a leaking shut-off valve.

would be no leakage through the second valve, even if it happened to not close tightly.

Manual reset valves are electrically latched safety valves that must be opened manually at the valve location. These valves are used when it is desirable to have an operator present when the boiler is lit.

Oxygen Trim Systems

In order to remedy the problem of reduced boiler efficiency at partial load, a modern technological advance (**oxygen trim**) constantly monitors the amount of oxygen in the flue gas. The combustion air flow is then controlled independently from the gas valve in order to maintain a low preset amount of excess air. Oxygen trimming can reduce the required amount of excess air from the 10 to 50 percent found on some boilers to less than 5 percent.

Boiler Inspection

Large boilers are inspected regularly in order to assure safety of operation, and to satisfy insurance company requirements. Safety controls are checked on a regular basis in order to be certain that they function at the correct set points, and that they, in fact, take the appropriate corrective action. Annually, internal inspections are made to ensure structural integrity. Heat transfer surfaces are inspected to make sure that they are not becoming coated with a buildup of solids from the boiler water.

To test the pressure-holding ability, the water or steam side of the boiler is pressurized with water to approximately $1\frac{1}{2}$ times the normal operating pressure (the boiler is cold, and the pressurization is accomplished only with pumps). The tubes are then inspected, one by one, from the fire side. Any leakage of water will indicate a defect in the boiler. During this pressurization, the relief valves must be gagged. A **gag** is a device designed to hold the relief valve closed, regardless of the pressure sensed. Extreme care must be taken to make sure that any gagged relief valves are returned to normal service before the boiler is placed back in operation.

Each tube is also inspected for signs of overheating. Potential failures are indicated by blisters or unevenness in the tube surface.

Refractory, which is the brick work that is used to contain the heat within the boiler, is also inspected during the annual inspection. It is quite normal for the refractory to require periodic repair.

Deareator

Corrosion in steam and condensate piping systems is a result of dissolved oxygen. Major boiler installations use a **deareator** tank in order to drive off the oxygen in the feedwater to the boiler

(Figure 18-29). The deareator tank (commonly called a DA tank) is a simple device. The feedwater is sprayed into the tank, and brought into close contact with live steam that is also admitted into the tank. By increasing the temperature of the feedwater, its ability to hold oxygen is reduced. The oxygen that is released is vented to atmosphere.

Economizer

High pressure steam boilers produce a high steam temperature, equal to the saturation temperature corresponding to the steam pressure. The stack temperature will be 50°F to 100°F higher than the steam temperature, resulting in flue gas temperatures in the range of 550°F to 650°F. This tremendous loss of heat up the stack has a negative effect on the boiler efficiency.

To improve the boiler efficiency, a boiler **economizer** can be used between the boiler and the stack (Figure 18-30). The economizer is similar in concept to a heating coil. Feedwater is supplied through tubes, and the flue gas is allowed to

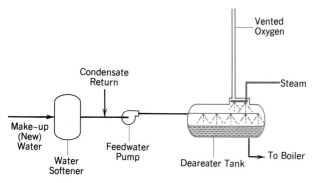

Figure 18-29 Deareator tank in a boiler feedwater system removes oxygen.

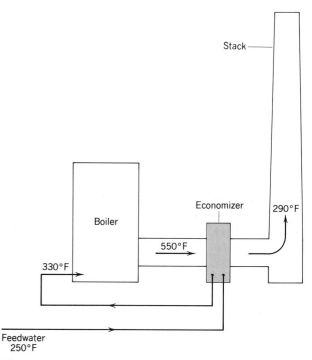

Figure 18-30 The boiler economizer preheats the feedwater by extracting heat from the flue gas.

pass over the tubes transferring heat into the feedwater. For average designs, the products of combustion are cooled to within 40°F of the inlet feedwater temperature. The flue gas temperature generally is reduced by 200°F to 300°F, and the feedwater temperature increases by 70°F to 100°F.

There are two potential operating problems that can be caused by the economizer reducing the flue gas temperature so drastically. First is the risk of condensation on the gas side of the heat exchanger, which can lead to corrosion. For fuel oils, there is some concentration of sulfur in the flue gas, and if condensation is allowed to occur, sulfuric acid may be formed. This problem is avoided by either using materials of construction in the economizer that will be unaffected by the condensation, or by keeping the flue gas temperature high enough so that the condensation will not form.

The second operational problem characteristic of economizers is the formation of soot on the gas side of the heat exchanger. A **soot blower** is a pipe that extends inside the economizer, that is supplied with steam or compressed air on a periodic basis. The soot is blown off the heat transfer surfaces, and is discharged out the stack with the flue gas.

Air Heaters

Air heaters are another improvement in boiler efficiency. A boiler economizer is limited in that the flue gas temperature cannot be any lower than the entering feedwater temperature. The air heater then transfers heat from the resulting flue gas into the entering combustion air (Figure 18-31). The air heater is a relatively light-duty device, as it is not designed to withstand boiler pressure. Figure 18-32 shows the fuel savings that are available by increasing the combustion air temperature with an air heater.

Air heaters are even more susceptible to corrosion due to condensation than the boiler economizer is due to the lower flue gas temperature. Either soot blowing or water washing is used to maintain the cleanliness of the exterior heat transfer surfaces.

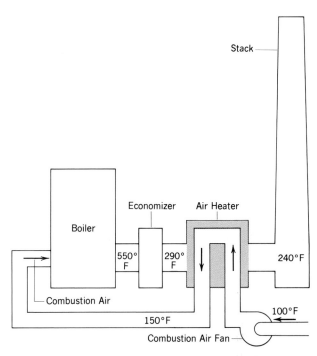

Figure 18-31 The air preheater preheats the combustion air by extracting heat from the flue gas.

Combustion Air Temperature	Fuel Savings (%)
100°F	—
200°F	2.36
300°F	4.73
400°F	7.10
500°F	9.47
600°F	11.80

Figure 18-32 Approximate fuel savings available by increasing the combustion air temperature.

Programmed Boiler Controllers

Figure 18-33 illustrates a programmed boiler controller that is designed to go through a complex set of checks before lighting off a boiler. These controllers are used on larger boiler systems to assure a safe light-off. The programs available vary from installation to installation, but are generally similar to the following.

Figure 18-33 Programmable boiler controller. *(Courtesy Weil-McLain Co.)*

1. Check for low water or other open safety controls.
2. Purge the firebox for 3 min. This involves starting the combustion air fan and opening combustion air dampers.
3. At the completion of the purge cycle, open the pilot gas valve and start the ignition system. If the pilot flame sensor does not prove a pilot flame within 30 sec, shut the system down.

Some controllers will return the system to step one and try again to light the pilot flame.

4. When the pilot flame is proved, open the main fuel valve. If the main flame is not proved within 30 sec, shut the system down.

There are many detailed checks that the programmed controller may make in addition to the major ones described above. Some of these might be as follows.

1. After the combustion air fan has started, prove that it is, in fact, moving air by the use of a pressure or sail switch.
2. After the combustion air dampers receive a signal from the programmed controller to open, check with a micro-switch to make sure that they have, in fact, opened fully. In this way, an adequate purge is assured.
3. When opening the main fuel valve, the controller sets it to a minimum fire position. A micro-switch senses the position of the fuel valve stem to assure that the valve is not opened to any more than minimum fire position.

In short, every output signal from the programmed controller requires an input signal from another sensor or switch to prove that the operation that was supposed to happen in fact happened. This level of redundancy is justified by the potentially large losses of life and property inherent with a malfunctioning boiler.

KEY TERMS AND CONCEPTS

Power Boiler	Steam Drum	Photocell
ASME	Mud Drum	Flame Rod
Low and High Pressure Boilers	Boiler Horsepower	Power Burner
Watertube	EDR	Steam Atomization
Firetube	Safety Relief Valve	Induced Draft
Packaged Boiler	Overpressure	Forced Draft
Field Erected Boiler	Overheating	Interruptible Rates
D-tube Boiler	Hartford Loop	Tankless Water Heater
Blowdown	Programmed Control	Compression Tank
Surface Blow	UV	Expansion Tank
TDS	IR	
	Make-up Water	

Zone Control

Turndown Ratio

Double Block and
   Bleed

Oxygen Trim

Hydrostatic Testing

Gag

Refractory

Deareator

Economizer

Air Heater

Soot Blower

QUESTIONS

1. What is a power boiler?

2. What is ASME, and how does it affect boiler manufacturing?

3. A hot water boiler is operating at 100 psi and a temperature of 260°F. Is this a high pressure boiler? Why?

4. What are the two drums called in a D-tube boiler?

5. What is the purpose of blowdown?

6. What is the heat input in Btu/hr of a 90-hp boiler?

7. What device is used to protect a boiler against overpressure?

8. What physical damage will occur to a boiler if it is fired without a sufficient level of water?

9. What happens during the purge cycle on a boiler start-up?

10. What type of flame detector detects ultraviolet radiation?

11. What type of flame detector detects infrared radiation?

12. What is the difference between a power burner and an atmospheric burner?

13. Name four methods of atomizing fuel oil.

14. Why are dual fuel burners used?

15. In a tankless water heater, what is losing heat? What is gaining heat?

16. What is the purpose of a compression tank? How is it different from an expansion tank?

17. When fuel flow is modulated, what other input to the boiler is also modulated?

18. What is the advantage of oxygen trimming systems?

19. Name two ways in which zone control is accomplished on small hot water boilers.

20. In a boiler economizer, what gains heat? What loses heat?

CHAPTER 19

WATER, STEAM, AND GAS PIPING

The purpose of this chapter is to demonstrate the relationship between pipe sizing and rate of flow, and to discuss the valves and specialty items used in piping systems. The examples used center around hot water and steam piping. The principles for chilled water piping are identical, and many of the aspects of refrigerant piping rely on the same physical concepts described in this chapter.

Flow and Friction

A basic principle of fluid flow is that a fluid (either a liquid or a vapor) will only flow from point A in a pipe to point B if the pressure is lower at point B than it is at point A. Whenever a fluid is flowing, there is **friction** between the fluid and the inside walls of the pipe. The effect of this friction is to reduce the fluid pressure along its path of flow. The amount of friction depends on three factors.

1. The velocity of the fluid in the pipe.

2. The roughness of the inside surface of the pipe.

3. The density, viscosity, and other characteristics of the fluid that is flowing.

For purposes of simplification, we will talk about the flow of water, and we will not deal with all the variables inherent in (3) above. In order to demonstrate the effects of velocity on friction, let's look at the example in Figure 19-1. A $\frac{3}{4}$-in. garden hose, 100 ft long, is attached to a hose bibb. Let's say that there is 40 psi of pressure available at the starting end of the hose, and we get a flow rate of 12 gal/min (gpm). If we were to place pressure gauges all along the length of the hose, we would find that the pressure becomes progressively lower as the water moves through the

hose. This is the **pressure drop** caused by friction. Just before the end of the hose nozzle, we only have 17 psi of pressure remaining. The last 17 psi of pressure in the hose is used up in friction through the nozzle, and the water pressure at the hose outlet is zero.

There are several factors that we can now change. Let's see what happens if we change the size of the hose, change the type of hose, and change the setting of the nozzle at the end of the hose.

If we reduce the size of the hose from $\frac{3}{4}$-in. diameter to $\frac{1}{2}$-in. diameter, the flow rate will be reduced. This is because in order to push the same 12 gpm through the more restrictive piping system, we would have higher velocities resulting in more friction. We would require more than the 40 psi that is available in order to deliver 12 gpm. Therefore, the flow rate is reduced until we reach, let's say, 7 gpm. At that lower flow rate through the smaller system, we obtain the same total 40 psi of friction loss through the hose and the nozzle.

If we change the type of $\frac{3}{4}$-in. hose to one that has a rougher inside surface, the effect will be the

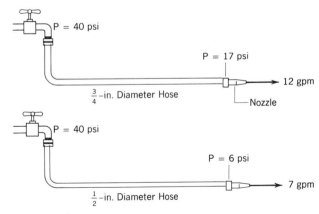

Figure 19-1 The pressure inside the hose is reduced along the length of the hose because of friction.

same as using the smaller diameter hose. The additional friction will cause additional pressure drop, and the flow rate will go down. Note, however, where the pressure drops are being taken in each example. In the first example, the 12 gpm had a pressure drop of 23 psi in the hose, and 17 psi in the nozzle. In the examples with the small hose and the rough hose, the pressure drop across the nozzle has been reduced to 6 psi, and the pressure drop through the hose has been increased to 34 psi. It makes sense that the pressure drop across the nozzle has been reduced. The nozzle configuration has not changed, so that with a reduced rate of flow, there will be reduced friction and reduced pressure drop.

If we return to our ¾-in. smooth surface hose and readjust our nozzle so that we get 7 gpm, we find that the pressure drop through the hose has been reduced from 23 psi to 8 psi, and the remaining 32 psi loss is taken across the readjusted nozzle.

From the previous discussion, it would appear that given enough pressure, we would be able to jam any amount of water through a hose or pipe (assuming that the hose strength is sufficient to withstand the pressure without bursting). Theoretically, this is almost true. In practice, however, there are physical and economic limitations to the amount of flow we can get through a pipe or hose.

Limitations to Flow Rate

In practice, the flow rate through a pipe is limited by the velocity of the water. When velocities become too high, the friction against the sidewall of the pipe becomes so high that the pipe actually wears out. The upper limit of velocity to prevent this **erosion** is a water velocity of 8 ft/sec (fps). For smaller pipes, or for pipes carrying silty or otherwise abrasive water that would aggravate the erosion problem, the velocity limitation is even lower.

The second limitation to pipe velocity is one of economics. Where we need to push water through a pipe, there needs to be a pump. The horsepower requirement of the pump depends on the flow rate and the pressure differential that the pump produces. As water velocity through a

gpm	2	4	8	17	25	50	80	140	280	850
Pipe size	½	¾	1	1¼	1½	2	2½	3	4	6
fps	2.2	2.5	3.0	3.4	4.0	5.0	5.5	6.4	7.2	9.5

Figure 19-2 Recommended water flow rates in pipe.

piping system is increased, the pressure drop (and therefore pump horsepower required) increases very rapidly.

The chart of maximum flow rate versus pipe size in Figure 19-2 represents a fairly good compromise between reliable design and economic pipe sizing and operating cost.

Friction Losses in Pipe

If the total friction loss is a piping system is known at any flow rate, we can calculate what the friction loss will be at any other flow rate. The table in Figure 19-3 gives you multiplication factors that relate the change in flow rate to the change in friction. Note that if you desire to double the flow rate, your pump will need to be capable of producing 3.6 times the pressure differential, and the horsepower requirement for the pump motor will increase by a factor of 7.2 times!

In order to size new piping, the engineer uses tables to determine the pressure drop per 100 ft of piping for the flow rate desired and the pipe size and type. Fittings, bends, valves, and size

New gpm Old gpm	Multiply Original Friction Loss By:	Multiply Original Horsepower By:
1.1	1.19	1.31
1.2	1.40	1.68
1.3	1.62	2.11
1.4	1.86	2.61
1.5	2.13	3.20
1.6	2.38	3.81
1.7	2.67	4.54
1.8	2.97	5.35
1.9	3.28	6.23
2.0	3.60	7.20

Figure 19-3 Change in friction loss and pump horsepower required caused by a change in the flow requirement.

changes are all handled by looking up an "equivalent length of straight pipe" for that particular fitting. Once the total pressure drop is determined for a piping system at any flow rate, the pressure drop at other flow rates may be determined by the factors given in Figure 19-3.

Piping Materials

The most commonly used **piping materials** for heating, air conditioning, and refrigeration systems in use today are as follows.

1. Black steel and galvanized steel.

2. Black iron and galvanized iron.

3. Copper.

4. PVC plastic.

Copper tubing is suitable for use in chilled water, condensate or make-up water systems, drain and condensate lines, steam and condensate, and hot water piping. The material cost for copper is higher than it is for iron or steel, but the quick installation makes copper an attractive material for use in sizes less than 2-in. Copper tubing is not recommended, however, for use in natural gas piping. There may be compounds in the gas that cause a scale to form on the inside of the tubing that can break loose and clog small orifices. Gas piping is black iron above ground, and either coated iron or plastic when run underground.

PVC plastic piping is gaining in popularity for applications permitted by local codes. Its advantages are low installed cost due to easy installation and high corrosion resistance. It also provides lower friction losses than iron pipe because of its smooth inside surface.

Determining the Required Water Flow

Before the piping sizes required in a hot or chilled water system can be determined, the required flow rate of water must be calculated. Hot water is normally supplied to heating coils at a temperature range of between 160°F and 200°F. As it passes through the coil, it will give up suffi-

cient heat to the air being heated so that the water leaves the coil 10°F to 30°F cooler than it entered. The relationship between the flow rate of the water, the temperature change it experienced, and the amount of heat it released is given by the formula

$$Q = 500 \times \text{gpm} \times \Delta T$$

Example

How much 180°F water must be supplied to a heating coil that will transfer 20,000 Btu/hr to the air passing through the coil (Figure 19-4)? The desired leaving water temperature is 160°F.

Solution

$$20,000 = 500 \times \text{gpm} \times 20°F$$
$$\text{gpm} = 20$$

It is very common to have a water ΔT of 20°F in heating coil applications. Where this is the case, the gpm required will always be the Btu/hr divided by 1000.

When investigating systems that do not provide sufficient heating, reading the inlet and outlet temperatures across the coil can provide clues as to the cause of the problem. For example, say that you find that the temperature of the water entering a heating coil is 180°F and leaving at

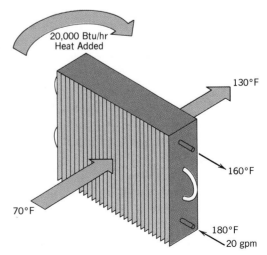

Figure 19-4   Typical water and air temperatures for a hot water heating coil.

120°F. The high ΔT would indicate that the water flow rate is too low. Similarly, if you found that the leaving water temperature was very close to the entering water temperature (very low ΔT), you would look for causes that would prevent the heat transfer rate from the water into the air from being as high as it should. Low airflow or dirty coils (inside or outside) could be the culprit.

Determining Water Flow from Pressure-Drop Readings

We described how the friction of water flow through piping causes a pressure drop. Pressure drop is also incurred when water passes through a heat exchanger, coil, or any other device. Manufacturers provide data that tell the pressure drop through their unit for any flow rate (Figure 19-5). If this manufacturer's data can be obtained for a unit, reading the inlet and outlet pressures can be used with the manufacturer's data in order to determine the actual flow rate through the unit.

Example

The heat exchanger whose pressure-drop characteristics are described in Figure 19-5 has an inlet pressure of 45 psi and an outlet pressure of 40 psi. How much water is flowing?

Solution

By subtracting the two pressures, we determine that the pressure drop through the heat exchanger is 5 psi. Enter the vertical scale of the graph at 5 psi, move horizontally to the right,

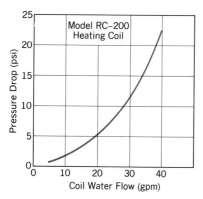

Figure 19-5   Example of manufacturer's data that may be supplied with any heat exchanger.

and when you reach the line, read vertically down. The flow rate is 20 gpm.

Valves

Valves serve two purposes. One is to control (adjust) the flow rate. The other is to provide isolation of a part of the system by shutting off a valve completely. The general types of valves that are available to perform these functions are

1. Globe valves.
2. Gate valves.
3. Ball valves.
4. Butterfly valves.
5. Balancing cocks.

Figures 19-6 and 19-7 show a **globe valve.** It may be used to throttle (adjust) flow rates, and may be left in a partially open position for normal operation. Globe valves must be installed with flow in the proper direction. This direction is marked with an arrow on the casting of the body. The correct flow direction is the direction that tends to push the valve plug away from the seat. If the valve is installed backwards, the flow will tend to close the valve when the plug is near the seat. A rapid chatter of the valve can result.

When globe valves are installed in a line, even when the valve is wide open, there is a significant pressure drop caused by the valve. The water must change directions twice just to make its way through the seat. If it is anticipated that a valve will only be used in either a wide open or completely closed position, a gate valve is used (Figures 19-8 and 19-9).

The **gate valve** presents almost zero obstruction to flow when the valve is open. And when the valve is closed, it provides tight shutoff. The gate valve uses a disc as a gate that can be lowered to completely block off the pipe, or raised to get completely out of the way. Gate valves have been known to fail when the gate separates from the stem of the valve. The stem gets screwed all the way out, but the gate remains obstructing the pipe.

Ball valves (Figure 19-10) consist of a ball that rotates tightly within a housing. A hole through the valve can be aligned with the pipe to allow

Figure 19-6 Globe valve for water and steam service. *(Courtesy Watts Regulator Co.)*

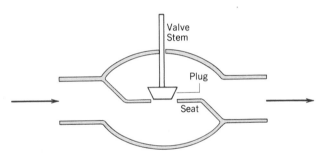

Figure 19-7 Cutaway of a globe valve with the correct flow direction indicated.

Figure 19-8 Gate valve. *(Courtesy Watts Regulator Co.)*

flow, or it can be turned one quarter of a turn to completely block off flow through the pipe. The handle of the ball valve indicates the position. If the handle lines up in the direction of the pipe, the hole in the ball is also aligned with the pipe, allowing full flow. Ball valves provide flow control characteristics that are as good as globe valves, and tight shutoff that is as good as gate valves. Ball valves are most commonly used in smaller sizes (2 in. and smaller). In the larger sizes, they tend to be more expensive than globe or gate valves.

Above the 2-in. size, **butterfly valves** are an attractive alternative (Figure 19-11). The butterfly valve is also a quarter-turn valve. Instead of a ball, a disc is rotated inside the valve. In the closed position, the disc may be seated against a rubber or other soft material to obtain tight shutoff.

A **balancing cock** (Figure 19-12) consists of a tapered cylinder that can rotate one-quarter of a turn. It has an opening in the cylinder that will permit flow when it is lined up in the direction of

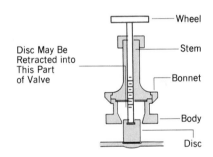

Figure 19-9 Gate valve cutaway. In the open position, the gate valve provides very low resistance to flow.

Figure 19-10 Ball valve. The position of the handle shows that the valve is in the open position. *(Courtesy Watts Regulator Co.)*

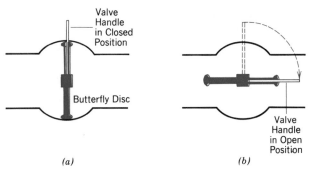

(a)   (b)

Figure 19-11   Butterfly valve. (*a*) Closed. (*b*) Open.

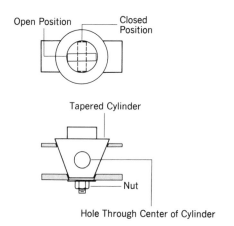

Figure 19-12   Balancing cock.

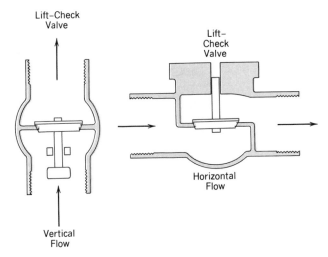

Figure 19-13   Lift-check valves.

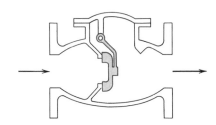

Figure 19-14   Swing-check valve.

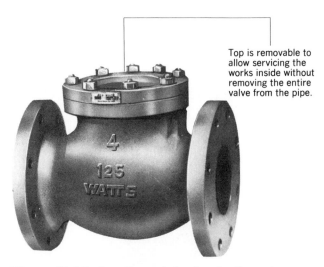

Figure 19-15   Iron flanged check valve for water or steam service, sizes 2 to 8-in. (*Courtesy Watts Regulator Co.*)

the pipe (similar to the ball valve). It is often supplied with no handle. It is set once by the installing technician (with a wrench), and remains in that position. If it is difficult to turn the balancing cock, it may be necessary to loosen the nut on the bottom, which holds the tapered barrel down. A sharp rap on the bottom of the nut will break the cylinder loose from the barrel. Balancing cocks are used to adjust the maximum flows in water systems, and are also common as shut-off valves on small gas lines.

Many valves have pressure ratings stamped on the body, for example, 125 WSP, 200 psi WOG. The abbreviation WSP is the maximum working steam pressure, and WOG is the maximum allowable pressure when the valve is used for water, oil, or gas.

Check Valves

Check valves used for water service are similar in concept to those used for refrigeration service.

Two common types of check valves are **swing checks** and **lift checks**.

Figures 19-13 through 19-16 show the differences between the two types. The swing check creates less friction in the direction of flow when there is flow through the pipe.

Figure 19-16 Bronze check valve for water or steam, "Y" pattern, sizes ⅜ to 2-in. *(Courtesy Watts Regulator Co.)*

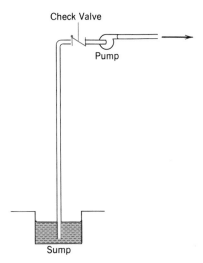

Figure 19-17 Check valve assures that water will remain in the pump during shutdown.

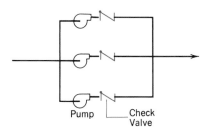

Figure 19-18 Check valves in the pump discharge lines prevent reverse flow through an idle pump.

Figure 19-17 shows a check valve installed in the suction line of a pump that has a suction lift. Without the check valve, whenever the pump turned off, all the water in the line would run back down into the sump, leaving the pump full of air. When the pump restarted, it would merely

spin, without moving any water at all. The check valve prevents the pump from losing its prime.

Figure 19-18 shows a check valve installed in the discharge line of pumps that are installed in parallel. Parallel piping of pumps is a wise idea that provides system stand-by. When only one of the pumps is required to run and the other is turned off, the check valve prevents the running pump from pumping backwards through the idle pump.

Pumps

Hot water and chilled water systems have pumps that move the water from the hot water boiler (or water chiller) through the heating or cooling coils. For normal service, **centrifugal pumps** are used. An impeller is driven by a motor. Water is thrown to the outside perimeter of the impeller, and its pressure is increased. Figures 19-19 and 19-20 show a small, close coupled booster pump that is mounted directly in the water line. Where larger flows are required, pumps such as those shown in Figures 19-21 and 19-22 may be used. They can be either **end-suction** pumps or **in-line** pumps.

End suction means that the water is brought into the side of the housing (center of the impeller), and the discharge is radial from the impeller. These pumps can be either close coupled (the impeller is fastened to the motor shaft), or there may be a separate motor and pump. Where the motor and pump each have their own shaft, they are connected by a flexible coupling. The flexible

Figure 19-19 In-line pump. *(Courtesy ITT Fluid Handling Division)*

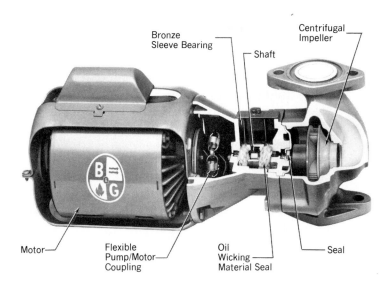

Figure 19-20 Cutaway of in-line pump. *(Courtesy ITT Fluid Handling Division)*

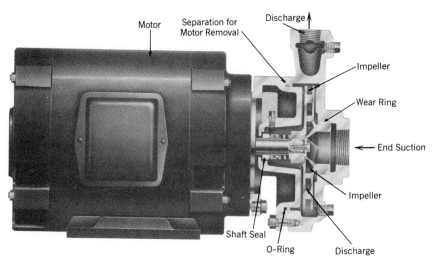

Figure 19-21 Close-coupled end-suction pump. *(Courtesy Aurora Pump)*

coupling allows for independent movement of the two shafts, allowing for slight misalignment or differential expansion as temperatures change. Where the pump shaft passes through the pump housing, there will be a shaft seal. This seal may be mechanical, such as the ones used for open compressors, or it may be a stuffing box with packing gland and packing. The mechanical seals do not permit any leakage, but they are susceptible to damage if the water being pumped is not clean. The stuffing box is padded with a soft material that is compressed to the shaft inside a closed box by the packing gland. The stuffing box depends upon a slight flow of the water being pumped to leak out through the seal packing to provide seal lubrication. The untrained

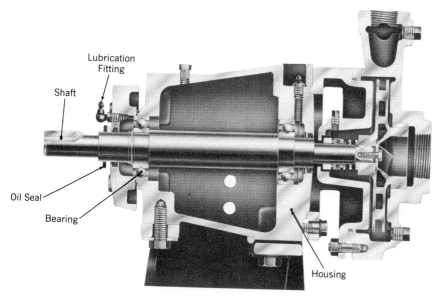

Figure 19-22 Flexible coupled pump – end suction. *(Courtesy Aurora Pump)*

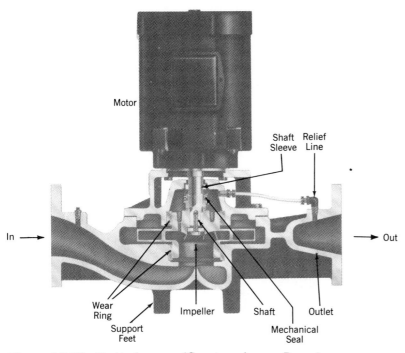

Figure 19-23 Vertical pump. *(Courtesy Aurora Pump)*

technician who tries to overtighten the packing gland to stop the leakage causes damage in two ways

1. The lack of lubrication ruins the seal packing.
2. The tightness of the seal packing around the shaft can cause the shaft to wear. The resulting unevenness (after a few weeks of opera-

tion) makes a good seal impossible, even if a new seal packing is installed (it will fail quickly). In severe cases, the increased friction of a too tight seal can cause the motor to overload.

The pumps shown to this point are classified as horizontal. Figure 19-23 shows a vertical pump. Where a sump is used as a water collection point, the vertical arrangement allows the pump to be located inside the sump, while the motor remains safely above.

Pump Flow Characteristics

Centrifugal pumps will pump an amount of water that depends upon the amount of pressure opposing the flow. Figure 19-24 shows a typical **pump curve,** which is available for all pumps from the manufacturer. It shows how the flow output (gpm) varies as the pressure difference across the pump changes. The pump curve is usually based on 1750 rpm, which is the most common motor speed. Some are based on 3500 rpm or other available motor speeds. The various curves describe the flow characteristics for different diameter impellers that can be used in the same pump housing. Impeller diameters may be obtained in any incremental size, merely by taking the maximum size impeller and turning down the blades on a lathe.

The horsepower drawn by the centrifugal pump depends on the following two factors.

1. The gpm being delivered.

2. The pressure against which the pump works.

The pump curve shows the actual horsepower required to drive the pump under all combinations of flow and pressure. The interesting observation here is that if you open a partially closed valve in the water line, the horsepower draw from the motor will actually go up! At first glance, this may seem contrary to common sense, because when you open a valve, you provide an easier path for the water to flow, and the pressure differential across the valve is reduced. In reducing the pressure drop, however, the pump now delivers more water. The increase in horsepower due to the increased flow more than offsets the reduction in horsepower due to the reduction in pressure drop.

Where no pump curve is available, the horsepower draw from a pump can be estimated as

$$hp = (gpm \times \Delta P)/2800$$

The ΔP in this formula must be in feet of water. If the pressure difference is known in psi, feet can be obtained by multiplying the psi by 2.3. The formula is based upon a pump that is 70 percent efficient. That is, 70 percent of the work being done by the motor is going into increasing the pressure of the water, while 30 percent is wasted in friction, turbulence, and heat.

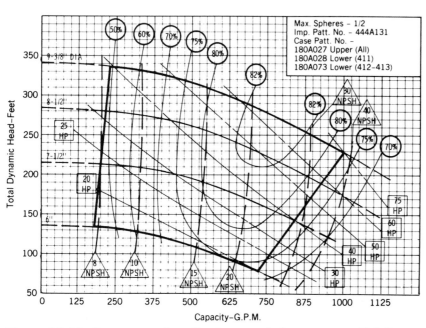

Figure 19-24 Pump curve describes how much flow a pump will deliver at any pressure differential. *(Courtesy Aurora Pump)*

Example

Estimate the flow being provided by a pump. Its discharge pressure is 90 psi and its suction pressure is 30 psi. The 10 hp motor that is driving the pump is operating at 80 percent of its design full load amps.

Solution

The motor horsepower output can be estimated as 80 percent of its rated horsepower.

Pump horsepower draw = 0.80 × 10 = 8 hp

The pressure differential across the pump is

$$(90 - 30) \text{ psi} \times 2.3 \text{ feet/psi}$$
$$= 138 \text{ feet}$$

The flow rate of the pump is

$$\text{gpm} = (\text{hp} \times 2800)/\Delta P$$
$$= (8 \times 2800)/138$$
$$= 162$$

Actual pump efficiencies can easily vary from 50 percent to 85 percent, so the previous formula must be used with the understanding that it is only an estimate.

Net Positive Suction Head

Net positive suction head (abbreviated **NPSH**) is the amount of pressure that is required on the suction side of the pump to prevent it from cavitating. **Cavitation** is an effect that occurs when the pump handles some vapor along with the liquid. This effect can be caused when there is insufficient liquid pressure feeding the pump suction. Cavitation causes severe damage to impellers. An impeller that has been cavitating for only a month or two may look like it has been severely beaten with a ball peen hammer. If an operating pump sounds as if it is pumping rocks inside the piping, it is cavitating. The vapor bubbles inside are collapsing as they go through the pump, causing thousands of implosions. The problem is most common when handling very hot water (near its boiling point) or where the pump must lift the water from a reservoir that is located below the level of the pump.

Example

You want to use a pump that requires 16 ft of NPSH in order to pump water out of a sump. The sump is located 20 ft below the pump suction. Will it work?

Solution

The pump will "see" atmospheric pressure, minus the pressure required to lift the water 20 ft. The net positive suction head available is

$$\text{NPSH} = (\text{atmospheric pressure}) - (\text{pressure losses})$$
$$= (14.7 \text{ psi} \times 2.3 \text{ psi/foot}) - (30 \text{ foot lift})$$
$$= 33.8 - 20$$
$$= 13.8 \text{ feet}$$

The pump will not work without cavitating. The friction loss of the water flow through the pump suction line should also be subtracted from the NPSH available, making the situation even worse. The options available to prevent this cavitation would be as follows.

1. Select a pump that requires less NPSH.
2. Mount the pump at a lower elevation.
3. Use a vertical pump located in the sump.

Piping Connections

Water or steam piping systems can be assembled with any of the following **piping connections.**

1. Screwed.
2. Flanged.
3. Welded.
4. Soldered.
5. Grooved.

Screwed pipe is commonly used in $\frac{1}{2}$-, $\frac{3}{4}$-, 1-, $1\frac{1}{4}$, $1\frac{1}{2}$-, and 2-in. sizes (larger sizes are available but not common). The sealing of a screwed joint relies on an interference fit. The male thread is actually a tapered thread. The female threaded part squeezes tighter onto the male pipe as it is threaded.

Prior to assembly of a screwed joint, a compound called **pipe dope** or pipe joint compound is applied to the male pipe end. The primary purpose of the pipe joint compound is to lubricate the threads, allowing them to be turned tighter. This lubrication also prevents the threads from being damaged as they are tightened. A secondary purpose for the pipe joint compound is to provide additional material that fills the space between the two mating fittings, thus providing some additional measure of sealing.

For convenience and speed of assembly, a specially formulated plastic tape may be used instead of the pipe dope. Sometimes called **tape dope,** the Teflon tape is wrapped tightly two turns around the male pipe, in the direction of the thread (Figure 19-25). If it is wrapped in the wrong direction, when the parts are assembled, the tape dope will tend to unwrap. Many installers prefer tape dope for its superior lubricating and sealing properties. Where tape dope is used on gas piping, it is imperative that the edge of the tape be wrapped no closer than $\frac{1}{8}$-in. to the end of the male pipe. Otherwise, there is a risk of the tape getting into the inside of the piping system where it may easily clog small orifices.

Figure 19-26 shows piping that is connected with flanges. One flange is welded to each pipe end, and the flanges are then bolted together, with a gasket in between. Flanged joints provide a convenient method for disassembling large diameter piping. Flanges are either flat face (ff) or raised face (rf). This refers to the mating surfaces. Most flanged joints are flat face. Piping that is subjected to unusually high pressure will use raised-face flanges.

Welded pipe joints (Figure 19-27) are the most permanent type of connection. They must be cut apart for removal of any portion of the system (unless flanges are provided at valves, etc.). Welded joints can be made to have a very smooth surface, and are widely used in foods processing and other industries. They are not common in ordinary hot water, chiled water, or steam systems (except for high pressure steam). The cost of field pipe welding makes this the most expensive of the piping methods discussed.

Soldered connections for hot or chilled water may be used with copper tubing and are prevalent in sizes of less than 2-in. The filler material may be 50/50 tin–lead, or 95/5 tin-antimony (except for localities where the use of lead-based solders is restricted).

Figure 19-25   Wrapping a pipe thread with tape dope.

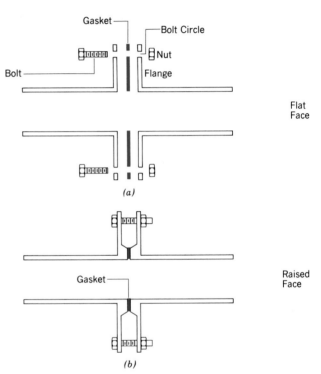

Figure 19-26   Piping connected with flanges.

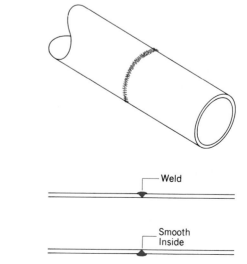

Figure 19-27   Welded pipe.

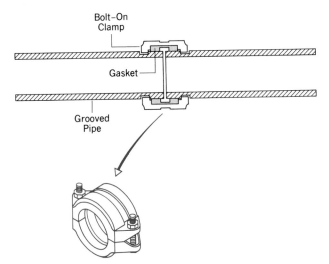

Figure 19-28 Grooved pipe connection uses a bolt-on clamp to hold a gasket in place. The pipes are pregrooved to accept the clamps.

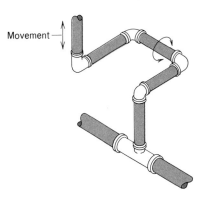

Figure 19-29 Piping detail to relieve pipe stress caused by thermal expansion.

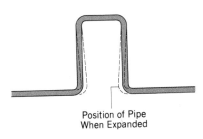

Figure 19-30 Piping detail of an expansion loop to allow for thermal expansion in a long, straight run of pipe.

Grooved connections (Figure 19-28) have gained popularity in recent years due to the fast assembly and ease of disassembly. They are available in sizes from 2- to 8-in. The pipe ends are prepared in the shop with a special grooving machine. In the field, a neoprene gasket is slipped over the joint, and a housing is then tightened around it. The housing holds the two pipes from separating, and squeezes the gasket tight to form a leaktight joint. These joints are sometimes referred to as Victualic couplings, although Victaulic Company is only one of several manufacturers.

For short runs of piping, pipe expansion may be accommodated by the use of expansion joints. These expansion joints require packing and lubrication, so they must be accessible for regular maintenance.

Pipe Expansion

When the temperature of a piping system increases, the pipe grows in length. This can occur upon start up of a steam or hot water system, or upon shut down of a chilled water system. If two ends of a long run of straight piping are anchored in place, something will fail when the piping expands. Steel piping will grow in length by $\frac{3}{4}$ in. per 100 ft for each 100°F increase in temperature. Copper tubing will expand by $1\frac{1}{8}$ in. per 100 ft for each 100°F temperature increase. To allow for this expansion, **expansion loops,** such as those in Figures 19-29 and 19-30, are used. The pipe expansion is allowed to occur by the bending of the legs of the expansion loop.

One-Pipe Hot Water Systems

The **one-pipe** hot water system (Figure 19-31) is used only in small residential systems. Water from a hot water boiler is supplied at 180°F. At the first room heater, a portion of the 180°F water is diverted into a branch pipe into the heater. This diversion is accomplished by use of a device such as the **monoflo** fitting (Figures 19-32 through 19-34). The velocity of the flow of water in the main through the venturi in the monoflo fitting causes a pressure reduction that draws the water up into the room heater and back into the main.

The disadvantage to this system is that the water temperature in the main becomes lower and lower as each successive heater removes

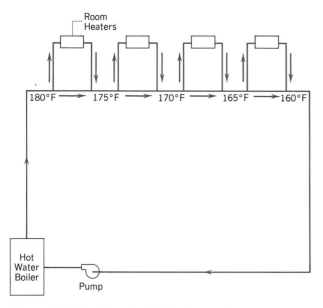

Figure 19-31 One-pipe hot water system.

Figure 19-32 Monoflow fitting for use with a one-pipe water system, threaded connections. (*Courtesy ITT Fluid Handling Division*)

some of the heat. The last heaters on the line may not get water at a sufficiently high temperature to do enough heating. In any event, the heaters at the end of the line must be physically larger than the heaters at the beginning of the line. The only advantage to the one-pipe water system is the low piping cost.

Two-Pipe Water System

The **two-pipe** water system shown in Figure 19-35 uses a supply main and a return main. The temperature in the supply main remains at 180°F, available to each of the heaters. When each heater uses some of the 180°F water, the cooler return water is collected in the return header. This system is called a two-pipe **direct-return** system. As each water flow exits from a heater, it is routed directly back to the hot water boiler. This presents some problems in obtaining balanced water flows to all the heaters. The heaters near the boiler have the shortest supply and return piping lengths, and they will tend to receive too much water. The heaters at the end of the line may be starved for sufficient water. Balancing valves would need to be installed in each of the heater branch lines in order to balance the flows.

Figure 19-36 shows another two-pipe system, but this one is **reverse return** instead of direct

Figure 19-33 Monoflo fitting with solder connections. (*Courtesy ITT Fluid Handling Division*)

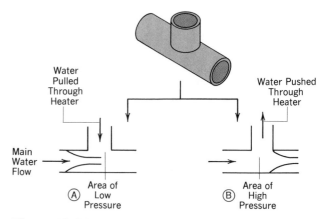

Figure 19-34 Monoflo fitting cutaway shown on return side (A), supply side (B), or both sides of a hot water heater.

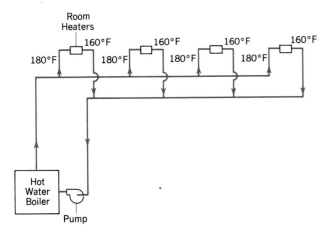

Figure 19-35 Two-pipe direct-return hot water system.

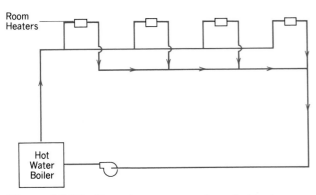

Figure 19-36 Two-pipe reverse-return hot water system.

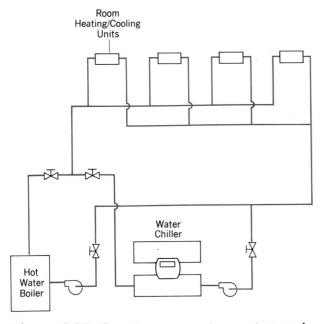

Figure 19-37 Two-pipe reverse-return system used to provide heating or cooling, but not both at the same time.

return. The heater closest to the boiler has the shortest supply piping length, but it has the longest return piping length. The piping lengths for each of the heaters tends to be more equal with the reverse-return system, and the flows tend to be more self-balancing. The cost for this advantage is the additional length of return piping required.

Two-pipe systems may be used either on hot water or chilled water systems. Some commercial installations will use a two-pipe system (Figure 19-37). The operator will choose the mode of operation, either heating or cooling, but not both. The shortcoming of this type of system is that if some areas need heating while others need cooling (which is very likely), somebody is going to be very uncomfortable.

Four-Pipe Water System

The **four-pipe** system shown in Figure 19-38 is in common use in large commercial buildings. The hot water boiler and water chiller provide hot and chilled water simultaneously. The individual HVAC units have both a hot water coil and a chilled water coil. The controls on these water coils will choose which valve to open. Obviously, this is the most expensive of the piping systems to install. It is the equivalent of two complete two-pipe systems. It is also, however, the only system that can adequately meet the needs of different spaces, some of which need heating when others need cooling.

Control of Water Coils

The amount of heating done by a water coil may be controlled in one of two methods.

1. Control the air flow across the coil.

2. Control the water flow through the coil.

Two methods are used to control the airflow across a heating coil. The unit heaters shown in Figures 19-39 and 19-40 have hot water that runs wild (does not have a control valve). The thermostat, when satisfied, will simply turn off the fan. The water flow through the coil will exit at virtually the same temperature at which it entered.

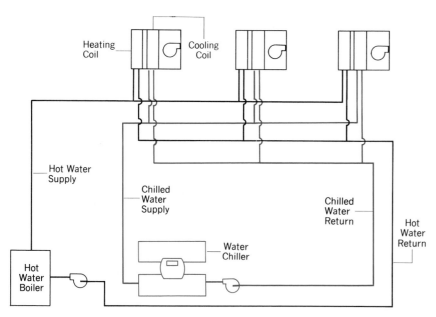

Figure 19-38 The four-pipe system can provide heating to one area and cooling to a different area at the same time.

Figure 19-39 Unit heater for horizontal air discharge. *(Courtesy The Trane Company, LaCrosse, Wis.)*

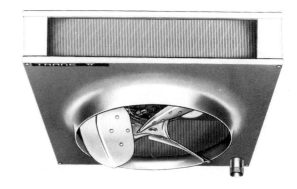

Figure 19-40 Unit heater for vertical air discharge. *(Courtesy The Trane Company, LaCrosse, Wis.)*

A second method of adjusting airflow through a **wild coil** is called **face-and-bypass** control (Figure 19-41). When heat is not required, the air is allowed to bypass around the coil.

Figure 19-42 shows control of a water coil (chilled water or hot water) by the use of a two-way control valve. The room temperature control system will cause the valve to modulate toward

the closed position as the demand for heating (or cooling) becomes reduced.

Three-way valves may also be used to control water flow through chilled water and hot water coils (Figure 19-43). The only advantage to using three-way valves is that each water user gets a constant flow of water, whether it goes through the coil or not. In this way, the control of one coil does not affect any part of the rest of the system. The disadvantage of this system is that the pump must handle the full flow of water all the time, even though full water flow may not always be required. Systems with three-way valves may be

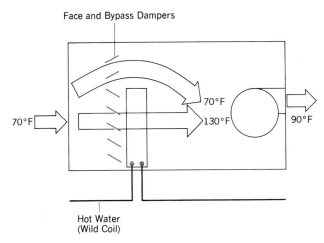

Figure 19-41 Discharge air temperature controlled by face-and-bypass dampers.

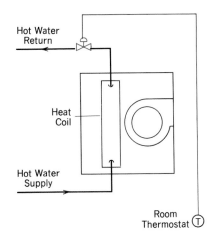

Figure 19-42 Two-way control valve on a heating coil.

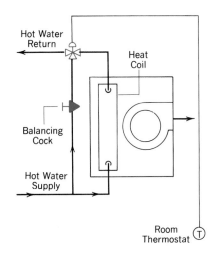

Figure 19-43 Three-way control valve on a heating coil. The balancing cock is set to match the bypass flow rate to the coil flow rate.

changed to two-way control by blanking off the bypass. It may also be necessary to install a pressure relief from the pump discharge into the suction if there is a possibility of all the users requiring near zero flow.

Steam Piping

The most common type of piping system used in steam systems is the two-pipe system shown in Figure 19-44. The steam main carries steam to each of the heating units. When the steam gives up its heat to the room air, it condenses and returns to the boiler through the condensate return line. The condensate return line is sized much smaller than the steam main. In order to prevent the steam from just blowing through the coil into the condensate return line, a steam trap is provided at the outlet from each heater. The steam trap is a device that will allow condensate to pass, but it will close if steam tries to pass. The operation of the various types of steam traps are discussed later in this chapter.

One-pipe steam systems in which the steam and condensate are both carried in the same pipe are limited to application in residential applications.

A good installation practice for steam systems is to pitch (slope) all supply pipes away from the boiler, and all condensate lines toward the boiler. A quarter inch of pitch for each 10 ft of horizontal run is sufficient.

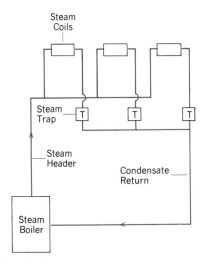

Figure 19-44 Two-pipe steam system.

Steam and Condensate Pipe Sizing

The criterion for sizing steam piping is that the piping must be sufficiently large to prevent excessive pressure drop, noise, and erosion of the piping. Figures 19-45 and 19-46 gives suggested steam carrying capacities for the different sizes of pipe. Oversizing of steam or condensate piping presents no particular operational problems, other than the increased first cost of installation.

The number of pounds of steam that must be carried to a heater depends upon the amount of heating capacity required, and the pressure of the steam. Obviously, the more heating capacity required, the more pounds of steam will be required. But surprisingly, the higher the pressure of the steam being supplied to the heating coil, the more pounds of steam will be required. This can be explained by looking at the table in Figure 19-47, which shows the latent heat of vaporization for steam at various pressures. At 15 psig, steam has a saturation temperature of 250°F, and a latent heat of vaporization of 946 Btu/lb. When 1 lb of 15 psig steam is allowed to condense in a heating coil, it will release 946 Btu of heat.

Now look at steam at 100 psig. Its saturation temperature of 344°F is much higher than that of the 15 psig steam, but its latent heat of vaporization is 881 Btu/lb. This means that the 100-psig steam will require a smaller heating coil in order to transfer the same amount of heat as a 15-psig coil, but more pounds of steam will be required.

The number of pounds of steam that is required may be calculated by dividing the heating requirement by the latent heat of vaporization for the steam pressure being used.

Example

Calculate the number of pounds of steam per hour that must be supplied to a heating coil that is to have a capacity of 30,000 Btu/hr. The steam supply pressure is 15 psig.

Solution

pounds per hour = (30,000 Btu/hr)/(946 Btu/lb)
= 31.7 lb/hr

Note that for the same amount of heating, it would require 34.1 lb/hr of 100-psig steam.

Pipe Carrying Capacity (lb/hr)

Pipe Size	Steam Supply	Condensate Return
$\frac{3}{4}$	20	140
1	40	250
$1\frac{1}{2}$	120	670
2	230	1400
3	660	3800
4	1400	7800

Figure 19-45 Approximate pipe capacities for low pressure steam systems.

Pipe Carrying Capacity (lb/hr)

	Pipe Size	Pressure Drop psi/100 ft		
		$\frac{1}{4}$	$\frac{3}{4}$	2
Steam Supply	$\frac{3}{4}$	40	80	180
	1	80	160	370
	2	580	1,170	2,000
	3	1,750	3,500	6,000
	4	3,700	7,400	12,000
	6	11,300	22,500	36,500
Condensate Return	$\frac{3}{4}$	230	460	890
	1	460	910	1,780
	$1\frac{1}{2}$	960	1,950	3,700
	2	3,300	6,400	12,300

Figure 19-46 Approximate pipe capacities for high pressure steam systems.

Steam Properties

Pressure (psig)	Saturation Temperature (°F)	Latent Heat of Vaporization (Btu/lb)
0	212	970
5	227	961
10	239	953
15	250	946
20	259	940
30	274	929
40	287	918
50	298	912
100	338	881
150	366	857
200	388	838

Figure 19-47 Saturated steam temperatures and latent heats of vaporization.

Once the number of pounds of steam required is determined, Figure 19-45 is used to select the pipe size (1-in.) that is capable of carrying more than the required quantity of steam.

Condensate Systems

The handling of condensate is probably the most important single factor in assuring the reliable operation of a steam system. Improper condensate piping can be the cause of

1. Water hammer.
2. Tube erosion.
3. Thermal shock.
4. Failure of control valves.
5. Freeze up.

Figure 19-48 shows how the condensate can build up in a steam main. As the steam loses heat to the air surrounding the pipe, condensate is formed. As the amount of condensate builds, waves begin to form. If sufficient condensate is allowed to collect, it will form a water seal across the pipe. At that point, the condensate will be propelled through the pipe at the velocity of the steam, which could easily be 10,000 to 20,000 fpm (over 200 mph)! You can imagine the energy that is carried by a slug of condensate traveling through a pipe at this velocity. When the piping makes a turn, the condensate is just as likely to keep going straight (taking the elbow along with it) as it is that it will make the turn.

If the condensate does not build up sufficiently to cause water hammer, erosion problems can still result from the droplets of condensate traveling at the speed of steam. When the steam is free of all condensate, it is said to be **dry steam.** If **wet steam** is allowed to be supplied to a heating coil, the tubes of the heating coil will become eroded. If wet steam passes through a control valve, the plug and seat of the valve will become worn.

When condensate is not drained freely from a heating coil, the condensate may remain in the coil during the off cycle. When steam is then reintroduced into the coil, it can cause the condensate to boil, thus causing a rapid thermal change. This rapid expansion can cause the joints of the heating coil to fail and begin to leak. Whenever failures of this type are found, it is wise to examine the condensate system for free drainage to assure that a replacement component will not also fail for the same reason as the original.

Where heating coils are called upon to heat outside air that is at temperatures below freezing, a special problem is presented. If the condensate is not removed quickly enough, it may be cooled by the outside air sufficiently to cause it to freeze. When the water freezes, it expands and causes the tubing to rupture.

In the section that follows, the required piping schemes to avoid all these failures are described.

Condensate Piping

In order to prevent water hammer or supplying wet steam to valves or coils, the condensate must be removed from the steam main. This is easily done by providing an end-of-the-main drip leg (Figure 19-49). The condensate that forms in the steam main is carried along the bottom of the pipe and allowed to drain out through a steam trap to the condensate system. As a further precaution against supplying condensate to the coil,

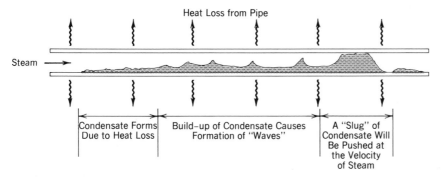

Figure 19-48 Condensate build-up in an uninsulated steam supply line.

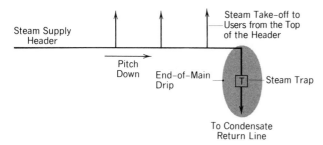

Figure 19-49 End-of-main drip leg prevents an accumulation of condensate from forming in the steam header.

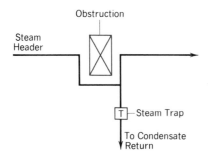

Figure 19-50 Correct method of preventing condensate accumulation when dropping the steam main below an obstruction.

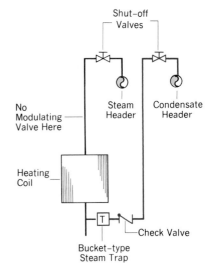

Figure 19-51 Lifting condensate to an overhead condensate return header.

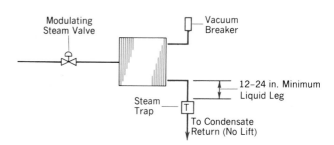

Figure 19-52 Modulated steam coil with vacuum breaker and 12 to 24-in. minimum liquid leg at the coil outlet.

the branch lines will take off from the top of the supply main. Drip legs may also be used at other places in the steam system. Just ahead of control valves and/or heating coils is a good location for a drip leg. Where a steam line must jog to avoid an obstruction (Figure 19-50), a drip leg should be provided to prevent the accumulation of condensate.

Once the dry steam is admitted to the heating coil, it will condense and must be quickly removed from the coil. If the coil is running wild (no control valve), there will be sufficient pressure to push the condensate out of the coil. There can also be sufficient pressure to lift the condensate, as shown in Figure 19-51. To avoid water hammer and erratic operation, the condensate should not be lifted more than 5 ft with a single-step **condensate lift** as shown. Also, the amount of steam pressure available to push the condensate uphill must theoretically exceed 1 psi for each 2.3 ft of lift required. In actual practice, 1 psi for each 2.0 ft of lift is recommended.

For steam coils that are controlled by a modulating valve, some additional problems must be solved. When the valve throttles to a partially

closed position, the pressure of the coil can easily go into a vacuum. This is caused by the condensing of the steam, and the corresponding tremendous reduction in volume occupied. If this vacuum is allowed to form, it will hold up the condensate in the coil, just as if it were soda pop being held up by the vacuum you form in a straw. Figure 19-52 shows a **vacuum breaker** installed in the coil. It will admit air into the system to prevent internal coil pressure from falling significantly below atmospheric pressure. Obviously, the condensate from a modulating system may not be lifted, as there may be insufficient pressure available at part load. The condensate must be allowed to drain vertically downward from this type of installation. In addition, the vertical distance between the coil outlet and the steam trap must be sufficient to provide enough pressure to overcome the pressure drop in the steam trap. Usually, 1 to 2 ft is sufficient.

COMBINED VENT TABLES

LEAST TOTAL HEIGHT	CONNECTOR RISE	MAXIMUM INPUT TO VENT CONNECTOR, Thousands of BTU Per Hour METALBESTOS VENT SIZE, INCHES													
		3"	4"	5"	6"	7"	8"	10"	12"	14"	16"	18"	20"	22"	24"
5'	1'	25	44	70	101	137	180	280	404	550	725	917	1135	1372	1635
	2'	30	53	83	120	162	213	333	480	626	817	1035	1277	1545	1840
	3'	33	60	93	135	183	240	375	540	730	950	1175	1450	1755	2090
6'	1'	26	46	72	104	142	185	289	416	577	755	955	1180	1425	1700
	2'	31	55	86	124	168	220	345	496	653	853	1080	1335	1610	1920
	3'	35	62	96	139	189	248	386	556	740	967	1225	1510	1830	2180
8'	1'	27	48	76	109	148	194	303	439	601	805	1015	1255	1520	1810
	2'	32	57	90	129	175	230	358	516	696	910	1150	1420	1720	2050
	3'	36	64	101	145	191	258	402	580	790	1030	1305	1610	1950	2320
10'	1'	28	50	78	113	154	200	314	452	642	840	1060	1310	1585	1890
	2'	33	59	93	134	182	238	372	536	730	955	1205	1490	1800	2150
	3'	37	67	104	150	205	268	417	600	827	1080	1370	1690	2040	2430
15'	1'	30	53	83	120	163	214	333	480	697	910	1150	1420	1720	2050
	2'	35	63	99	142	193	253	394	568	790	1030	1305	1610	1950	2320
	3'	40	71	111	160	218	286	444	640	898	1175	1485	1835	2220	2640
20'	1'	31	56	87	125	171	224	347	500	740	965	1225	1510	1830	2190
	2'	37	66	104	149	202	265	414	596	840	1095	1385	1710	2070	2470
	3'	42	74	116	168	228	300	466	672	952	1245	1575	1945	2350	2800
30'	1'	33	59	93	134	182	238	372	536	805	1050	1330	1645	1990	2370
	2'	39	70	110	158	215	282	439	632	910	1190	1500	1855	2240	2670
	3'	44	79	124	178	242	317	494	712	1035	1350	1710	2110	2550	3040
40'	1'	35	62	97	140	190	248	389	560	850	1110	1405	1735	2100	2500
	2'	41	73	115	166	225	295	461	665	964	1260	1590	1965	2380	2830
	3'	46	83	129	187	253	331	520	748	1100	1435	1820	2240	2710	3230
60'	1'	37	66	104	150	204	266	417	600	926	1210	1530	1890	2280	2720
	2'	44	79	123	178	242	316	494	712	1050	1370	1740	2150	2590	3090
	3'	50	89	138	200	272	355	555	800	1198	1565	1980	2450	2960	3520

COMMON VENT TABLE

LEAST TOTAL HEIGHT	VENT TYPE*	MAXIMUM COMBINED INPUT TO EACH SECTION OF COMMON VENT, Thousands of BTU Per Hour METALBESTOS VENT SIZE, INCHES													
		3"	4"	5"	6"	7"	8"	10"	12"	14"	16"	18"	20"	22"	24"
5'	L	—	48	76	109	149	195	310	455	630	830	1060	1330	1600	1920
	V	—	60	95	137	186	242	383	548	752	982	1240	1530	1815	2205
6'	L	—	52	82	117	160	210	325	468	708	925	1170	1445	1680	2080
	V	—	65	103	147	200	260	410	588	815	1065	1345	1660	1970	2390
8'	L	—	58	91	130	178	230	365	520	793	1035	1310	1620	1920	2330
	V	—	73	114	163	223	290	465	652	912	1190	1510	1860	2200	2680
10'	L	—	63	98	142	193	250	395	568	865	1130	1430	1765	2090	2540
	V	—	79	124	178	242	315	495	712	995	1300	1645	2030	2400	2920
15'	L	—	73	114	164	224	290	460	656	1008	1315	1665	2060	2430	2960
	V	—	91	144	206	280	365	565	825	1158	1510	1910	2360	2790	3400
20'	L	—	81	127	182	250	325	510	728	1126	1470	1860	2300	2720	3310
	V	—	102	160	229	310	405	640	916	1290	1690	2140	2640	3120	3800
30'	L	—	94	147	211	290	375	590	844	1327	1735	2190	2710	3210	3900
	V	—	118	185	266	360	470	740	1025	1525	1990	2520	3110	3680	4480
40'	L	—	105	164	236	320	420	660	945	1492	1950	2470	3050	3610	4390
	V	—	131	203	295	405	525	820	1180	1715	2240	2830	3500	4150	5050
60'	L	—	—	178	259	352	460	720	1100	1750	2280	2890	3570	4230	5050
	V	—	—	224	324	440	575	900	1380	2010	2620	3320	4100	4850	5900
80'	L	—	—	—	275	374	488	765	1232	1950	2550	3230	3980	4720	5750
	V	—	—	—	344	468	610	955	1540	2250	2930	3710	4590	5420	6600
100'	L	—	—	—	—	383	500	780	1335	2140	2790	3530	4360	5160	6290
	V	—	—	—	—	479	625	975	1670	2450	3200	4050	5000	5920	7200

*For definition of L and V, see section 2-1.

Figure 20-6 Combined flue sizes. (*Courtesy Selkirk Metalbestos*)

b. Connector rise is the vertical distance from the draft hood outlet to the point at which the next higher connection from another appliance ties into the combined vent. Each heater may have a different connector rise dimension. For the heater that has the highest tie-in point to the combined vent, the connector rise is the vertical distance from that heater draft hood to the point where it ties into the combined vent.

c. A **common vent** is the portion of the flue stack starting with the lowest heater connection and ending at the discharge end of the flue stack. When this portion of the venting system is vertical from beginning to end, it is said to be a **V-type** (for vertical) common vent. If there are any bends in the common vent, then it is an **L-type** (lateral).

Once the system dimensions have been sketched out, determine the least-total-height, the connector rise for each heater, and determine whether it is a type V or a type L combined vent. Enter the table in Figure 20-6 at the least-total-height dimension for the system. For heater number 1, determine the connector rise, and then read horizontally to the right to determine the vent size with a capacity greater than the input rating for heater number 1.

Enter the table again for each appliance, using the same least-total-height dimension. Determine a vent size for each appliance in the system. Where the table value indicates a vent size smaller than the draft hood connector, use the same size as the vent hood connector.

Once the vent connectors are all sized, the common vent size must be determined. Add the heat input ratings (Btu/hr) for all of the heaters that will be discharging into the common vent. Enter the table in Figure 20-6 at the same least-total-height used for the individual connectors sized previously. Read across to the right on either the L line or the V line, depending on the physical configuration of the common vent, and select the vent size that will handle more than the combined inputs. Regardless of the table value, never use a combined vent that is any smaller than the largest connection into the common vent. If two or more connectors of the same size enter the common vent, the common vent must be at least one size larger than the connectors.

Example

Size the vent connectors and the common vent for the combination of a water heater and furnace shown in Figure 20-7.

Solution

Enter Figure 20-6 at a least-vertical-height of 20 ft and a connector rise of 2 ft. Read horizontally to the right, stopping at the 4-in. size, which has a capacity of 66,000 Btu/hr. The water heater connector will be 4 in. Enter the same table again, at the same least-vertical-height of 20 ft, and a connector rise of 3 ft. Read horizontally to the right, stopping at 5 in., which has a capacity of 116,000 Btu/hr. The furnace vent will be 5 in. To size the combined vent, enter the table in Figure 20-6 at a least-total-height of 20 ft. Read across on the V line horizontally to the right (the common vent rises vertically from the bottom connector to the top of the combined vent, stopping at 5-in., which has a capacity of 160,000 Btu/hr. The 5-in. combined vent size will be sufficient to handle the total 140,000 Btu/hr input of our water heater and furnace.

Note that if there had been a third appliance entering the common vent above the water heater, the connector rise for the water heater would have been greater. We would then have needed to size the portion of the common vent that carried the flue gas from all three devices.

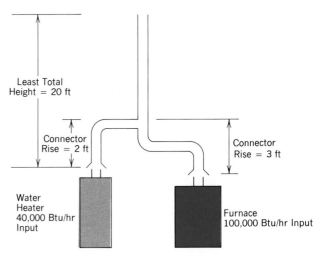

Figure 20-7 Example flue sizing problem.

Location of the Top of the Flue Stack

The top of the flue stack must not be close enough to any obstructions where the effect of wind will offset the chimney effect. Where the flue stack penetrates a flat roof, the stack must continue upward for at least one foot before terminating. The same applies for roofs with a relatively flat pitch (less than a 7-in. rise over a 12-in. run). For steeply pitched roofs Figure 20-8 shows the rise required past the roof penetration for various roof pitches.

Where a flue stack penetrates a roof, but is still

less than 8 ft away from another wall (perhaps from a parapet or a taller neighboring building), the end of the flue stack must rise 2 ft above the top of the obstruction (Figure 20-9). Common sense dictates the location of flue stacks that are not described here.

Power Venting

For many years, engineers have used fans on boilers to either push combustion air through the boiler and flue stack, or to pull the products of combustion out of the boiler. These are referred to as forced-draft or induced-draft combustion air fans.

In recent years, rooftop furnaces and residential heating units have gone to the technology of **power venting.** A small fan motor is energized whenever the primary heating element is energized. For some rooftop units, this has the advantage of allowing proper venting of the products of combustion, without needing to erect a tall, unsightly flue stack. Figure 20-10 shows one arrangement of a power-vented rooftop furnace. A small sail switch closes a microswitch when the combustion air fan airflow is proved. The combustion air fan pressurizes a section of the furnace, assisting the exit of the flue gas. A panel that is replaced carelessly by a service technician can destroy this pressurization, and seriously impair the draft through the unit.

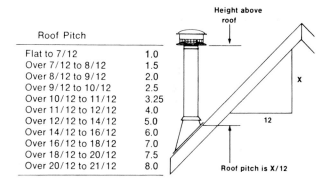

Roof Pitch	
Flat to 7/12	1.0
Over 7/12 to 8/12	1.5
Over 8/12 to 9/12	2.0
Over 9/12 to 10/12	2.5
Over 10/12 to 11/12	3.25
Over 11/12 to 12/12	4.0
Over 12/12 to 14/12	5.0
Over 14/12 to 16/12	6.0
Over 16/12 to 18/12	7.0
Over 18/12 to 20/12	7.5
Over 20/12 to 21/12	8.0

Figure 20-8 Location of vent termination above a sloped roof penetration. *(Courtesy Selkirk Metalbestos)*

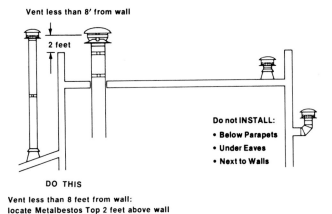

Figure 20-9 Location of vent termination above a flat roof penetration. *(Courtesy Selkirk Metalbestos)*

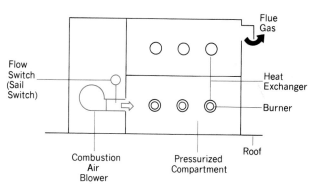

Figure 20-10 Power venting on a rooftop furnace.

KEY TERMS AND CONCEPTS

Products of
  Combustion

Excess Air

Dilution Air

Induced Draft

Forced Draft

Draft Diverter

Downdraft

Stack Effect

Condensation

Double-wall Vent

Sizing Vents

Combination Vents

Type L and V Vents

Least-total-height

Power Vents

QUESTIONS

1. What are the normal products of combustion that are carried by the vent system?

2. In a naturally vented heater, what should the pressure be just inside the draft diverter?

3. How does the vertical height of the vent duct affect the velocity of the flue gas in the duct?

4. What is the limiting factor in how much the flue gas can be allowed to cool in the flue stack?

5. In what locations may single-wall vent ducting be used?

6. What is the maximum allowable lateral length for a 120,000 Btu/hr heater vented with 5-inch vent pipe, 15 feet high?

7. A furnace with 200,000 Btu/hr input has a stack that runs vertically for 4 ft, horizontally for 6 ft, and then vertically again for another 6 ft. What diameter vent should be used?

8. If the furnace in Question 7 comes equipped with an 8-in. vent connector, then what diameter vent should be used?

9. A wall heater with an input of 30,000 Btu/hr has a vent with a total height of 16 ft and a lateral length of 10 ft. What diameter vent should be used?

10. A wall heater with a 20,000 Btu/hr input has a vent with a 30-ft vertical vent. The vent connector furnished with the heater is 4 in. in diameter. What diameter vent line should be used?

11. For the combination vent system shown in Figure 20-7, determine the required vent sizes if the water heater is 40,000 Btu/hr and the furnace is 200,000 Btu/hr.

12. A vent penetrates a pitched roof, which has a 45-degree angle. How much higher past the point of penetration must the vent line be run before it ends?

13. Why is a sail switch used in rooftop heaters with a combustion air fan?

CHAPTER 21

ELECTRIC HEAT

This chapter explains the uses for electric resistance heating, and describes the components that are unique to the electric furnace.

Resistance Heating

When an electric current is passed through a resistance, the electrical energy is converted into heat energy. The principle of **resistance heating** is used in toasters, electric ovens and stoves, electric water heaters, and electric furnaces. All electric heat is, by definition, 100 percent efficient. That is, 100 percent of the electrical energy that is consumed is converted into heat. For each kilowatt-hour of electricity that is consumed in an electric heater, 3415 Btu of heat is produced. Some electric companies tout this efficiency in order to imply that operating costs are low. This is not the case. Even though there is no waste of the electricity consumed by the electric heater, the cost per Btu purchased is much higher for electricity than for gas or oil. Electric resistance heat is low in first cost, units are relatively small and compact, and are sometimes used where gas or oil are simply not available. Where operating costs are important and electricity is the only available source of energy, the first cost and operating costs of an electric furnace should be compared to a heat pump system.

The Electric Furnace

Figure 21-1 shows a cutaway view of an upflow electric furnace. Its configuration is similar to that of a gas furnace, except for the following important major differences.

1. Instead of a heat exchanger, the blower sends the room air through an electric heating element.

2. There is no flue stack. There are no products of combustion to remove, and 100 percent of the heat from the elements is transferred to the air.

3. The furnace uses a 230-V power supply instead of a 115-V power supply.

4. When the thermostat calls for heat, the electric heating element and the fan motor are energized simultaneously. The heating element can never be allowed to operate without the fan, or the element will overheat and burn out.

The Heating Element

Figure 21-2 shows a **heating element** used in an electric furnace. It consists of a long spiral wound wire made of a nickel and chromium alloy commonly called **Nichrome.** A furnace will use several

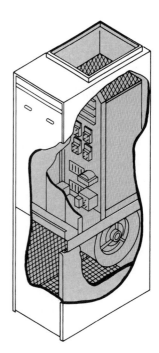

Figure 21-1 Electric furnace.

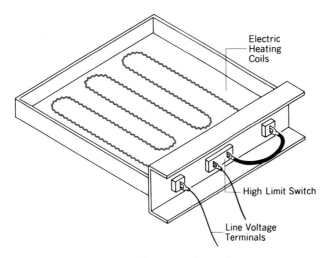

Figure 21-2 Electric heating element.

of these heating elements. Each element will draw between 3 and 8 kilowatts (kW). The framework provides an insulated support for the Nichrome element and prevents the wires from touching. When the heating element is operating, there is a 15- to 25-amp current passing through the uninsulated wire at 230 V. IT IS IMPERATIVE THAT YOU DO NOT TOUCH THE HEATER WIRE WHILE IT IS OPERATING. As a safety precaution, electric furnaces have an electrical interlock on the access panel to the heater elements. When the access panel is opened, the heater element is deenergized.

The furnace control panel will have a 20- to 40-amp cartridge fuse on each of the two legs of each heater. Therefore, an electric furnace with three 5-kW heaters would have six 30-amp cartridge fuses.

In addition to the large fuses that protect the furnace wiring against a shorted heating element, there is also a thermal cutout and a limit switch mounted on the heater element. The limit switch senses the air temperature around the heating element, and will deenergize the element if the limit set point is reached (usually between 140 to 190°F). The limit switch is a bimetal type of element that will open on a rise in temperature, and will reset itself when the element cools down. There is a nonadjustable differential between the cutout and cut-in of 25°F to 40°F. This prevents the heater from short cycling.

The **thermal fuse** will also open if it detects an abnormally high temperature around the heating element. Its function is the same as that of the limit switch, and acts as a backup safety switch.

Operating Sequence

When the thermostat calls for heat, it energizes a small resistance heater in a **heat relay** (Figure 21-3). This heat relay is identical in operation to the timed start fan control used in downflow gas furnaces. The difference is that the contacts are rated for the higher amp draw of the electric heating elements as compared to a blower motor. When the heat generated in the heat-relay resistance heater is enough to warp a bimetal switch, the normally open contacts of the relay will close, energizing the heater element.

Where there are two or more heater elements, the control may be staged. A two-stage thermostat, which operates as follows, may be used.

1. When the room temperature drops below the thermostat set point, one heater element is energized.

2. If the heat produced by the first stage of heating is insufficient, the room temperature will continue to drop. When it drops 2°F below the first-stage set point, a second heater element is energized.

In this way, the furnace can operate at either

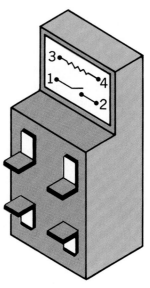

Figure 21-3 Heat relay used for sequencing electric heating elements.

half or full capacity, depending upon the heating requirements of the room.

Element Staging—Single-Stage Thermostat

Electric furnaces will have between two and six heating elements. In order to reduce the surge of electric current when the thermostat calls for heat, this scheme was devised to sequence the elements. A single-stage thermostat energizes a 24-V heater in heat relay number 1. When the contacts in heat relay number 1 close, 230 V pass to heating element number 1.

In series with the first heating element is a second heat relay. It is different from the first heat relay in that its heater element is rated at 230 V. When heating element number 1 comes on, the resistance heater in heat relay number 2 is also energized. After a time delay, the contacts in heat relay number 2 close and bring on the second heating element. If there is a third heating element, there would be a 230-V heater in heat relay number 3, which powers up when heating element number 2 is energized. There could be as many as six different heat relays in an electric furnace with six heating elements.

Blower Control

Figure 21-4 shows a blower control that is used on electric furnaces. This control will energize the blower motor whenever the heating element in the furnace is energized. It is a solid state device that has an electronic circuit connected to a **current-sensing loop** mounted on the outside of the blower control. A wire that supplies current to the heater element passes through the current-sensing loop. Whenever this element is energized, the current-sensing loop will cause a set of normally open contacts in the blower relay to close, thus energizing the fan motor.

Electric Baseboard

Figure 21-5 shows an electric element mounted in a baseboard cabinet on a wall. The most effective location for a baseboard heater is under a window where the greatest heat loss occurs. The element in the **electric baseboard heater** is designed to operate reliably, even without a fan to force the air through the element. These elements, however, do rely on a free flow of air due to convection. Restrictions to the free convection of air will cause the element to overheat and burn out. A thermal element is provided on many units to interrupt the power supply when overheating occurs, thus protecting the element.

The cabinets of electric baseboard heaters can run extremely hot. They should not be installed where there is a high risk of injury to children. The recommended clearance distances to floors and drapes must also be strictly observed to eliminate any possibility of fire.

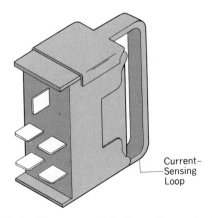

Figure 21-4 Blower control closes its contacts when the current-sensing loop senses that the power wiring to the blower is energized.

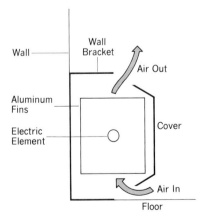

Figure 21-5 Cross section of an electric baseboard heater. Airflow is accomplished by natural convection.

The capacity of the baseboard will depend upon the length of the unit and the voltage applied. A heater may be installed with any voltage supply as long as it does not exceed the rated voltage supply. A unit rated for 240 V may be used on a 120-V power supply. It will not run as hot, and it will produce only one quarter of its rated capacity. A unit rated at 120 V, however, will quickly burn out if used in a 240-V system. On the average, electric baseboard will consume 250 watts per foot (W/ft) of element when operating at the rated voltage. Where units are rated at dual voltage, such as 208/240, the unit capacity must be derated for the lower voltage. The watts (heating capacity) varies with the square of the voltage applied.

$$\text{Actual watts} = \left(\frac{\text{actual volts}}{\text{rated volts}}\right)^2$$

Each watt of electrical consumption will translate into 3.414 Btu/hr of heating capacity.

Example
Estimate the heat output of a 5-ft section of electric baseboard, rated at 240 V, and installed on a 208-V system.

Solution
Estimating a capacity of 250 W/ft when operating at the rated voltage of 240 V, the heater will consume 1250 W. Multiplying by 3.414 Btu/hr for each watt, the rated capacity is 4267 Btu/hr. To derate the unit to 208 V, multiply by the ratio of $(208/240)^2$. The estimated capacity of the 5-ft section operating at 208 V is

$$4267 \times \left(\frac{208}{240}\right)^2 = 3205 \text{ Btu/hr}$$

The baseboard unit comes prewired, with junction boxes at each end. Units may be installed end to end in order to make up long runs of continuous heat. Standard units are available to 10 ft.

Figure 21-6 shows a thermostat installed in one of the junction boxes. The thermostat is wired in series with the element and switches the line voltage. The ability of each heater to have its own thermostat is a major advantage associated with electric baseboard heat. It is easy to keep different building areas at different temper-

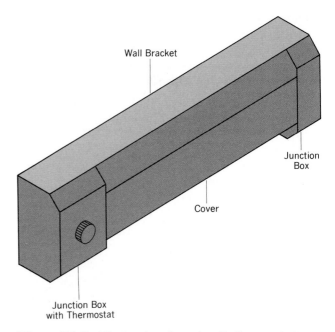

Figure 21-6 Electric baseboard with thermostatic control.

atures, or even shut off the heat to unused spaces. If thermostat settings are carefully controlled, a significant portion of the high operating cost can be offset.

Electric Duct Heaters

A **duct heater** is an electric heater element that is installed in a run of duct. It may be used on new installations in order to provide individual temperature control for different occupied areas. It is also often used to fix systems that have been improperly designed or installed. When one area is too cold while all other areas are comfortable, a duct heater may be used to "boost" the capacity of the one cold area.

The duct heater is supplied with an element that goes inside the duct, and an attached control compartment, which remains outside the duct (Figure 21-7). Most times, the heater is designed for slip-in installation. That is, a rectangular slot is cut in the side of the duct, and the element is simply slipped in. The control box is then attached to the side of the duct with sheet-metal screws. On some larger installations, the heater will be flanged to the same dimensions as the duct. The heater is then bolted to the duct around the perimeter.

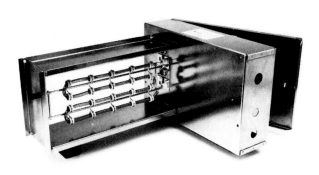

Figure 21-7 Electric duct heater. Element is inside the duct; control box hangs on the outside. *(Courtesy Tutco Inc.)*

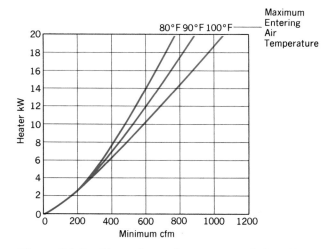

Figure 21-8 Chart shows the maximum kilowatts that should be installed in a duct with a given airflow, or the minimum airflow required to accommodate a given heater.

The heat output of the duct heater is constant whenever the element is energized. It is designed so that a minimum air velocity must be provided to prevent element overheating. Figure 21-8 shows how much airflow is required for heaters of each wattage.

Example

What is the largest duct heater that may be safely installed in a duct carrying 600 cfm of air at 80°F?

Solution

Enter the table in Figure 21-8 at 600 cfm of airflow. Read vertically upward to a maximum entering air temperature of 80°F. Read to the left a maximum of 14 kW. This is the highest capacity heater that can be installed without danger of overheating the elements.

For electric heaters in ducts, it is important to understand that the average velocity is not always the same as the velocity of the air across the heater at each point. Duct heaters should not be installed where there would be **nonuniform airflow** (Figure 21-9). They should be installed several duct diameters away from any elbow or change in duct size where uneven airflow may be present.

Duct heaters can be supplied with two different kinds of elements. By far the most popular is the **open** type of element, rather than the **enclosed** or **sheathed element.** The open element has Nichrome wires exposed to the airstream. These heaters have the following advantages.

1. Lower cost than enclosed heaters.

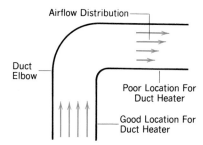

Figure 21-9 Avoid installation of a duct heater where there is a nonuniform air distribution inside the duct.

2. Cooler coil temperatures, yielding longer coil life and more reliable operation.

3. The open coil design presents virtually no resistance to airflow.

4. Quick control response. When the heater element is energized, it immediately begins heating the air. The enclosed element must first heat the enclosure before the heat is transferred to the airstream.

The only disadvantage of the open type of coil is the potential of igniting potentially explosive atmospheres. Where chemicals, dust, or other contaminants may be present in the airstream, an enclosed-type element would be used.

Control of Duct Heaters

The electric duct heater can have single or multiple elements. The elements are staged on by the room thermostat, through a relay. The relay uses the room low voltage switch to energize, in sequence, the line voltage switches that control the heating elements. The major difference between the control of a duct heater and the control of an electric furnace is that the duct heater has no control over the operation of the fan. The duct heater may use a **sail switch** that closes a set of contacts when it senses airflow in the duct. In the absence of sufficient airflow, the duct heater will not be allowed to become energized.

In unusual applications where a noncycling temperature leaving the heater is required, step control is not adequate. For these applications, a silicon-controlled rectifier (commonly abbreviated to SCR) is used to modulate the output of the heater element.

Electric Unit Heaters

Figures 21-10 and 21-11 show a unit heater that combines an electric heating element and a blower fan. These **electric unit heaters** are available in both horizontal and vertical models (designated by the direction in which the air is discharged). Unit heaters are widely used in commercial and warehouse applications where ap-

Figure 21-10 Electric unit heater for horizontal air discharge. *(Courtesy The Trane Company, LaCrosse, Wis.)*

Figure 21-11 Electric unit heater for vertical air discharge. *(Courtesy The Trane Company, LaCrosse, Wis.)*

pearance is of secondary importance. The units are suspended from the ceiling or supported from a wall. They are supplied with open or enclosed heating elements and automatic overheating protection. They can operate on single-phase or three-phase power, usually 240 V.

A thermostat is used to cycle the heater ele-

| Nominal kW | Fan cfm | Maximum Mounting Height | | Diameter of Floor Area Covered (Vertical Units) (ft) |
		Horizontal (ft)	Vertical (ft)	
3.0	400	9	9	15
5.0	400	9	9	15
7.5	700	10	12	30
10	700	10	12	30
15	1100	11	16	40
20	1100	11	16	40

Figure 21-12 Suggested mounting dimensions for electric unit heaters.

ment and the fan, or the fan may be operated continuously. Continuous fan operation has the advantage of reducing the stratification of warm air in the unoccupied areas near the ceiling. The circulation of unheated air, however, may be perceived by some occupants as an objectionable draft.

Heating requirements can vary widely, but for average commercial construction (8-ft ceilings), a heating requirement of 2 W/cu ft is usually adequate. Manufacturer's data should be checked to be certain that the throw of the unit will be adequate to supply heated air to the required areas. This is especially critical with units that are required to deliver hot air from an elevated unit. If improperly selected, the buoyancy of the warmed air will cause it to turn around and rise toward the ceiling before it has ever reached the occupied zone. In the absence of manufacturer's data, Figure 21-12 gives suggested maximum mounting heights for both vertical and horizontal unit heaters.

KEY TERMS AND CONCEPTS

Resistance Heating

Heating Element

Thermal Fuse

Heat Relay

Staged Control

Current-Sensing Loop

Electric Baseboard Heater

Duct Heater

Slip-In Heater

Nonuniform Airflow

Open Element

Enclosed Element

Sail Switch

Electric Unit Heater

QUESTIONS

1. What are the advantages of using resistance heating compared to a heat pump?

2. What are the disadvantages of using resistance heating compared to a heat pump?

3. How much will it cost each hour to heat a room with a heat loss of 10,000 Btu/hr using a 5-kW electric resistance heater? Electricity costs $0.12 per kilowatt-hour.

4. For the room in Question 3, how much will it cost each hour if the heating is done with a 10-kW heater?

5. What is Nichrome wire?

6. What is a heat relay?

7. What is the purpose of an outdoor-air thermostat in an electric furnace?

8. What is sensed to determine when to energize the furnace blower fan?

9. Where is the most effective mounting location for electric baseboard?

10. How many Btu/hr will be delivered by a 6-ft-long section of electric baseboard rated at 240 V and operating on a 220-V system?

11. What is the minimum air quantity that must pass through a 10-kW duct heater? Air will enter the heater at 100°F.

12. What is the maximum allowable mounting height for a 10-kW unit heater that discharges vertically downward?

13. What is the advantage of continuous fan operation on a unit heater? What is the disadvantage?

14. What does an SCR do?

CHAPTER 22

SOLAR HEATING

The objective of this chapter is to describe the nature of solar energy, devices used to collect this energy, and systems that convert this energy to useful purposes.

Energy from the Sun

A tremendous amount of energy reaches the earth each day from the sun. The total amount of **solar radiation** striking the earth is equivalent to over 400 times the total energy consumption of the United States. The energy is in the form of visible light, ultraviolet radiation, and infrared radiation.

Not all of the solar energy that reaches the earth's atmosphere reaches the earth. The energy reaching the upper atmosphere is called the solar constant, and is equivalent to 430 Btu/hr for each square foot. The amount of radiation that reaches the earth is consumed by the following.

1. Twenty-five percent is absorbed by the earth's atmosphere.

2. Twenty percent is reflected away by clouds.

3. Five percent is reflected away by the ground.

Of the 50 percent that remains, half reaches the ground in diffuse, nondirectional rays that do not lend themselves to capture by solar collectors. The remaining half reaches the ground in direct rays that may be easily converted to heat energy.

The amount of solar energy that reaches the earth in a form suitable for use by solar collectors varies according to geography. It depends upon **latitude,** atmospheric clarity, time of the year, and climate. A flat, horizontal surface of one square foot in Arizona receives an average of 1900 Btu each day. The same surface in the northeast United States would receive an average of only 1100 Btu per day. The amount of radia-tion striking a surface each day is called the **insolation** value.

The amount of solar radiation that strikes the roof of the average residence exceeds the annual heating demand for the residence by a wide margin. Unfortunately, the times at which the solar insolation is at its greatest does not always correspond with the times at which the need for heat is at its greatest. **Heat storage** systems must be made part of solar collection systems so that heat collected during the day can be used at night.

Flat-Plate Solar Collectors

Devices that gather the sun's energy and convert it to heat are quite simple in nature. An automobile whose inside temperature has soared much higher than the outside air temperature while parked at the beach is a good example of a solar collector. The sun's rays enter through the windows, and strike the dark surfaces inside. This radiant energy causes these surfaces to warm, and the warm surfaces in turn heat the air inside the car.

The solar collector in Figure 22-1 is quite common. It is called a **flat-plate collector,** and it

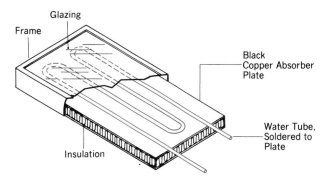

Figure 22-1 Flat-plate solar collector.

works just like our hot automobile at the beach. The sun's rays enter through the glass (called glazing) and strikes the black copper surface. It is painted because the black surface will absorb more solar energy than any other color. When collectors are on the jobsite, they should remain covered until they are installed. The temperatures inside a collector with no water to remove the heat can get high enough to damage the black absorbing surface.

A copper tube is soldered to the copper surface. Water is allowed to flow through the tube and draw heat away from the collector surface. The function of the glazing is to prevent the copper surface from losing heat to the surrounding air. The collector may be either single glazed or double glazed for application in extremely cold climates. The glass used is specially formulated with a low iron content that absorbs only 3–4 percent of the total energy passing through. This compares with 7–8 percent absorption for conventional glass. Glass is also available with a special selective coating that will reduce the amount of energy that is **reradiated** back to the atmosphere.

Some collectors are glazed with materials other than glass. Plastic materials are used that are lighter and stronger than glass, but they absorb and reradiate much higher percentages of the energy striking the surface. They are also subject to discoloration and are less resistant to scratching than glass panels.

The rear side of the copper plate is insulated to prevent losing heat from that direction.

Focusing Collectors

Figure 22-2 shows another type of collector called a **focusing collector.** The shape of the curved reflecting surface is a carefully calculated design. It works very much like an automobile headlight reflector in reverse. In cars, a light source for the headlight (bulb) is located at a point called the focus of the curve. The light from the bulb is emitted in all directions. But all light that strikes the curved reflector is reflected out in the same direction, resulting in a beam of light.

The solar collector receives parallel light rays from the sun. They strike the reflector and are all

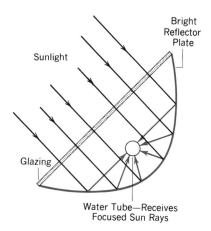

Figure 22-2 Focusing collector concentrates the collected solar energy along a single line.

focused through one point, and a tube is placed lengthwise through the collector along that point. The focusing of the sun's rays on a single point can result in very high temperatures. Water flowing through the tube can actually be made to boil!

The focusing collector must be pointed directly at the sun in order to work effectively. These systems are used with **tracking systems** that constantly monitor the position of the sun, and move the collector orientation accordingly.

Focusing collectors work extremely well when exposed to direct sunlight. Their effectiveness, however, is rapidly reduced under conditions of cloudy or hazy skies. The focusing collector is less popular than the flat-plate collector due to its increased complexity, cost, and susceptability to mechanical failure.

Thermosiphon System

Figure 22-3 shows a very simple solar hot water heating system called a **thermosiphon** system. Flat-plate collectors absorb energy from the sun and heat the water in the collector tubes. The warm water has a slightly lower density than the cooler water, and therefore tends to rise toward the insulated storage tank. Water that is displaced from the tank falls back to the collectors, setting up a natural circulation of water. When there is a demand for hot water, new cold water is admitted into the bottom of the tank, while the

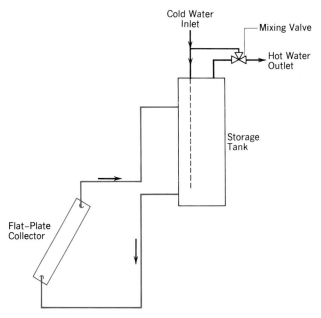

Figure 22-3 Thermosiphon system relies on natural convection for water circulation.

hot water is drawn from the top of the tank. The hot water in the tank could potentially be hotter than desired, so a mixing valve is provided. The mixing valve is set for the highest allowable desired hot water temperature. If the tank temperature exceeds this temperature, cold water will be mixed with the hot water through the mixing valve.

This system has the advantage of simplicity, lowest initial cost, reliability, and no moving parts. It has, however, the following disadvantages.

1. The collectors must be located at a lower elevation than the storage tank. This does not fit well with the most popular location for collectors, which is on a roof.

2. No provision is made to protect the system against freezing.

Open-Loop Recirculating System

Figure 22-4 shows a system similar to the thermosiphon system, except for one major addition — a circulating pump. The addition of a circulating pump allows the collectors and the storage tank to be located any place with respect to each other. A controller has been added to operate the pump whenever the collector is hotter than the storage tank. The operation of the controller is discussed in a later section. This is called an **open loop recirculating system** because the water that goes to the hot water users is the same water that circulates through the collectors.

An **auxiliary electric heating** element has been added in the side of the storage tank. This heater serves two different purposes.

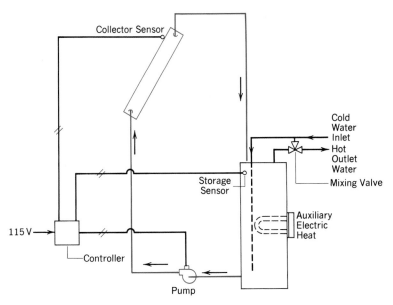

Figure 22-4 Open-loop recirculating system.

1. When the temperature of the storage tank is not high enough to satisfy the need for hot water, the auxiliary heater will become energized.

2. In the event that the water in the collectors approaches freezing, the controller will cause the pump to operate. The auxiliary heater will then come on if required in order to maintain the water temperature in the collector above freezing.

Draindown System

Figure 22-5 is a **draindown system,** designed to drain all the water in the collectors whenever there is a potential danger of freeze-up. The frost cycle sensor initiates the operation of the two solenoid valves, draining the water from the collector. The vacuum breaker at the top of the collector system permits air to enter the system, allowing free drainage of the water. During normal operation of the system, both solenoid valves are energized. In the event of a power failure, the normal (deenergized) position of the solenoids is in the draindown position.

Drainback System

The **drainback system** in Figure 22-6 overcomes the potential freezing problem without draining water each time the system shuts down. It is a **closed-loop** system in which a segregated water loop circulates water through the collectors and a heat exchanger. Heat is then taken from the heat exchanger from the open water system. Each time the system shuts down, the water from the collectors drains back into the heat exchanger tank. A level gauge is provided on the heat exchanger tank so that the system may be filled, leaving sufficient room for the water in the collector to drain.

Closed-Loop Antifreeze System

Figure 22-7 is a system that relies on the use of an antifreeze such as ethylene glycol to prevent freezing of the collectors. Cold water is then heated from the warmed antifreeze. If this type of system is used, many local codes require a special double separation of the fluids in the heat exchanger. This is to prevent accidental con-

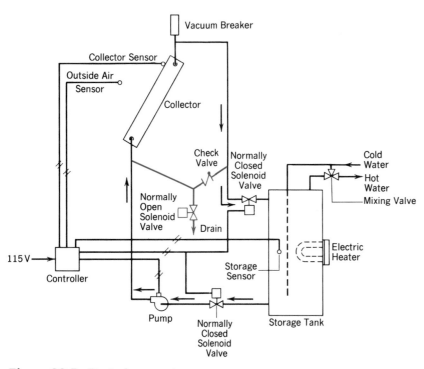

Figure 22-5 Draindown system.

tamination of the water system from the antifreeze system in the event of a leak between the systems.

Differential Temperature Controllers

The brains of the solar system is the **differential temperature controller** (Figure 22-8). The out-

put of this controller goes to the system circulating pump. The output may be On–Off, or it may be a proportional type that varies the pump speed. For **proportional controllers,** the greater the difference in temperature between the collector and the water storage tank, the more heat is available and the higher the pump rpm.

For **On–Off controllers,** there are two set points. One set point determines how much warmer the collector must be than the storage

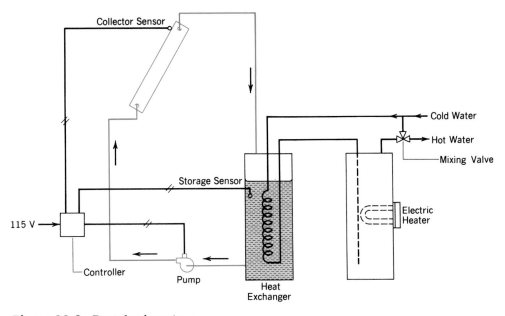

Figure 22-6 Drainback system.

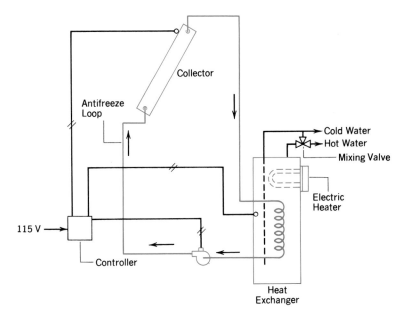

Figure 22-7 Closed-loop antifreeze system.

tank before the pump turns on. This is called the delta T on (ΔT on). When the pump runs, the temperature of the collector drops. When it drops to within, say, 2°F of the storage tank temperature, the pump cycles off. This is called the delta T off (ΔT off) set point. These set points may be adjustable or fixed, depending upon the brand of controller.

Figure 22-9 shows a sample daily operating cycle for a collector and storage system. Sometime in the morning, the difference between the collector and storage temperatures reaches the

ΔT on set point, and the pump turns on. The collector temperature immediately drops. It may even drop far enough to equalize the collector and storage temperatures, causing the pump to stop. If that happens, it will try again later.

As the day wears on, the collector continues to absorb more heat. The pump is on, so the temperature of the storage tank slowly warms. When the sun starts moving away from directly in front of the collector, the collector temperature drops causing the pump to stop. When it does, the collector temperature immediately rises and may

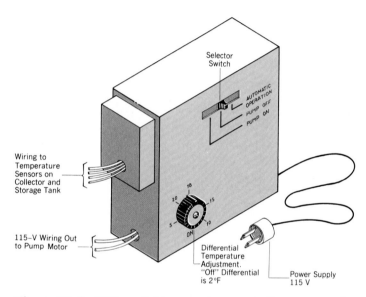

Figure 22-8   Differential temperature controller senses collector temperature and tank temperature, and operates a switch that controls a water-circulating pump.

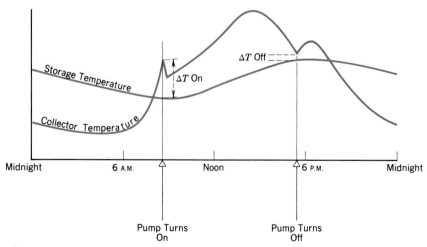

Figure 22-9   Temperature of storage tank and collector through a one-day cycle for an On–Off control system.

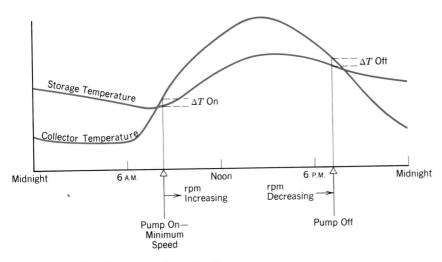

Figure 22-10 Storage tank and collector temperatures for a modulating control system.

even turn the pump back on a couple of times. Soon, though, the collector temperature will drop to lower than the storage temperature, and the pump will remain off until the morning.

Figure 22-10 shows a similar system, operating with a proportional controller. It will be slightly more efficient than the On–Off controller. The pump may be started when the collector is only a few degrees warmer than the storage temperature. The danger of cycling the pump off is minimized, because at this low ΔT the pump will run only very slowly. And at the end of the day, the pump will gradually slow as the collector temperature reaches the storage temperature. The pump will stop when the collector is within a couple of degrees of the storage tank. Note the time lines. The proportional control system will start up earlier in the day, and it will remain on until later in the day than the On–Off system.

Sensors

The **sensors** used to sense the collector temperature and the storage tank temperature (Figure 22-11) are connected to the controller by a single pair of wires for each sensor. The electrical resistance of the sensor changes with the temperature it senses (thermistors). Depending upon the manufacturer, the thermistor may be either **positive-responding** or **negative-responding.** The resistance of a positive-responding ther-

mistor will increase as temperature rises. A negative-responding thermistor will decrease in resistance as the temperature rises. Figure 22-12 shows a typical resistance versus temperature chart for a positive-responding thermistor.

Collector Placement

The placement of the flat-plate collector will determine to a large extent how much of the available energy is collected. The efficiency of the collector will be much higher when it is pointed directly at the sun. The sun is always in motion,

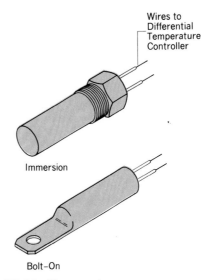

Figure 22-11 Temperature sensors.

Temperature (°F)	Resistance
32°	32.6 K
40°	26.0 K
50°	19.9 K
60°	15.3 K
70°	11.9 K
80°	9.2 K
90°	7.3 K
100°	5.8 K
110°	4.6 K
120°	3.7 K
140°	2.5 K
160°	1.7 K
180°	1.2 K
200°	835 Ω

Figure 22-12 Typical temperature versus resistance for a temperature sensor.

of course, so a fixed-plate collector must be placed so that it will receive the maximum amount of radiation throughout the entire day.

As the earth rotates, the apparent position of the sun is that it rises in the east and sets in the west. At noon it is at the center of its daily journey across the sky. The arc traveled by the sun does not take it directly overhead, even at noon. The actual angle compared to horizontal depends upon the latitude of the location where the system is installed, and the time of year (Figure 22-13). During the year, the sun will rise higher in the sky during the summer, and will stay lower in the sky during the winter for all locations north of the equator. The opposite is true for locations south of the equator. A well-designed awning will be large enough for shading a window in the summer, but it will be small enough to allow the rays of the sun to shine under the awning during the winter heating months (Figure 22-14).

So with the sun moving around the sky each day, and changing each day from the day before, where should the flat collector be pointed? The answer can be given simply as follows:

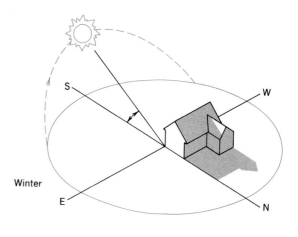

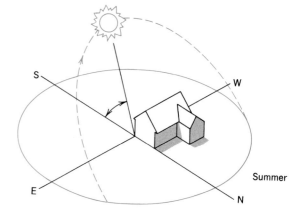

Figure 22-13 The angle of the sun relative to horizontal will vary with the seasons.

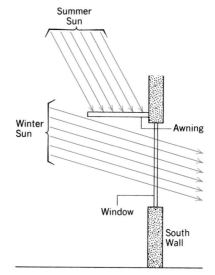

Figure 22-14 Awning sized for maximum year-round energy efficiency. The summer sun is shaded out, but the winter sun is allowed to pass under the awning.

1. For collectors used for winter use only (space heating), the tilt angle of the collector should be equal to the latitude plus 15 degrees.

2. For collectors used for year-round heating (such as for domestic hot water heating), the angle of the collector should be equal to the latitude.

3. Collectors for all applications should face due south.

4. Maximum variations from these recommendations should not be more than plus or minus 10 degrees in any direction.

The **collector tilt angle** is the angle formed between the collector surface and the horizontal. A vertical collector (pointing horizontal) has a tilt angle of 90 degrees. A collector laying on a flat roof pointing straight up has a tilt angle of zero degrees. Note that for a collector that is facing the sun directly, the collector tilt angle plus the solar tilt angle always equals 90 degrees. It makes sense that for systems operating during the winter only, the recommended tilt angle is higher than for year-round collectors. For a greater percentage of the operating hours, the collector would then be more closely pointing toward the actual position of the sun.

Figure 22-15 shows an insolation table that gives the solar radiation striking collectors at various locations and at various angles. These tables are available with data for specific cities. The **solar tilt angle** is the angle of the collector, as measured from vertical. The azimuth is the number of degrees that the collector is facing away from true south (a collector facing true south has an azimuth of zero).

Example

For the sample city for which the insolation data are given in Figure 22-15, how much heat will be collected (on the average) for a collector with a tilt angle of 50 degrees during the months of June, July, and August? The collector is facing due south.

Solution

Using Figure 22-15, read the following insolation values.

June   1830 Btu/day/ft^2

July   1860 Btu/day/ft^2

Aug.   1940 Btu/day/ft^2

The average of the three values is 1877 Btu/day for each square foot of collector area.

If you are interested only in determining the best tilt angle, without regard for the insolation value, use the chart in Figure 22-16. This gives the best tilt angle for each month for various latitudes. Note that the average tilt angle for each latitude when figured for year-round operation just happens to equal the latitude angle. This

Average Radiation—BTU/DAY-SQ. FT.

Slope	Azimuth	Jan.	Feb.	Mar.	Apr.	May	June	July	Aug.	Sept.	Oct.	Nov.	Dec.
30	0	1520	1770	2150	2120	2170	2250	2250	2210	2180	2050	1770	1530
40	0	1630	1850	2160	2040	2020	2060	2070	2100	2150	2130	1890	1660
50	0	1700	1880	2120	1910	1830	1830	1860	1940	2080	2140	1960	1740
60	0	1730	1860	2030	1750	1600	1570	1600	1740	1950	2100	1980	1780
30	15	1500	1750	2130	2120	2180	2270	2260	2220	2170	2030	1740	1510
40	15	1610	1820	2150	2050	2030	2080	2090	2110	2150	2100	1860	1630
50	15	1670	1850	2110	1930	1850	1860	1880	1960	2080	2110	1920	1700
60	15	1690	1830	2020	1770	1640	1610	1640	1770	1960	2070	1940	1740
30	30	1440	1700	2100	2120	2200	2310	2300	2230	2150	1970	1670	1440
40	30	1520	1750	2110	2060	2080	2140	2150	2140	2130	2030	1960	1540
50	30	1570	1770	2070	1950	1920	1940	1970	2010	2060	2030	1810	1600
60	30	1580	1740	1980	1810	1730	1720	1750	1840	1950	1980	1820	1620

Figure 22-15 Insolation table for a sample city.

LATITUDE→ MONTH ↓	32°	36°	40°	44°	48°
Dec.	55°	59°	63°	67°	71°
Jan., Nov.	52°	56°	60°	64°	68°
Feb., Oct.	43°	47°	51°	55°	59°
Mar., Sept.	32°	36°	40°	44°	48°
Apr., Aug.	21°	25°	29°	33°	37°
May, July	12°	16°	20°	24°	28°
June	8°	12°	16°	20°	24°

Figure 22-16 Best collector tilt angle for maximum collection each month.

chart is useful for solar applications that will only be used for a few specific months during each year.

True South Compared to Magnetic South

The south direction indicated on a magnetic compass can be significantly different from the true south direction that is recommended for solar panels. The magnetic lines of force around the world do not follow exactly perfect north–south directions. Figure 22-17 shows a map of the United States with lines indicating differences between true north and magnetic north. If, for example, you are interested in a city that is near a line labeled 10 degrees east, that means that true south is actually 10 degrees to the east of the reading you obtained on a compass.

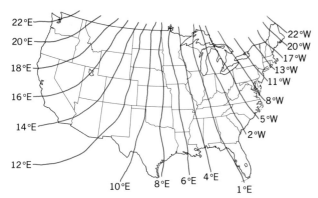

Figure 22-17 Difference between true south and magnetic south.

Air-Heating Collectors

Some systems have been built in which the collector is used to heat air instead of water. The heated air is then used directly into an air handler. Storage for the solar air heater is accomplished by passing the air through a tank filled with rocks. The rocks provide a lot of mass that is heated during the day. The room air may then be passed over the rock storage bed during the cooler evening hours.

Air-heating systems have not gained the same popularity as water-heating systems for several reasons.

1. The heat transfer coefficients in the collector are poorer for air than for water.
2. Installation of the ductwork and the rock storage is more expensive than a comparable piping and water-storage system.
3. Heat losses and leaks from the ductwork impair the system efficiency.

Solar Angles

The position of the sun in the sky is predictable for any location, and for any time of the year. The solar tilt angle or altitude is the angle formed between the sun and the earth. It starts at zero each day and rises to its daily peak each day at noon. During the winter and spring months, its daily peak increases each day. During the summer and fall, its daily peak is a little lower each day. June 21 is the first day of summer, and it represents the day on which the solar tilt angle is the highest it will be for the entire year. December 21, the first day of winter, is the day on which the solar tilt angle is lower than for the same time of the day on any other day.

The **azimuth angle** is the number of degrees away from a due south direction at which the sun is located at any time during the day. The azimuth is always zero at noon.

The graphs in Figures 22-18 through 22-20 allow you to determine the position of the sun (tilt angle and azimuth) for any day of the year, for any time of the day, and for three different latitudes.

Knowing the position of the sun at any point

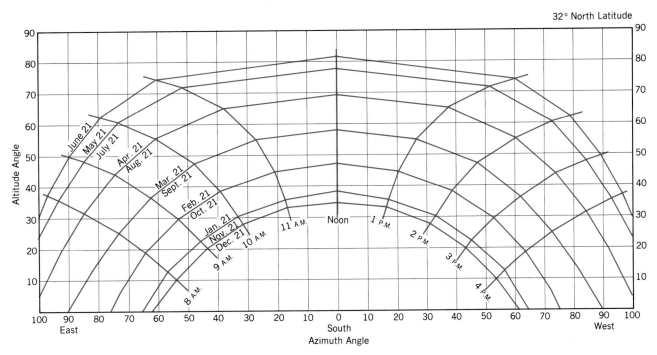

Figure 22-18 Solar angles for 32° north latitude.

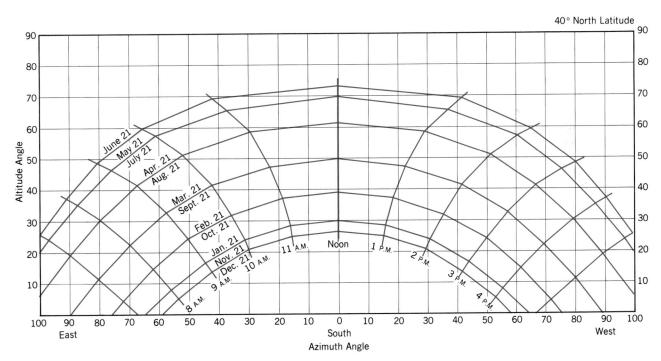

Figure 22-19 Solar angles for 40° north latitude.

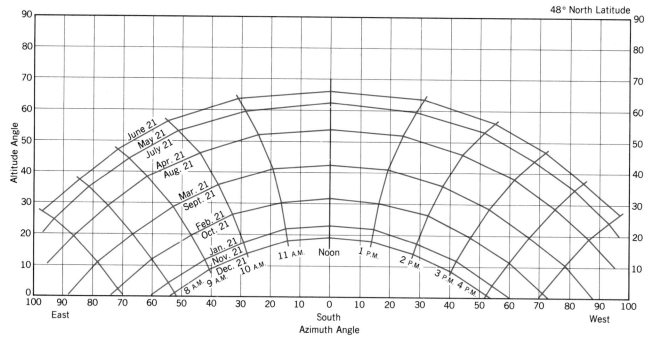

Figure 22-20　Solar angles for 48° north latitude.

in time allows you to calculate the **length of shadows** that will be cast by a stationary object. There are two important factors concerning shadows that you will be interested in.

1. When collectors are installed around nearby obstructions, such as buildings or trees, you will need to know if the obstruction will be casting a shadow on the collector.

2. For determining if the sun will shine in a window that has an overhang, or for selecting an awning that will provide a certain area of shade, the dimensions of the shadow must be determined.

Figure 22-21 presents factors that may be used to determine the length of shadows for any solar tilt angles.

Example

For a city with a location of 32 degrees north latitude, it is desired to install an awning over a sliding glass door that provides a shadow on the wall at least 8 ft below the awning for all hours between 10:00 A.M. and 2:00 P.M. every day between June 21 and August 21. How far must the awning project from the wall?

Solution

Make a sketch of the problem as shown in Figure 22-22. From Figure 22-18 you can determine that the lowest solar tilt angle during the period being investigated is 56 degrees. The ratio X/Y for a solar angle of 56 degrees is 1.48. We can set up a simple equation

$$8/Y = 1.48$$
$$Y = 5.4 \text{ ft}$$

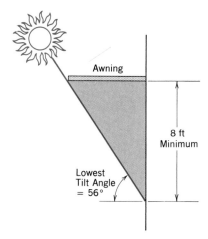

Figure 22-22　Example problem.

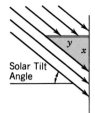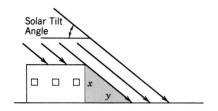

Shading on a Vertical Surface		Shading on a Horizontal Surface			
Solar Tilt Angle	x/y	Solar Tilt Angle	x/y	Solar Tilt Angle	x/y
1	.018	31	.601	61	1.80
2	.035	32	.625	62	1.88
3	.052	33	.649	63	1.96
4	.070	34	.675	64	2.05
5	.088	35	.700	65	2.14
6	.105	36	.727	66	2.25
7	.123	37	.751	67	2.36
8	.141	38	.781	68	2.48
9	.158	39	.810	69	2.61
10	.176	40	.839	70	2.75
11	.194	41	.869	71	2.90
12	.213	42	.900	72	3.08
13	.231	43	.933	73	3.27
14	.249	44	.966	74	3.49
15	.268	45	1.00	75	3.73
16	.287	46	1.04	76	4.01
17	.306	47	1.07	77	4.33
18	.325	48	1.11	78	4.70
19	.344	49	1.15	79	5.14
20	.364	50	1.19	80	5.67
21	.384	51	1.23	81	6.31
22	.404	52	1.28	82	7.12
23	.425	53	1.33	83	8.14
24	.445	54	1.38	84	9.51
25	.466	55	1.43	85	11.4
26	.488	56	1.48	86	14.3
27	.510	57	1.54	87	19.1
28	.532	58	1.60	88	28.6
29	.554	59	1.66	89	57.3
30	.577	60	1.73	90	∞

Figure 22-21 Factors to determine minimum required shading lengths.

Whenever the solar tilt angle is above 56 degrees, the awning will provide more shading than is required. During the winter months when the solar tilt angle is lower, the sun will be able to shine under the awning to help reduce the winter heating load.

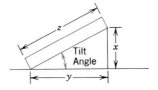

Sizing the Collector Frame

Figure 22-23 gives factors to be used in figuring the frame dimensions required to mount a collector at a predetermined angle. With the dimension of the collector known, either X/Z or Y/Z may be used to calculate the length of the horizontal or vertical support members.

Example
A 4-ft wide collector is to be mounted on a flat roof at a tilt angle of 50 degrees. How much higher must the top of the collector be than the bottom?

Solution
From Figure 22-23, for a 50 degree tilt angle, read

$$X/Z = 0.766$$

Substituting 4 ft for the collector dimension Z,

$$X/4 = 0.766$$
$$X = 3.06 \text{ ft}$$

Tilt Angle	x/z	y/z
24	0.407	0.914
26	0.438	0.899
28	0.470	0.883
30	0.500	0.866
32	0.530	0.848
34	0.559	0.829
36	0.588	0.809
38	0.616	0.788
40	0.643	0.766
42	0.669	0.743
44	0.695	0.719
46	0.719	0.695
48	0.743	0.669
50	0.766	0.643
52	0.788	0.616
54	0.809	0.588
56	0.829	0.559
58	0.848	0.530
60	0.866	0.500
62	0.883	0.470
64	0.899	0.438
66	0.914	0.407

Figure 22-23 Factors to determine support-frame dimensions.

KEY TERMS AND CONCEPTS

Solar Radiation

Latitude

Insolation

Heat Storage

Flat-plate Collector

Reradiation

Focusing Collector

Tracking Collector

Thermosiphon

Open-Loop Recirculating

Auxiliary Electric Heat

Draindown System

Drainback System

Closed Loop

Differential Temperature Controller

On–Off Controller

Proportional Controller

Sensor

Positive and Negative Responding

Collector Placement

Collector Tilt Angle

Solar Tilt Angle

Magnetic South

Air-Heating Collector

Azimuth Angle

Length of Shadows

Sizing Collector Frames

QUESTIONS

1. What is meant by the term insolation value?

2. Why is heat storage provided on many solar heating systems?

3. Why must flat-plate collectors remain protected from the sun until the installation is completed?

4. What is glazing?

5. What are the advantages of a focusing-type collector?

6. What are the disadvantages of a focusing-type collector?

7. How is circulation of the water accomplished with a thermosiphon system?

8. How is the open-loop recirculating system protected from freezing?

9. How is the draindown system protected from freezing?

10. How is the drainback system protected from freezing?

11. What two temperatures are sensed by the controller in order to determine the ΔT?

12. What is the difference betweeen an On–Off controller and a proportional-type controller?

13. What is a thermistor?

14. What is the difference between a positive-responding thermistor and a negative-responding thermistor?

15. During which season does the sun move across the sky in its highest arc?

16. During which season will a horizontal awning throw the longest shadow upon a wall?

17. During which season will a vertical wall throw the longest shadow upon the ground?

18. Which collector would be placed flatter (more horizontal), one to be used for water heating or one to be used for space heating?

19. In the northern hemisphere, what direction (north, east, south, west) should the flat collector plate be facing?

20. Which type of system uses a tracking system that allows the collector to follow the sun across the sky?

PART THREE

AIRSIDE SYSTEMS

Part 3 describes how the heating and refrigeration equipment of Parts 1 and 2 are used to condition air for personnel comfort. Included are practical applications of psychrometrics (the science of air), load calculation methods, the air distribution systems used in building air conditioning, and the control and zoning of these systems. One chapter is devoted to the special air-measuring tools and techniques used in air balancing.

OBJECTIVES

After completing Part 3, the student should be able to do the following:

1. Use the psychrometric chart to check the operation of air conitioning systems.
2. Understand the operation of different types of building air systems.
3. Use air measuring instruments.
4. Read pneumatic controls diagrams.
5. Do building heating and cooling load calculations.

CHAPTER 23

PSYCHROMETRICS

The psychrometric (si-kro-metric) chart is a graph that shows how the properties of air are related to each other. Those properties we will be concerned with are the air temperature, the amount of water vapor contained in the air, and the amount of heat contained in the air.

The engineer uses the psychrometric chart in order to select air conditioning equipment to be installed. The service technician can use the psychrometric chart to recognize when the air conditioning system is not operating properly, or when the air conditioning load is greater than the system cooling capacity. This chapter explains the development and use of the psychrometric chart.

Dry-Bulb Temperature

When you measure air temperature with a standard thermometer, you are measuring the property of the air called its **dry-bulb** temperature. Unless air temperature measurements are specifically referred to otherwise, the measurements are dry-bulb temperatures. In this text, the units for measuring dry-bulb temperature will be degrees Fahrenheit (°F) and dry bulb will be abbreviated to db.

Moisture Content

We are familiar with terms like dry desert heat or a hot, muggy day. In the desert areas of California 80°F db might be quite comfortable, but the same 80°F db could be quite uncomfortable on a muggy day in New Orleans. The difference in these two conditions is the amount of moisture contained in the air.

The ability of air to hold water vapor depends upon the dry-bulb temperature of the air. For example, consider a roomful of air. An average room contains about 100 lb of air. If that air is at a temperature of 60°F db, it is capable of holding about 1.1 lb of water vapor. If the air in the room is 80°F db, 100 lb of that air is capable of holding about 2.2 lb of water vapor. The higher the dry-bulb temperature, the more water vapor the air is capable of holding. This relationship is shown in Figure 23-1.

For the general case, we don't deal with a roomful of air. Rather, we state the amount of water vapor that can be held per pound of dry air. For example, in the previous example, the 60°F room can hold about 0.011 lb of water vapor per pound of dry air. More commonly, the amount of water in the air is expressed in grains rather than in pounds. There are 7000 grains in 1 lb, so 0.011 lb of water vapor is the same as 77 grains. The relationship between dry-bulb temperature and its moisture-holding capability, as it will be shown on the psychrometric chart, is shown in Figure 23-2.

When a sample of air is holding as much moisture as it is capable, it is said to be **satu-**

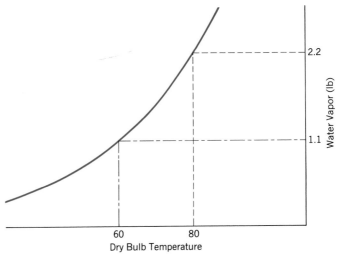

Figure 23-1 If 100 pounds of air at 60°F can hold 1.1 lb of water vapor, it can hold 2.2 lb of water vapor at 80°F.

369

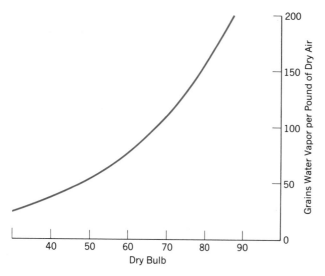

Figure 23-2 As dry-bulb temperature increases, the air is capable of holding an increased amount of water vapor.

rated. Figure 23-2 shows the amount of moisture that can be held by the air at saturation, for any dry-bulb temperature.

Example
Fifteen pounds of dry air at 80°F is allowed to absorb moisture until it reaches a saturated condition. How much moisture does it contain at this saturated condition?

Solution
See Figure 23-2. Locate 80°F on the dry-bulb scale at the bottom of the chart. Move vertically up to the saturation curve. Then move horizontally to the right and read 156 grains of moisture per pound of dry air. Therefore, 15 lb of dry air at 80°F and at a saturated condition holds:

$$15 \text{ lb dry air} \times \frac{156 \text{ grains moisture}}{\text{lb dry air}}$$
$$= 2340 \text{ grains}$$

Relative Humidity

Usually, air contains less than the maximum amount of moisture that it is capable of holding. When we compare the amount of moisture contained in an air sample to the maximum it can possibly hold, we are describing **relative humid-**

ity. For example, if an air sample of 80°F air contains 70 grains of moisture per pound of dry air, the air is at 46% relative humidity (abbreviated rh). That is because its actual moisture content (70 grains) is only 46% of the moisture the air could hold if it were saturated (156 grains).

When a sample of air is heated, its relative humidity decreases. Even though the total pounds of moisture does not change, when the air is heated, it is *capable* of holding more moisture. Therefore the relative humidity decreases. Similarly, when air is cooled, it is less capable of holding moisture. Therefore, with no change in the total pounds of water vapor contained, the relative humidity is increased.

Relative humidity is a popular term used on weather reports. But to the service technician, it can be misleading. Air at 40°F db and 100% relative humidity actually contains less moisture than air at 70°F and 50% relative humidity. When we begin exploring humidification (adding moisture to air) or dehumidification (removing moisture from air), it becomes apparent that we need a term to accurately describe the absolute amount of moisture contained in the air.

Absolute Humidity

Absolute humidity (abbreviated H) is the actual weight of moisture contained in each pound of dry air. In Figure 23-2, the horizontal axis is dry-bulb temperature, and the vertical axis is absolute humidity, expressed as grains of water vapor per pound of dry air.

When air is heated or cooled with no change in the actual quantity of moisture contained, the relative humidity changes, but the absolute humidity remains the same. Other terms that are used interchangeably with absolute humidity are humidity ratio and specific humidity.

Wet-Bulb Temperature

A **wet-bulb** thermometer is illustrated in Figure 23-3. It is simply a standard thermometer that has been outfitted with a cloth sock or wick around the thermometer bulb. When the sock is

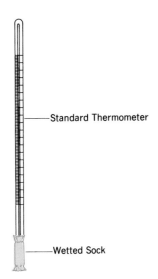

Figure 23-3 Wet-bulb thermometer.

wetted, the evaporation that then takes place causes a cooling effect on the bulb. The lowest temperature reached by this thermometer while the moisture is being allowed to evaporate is the wet-bulb temperature of the air (abbreviated wb). The wet-bulb temperature will never be higher than the dry-bulb temperature. For air that is at 100% relative humidity, no evaporation will take place, and the dry- and wet-bulb temperatures will be equal.

It is interesting to note that the final temperature reached by the wet-bulb thermometer does not depend on the temperature of the water used to wet the sock. If warm water is used, it will keep evaporating and cooling until the wet-bulb temperature is reached. Similarly, cool water will be warmed by the surrounding air until it reaches the same wet-bulb temperature.

In order to maximize the rate of evaporation from the sock, a **sling psychrometer** is used (Figures 23-4 and 5). The dry- and wet-bulb thermometers are both mounted on an apparatus that is spun around on a swivel handle. The velocity past the wet-bulb thermometer causes the maximum rate of evaporation to occur, but it does not affect the reading of the dry-bulb thermometer. With a dry- and wet-bulb temperature, all the other properties of the air may be determined from the psychrometric chart.

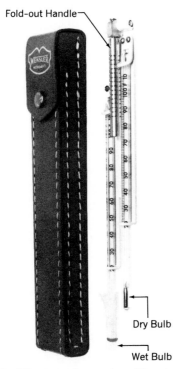

Figure 23-4 Sling psychrometer. *(Courtesy Robinair)*

Figure 23-5 Pocket-sized sling psychrometer. *(Courtesy Airserco Manufacturing Company)*

Dew Point Temperature

When air is cooled, its ability to hold water vapor is decreased. If we cool a sample of air enough, the water vapor in that sample will condense onto the surface that is cooling the air. The temperature of the cold surface that will cause the water vapor to just begin condensing is called the **dew point** temperature.

Heat Content

The term heat content or **enthalpy** describes how much heat is contained in an air sample, per pound of dry air. It is expressed as Btu/lb. For convenience, we define the amount of heat contained in air at 0°F as zero, even though air at that condition actually does contain some heat. Air at higher temperatures and absolute humidities have a higher heat content. Enthalpy is an extremely important property of air. The quantity of air being heated or cooled, and the difference in enthalpy between the beginning and ending air conditions determine the size of the heating or cooling equipment required. For example, suppose we need to cool 1000 lb per hour of air from a heat content of 34.0 Btu/lb to a lower heat content of 22.0 Btu/lb. We would need an air conditioning system that can remove 1000 lb/hr × (34.0 − 22.0) Btu/lb, or 12,000 Btu/hr. This is equivalent to 1 ton of cooling capacity.

Specific Volume

We usually don't think about air as having any weight, but it does. In fact, the amount of air contained in an average size room would weigh about 50 lb! Helium balloons float in air because they are lighter than air. For average conditions (between 60° and 70°F), air weighs approximately 0.075 lb for each cubic foot. Said another way, each pound of dry air occupies approximately 13.3 cu ft. This is called the **specific volume** of the air. It is affected only slightly by changes in temperature and humidity. Air at a temperature of 35°F has a specific volume of 12.5 cu ft/lb. Air at a temperature of 100°F has a specific volume of about 14.4 cu ft/lb.

The Psychrometric Chart

We are now ready to develop the **psychrometric chart.** Your understanding of this chart will give you a powerful tool in understanding what happens in an air conditioning, heating, or refrigeration system.

The graph in Figure 23-2 provides the framework for the psychrometric chart. It is simply a graph of dry-bulb temperature versus the amount of water vapor the air is capable of holding. The curved line indicates the maximum water-holding capability, and therefore is referred to as the saturation line or the 100% relative humidity line.

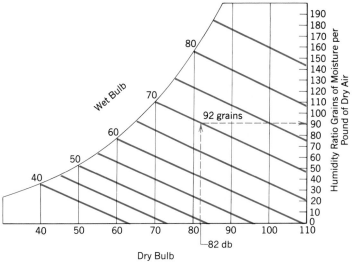

Figure 23-6 Relationship between dry bulb, wet bulb, and absolute humidity.

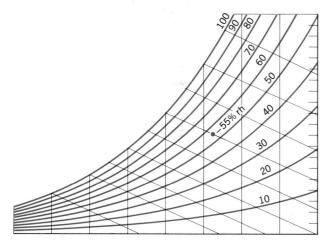

Figure 23-7 Relative humidity lines.

In Figure 23-6, we have added wet-bulb lines to the graph. Wet bulb is read along the diagonal lines, dry bulb is read along the vertical lines, and absolute humidity is read along the horizontal lines.

Example

Using a sling psychrometer, you have obtained a dry-bulb reading of 82°F and a wet-bulb reading of 70°F. What is the absolute humidity?

Solution

On Figure 23-6, locate 82°F along the dry-bulb scale. Proceed vertically upward until you reach

the diagonal 70°F wet-bulb line. Then proceed horizontally to the right and read 0.013 lb/lb on the absolute humidity scale.

Note that there are many different combinations of dry-bulb temperature and moisture content that will have a 70°F wet-bulb reading. For example, air at 91°F db and 78 grains of absolute humidity has a wet bulb of 70°F. The higher dry-bulb temperature is exactly offset by a higher rate of evaporation attributable to the lower moisture content.

In Figure 23-7 we have developed the chart one step further by adding relative humidity lines. Continuing with the previous example, if you place a dot on the graph at 82°F db and 70°F wb, note that it falls between the relative humidity lines labeled 50% and 60%. We can therefore estimate the relative humidity at 55%.

In Figure 23-8, the enthalpy lines have been added, and are almost parallel to the wet-bulb lines. Using the wet-bulb lines to read enthalpy will introduce only a small error. If exact enthalpy readings are desired, the enthalpy scales on both sides of the chart are used.

In Figure 23-9, we have completed the psychrometric chart by the addition of specific volume lines, dew point, and a sensible heat ratio scale. The specific volume lines are normally not of interest to the service technician for troubleshooting. They are sometimes used by the en-

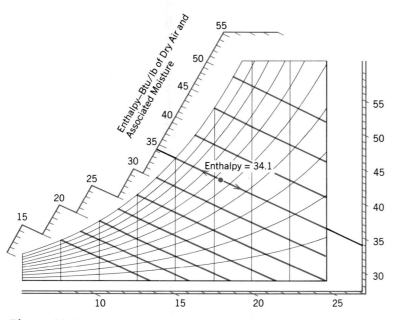

Figure 23-8 Enthalpy lines.

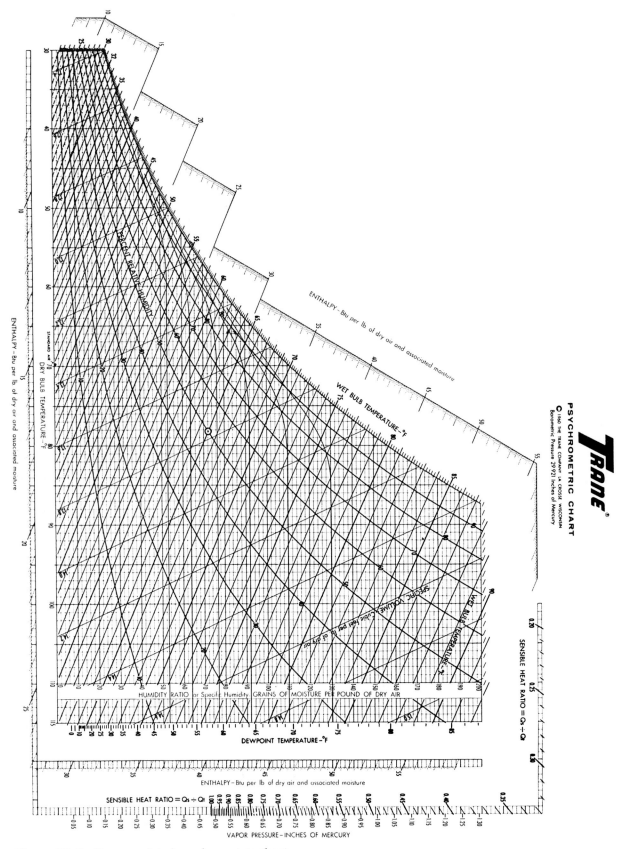

Figure 23-9 The completed psychrometric chart.
(Courtesy The Trane Company, LaCrosse, Wis.)

gineer for design calculations. The sensible heat ratio scale is used in selecting air conditioning equipment. Its use will be described in a later section.

The Psychrometric Process

A **psychrometric process** is the name we give to any process that changes the temperature or moisture content of air. Generally, these processes involve heating, cooling, humidification (adding moisture), and dehumidification (removing moisture). The following sections describe the most common of the psychrometric processes.

Sensible Heating

Consider a closed room that contains air at 70°F db, 62°F wb. From the psychrometric chart, we can determine all of the other properties of that air. This air is shown as point A on Figure 23-10, from which we can read the other properties as follows.

$$\text{absolute humidity} = 72 \text{ grains/lb}$$
$$\text{relative humidity} = 65\%$$
$$\text{dew point} = 57.2°F$$
$$\text{enthalpy} = 27.9 \text{ Btu/lb}$$

Now if we turn on an electric heater in the room to increase the room temperature by 10°F, the new room condition will be 80°F. We have not added or removed any moisture, so the absolute humidity must remain the same at 72 grains/lb. Locating the point 80°F db 72 grains/lb on the psychrometric chart (point B on Figure 23-10), we can determine the remaining properties at the warmer air condition. They are as follows.

$$\text{wet bulb} = 65.3°F$$
$$\text{relative humidity} = 46\%$$
$$\text{dew point} = 57.2°F$$
$$\text{enthalpy} = 30.3 \text{ Btu/lb}$$

This process is called **sensible heating** because there is a change in temperature with no addition or removal of water vapor. On the psychrometric chart, the psychrometric process ap-

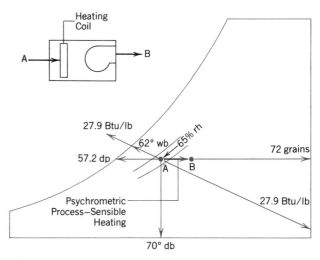

Figure 23-10 Sensible heating on the psychrometric chart. Dry-bulb temperature increases, while absolute humidity remains constant.

pears as a horizontal line connecting the initial air condition with the final air condition. It is interesting to note that even though no moisture is added or removed during a sensible heating process, the wet-bulb temperature increases and the relative humidity decreases. During the cold winter months, when outside air leaks into a space and is then sensibly heated to 70°F or 75°F, it is easy to see how extremely low (and uncomfortable) relative humidity conditions can result.

The sensible heating process is shown as line A-B on the psychrometric chart in Figure 23-10. Some common examples of sensible heating are as follows.

1. Lights, electrical appliances, or motors within a room.

2. Sun shining through windows.

3. Heat that is transmitted through walls.

4. Air in a duct passing over a heating coil or any indirect fired furnace.

Sensible Cooling

Returning to our 70°F db, 62°F wb room, suppose we have a cooling coil at 58°F inside the room, as in Figure 23-11. The air is cooled, with no change in absolute humidity. If the air is cooled to 60°F db, we can read the following properties at point B.

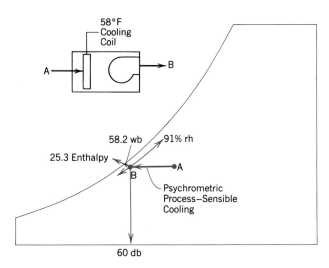

Figure 23-11 Sensible cooling. Dry-bulb temperature decreases, while absolute humidity remains constant.

$$\text{wet bulb} = 58.2°F$$
$$\text{relative humidity} = 91\%$$
$$\text{dew point} = 57.2°F$$
$$\text{enthalpy} = 25.3 \text{ Btu/lb}$$

Note that as the air is cooled, the relative humidity increases and its dew point remains the same. In this process, called **sensible cooling**, we would say that the cooling coil is a dry coil. No condensation of moisture is taking place on the coil surface. The coil will always be dry when its surface temperature is higher than the dew point temperature of the entering air.

Cooling with Dehumidification

Cooling with dehumidification is the most important process in the air conditioning and refrigeration industry. Let's use our 70°F db 62°F wb room as our example once again. But this time, suppose our cooling coil surface temperature is at 48°F. As the air sample passing through the coil cools, it will reach its dew point temperature of 57.2°F. Any further cooling of the air will result in moisture condensing out of the air, causing the absolute humidity to decrease. It may seem odd, but in order to dehumidify the air, we first need to cool it until it reaches 100% relative humidity.

In actual practice, as the air passes through a dehumidifying cooling coil, its relative humidity may be less than 100%. Shallow coils of two

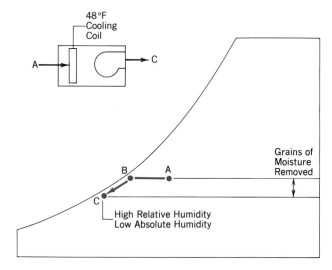

Figure 23-12 Cooling plus dehumidification. Absolute humidity decreases, while relative humidity increases.

or three rows deep may have leaving air conditions of around 90% rh. Deeper coils of six or eight rows will have leaving air that approaches 100% rh.

In Figure 23-12, the line A-C represents cooling with dehumidification. As the air passes through the coil, its dry bulb decreases, it is dehumidified (absolute humidity decreases), and its relative humidity increases, approaching 100%. In this case, some of the heat removed from the air causes a reduction in dry-bulb temperature. This portion is called sensible cooling. Some of the heat that is removed from the air, however, is required to condense the water vapor into a liquid without changing its temperature. This portion is called **latent cooling** (latent means hidden). It is latent because its effect cannot be sensed with a thermometer.

The most common example of cooling with dehumidification occurs in air conditioners or air handlers. In the air conditioner, air is cooled and dehumidified as it passes over a coil containing cold refrigerant. This is called a direct expansion or D-X system. In the air handler, the air is cooled and dehumidified as it passes over a coil containing cold water. This is known as a chilled water coil. Some very common design conditions for this type of equipment are shown as follows.

1. For D-X coils, refrigerant enters the coil at 35°F–45°F saturated suction temperature. It leaves the coil with 7°F–15°F superheat, or at 42°F–52°F.

2. For chilled water coils, water enters the coil at a temperature of between 40°F and 48°F, and leaves the coil at a temperature of 10°F–12°F higher than the entering water temperature.

3. Air leaves the coil at a dry-bulb temperature of between 50°F and 58°F. For residential air conditioners, a temperature drop across the evaporator coil of 18°F–20°F is considered normal.

Disposal of the condensate that collects on the coil may be handled in one of several ways. Most commonly, it is collected in a drain pan as it runs down the face of the coil and is piped away to a sanitary or storm drain (depending on local building codes). Some small packaged air conditioners will get rid of the condensate by allowing it to flow toward the condenser where it is picked up by a slinger ring on the condenser fan. It is then reevaporated on the surface of the condenser coil.

Evaporative Cooling

Figure 23-13 shows a process in which outside air at 95°F db, 65°F wb is passed through a 65°F recirculating water spray. It leaves the spray banks at 70°F db, 65°F wb, and then enters the

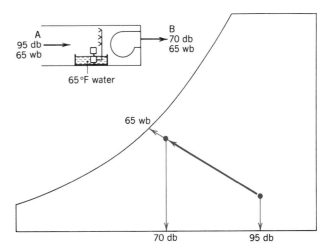

Figure 23-13 Evaporative cooling. Dry-bulb temperature decreases, while absolute humidity increases. Enthalpy and wet bulb remain constant.

room having been cooled by 25°F. In this process, the water will always reach the same temperature as the air wet bulb (through evaporation). The water that evaporates is absorbed by the air, increasing the humidity ratio.

The **evaporative cooling** (Figure 23-14) process is most commonly used in hot, dry climates. There is no mechanical cooling involved, so purchase costs and operating costs are quite low. Most evaporative cooling units use a plastic media pad instead of sprays. The pads are wetted

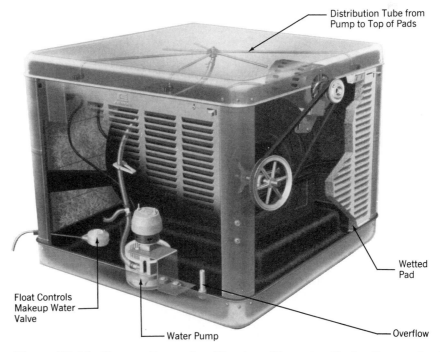

Figure 23-14 Evaporative cooler. (*Courtesy Champion Cooler Corporation*)

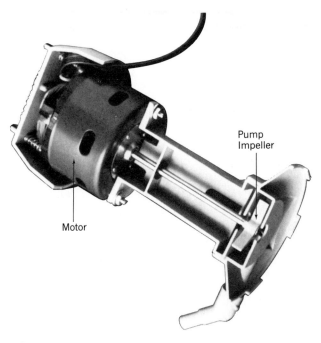

Figure 23-15 Cutaway of a sump pump from an evaporative cooler. *(Courtesy Champion Cooler Corporation)*

with water drawn from the sump by a submersible pump (Figure 23-15). Outside air is then drawn through the pad, where the evaporative cooling effect takes place. Evaporative coolers are commonly referred to as swamp coolers, because of their tendency to create room conditions of high humidity.

The efficiency of an evaporative cooler is determined by dividing the actual drop in dry-bulb temperature by the maximum possible theoretical drop in dry-bulb temperature.

Example

Calculate the efficiency of an evaporative cooler whose performance is shown in Figure 23-13.

Solution

The actual reduction in dry-bulb temperature is (95°F − 70°F) = 25°F. The maximum possible theoretical drop in dry-bulb temperature would occur if the air became totally saturated. That would occur at a db and wb of 65°F. Therefore, the maximum possible drop in air temperature would be (95°F − 65°F) = 30°F. The efficiency of this evaporative cooler is $\frac{25}{30}$ = 0.833, or 83.3 percent.

Mixing

A room or space can be kept at a desired temperature and humidity by constantly recirculating the same air and adjusting the temperature or humidity as desired. But the occupants would soon complain of the air being stale or the room feeling stuffy. In order to avoid this problem, only a portion of the air supplied to the room is returned to the air handler. This return air is then **mixed** with outside air, and that mixture passes through the cooling coil (Figure 23-16). The excess air, which is not returned to the air handler, leaks out through doors or cracks. This out-leakage is called exfiltration.

The dry-bulb and wet-bulb temperature of the mixed air may be predicted using the psychrometric chart. The outside air and return air conditions are plotted on the psychrometric chart, and a straight line is drawn between those two points. The mixed-air condition will fall somewhere on this line, depending on the percentage of outside air versus the percentage of the return air. The point along this line corresponding to the mixed-air condition may be determined by

$$MAT = RAT + \%OA \times (OAT - RAT)$$

where

MAT = mixed air dry-bulb temperature.

RAT = dry-bulb temperature of the return air.

OAT = dry-bulb temperature of the outside air.

%OA = percentage of the total air flow that is from the outside air.

Example

In this case, 800 cfm of return air is at 75°F db, and 50% rh is mixed with 200 cfm of outside air at 95°F db, 78°F wb. What is the dry-bulb and wet-bulb temperature of the mixture?

Solution

$$RAT = 75°F$$
$$OAT = 95°F$$
$$\%OA = 200/(200 + 800)$$
$$= 0.200$$

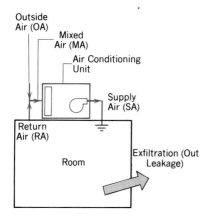

Figure 23-16 A common air-circulation pattern for air conditioning.

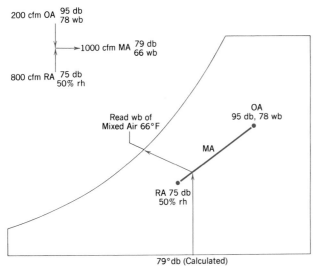

Figure 23-17 Mixing of two air streams. The condition of the mixed air will lie on the line connecting the conditions of the original air streams.

$$MAT = 75 + 0.200\ (95 - 75)$$
$$= 75 + 4$$
$$= 79°F$$

From the psychrometric chart in Figure 23-17, the web-bulb temperature is read as 66°F.

The designer uses this technique to predict the mixed-air temperature so that a cooling coil may be correctly selected. The service technician can use this technique in reverse to determine mixed-air percentages.

Determining the Supply Air Quantity

An air conditioned space has a constant flow of heat into it from many sources. Chapter 26 describes how to calculate the rate of heat flow into a conditioned area. The air conditioned room is maintained at a cool temperature only when heat is being removed by the system at a rate equal to all the heat gains. In a 75°F room, air is introduced at approximately 55°F, and absorbs heat as it warms to 75°F. In order to absorb sufficient heat, there must be a sufficient quantity of air supplied to the space at a sufficiently low temperature. There is a relationship between the quantity of air supplied, the amount of heat it absorbs, and the change in temperature experienced by that air. The relationship is

$$Q = 1.08 \times \text{cfm} \times \Delta T$$

where

Q = sensible heat Btu/hr.

cfm = air flow ft^3/min.

ΔT = change in db temperature.

The 1.08 is a constant. It converts the air flow from cfm to lb per hour, and it factors in the specific heat of the air, which is 0.24 Btu/lb-°F. It can be derived as follows.

$$\frac{0.075\ \text{lb}}{\text{ft}^3} \times \frac{60\ \text{min}}{\text{hr}} \times \frac{0.24\ \text{Btu}}{\text{lb-°F}} = 1.08$$

Example
A room is being maintained at 76°F by an air conditioning system that supplies air at 56°F. If the sensible heat load in the room is 60,000 Btu/hr, how much air is being supplied?

Solution
$$Q = 1.08 \times \text{cfm} \times \Delta T$$
$$60,000 = 1.08 \times \text{cfm} \times (76 - 56)$$
$$\text{cfm} = 2778$$

Note that the same amount of sensible cooling can be done on this room by supplying 3086 cfm at 58°F, or by supplying 2525 cfm at 54°F.

Sensible Heat Ratio

In the previous section, we only dealt with the portion of room heat gain that is sensible. Most air conditioned or refrigerated spaces, however, also have some latent heat gain. Latent loads are those that add moisture to the room. Examples of latent loads are people, plants, coffee urns, and steam tables. If cool air is introduced into a room that has no latent load, it will absorb only sensible heat. Its dry bulb will increase, but its absolute humidity will be unchanged (line A-B in Figure 23-18). If the supply air picks up moisture from latent loads in the room, however, it will gain in db temperature and absolute humidity. An example is shown as line A-C in Figure 23-18. If the percentage of latent load increases compared to the sensible load, line A-C will be steeper. Once the sensible and latent room loads have been determined, it is possible to determine the slope of line A-C by calculating the **sensible heat ratio (SHR)**. This is calculated as follows.

$$SHR = \frac{room\ sensible\ load}{room\ total\ load}$$

Example

A room has a sensible heat load of 110,000 Btu/hr and a latent heat load of 5700 Btu/hr. What is the sensible heat ratio of the room?

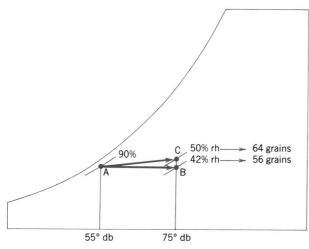

Figure 23-18 With supply air at condition A, a room with no latent heat gain (A-B) will have lower relative humidity than a room that has latent heat loads (A-C).

Solution

The SHR is calculated as

$$SHR = \frac{110,000}{110,000 + 5700} = \frac{110,000}{115,700} = 0.95$$

Stated in words, of the total heat gain in the room, 95 percent is sensible and 5 percent is latent.

The sensible heat ratio will never be greater than 1.00. Most normal air conditioning applications will have a room SHR of between 0.92 and 1.00. Below 0.85, the room latent load would be considered to be quite heavy.

The SHR may be represented graphically as shown by the dotted line in Figure 23-19. This line is drawn parallel to line A-C, and through the reference point at 78°F db and 65°F wb. This line is then extended through the SHR scale at the right.

The designer uses the room heat gain calculations to determine the SHR. Then, knowing the desired room conditions, the required room supply air temperature is determined.

Example

A room is to be maintained at 72°F db and 50% rh. The SHR (determined from the heat gain calculations) is 0.95. What is the required supply air condition?

Solution

See Figure 23-19. Line X-X is drawn through the reference point and 0.95 on the SHR scale. Line A-C is then drawn parallel to line X-X, and passing through the room design condition of 72°F db 50% rh. We will assume an air condition of 90% rh leaving the coil. The intersection of line A-C with the 90% rh line pinpoints the required air supply. Air should be supplied to the room at 54.5°F db, 52.8°F wb. As it picks up sensible and latent load from the room, it will warm up to 72°F db and 50% rh along the SHR line A-C. Note that if we choose to select a supply air temperature higher than 54.5°F db, 52.8°F wb, when it warms to 72°F the relative humidity will be higher than 50%. If we choose a supply air temperature lower than 54.5°F db, 52.8°F wb, the room relative humidity will be less than 50% and the operating cost of the system will be higher than necessary.

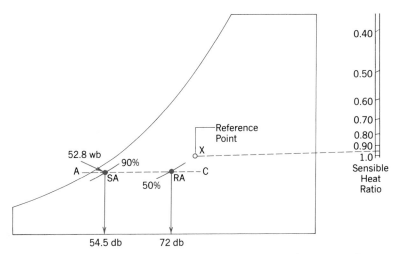

Figure 23-19 Using the SHR scale to determine the required condition of the supply air.

In this problem we assumed a 90% rh leaving the coil. This is a conservative assumption that actually provides some safety factor in the selection of the coil. Without this simplifying assumption, the exact selection process becomes a tedious process that many coil manufacturers have computerized.

Calculating the Air Conditioning Load

In a previous section, you were introduced to the formula

$$Q = 1.08 \times \text{cfm} \times \Delta T$$

That formula described the relationship between sensible heat added or removed, the quantity of airflow, and the change in temperature experienced by the air. But when air is cooled in an air conditioning or refrigeration system, it usually is also dehumidified. Therefore, we need a formula that describes air that is both cooled and dehumidified at the same time. That relationship is

$$Q = 4.45 \times \text{cfm} \times \Delta h$$

where

Q = sensible plus latent cooling, Btu/hr.

cfm = airflow, ft³/min.

Δh = the change in enthalpy of the air, Btu/lb.

The following example demonstrates the use of this formula. It also pulls together much of the information previously described in this chapter.

Example

Calculate the required cooling capacity of an air conditioning unit to maintain a building at 72°F db, 50% rh. The sensible load in the building is 110,000 Btu/hr. The latent load is 5700 Btu/hr. Twenty percent of the air supplied to the room will come from outside. The outside air design condition is 95°F db and 78°F wb.

Solution

See Figure 23-20. The room air and **outside air** are plotted and labeled as OA and RA. The **return air** from the building and outside air are mixed together before entering the air conditioning unit. The temperature of this mixture is calculated as

$$\text{MAT} = 72 + (0.20) \times (95 - 72)$$
$$\text{MAT} = 72 + 4.6$$
$$\text{MAT} = 76.6°\text{F db}$$

A dot is placed on the line connecting OA and RA, at 76.6°F db, and it is labeled MA. This point represents the condition of the air which will enter the cooling coil. From the psychrometric chart, the **mixed air** wet bulb is 64°F.

Next, the SHR line is drawn through the room air temperature to allow us to determine the supply air condition. The SHR is calculated as

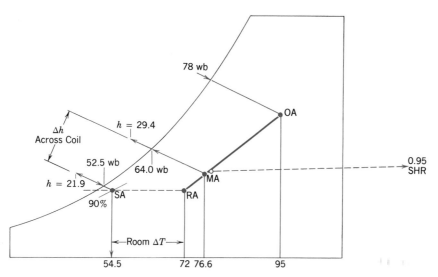

Figure 23-20 A complete air conditioning problem.

$$\text{SHR} = \frac{110,000}{110,000 + 5700} = 0.95$$

A line of 0.95 SHR is drawn through the RA, and a dot is placed on the psychrometric chart where the SHR line crosses the 90 percent rh line. This point (54.5°F db/52.5°F wb) represents the required supply air condition, and is labeled SA. The line from MAT to SA represents the process as the air passes through the cooling coil in the air conditioner. The line from SA to RA represents the process as the air picks up heat and moisture in the room.

The quantity of air required in order to pick up the room sensible load is calculated as

$$Q \text{ sensible} = 1.08 \times \text{cfm} \times \Delta T$$
$$110,000 = 1.08 \times \text{cfm} \times (72 - 54.5)$$
$$\text{cfm} = 5820$$

And finally, we can calculate the refrigeration capacity required. We know that we need to be able to cool 5820 cfm of air from 76.6°F db, 64.0°F wb to 54.5°F db, 90% rh. We can read the enthalpy of the mixed air as 29.4 Btu/lb, and the enthalpy of the supply air as 21.9 Btu/lb.

The **coil load** is now determined by

$$Q = 4.45 \times \text{cfm} \times \Delta h$$
$$Q = 4.45 \times 5820 \times (29.4 - 21.9)$$
$$Q = 194,000 \text{ Btu/hr.}$$

Note that a difference of 0.1 or 0.2 Btu/lb in reading the enthalpy, or a slight error in drawing

the slope of the SHR, will make a significant difference in coil load. Therefore, it is appropriate to round off the calculated load to the nearest thousand.

This example showed how to determine the total amount of cooling required, starting with a known **room load.** The service technician can use this same technique to determine the actual room load, or to determine the actual cooling being done by the cooling coil. In order to determine the room load, the technician can measure supply air db and wb, return air db and wb, and supply cfm. If it is desired to determine the actual cooling output of a unit, the technician can measure supply air db and wb, mixed air db and wb, and supply cfm.

Techniques for measuring cfm are discussed in Chapter 24. For quick estimating purposes for most standard air conditioning systems, the cfm can be determined by multiplying the nameplate tonnage by a factor of 400.

Troubleshooting

The end result of studying psychrometrics must be either to allow us to design or troubleshoot a system. The following examples are designed to demonstrate how psychrometrics are used to detect troubles that are not the routine broken belt or faulty switch type problems.

Example

Figure 23-21 shows a system that includes an air handler with a chilled water coil and a steam heating coil. A room thermostat controls the valves on the coils to maintain room temperature between 70°F and 74°F in a laboratory room where water is being used in some of the customer's testing procedures. The customer is complaining of condensation forming on the inside surface of the windows whenever the outside temperature falls to around 45°F.

Solution

A sling psychrometer is used to take the following readings.

Room air	72°F db, 61°F wb
Outside air	36°F db, 34°F wb
Mixed air	55°F db, 50°F wb
Supply air	90°F db, 66°F wb

When these data are plotted on a psychrometric chart as shown in Figure 23-21, the problem becomes immediately apparent. With the room at 72°F db/61°F wb, the dew point is 54.4°F. It is therefore not surprising that the windows sweat. Now we must determine why the room dew point is so high. In the mixed air, the moisture content is 45 grains/lb; in the supply air, the moisture content is 56 grains/lb. We have found that not only is the heating coil adding heat, but it is also adding moisture. The only way this is possible is

if there is a leak in the coil and steam is being injected directly into the airstream.

Example

Consider the same system and same complaint as in the previous example. This time, however, the psychrometric data are as follows:

Room air	72°F db, 61°F wb
Outside air	36°F db, 34°F wb
Mixed air	55°F db, 50°F wb
Supply air	90°F db, 63.5°F wb

Solution

This problem is plotted on the psychrometric chart shown in Figure 23-22. The absolute humidity of the supply air is the same as the absolute humidity of the mixed air. After the air enters the lab, however, its moisture content increases from 46 grains to 64 grains. In this case, there is nothing wrong with the HVAC system. The technician should suggest to the customer the following potential solutions to the problem.

1. Identify the source of the moisture within the lab, and reduce the amount being allowed to enter the room.

2. If the moisture being released cannot be reduced, install an exhaust system very close to the moisture source. In this way, the moisture

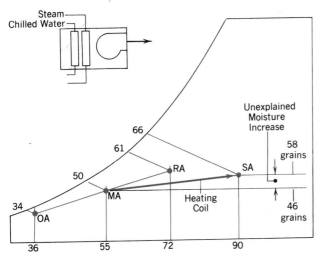

Figure 23-21 Example problem—a leaking steam coil.

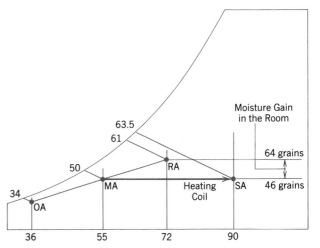

Figure 23-22 Example problem—high latent room load.

is removed before it is dispersed into the room.

3. Replace the windows with double (or triple) pane glass. The inside window surface temperature will be increased so that it will not be below the dew point of the room air.

4. Increase the percentage of dry outside air. This will have an adverse effect on operating costs, but it will reduce the room humidity when it is cold outside.

Example

An air conditioning unit is mounted on a rooftop, and is cooling a computer room located inside a warehouse as shown in Figure 23-23. The complaint is that the room temperature gets too high whenever it gets hot outside.

Solution

The air conditioning unit is checked for proper mechanical functioning and found to be working properly. The following psychrometric data are taken.

Room air	75°F db, 62.5°F wb
Outside air	85°F db, 70°F wb
Mixed air	80°F db, 66.3°F wb
Supply air	56°F db, 55°F wb

The room is cool enough now, because it is not too hot outside. Plotting the system on the psychrometric chart in Figure 23-23, however, re-

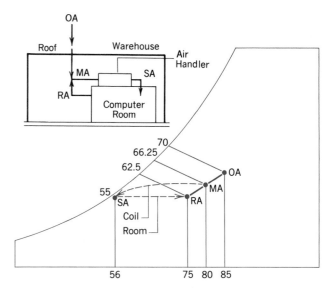

Figure 23-23 Example problem—too much outside air.

veals the problem. With a room air temperature at 75°F, outside air at 85°F, and mixed air at 80°F, we are drawing 50 percent outside air into the system. The computer room has mostly sensible load, only a few people, and there is not a requirement for a lot of outside air. When it becomes warm outside, the unit runs out of capacity just as surely as if all the doors and windows had been left open in the room. The solution is to readjust the damper in the outside air intake so that the unit draws only about 10 percent outside air, or the minimum as required by local building codes.

KEY TERMS AND CONCEPTS

Dry Bulb	Psychrometric Chart	Supply Air Quantity
Saturated Air	Psychrometric Process	SHR
Relative Humidity		Return Air
Absolute Humidity	Sensible Heating	Mixed Air
Wet Bulb	Sensible Cooling	Outside Air
Sling Psychrometer	Latent Cooling	Coil Load
Dew Point	Evaporative Cooling	Room Load
Enthalpy	Mixing	
Specific Volume		

QUESTIONS

1. The air in a room is determined to be at 75°F db and 63°F wb. Using the psychrometric chart, determine the
   (a) rh.   (c) enthalpy.
   (b) dew point.   (d) absolute humidity.

2. As air passes through a heating coil, what happens to the db temperature (goes up, goes down, or remains the same)? The wb temperature? The relative humidity? The absolute humidity? The enthalpy?

3. What is saturated air?

4. What is dew point?

5. As air passes through a sensible cooling coil, what happens to the db? wb? rh? absolute humidity? enthalpy?

6. By looking at a cooling coil, how can you tell if it is dehumidifying the air?

7. As air passes through an evaporative cooler, what happens to the db? wb? rh? absolute humidity? enthalpy?

8. How much moisture can be held by 40°F air? 60°F air? 80°F air?

9. Air at 80°F db and 40% rh enters the evaporator coil of a system operating at 50°F saturated suction temperature. Will the coil be wet or dry? Why?

10. Air enters a cooling coil at 80°F db and 70°F wb. It leaves the coil at 55°F db and 90% rh.

    (a) What are the entering psychrometric properties other than db and wb?

    (b) What are the leaving psychrometric properties other than db and rh?

11. Suppose 8000 cfm of air at 75°F mixes with 2000 cfm of air at 35°F. What is the temperature of the mixture?

12. Outside air at 96°F is mixing with return air at 72°F. The mixed air temperature is 76°F. What is the percentage of outside air?

13. Air at 94°F db, 30% rh passes through an evaporative cooler. If the cooler is 100 percent efficient, what will be the db and wb of the leaving air? What will the leaving db and wb be if the cooler is only 80 percent efficient?

14. Suppose 1500 cfm of outside air at 38°F db and 50% rh is to be introduced into a building at 72°F db and 30% rh.

    (a) How much moisture (gal per day) must be provided by a humidifier?

    (b) What is the difference in enthalpy between the outside air and the air entering the room?

    (c) How much heat (Btu/hr) must be added to the outside air?

15. A room is at 78°F db, 50% rh. There is moisture (condensation) form-

ing on the inside surface of the wall. What can you say about the temperature of the inside surface of the wall?

16. Suppose 4000 cfm of air is cooled from 82°F db, 70°F wb to 54°F db.

   (a) How much sensible cooling was done?

   (b) How much total cooling was done?

   (c) How much latent cooling was done?

17. The engineer's duct drawings specify 1000 cfm of outside air and 9000 cfm of return air. On start-up of the unit, you measured 94°F db and 76°F wb for the outside air condition and return air from the room at 72°F db and 60°F wb. What mixed air db and wb should you measure?

18. A return air duct is drawing 9500 cfm of air at 76°F db and 40% rh from a room. The duct from the room to the air-handling unit passes through a warehouse that is at 100°F db and 76°F wb. The air that is entering the air handler is at 79°F db and 61.5°F wb. Is the warehouse air leaking into the return air duct? How do you know?

19. Suppose 3000 cfm of air at 80°F db and 70°F wb enters an air conditioning unit. It leaves at 53°F db. How much cooling is being done to the air?

20. Air is entering an air handler at 72°F db and 60°F wb. It passes through a steam coil where it is heated to 94°F db and 72°F wb. Is there a leak in the steam coil? How do you know?

CHAPTER 24

MEASURING AIRFLOW AND AIR PRESSURE

The service technician is often called upon to measure air quantities in order to set up or check the operation of a forced-air system. The following sections will acquaint you with methods and instruments used in the field to calculate airflows.

Velocity and CFM

When we talk about airflow quantities, we are talking about **cfm** or cubic feet per minute. When we talk about velocity, we are describing how fast a quantity of air is moving, in feet per minute or **fpm.** Consider the opening in a wall shown in Figure 24-1. The opening is one foot wide and one foot high, and if the air moving through the opening is moving at a velocity of 1500 ft/min,

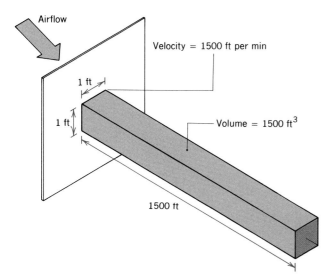

Figure 24-1 Velocity is measured in feet per minute, and flow rate is measured in cubic feet per minute.

the total air that moves through the opening in 1 min would be able to fill a space with a volume of 1 ft × 1 ft × 1500 ft, or 1500 cu ft. Therefore, we could say that if we measured a velocity of 1500 fpm through an opening that had a cross-section area of 1 sq ft, the air quantity woud be flowing at a rate of 1500 cfm. It is easy to extrapolate from this example that if the opening had been 2 ft × 2 ft, a velocity of 1500 fpm would mean an airflow rate of 6000 cfm. Or if we had a velocity of 2000 fpm through an opening of 2 ft × 3 ft, the airflow would be 12,000 cfm. The general relationship between velocity and air quantity is

$$Q = A \times V$$

where

Q = flow in cfm.

A = cross-section area, sq ft.

V = velocity in fpm.

In most cases, airflow is not measured directly with a meter that reads out in cfm. Rather, we usually use a meter that measures velocity, and we use the previous formula to calculate the actual air quantity.

The Velocity Traverse

In the previous example, we assumed that the air velocity through the opening was the same at every point in the opening. In fact, this is rarely the case. If you have even been on a river, you know that the water velocity in the center is much higher than the velocity near the banks. If you wanted to measure the water flow in the river, you would need to measure the velocity at different distances from the bank, and arrive at

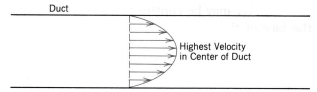

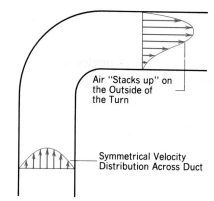

Figure 24-2 Velocity distribution of air flowing inside a duct.

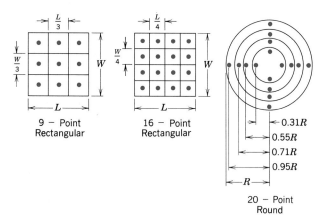

9 – Point Rectangular

16 – Point Rectangular

0.31R
0.55R
0.71R
0.95R
R

20 – Point Round

Figure 24-3 Measurement locations for a velocity traverse.

some sort of average velocity. We follow the same procedure in measuring air velocity. It is called taking a **velocity traverse.**

For air flowing through a duct, the velocity distribution might look something like Figure 24-2. If the length of the arrows represent the actual air velocity, you can see that the air velocity is highest in the center of the duct and drops off very quickly near the sides of the duct. In order to arrive at an average, the technician mentally divides the duct cross section into **equal areas** and measures the velocity at the center of each of the equal areas (Figure 24-3). In a straight duct, some service technicians get rough estimates of the average velocity by measuring just the maximum velocity in the center of the duct and multiplying that value by 0.85.

When selecting a location to take a duct traverse, a location that is preceded by a long straight duct is ideal. If measurements are made close to a fan or an elbow, a very odd and non-uniform set of readings will result (Figure 24-4). When measuring velocities from a supply air or return air grille, the face area is mentally divided and individual velocity readings are taken and averaged.

Air "Stacks up" on the Outside of the Turn

Symmetrical Velocity Distribution Across Duct

Figure 24-4 Uneven velocity distribution downstream from a turn in a duct.

	32 in.			
16 in.	1180	1220	1200	1190
	1210	1370	1350	1230
	1210	1360	1360	1220
	1180	1240	1280	1200

$$\text{Average velocity} = \frac{20,000}{16} = 1250 \text{ fpm}$$

$$\text{Area} = \frac{32 \times 16}{144} = 3.55 \text{ ft}^2$$

$$\text{Airflow} = 3.55 \text{ ft}^2 \times 1250 \frac{\text{ft}}{\text{min}}$$

$$= 4444 \frac{\text{ft}^3}{\text{min}}$$

$$= 4444 \text{ cfm}$$

Figure 24-5 Sample duct traverse.

Example

A velocity traverse is taken across the face of a 16 in. × 32 in. return air register as shown in Figure 24-5. How much air is passing through this return register?

Solution

$$Q = A \times V$$

The average velocity is calculated by averaging all the individual readings.

$$\text{Average velocity} = \frac{20,000}{16} = 1250 \text{ fpm.}$$

and then

$$Q = \frac{(16 \times 32)}{144 \times 1250}$$

$$= 4444 \text{ cfm}$$

the side, which is reading pressure in the area between the two tubes. The pressure is created by a series of small holes around the circumference of the outer tube. As each of these holes is perpendicular to the direction of airflow, they will sense static pressure only.

The Pitot tube may be used with a manometer to measure velocity pressure in a duct by hooking it up as in Figure 24-14a. It does not matter if the static pressure within the duct is positive or negative. The Pitot tube may also be used to measure static pressure in a duct as in Figure 24-14b.

in order to measure a static or velocity pressure of 1.0 in. w.c., a vertical displacement of 1.0 in. in a tube must be read. In reading commonly encountered velocity pressures in air conditioning ducts, however, pressures of only a few hundredths of an inch must be read. With a U-tube manometer, it is impossible to read hundredths of an inch with any degree of accuracy. With the **inclined manometer,** shown in Figure 24-18, however, the situation changes. This manometer essentially spreads out the one inch of vertical column into a much longer horizontal length. The actual vertical rise is still one inch, but it becomes much easier to read accurately when the scale has been expanded.

Inclined Manometer

The manometers discussed up to this point have been **U-tube manometers** (Figure 24-16) or **slack-tube manometers** (Figure 24-17). That is,

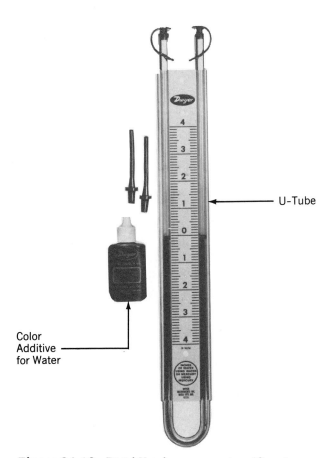

Figure 24-16 Rigid U-tube manometer. *(Courtesy Dwyer Instruments, Inc.)*

Figure 24-17 Slack-tube manometer rolls up for easy storage. When unrolled, it functions like a U-tube manometer. *(Courtesy Dwyer Instruments, Inc.)*

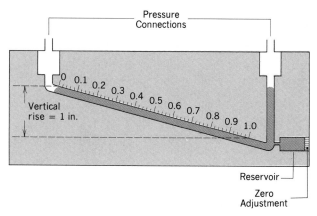

Figure 24-18 Inclined manometer spreads out the pressure scale.

Some inclined manometers use a red gauge oil instead of water because it is easier to read. There is also a problem with using water in that when it evaporates, the solids that are left behind can cloud the inside of the manometer tube. Manometers that use red gauge oil will not have an actual vertical rise that matches the numbers on the gauge. This is explained by the fact that the gauge oil has a specific gravity of 0.86 compared to that of water. Water and red gauge oil may not be used interchangeably. The correct fluid to match the actual calibration of the manometer must be used.

When using inclined manometers (Figure 24-19), it is important that they be mounted in a level position when taking readings. Many units are supplied with an integral water level, and magnetic mounts or screw-type feet for attaining a level position. Once leveled, the gauge must be set to zero by adjusting the amount of fluid allowed to remain in the reservoir. When using the inclined manometer, it is important to anticipate which pressure-sensing port will see the greater pressure and always have the greater pressure pushing the fluid down the scale. When hooked up backwards, the technician will face the em-

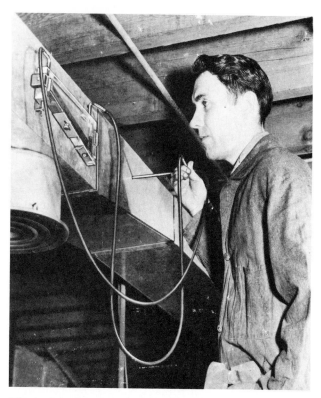

Figure 24-19 Technician measuring velocity pressure with a Pitot tube and an inclined manometer. *(Courtesy Dwyer Instruments, Inc.)*

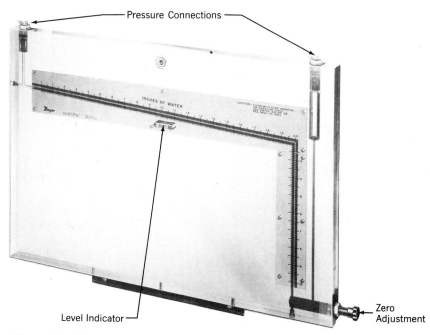

Figure 24-20 Manometer is inclined from 0–2in.w.c. and vertical from 2–10in.w.c. *(Courtesy Dwyer Instruments, Inc.)*

Ratio of (new cfm/old cfm)	New pressure drop = (old pressure drop) × (this factor)
0.50	0.25
0.55	0.30
0.60	0.36
0.65	0.42
0.70	0.49
0.75	0.56
0.80	0.64
0.85	0.72
0.90	0.81
0.95	0.90
1.00	1.00
1.05	1.10
1.10	1.21
1.15	1.32
1.20	1.44
1.25	1.56
1.30	1.69
1.35	1.83
1.40	1.96
1.45	2.10
1.50	2.25

Figure 25-8 Factors to predict the new pressure drop through a duct system when the flow rate is changed.

sense tells us that when a fan is placed in a duct system to move air, the duct cfm will equal the cfm being carried in the duct, and the change in static pressure being produced by the fan will be exactly the same as the static pressure lost in the duct. On the **fan/system curve,** the intersection represents that point where the static pressure and cfm for the fan and the duct match.

Figure 25-10 shows a fan/system curve for the same fan and the same system (labeled system A), and it also shows another system (labeled B). System B represents a more restrictive system than system A. It might be smaller ductwork, longer runs, more fittings, or a cooling coil with higher pressure drop than system A. But at any cfm, you can see that system B will experience more static pressure than system A.

If our original fan is matched with system B, the operating point will ride up the fan curve. The fan will not move as much air as it was able to move through system A, and it will produce a higher static pressure differential.

Figure 25-11 shows a smaller fan (fan B) now added to the fan/system curves. This may be either a smaller diameter fan or the same fan operating at a lower rpm. With two different fans matched against two different systems, there are four possible combinations for matching the components. The interesting comparison is for the intersections labeled Q and R. Each point produces the same cfm. Point Q gets there with system A and a small fan. Point R gets there with smaller (less expensive) ductwork, but must use the larger fan with the higher operating cost.

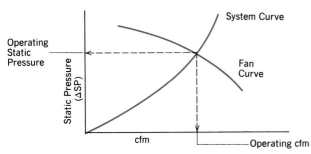

Figure 25-9 Fan curve matched with a system curve. Where they match is the point at which the fan and duct will balance.

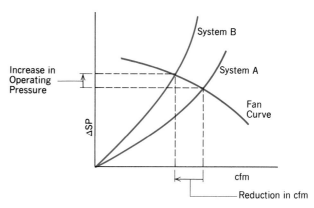

Figure 25-10 System B is more restrictive (smaller duct) than system A, causing the operating point to move upward on the fan curve.

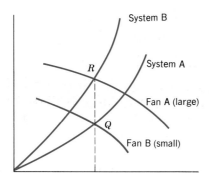

Figure 25-11   Two possible fans matched with two possible systems provide four possible operating points.

Which match is best depends upon the actual numbers, and an analysis of how much the operating cost penalty will be for the potential reduction in first cost.

Fan Types

The two general categories of fans are **Centrifugal** and **Axial.**

The centrifugal fan (Figure 25-12) consists of an impeller and a housing. The overall direction of airflow is perpendicular to the direction of the fan shaft. Air is drawn into the center of the impeller and discharged through the housing. The fan may have the inlet on just one side; a double fan width may be used with inlets on both sides. These two arrangements are commonly called SWSI (single-width single inlet) or DWDI (double-width double inlet).

The impeller is driven either through a direct-drive or belt-drive arrangement (Figure 25-13). Fans that are direct drive, with the motor housing inside the impeller (Figure 25-14) are sometimes referred to as squirrel cage fans.

There are several different designs in use for the centrifugal fan impeller. They are

1. Forward curved (FC).

2. Backward inclined (BI).

3. Airfoil (AF).

4. Radial.

The blade configuration for each of the impeller types is shown in Figure 25-15. Of all the different designs, the **FC fan** is by far the most commonly used in residential and commercial

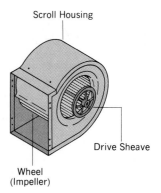

Figure 25-12   Centrifugal fan.

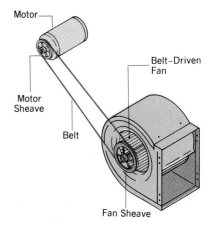

Figure 25-13   Belt-drive fan.

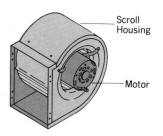

Figure 25-14   Direct-drive squirrel cage fan.

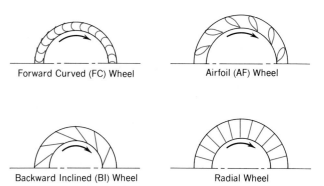

Figure 25-15   Impeller types used in centrifugal fans.

applications. For a given cfm requirement, the FC fan will be the smallest wheel and will operate at the lowest rpm. The FC wheel is used where the static pressure requirements are moderately low (up to 2 to 3 in. w.c.) The forward-looking cup of the FC wheel can become filled with dirt, and thus reduce the fan's effectiveness. In clean applications, this is not a problem.

The **BI** and **AF fans** are more efficient and more expensive than the FC wheel. The rpm for these fans is much higher than for the FC fans. Note the direction of rotation indicated. It is opposite from what many people would suspect. You must be able to recognize the correct rotation direction for a fan. It is not uncommon for the rotation direction to be backwards when troubleshooting a new installation. A fan that is rotating backwards will not move air in a backwards direction. It will move air in the proper direction through the ductwork, but at a drastically reduced flow rate.

The radial design fan blade is not generally used in normal air conditioning applications. It is designed for high-strength, rugged construction. It is more commonly used in material-handling applications.

Axial fans are fans in which the airflow is in the same direction as the fan shaft. Common types of axial flow fans are

1. Propeller
2. Tubeaxial
3. Vaneaxial.

The **propeller fan** has relatively low efficiency, inexpensive construction, and is not capable of producing static pressures in excess of a few tenths of an inch. A household window fan would be an example of a propeller fan. Propeller fans can move lots of air with a free discharge. But they have a very flat fan curve. As the static pressure requirement increases, the cfm decreases very dramatically (Figure 25-16).

The **tubeaxial fan** (Figure 25-17) consists of propeller-type fan blades inside a cylindrical tube. The clearance between the blades and the tube is close, increasing the efficiency. The tubeaxial is capable of developing enough static pressure to move air through low resistance duct systems.

The **vaneaxial fan** has a set of stationary vanes placed ahead of or after the propeller (Figure 25-18). These vanes help straighten the air and convert the rotation energy to static pressure energy. The fan blades may be fixed or adjustable (Figure 25-19) to meet the specific application. Vaneaxial fans can develop static pressures comparable to those produced by the centrifugal fans. They have the advantage of a straight-through design, and are well suited to installation in a straight run of ductwork. Some vaneaxial fans have higher noise characteristics than comparable-capacity centrifugal fans.

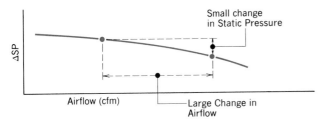

Figure 25-16 Flat fan curve for a propeller fan. This fan is not capable of delivering air through ductwork.

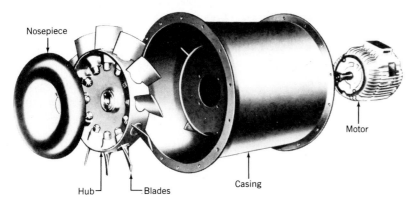

Figure 25-17 Tubeaxial fan. *(Courtesy Joy Mfg. Co.)*

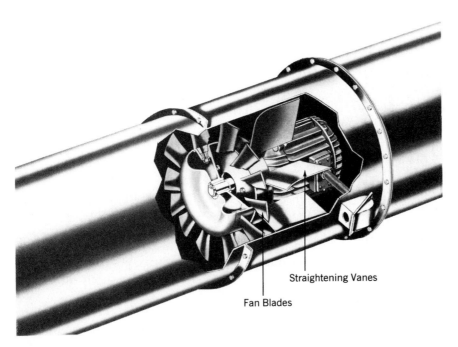

Straightening Vanes

Fan Blades

Figure 25-18 Vaneaxial fan. *(Courtesy Joy Mfg. Co.)*

Figure 25-19 Internal view showing the internal mechanism for adjustment of controllable pitch blades. *(Courtesy Joy Mfg. Co.)*

Fan-to-Shaft Connection

The fan is mechanically connected to its shaft by one of three different methods

1. Set screw.

2. Set screw and key.

3. Expanding hub.

The same methods are used for fastening pulleys (sheaves) to shafts.

Figure 25-20 shows the simplest type of **fan-to-shaft attachment.** The set screw threads through the fan hub and bottoms out on the fan shaft. When installing a fan with a set screw, there will be one flat side on the shaft. It is imperative that the set screw be lined up with this flat spot. If it is not, the fan may be secured to the shaft, but it will soon work its way loose.

Figure 25-21 shows a key fastening arrangement. This is much stronger than the simple set screw method. It is used for more severe fan duty. When removing a fan that is fastened in this manner, take care not to lose the key.

Figure 25-22 shows a fan with a separate tapered fan hub. The hub is tightened against the side of the fan, forcing the tapered section in between the fan hub and the shaft.

Belt-Drive Fans

The general arrangement for a belt-drive fan is shown in Figure 25-13. Generally, the motor speed is 1750 rpm and the fan speed is somewhat lower. The relationship between the fan rpm and the motor rpm is

Fan rpm = (PD motor/PD fan) × motor rpm

where

PD motor = the pitch diameter of the pulley on the motor.

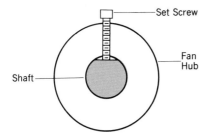

Figure 25-20 Fan hub fastened to shaft with a set screw.

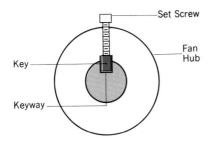

Figure 25-21 Fan keyed to a fan shaft.

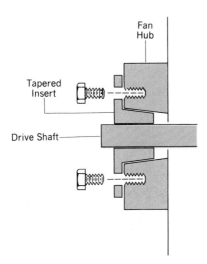

Figure 25-22 Tapered fan hub.

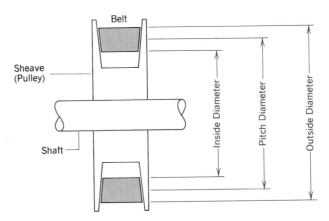

Figure 25-23 Pitch diameter of a sheave.

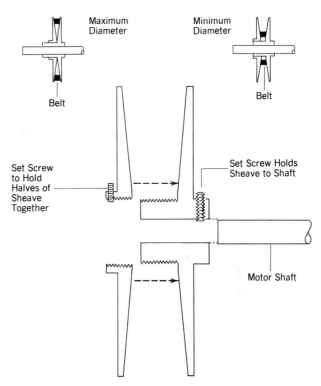

Figure 25-24 Variable-pitch sheave.

PD fan = the pitch diameter of the pulley on the fan.

Pitch diameter (PD) is a term used to describe a dimension in between the outside diameter of the belt and the inside diameter of the belt (Figure 25-23).

Example

A 1750 rpm motor with a 3-in. pitch-diameter pulley needs to drive a fan at 950 rpm. What size fan pulley should be used?

Solution

$$950 \text{ RPM} = (3/\text{PD fan}) \times 1750$$
$$\text{PD fan} = (3) \times (1750/950)$$
$$= 5.52 \text{ in.}$$

When a fan speed needs to be changed in order to change the airflow, the motor pulley size is increased to increase the flow, and decreased to decrease the flow. Because the motor pulley is usually smaller than the fan pulley, it is less expensive to change than the fan pulley.

Figure 25-24 shows a **variable-pitch sheave.**

The two halves of the sheave are two different pieces, with one threaded onto the shaft of the other. How closely the halves are positioned to each other determines the actual pitch diameter of the sheave. Variations of plus or minus 30 percent from the nominal pitch diameter are commonly available with this type of sheave. The variable-pitch sheave is only slightly more expensive than the fixed-pitch sheave in the smaller sizes. In the larger sizes, however, or for multiple-groove sheaves, the variable-pitch sheave can be two or three times the price of a fixed sheave.

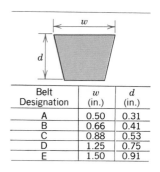

Belt Designation	w (in.)	d (in.)
A	0.50	0.31
B	0.66	0.41
C	0.88	0.53
D	1.25	0.75
E	1.50	0.91

Drive Belts

Drive belts, or **V-belts,** are designated by their cross-section dimension and their length, in inches. The standard cross-section dimensions of belt width and depth are given in Figure 25-25. A model for a belt might be, for example, B48. This would mean that it has a cross section of 0.66 in. width, 0.41 in. depth, and its length is 48 in. The length is at its pitch diameter. The actual outside dimension of the belt will be slightly longer. When replacing a fan belt, the new belt must be of the same cross section as the original. If it is different, it will ride higher or lower in the groove of the sheave and change the effective pitch diameter of the sheave. New belts will normally ride much higher in the sheave groove than an old belt. When the old belt wears, its w dimension can become smaller, causing the belt to sink deeper into the groove of the sheave.

Large fans may be driven with multiple belts. Multiple belts are required when the horsepower to be transmitted from the motor to the fan shaft is too much to be handled with a single belt. Mutliple belts are purhased in **matched sets.** Each of the belts in the set is very close to identical. When one belt breaks, the entire set should be replaced with a new matched set. If old belts are mixed with new, or if nonmatched belts are used, one belt will wind up carrying more than its proportional share of the load. It will fail prematurely, leaving all of the remaining belts overloaded.

When replacing a drive belt, it is necessary to move the motor to relieve the **belt tension.** After the belt is slipped over the sheaves, the belts are

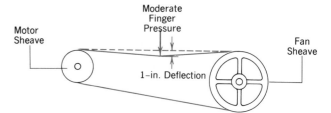

Figure 25-26   Setting belt tension.

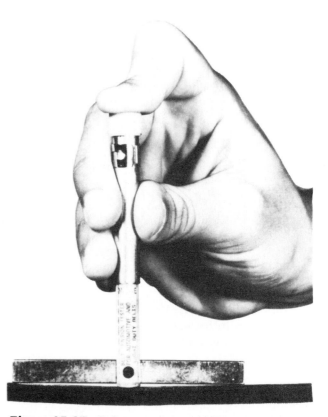

Figure 25-27   Belt tension tester. *(Courtesy Airserco Manufacturing Company)*

retensioned. One guideline is to make it tight enough so that the center of the span will deflect 1 in. when moderate finger pressure is applied (Figure 25-26). This is not very exact. There are tools available for measuring belt tension (Figure 25-27), but the required belt tension is not always known. A simple guideline is to make the belt as loose as possible without it slipping on the sheave or flopping around when it is operating. A belt that is tightened too much puts extra force on the fan and motor bearings, and can cause bearing failure.

Fan Laws

Fan laws are a set of mathematical relationships that predict the changes in a fan's operating characteristics when its speed is changed. When the speed of a fan is increased, the cfm increases, the SP produced by the fan (and used up in the system ductwork friction) increases, and the hp required to drive the fan increases. The following formulas predict how much each of these variables will increase.

$$cfm_2 = cfm_1 \times (rpm_2/rpm_1)$$
$$SP_2 = SP_1 \times (rpm_2/rpm_1)^2$$
$$hp_2 = hp_1 \times (rpm_2/rpm_1)^3$$

In words, the cfm increases in direct proportion with the rpm, the static pressure increases with the square of the rpm, and the horsepower consumed by the fan varies with the cube of the rpm.

Example

A fan is delivering 4000 cfm against a static pressure of 1.6 in. w.c. It is driven by a motor that has a 3.5-in. sheave. The fan horsepower is 1.4. If the motor sheave were changed to 4.5 in., what would be the new cfm, static pressure, and horsepower?

Solution

The diameter of the motor sheave has been increased by a factor of (4.5/3.5) = 1.29. Therefore, the ratio of the old fan rpm to the new fan rpm is $(rpm_2/rpm_1) = 1.29$. The new cfm being delivered will be

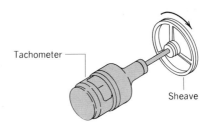

Figure 25-28   Direct reading tachometer. *(Courtesy Biddle Instruments)*

$$cfm_2 = (4000) \times (1.29)$$
$$= 5160$$

the new static pressure will be

$$SP_2 = (1.6) \times (1.29)^2$$
$$= 2.66 \text{ in. w.c.}$$

and the new horsepower will be

$$hp_2 = (1.4) \times (1.29)^3$$
$$= 3.0 \text{ hp}$$

The most important facet of the fan laws to the technician is to understand how fast the horsepower draw increases with a relatively small increase in fan speed. In the previous example the airflow and fan rpm were increased by 29 percent, and the horsepower requirement more than doubled.

When changing a sheave size, or increasing the setting on a variable-pitch sheave, be careful not to overload the motor. With reasonable accuracy, you can substitute "motor amps" in the previous equation to replace horsepower. Measurement of fan rpm is done with a tachometer (Figure 25-28).

Example

A motor with a 9.0 amp nameplate rating is using a 5-in. sheave to drive a fan. The motor is

If the New Fan rpm Divided by the Original Fan rpm is	Multiply the Original Fan cfm by	Multiply the Original Fan SP by	Multiply the Original hp and amps by
1.05	1.05	1.10	1.16
1.10	1.10	1.21	1.33
1.15	1.15	1.32	1.52
1.20	1.20	1.44	1.73
1.25	1.25	1.57	1.95
1.30	1.30	1.69	2.20
1.35	1.35	1.82	2.46
1.40	1.40	1.96	2.74
1.45	1.45	2.10	3.05
1.50	1.50	2.25	3.37

Figure 25-29   Tabulated data representing the result of application of the fan laws.

slightly oversized and the actual amperage draw is 6.2 amps. If the motor sheave size is increased to 6 in., will the motor be able to handle the increase in horsepower?

Solution
The increase in fan rpm is the same proportion as the increase in the motor sheave diameter, so

$$(\text{rpm}_2/\text{rpm}_1) = 6.0/5.0 = 1.2$$

The new amperage draw on the motor will be

$$\text{amps}_2 = 6.2 \times (1.2)^3$$
$$= 10.7 \text{ amps}$$

The increase in motor sheave diamater will require that the motor be changed to the next larger horsepower size.

Figure 25-29 is provided to simplify the process of figuring the rpm ratio squared and cubed.

Sizing the Ductwork

In the discussion on how a fan and a ductwork system balance with each other, it was demonstrated that there were a wide range of **ductwork** sizes that would work for a given airflow. Ductwork that is too small makes the sizing of the fan impractical, however, and ductwork that is too large causes a high installed first cost, with no significant operating benefit. This section provides some practical sizing guidelines for ductwork.

Figures 25-30 and 31 give the pressure drop due to friction for different airflow rates through different size ducts. The units for the friction rate are given in in. w.c. per 100 ft.

Example
What will be the static pressure loss due to friction of 1000 cfm flowing through a duct that is 12 in. in diameter and 250 ft long?

Solution
Enter the vertical axis of Figure 25-30 at 1000 cfm. Move horizontally to the right until you reach the diagonal line indicating 12-in. duct size. Proceed vertically down to read a friction rate of 0.2 in. per 100 ft. For a 250-ft run of ductwork, the friction loss will be

$$\text{Friction} = (0.2 \text{ in.}/100 \text{ ft}) \times 250 \text{ ft}$$
$$= 0.5 \text{ in. w.c.}$$

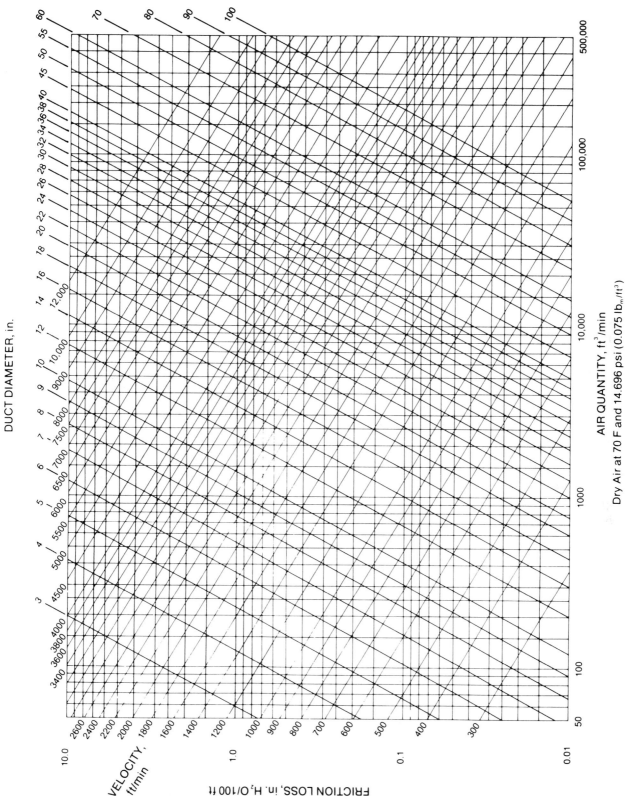

Figure 25-30 Friction rate of air in straight duct. *(Courtesy ASHRAE)*

Lgth Adj.	Length of One Side of Rectangular Duct (a), in.																
	4.0	4.5	5.0	5.5	6.0	6.5	7.0	7.5	8.0	9.0	10.0	11.0	12.0	13.0	14.0	15.0	16.0
3.0	3.8	4.0	4.2	4.4	4.6	4.7	4.9	5.1	5.2	5.5	5.7	6.0	6.2	6.4	6.6	6.8	7.0
3.5	4.1	4.3	4.6	4.8	5.0	5.2	5.3	5.5	5.7	6.0	6.3	6.5	6.8	7.0	7.2	7.5	7.7
4.0	4.4	4.6	4.9	5.1	5.3	5.5	5.7	5.9	6.1	6.4	6.7	7.0	7.3	7.6	7.8	8.0	8.3
4.5	4.6	4.9	5.2	5.4	5.7	5.9	6.1	6.3	6.5	6.9	7.2	7.5	7.8	8.1	8.4	8.6	8.8
5.0	4.9	5.2	5.5	5.7	6.0	6.2	6.4	6.7	6.9	7.3	7.6	8.0	8.3	8.6	8.9	9.1	9.4
5.5	5.1	5.4	5.7	6.0	6.3	6.5	6.8	7.0	7.2	7.6	8.0	8.4	8.7	9.0	9.3	9.6	9.9

Lgth Adj.	Length of One Side of Rectangular Duct (a), in.																				Lgth Adj.
	6	7	8	9	10	11	12	13	14	15	16	17	18	19	20	22	24	26	28	30	
6	6.6																				6
7	7.1	7.7																			7
8	7.6	8.2	8.7																		8
9	8.0	8.7	9.3	9.8																	9
10	8.4	9.1	9.8	10.4	10.9																10
11	8.8	9.5	10.2	10.9	11.5	12.0															11
12	9.1	9.9	10.7	11.3	12.0	12.6	13.1														12
13	9.5	10.3	11.1	11.8	12.4	13.1	13.7	14.2													13
14	9.8	10.7	11.5	12.2	12.9	13.5	14.2	14.7	15.3												14
15	10.1	11.0	11.8	12.6	13.3	14.0	14.6	15.3	15.8	16.4											15
16	10.4	11.3	12.2	13.0	13.7	14.4	15.1	15.7	16.4	16.9	17.5										16
17	10.7	11.6	12.5	13.4	14.1	14.9	15.6	16.2	16.8	17.4	18.0	18.6									17
18	11.0	11.9	12.9	13.7	14.5	15.3	16.0	16.7	17.3	17.9	18.5	19.1	19.7								18
19	11.2	12.2	13.2	14.1	14.9	15.7	16.4	17.1	17.8	18.4	19.0	19.6	20.2	20.8							19
20	11.5	12.5	13.5	14.4	15.2	16.0	16.8	17.5	18.2	18.9	19.5	20.1	20.7	21.3	21.9						20
22	12.0	13.0	14.1	15.0	15.9	16.8	17.6	18.3	19.1	19.8	20.4	21.1	21.7	22.3	22.9	24.0					22
24	12.4	13.5	14.6	15.6	16.5	17.4	18.3	19.1	19.9	20.6	21.3	22.0	22.7	23.3	23.9	25.1	26.2				24
26	12.8	14.0	15.1	16.2	17.1	18.1	19.0	19.8	20.6	21.4	22.1	22.9	23.5	24.2	24.9	26.1	27.3	28.4			26
28	13.2	14.5	15.6	16.7	17.7	18.7	19.6	20.5	21.3	22.1	22.9	23.7	24.4	25.1	25.8	27.1	28.3	29.5	30.6		28
30	13.6	14.9	16.1	17.2	18.3	19.3	20.2	21.1	22.0	22.9	23.7	24.4	25.2	25.9	26.6	28.0	29.3	30.5	31.7	32.8	30
32	14.0	15.3	16.5	17.7	18.8	19.8	20.8	21.8	22.7	23.5	24.4	25.2	26.0	26.7	27.5	28.9	30.2	31.5	32.7	33.9	32
34	14.4	15.7	17.0	18.2	19.3	20.4	21.4	22.4	23.3	24.2	25.1	25.9	26.7	27.5	28.3	29.7	31.0	32.4	33.7	34.9	34
36	14.7	16.1	17.4	18.6	19.8	20.9	21.9	22.9	23.9	24.8	25.7	26.6	27.4	28.2	29.0	30.5	32.0	33.3	34.6	35.9	36
38	15.0	16.5	17.8	19.0	20.2	21.4	22.4	23.5	24.5	25.4	26.4	27.2	28.1	28.9	29.8	31.3	32.8	34.2	35.6	36.8	38
40	15.3	16.8	18.2	19.5	20.7	21.8	22.9	24.0	25.0	26.0	27.0	27.9	28.8	29.6	30.5	32.1	33.6	35.1	36.4	37.8	40
42	15.6	17.1	18.5	19.9	21.1	22.3	23.4	24.5	25.6	26.6	27.6	28.5	29.4	30.3	31.2	32.8	34.4	35.9	37.3	38.7	42
44	15.9	17.5	18.9	20.3	21.5	22.7	23.9	25.0	26.1	27.1	28.1	29.1	30.0	30.9	31.8	33.5	35.1	36.7	38.1	39.5	44
46	16.2	17.8	19.3	20.6	21.9	23.2	24.4	25.5	26.6	27.7	28.7	29.7	30.6	31.6	32.5	34.2	35.9	37.4	38.9	40.4	46
48	16.5	18.1	19.6	21.0	22.3	23.6	24.8	26.0	27.1	28.2	29.2	30.2	31.2	32.2	33.1	34.9	36.6	38.2	39.7	41.2	48
50	16.8	18.4	19.9	21.4	22.7	24.0	25.2	26.4	27.6	28.7	29.8	30.8	31.8	32.8	33.7	35.5	37.2	38.9	40.5	42.0	50
52	17.1	18.7	20.2	21.7	23.1	24.4	25.7	26.9	28.0	29.2	30.3	31.3	32.3	33.3	34.3	36.2	37.9	39.6	41.2	42.8	52
54	17.3	19.0	20.6	22.0	23.5	24.8	26.1	27.3	28.5	29.7	30.8	31.8	32.9	33.9	34.9	36.8	38.6	40.3	41.9	43.5	54
56	17.6	19.3	20.9	22.4	23.8	25.2	26.5	27.7	28.9	30.1	31.2	32.3	33.4	34.4	35.4	37.4	39.2	41.0	42.7	44.3	56
58	17.8	19.5	21.2	22.7	24.2	25.5	26.9	28.2	29.4	30.6	31.7	32.8	33.9	35.0	36.0	38.0	39.8	41.6	43.3	45.0	58
60	18.1	19.8	21.5	23.0	24.5	25.9	27.3	28.6	29.8	31.0	32.2	33.3	34.4	35.5	36.5	38.5	40.4	42.3	44.0	45.7	60
62		20.1	21.7	23.3	24.8	26.3	27.6	28.9	30.2	31.5	32.6	33.8	34.9	36.0	37.1	39.1	41.0	42.9	44.7	46.4	62
64		20.3	22.0	23.6	25.1	26.6	28.0	29.3	30.6	31.9	33.1	34.3	35.4	36.5	37.6	39.6	41.6	43.5	45.3	47.1	64
66		20.6	22.3	23.9	25.5	26.9	28.4	29.7	31.0	32.3	33.5	34.7	35.9	37.0	38.1	40.2	42.2	44.1	46.0	47.7	66
68		20.8	22.6	24.2	25.8	27.3	28.7	30.1	31.4	32.7	33.9	35.2	36.3	37.5	38.6	40.7	42.8	44.7	46.6	48.4	68
70		21.1	22.8	24.5	26.1	27.6	29.1	30.4	31.8	33.1	34.4	35.6	36.8	37.9	39.1	41.2	43.3	45.3	47.2	49.0	70
72			23.1	24.8	26.4	27.9	29.4	30.8	32.2	33.5	34.8	36.0	37.2	38.4	39.5	41.7	43.8	45.8	47.8	49.6	72
74			23.3	25.1	26.7	28.2	29.7	31.2	32.5	33.9	35.2	36.4	37.7	38.8	40.0	42.2	44.4	46.4	48.4	50.3	74
76			23.6	25.3	27.0	28.5	30.0	31.5	32.9	34.3	35.6	36.8	38.1	39.3	40.5	42.7	44.9	47.0	48.9	50.9	76
78			23.8	25.6	27.3	28.8	30.4	31.8	33.3	34.6	36.0	37.2	38.5	39.7	40.9	43.2	45.4	47.5	49.5	51.4	78
80			24.1	25.8	27.5	29.1	30.7	32.2	33.6	35.0	36.3	37.6	38.9	40.2	41.4	43.7	45.9	48.0	50.1	52.0	80
82				26.1	27.8	29.4	31.0	32.5	34.0	35.4	36.7	38.0	39.3	40.6	41.8	44.1	46.4	48.5	50.6	52.6	82
84				26.4	28.1	29.7	31.3	32.8	34.3	35.7	37.1	38.4	39.7	41.0	42.2	44.6	46.9	49.0	51.1	53.2	84
86				26.6	28.3	30.0	31.6	33.1	34.6	36.1	37.4	38.8	40.1	41.4	42.6	45.0	47.3	49.6	51.7	53.7	86
88				26.9	28.6	30.3	31.9	33.4	34.9	36.4	37.8	39.2	40.5	41.8	43.1	45.5	47.8	50.0	52.2	54.3	88
90				27.1	28.9	30.6	32.2	33.8	35.3	36.7	38.2	39.5	40.9	42.2	43.5	45.9	48.3	50.5	52.7	54.8	90
92					29.1	30.8	32.5	34.1	35.6	37.1	38.5	39.9	41.3	42.6	43.9	46.4	48.7	51.0	53.2	55.3	92
96					29.6	31.4	33.0	34.7	36.2	37.7	39.2	40.6	42.0	43.3	44.7	47.2	49.6	52.0	54.2	56.4	96

Figure 25-31 Circular equivalents of rectangular ducts. *(Courtesy ASHRAE)*

Lgth Adj.	32	34	36	38	40	42	44	46	48	50	52	56	60	64	68	72	76	80	84	88	Lgth Adj.
								Length of One Side Rectangular Duct (a), in.													
32	35.0																				32
34	36.1	37.2																			34
36	37.1	38.2	39.4																		36
38	38.1	39.3	40.4	41.5																	38
40	39.0	40.3	41.5	42.6	43.7																40
42	40.0	41.3	42.5	43.7	44.8	45.9															42
44	40.9	42.2	43.5	44.7	45.8	47.0	48.1														44
46	41.8	43.1	44.4	45.7	46.9	48.0	49.2	50.3													46
48	42.6	44.0	45.3	46.6	47.9	49.1	50.2	51.4	52.5												48
50	43.6	44.9	46.2	47.5	48.8	50.0	51.2	52.4	53.6	54.7											50
52	44.3	45.7	47.1	48.4	49.7	51.0	52.2	53.4	54.6	55.7	56.8										52
54	45.1	46.5	48.0	49.3	50.7	52.0	53.2	54.4	55.6	56.8	57.9										54
56	45.8	47.3	48.8	50.2	51.6	52.9	54.2	55.4	56.6	57.8	59.0	61.2									56
58	46.6	48.1	49.6	51.0	52.4	53.8	55.1	56.4	57.6	58.8	60.0	62.3									58
60	47.3	48.9	50.4	51.9	53.3	54.7	56.0	57.3	58.6	59.8	61.0	63.4	65.6								60
62	48.0	49.6	51.2	52.7	54.1	55.5	56.9	58.2	59.5	60.8	62.0	64.4	66.7								62
64	48.7	50.4	51.9	53.5	54.9	56.4	57.8	59.1	60.4	61.7	63.0	65.4	67.7	70.0							64
66	49.4	51.1	52.7	54.2	55.7	57.2	58.6	60.0	61.3	62.6	63.9	66.4	68.8	71.0							66
68	50.1	51.8	53.4	55.0	56.5	58.0	59.4	60.8	62.2	63.6	64.9	67.4	69.8	72.1	74.3						68
70	50.8	52.5	54.1	55.7	57.3	58.8	60.3	61.7	63.1	64.4	65.8	68.3	70.8	73.2	75.4						70
72	51.4	53.2	54.8	56.5	58.0	59.6	61.1	62.5	63.9	65.3	66.7	69.3	71.8	74.2	76.5	78.7					72
74	52.1	53.8	55.5	57.2	58.8	60.3	61.9	63.3	64.8	66.2	67.5	70.2	72.7	75.2	77.5	79.8					74
76	52.7	54.5	56.2	57.9	59.5	61.1	62.6	64.1	65.6	67.0	68.4	71.1	73.7	76.2	78.6	80.9	83.1				76
78	53.3	55.1	56.9	58.6	60.2	61.8	63.4	64.9	66.4	67.9	69.3	72.0	74.6	77.1	79.6	81.9	84.2				78
80	53.9	55.8	57.5	59.3	60.9	62.6	64.1	65.7	67.2	68.7	70.1	72.9	75.4	78.1	80.6	82.9	85.2	87.5			80
82	54.5	56.4	58.2	59.9	61.6	63.3	64.9	66.5	68.0	69.5	70.9	73.7	76.4	79.0	81.5	84.0	86.3	88.5			82
84	55.1	57.0	58.8	60.6	62.3	64.0	65.6	67.2	68.7	70.3	71.7	74.6	77.3	80.0	82.5	85.0	87.3	89.6	91.8		84
86	55.7	57.6	59.4	61.2	63.0	64.7	66.3	67.9	69.5	71.0	72.5	75.4	78.2	80.9	83.5	85.9	88.3	90.7	92.9		86
88	56.3	58.2	60.1	61.9	63.6	65.4	67.0	68.7	70.2	71.8	73.3	76.3	79.1	81.8	84.4	86.9	89.3	91.7	94.0	96.2	88
90	56.8	58.8	60.7	62.5	64.3	66.0	67.7	69.4	71.0	72.6	74.1	77.1	79.9	82.7	85.3	87.9	90.3	92.7	95.0	97.3	90
92	57.4	59.3	61.3	63.1	64.9	66.7	68.4	70.1	71.7	73.3	74.9	77.9	80.8	83.5	86.2	88.8	91.3	93.7	96.1	98.4	92
94	57.9	59.9	61.9	63.7	65.6	67.3	69.1	70.8	72.4	74.0	75.6	78.7	81.6	84.4	87.1	89.7	92.3	94.7	97.1	99.4	94
96	58.4	60.5	62.4	64.3	66.2	68.0	69.7	71.5	73.1	74.8	76.3	79.4	82.4	85.3	88.0	90.7	93.2	95.7	98.1	100.5	96

Figure 25-31 (continued)

For ductwork that is rectangular instead of round, use Figure 25-31 to determine the equivalent round duct size, and then figure the friction as though the duct is round. Fan manufacturers have produced a variety of nomographs and other devices to simplify the friction graphs for both round and rectangular ducts. One of the more popular models is shown in Figure 25-32.

To determine what duct size should be used for a given airflow, there are velocity guidelines that minimize airflow noise through the duct, and represent an acceptable balance between first cost and operating cost. For general residential and commercial applications, choose a duct size that will create a friction rate of 0.08 to 0.10 in. w.c. Where noise is not a consideration, the velocity used may be somewhat higher, up to a friction rate of 0.2 to 0.4 in. w.c.

The allowable variation from these guidelines is quite large. In a theater, where noise is of much higher concern than for the average application, a much larger duct size might be chosen. For an industrial application where no amount of duct noise will be noticeable, and where the duct run is quite short, a very small, very high velocity duct may be the most practical.

Friction Losses in Duct Fittings

Every time air changes direction or passes through a partial obstruction, turbulance is produced and an additional friction loss is encountered. For example, for a round duct with 24-in. diameter, a smooth 90-degree elbow will add the equivalent friction as an additional 18 ft of straight duct. If the elbow is mitered, with no internal vanes to reduce the turbulance, the same 90-degree turn would impose friction equivalent to 130 ft of straight duct. Designers use tables of friction losses for each different type of fitting through which the air must pass, and determine the **equivalent length** for each fitting.

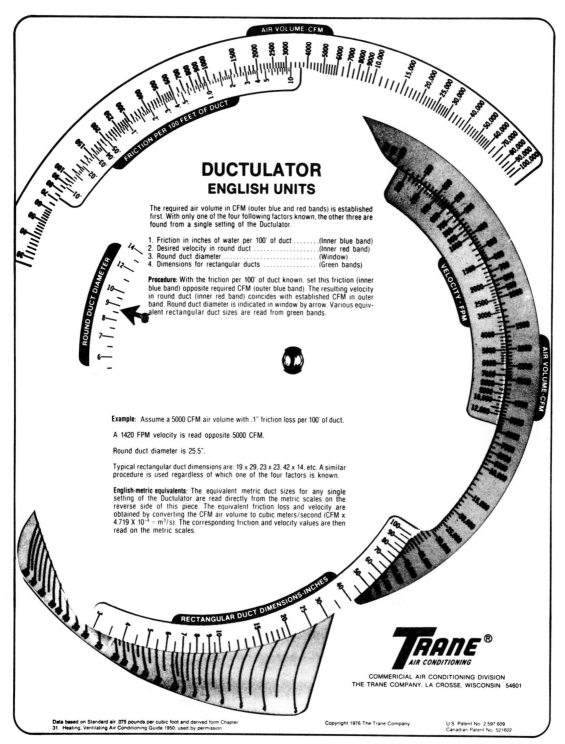

Figure 25-32 Selector to correlate duct size, flow rate, and friction rate. *(Courtesy The Trane Company, LaCrosse, Wis.)*

When these equivalent lengths are added to the actual length of ductwork, the friction is figured through the total equivalent length of the system.

Aspect Ratio

For rectangular ducts, the **aspect ratio** (Figure 25-33) is defined as the length of the larger dimension (width or depth) divided by the shorter dimension. A duct that is 36 in. × 12 in. has an aspect ratio of 3:1 (read three to one). Aspect ratios closer to 1:1 (a square duct) are more desirable than ducts with high aspect ratios. A high aspect ratio duct will require the use of more sheet metal to construct the duct of an equal cross-sectional area. In addition, even though the cross-sectional areas may be the same, a duct with a high aspect ratio will incur a greater loss than a square duct. The use of high aspect ratio ducts should be limited to areas where physical space limitations are such that there is no alternative.

Duct Fittings

Elbows are designated by the number of degrees turned, usually 90 degrees or 45 degrees. For round duct, Figure 25-34 shows a smooth elbow, a three-piece elbow, and a five-piece elbow. The three- and five-piece elbows are also called **mitered elbows.** For rectangular duct, Figure 25-35 shows a **radius elbow,** a short radius vaned elbow, and a square vaned elbow. The use of **turning vanes** in elbows prevents the air from stacking up on the outside of the turn and reduces the friction loss tremendously. Turning vanes may be either single thickness or double thickness, with the double thickness providing a somewhat lower pressure drop.

Transitions are duct fittings where the duct must change size or shape. For minimum pressure drop, the angles of change should be limited to 30 degrees from the direction of airflow, and never more than 45 degrees (Figure 25-36). The higher the velocity of the airflow in the duct, the more critical it is to have gradual instead of sudden size transitions.

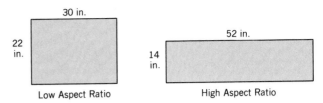

Figure 25-33 Equivalent ducts with very different aspect ratios.

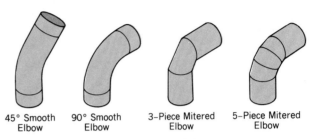

Figure 25-34 Elbows in round duct.

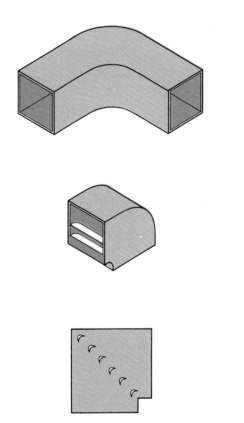

Figure 25-35 Elbows used in rectangular ductwork.

Takeoffs are important for getting air out of a main duct and into a branch duct (Figure 25-37). There have been many systems constructed where there is no fitting provided to turn the air into the branch duct. As a result, all the air tends

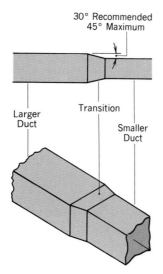

Figure 25-36 Duct transition.

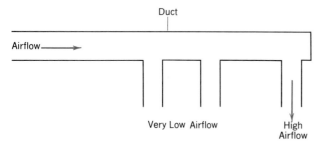

Figure 25-37 A poorly designed system of takeoffs. Air will stack up at the end of the main duct run.

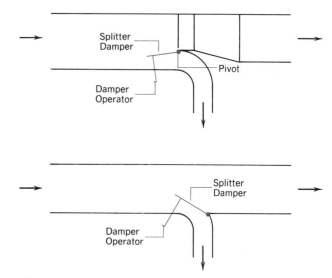

Figure 25-38 Splitter dampers used to turn air out of the main duct and into the branch duct.

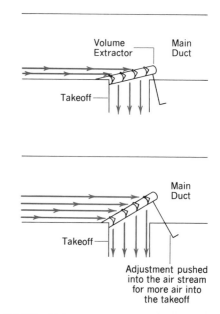

Figure 25-39 Volume extractor used to cause air to turn into the branch duct from the main duct.

to continue moving straight through the main duct until it stacks up at the end. The velocity pressure is converted to static pressure, and the takeoffs at the end of the main duct receive too much airflow, while the upstream takeoffs are starved for air.

Figure 25-38 shows two different methods for using a **splitter damper.** The amount of air that goes into the takeoff may be adjusted by positioning the splitter rod. Figure 25-39 is a volume extractor. This is an adjustable turning vane that pulls air out of the main duct and turns it into the branch. The further the extractor is pushed into the main duct, the more air will be extracted. The turning vanes are mounted on a linkage that keeps the angle of the vanes the same relative to the main duct, regardless of the position of the extractor.

Dampers installed in a duct may be either **parallel-blade, opposed-blade,** or **leaf dampers** (Figure 25-40). Leaf dampers are the least expensive, but suitable for only relatively small applications. Opposed-blade dampers are the best for adjusting air flows, and they are also the most expensive. Any of these damper types may be motor driven. The parallel-blade dampers may also be operated by gravity to close and airflow to open.

Figure 25-41 shows another damper, called a **fire damper.** It is not installed for any heating or air conditioning purpose, but is a safety device

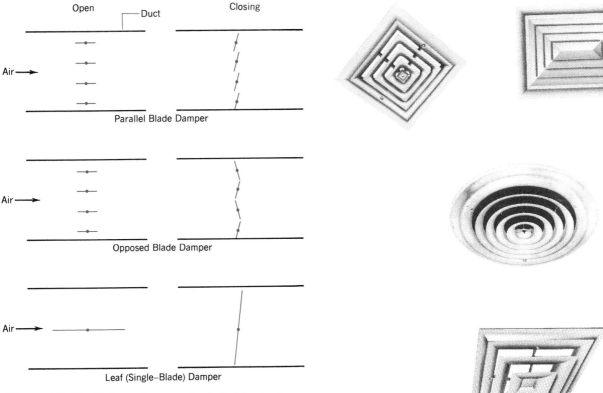

Figure 25-40 Volume dampers installed in ducts.

Open Duct Closing

Air →

Parallel Blade Damper

Air →

Opposed Blade Damper

Air →

Leaf (Single–Blade) Damper

Fire Wall

Accordian Damper

Access Door

Figure 25-41 Fire damper installed where a duct passes through a wall that is designed as a fire separation.

Figure 25-42 Ceiling diffusers deliver air to the room with maximum mixing with room air and minimum draft to the occupants.

only. In large industrial buildings, fire walls are provided to prevent the potential spread of fire from one area to another. When ducts penetrate these walls, the integrity of the fire separation is threatened. Fire codes require that fire dampers be installed at these fire walls. In the event of fire, the fusible link melts and the damper automati-

cally closes. This is of interest to the service technician because if the fusible link breaks, airflow will be interrupted for no apparent reason. The existence of a fire damper is signaled by an access door in the side of the duct where it passes through the wall. The access door is also required by fire codes to allow inspection of the fire damper.

A number of ductwork accessories are available for the termination of the ductwork at the room. **Ceiling diffusers** (Figure 25-42) are used to supply air into the room. A well-designed, properly selected diffuser will cause the supply air to mix quickly with the room air, equalizing its temperature with the room so that the occupants do not perceive a draft. The diffuser may be supplied with a balancing damper behind (Figure 25-43). A **linear diffuser** (Figure 25-44) may be used where long areas must be blanketed very evenly. A **grille** is simply a visual screening used on a supply or a return duct, although a supply grille may have some vanes to somewhat direct

Figure 25-43 Adjustable balancing damper used behind a ceiling diffuser. *(Courtesy Hart & Cooley)*

Figure 25-44 Linear diffuser. *(Courtesy Hart & Cooley)*

the supply air (Figure 25-45). The vanes behind the supply grille may be either horizontal or vertical or both. A **register** (Figure 25-46) is a grille with a damper behind it. A **louver** is used where a duct terminates at an exterior building wall, and may be used for outside air intake or exhaust air. The exhaust air louver function is only decorative. The intake louver is designed to prevent rain from entering with the air. When referring to these duct fittings, the terminology of diffuser, grille, register, and louver have specific meaning, and should not be referred to as vents.

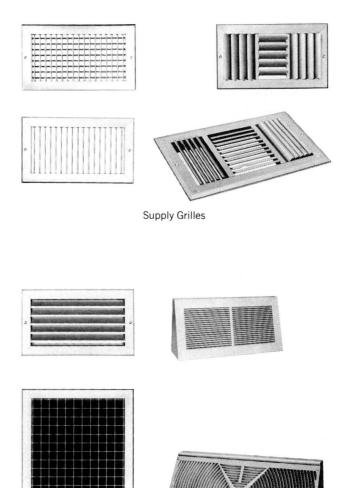

Supply Grilles

Return Grilles

Figure 25-45 Supply grilles and return grilles. *(Courtesy Hart & Cooley)*

Duct Construction

Most ductwork installations consist of sheet-metal pieces fabricated in a shop and delivered to

Figure 25-46 Supply air registers. *(Courtesy Hart & Cooley)*

the jobsite in sections for assembly. The duct construction details are dictated by the size of the ductwork and the internal pressure that it must withstand without distorting. Systems are classified as low velocity, medium velocity, and high velocity, according to the service they are performing (a combination of velocity and static pressure). For each velocity classification, there are construction details for each size classification. Many of these recommendations are spelled out in manuals available from an organization called Sheet Metal and Air Conditioning Contractors of North America (SMACCNA).

Other materials are used for duct construction due to lower cost and or easier installation. Fiberglass board of approximately 1 in. of thickness can be formed into ductwork sections and taped together. It is already insulated, and the fiberglass also attenuates ductwork noise. For cold ducts, a separate vapor barrier material (foil) is provided for the outside surface.

Flexible ductwork is popular in small sizes. It may be used in long lengths, minimizing the number of joints required. Its flexibility allows a single piece to be used to connect a main duct to a diffuser. The construction is a wire springlike core, covered with soft fiberglass and then covered with a plastic vapor barrier.

Construction Drawings

Construction drawings are prepared by the architects and engineers. They serve several purposes.

1. The owner reviews the construction drawings in order to make sure that the building and systems being designed are, in fact, the building and systems that are wanted.

2. The contractor uses the construction drawing in order to estimate the labor and material required in order to bid on the work.

3. The installer uses the construction drawings as a road map to make sure that the system is installed as bid by the contractor, and as designed by the architects and engineers.

4. The service technician uses the construction drawings (when available) for assistance in determining how the system is supposed to

work, and in locating parts of the system that are not readily visible from the occupied portions of a completed building.

A complete set of construction drawings will be divided into categories that correspond to the various trades. Each category will have drawings numbered consecutively. For example, the ductwork will appear on the mechanical drawings, which will be numbered M-1, M-2, M-3, etc. The electrical work will appear on drawings numbered E-1, E-2, etc. Figure 25-47 shows a common **drawing numbering system,** and the type of information contained in each category.

In addition to the construction drawings (called plans), there may be written material that describes methods of construction, materials, responsibilities of each of the parties, quality control and inspection, bidding requirements, and hundreds of other items describing how the job will be done. These documents are known as the specifications (specs). The plans and specs completely describe every aspect of the job and how it is to be constructed.

Drawing Category	Information Contained
A	**Architectural:** Construction materials and methods for walls, partitions, ceilings, floors, roofs, windows, doors and hardware, finishes, and furnishings. Ceiling plans will show ceiling diffuser locations
S	**Structural:** Beams and columns that hold the structure together. Includes framing and support details for HVAC equipment.
M	**Mechanical:** Shows all details of layout and sizing for the HVAC system, including the required airflows
P	**Plumbing:** Domestic water supplies and waste piping as well as storm drainage piping. Also, water supply or drainage for HVAC equipment is included
E	**Electrical:** Power wiring, control wiring, electrical panels, and lighting
Y	**Yard:** Outside work, including grading, landscaping, utilities, fencing, and a plot plan

Figure 25-47 Drawing categories for a set of construction drawings.

Reading the Mechanical Drawings

Drawings may show the items being represented from several points of view. Most common is the **plan view** (Figure 25-48), which draws the equipment as if you were looking down from directly above. Figure 25-49 depicts an **elevation,** that is, a view looking from the side. If it is a view from the west looking eastward, it is called the west elevation. Similarly, there can be north, east, and south elevations.

Less frequently used is the **isometric drawing.** The three-dimensional sketches used throughout the chapter are isometric drawings. While they are the most descriptive and easiest to visualize, they are somewhat more difficult to produce, and they are not accurate dimensionally.

The scale of the drawing is given on each sheet, or on each section where different scales are used. A scale of $\frac{1}{8}$ in. = 1 ft − 0 in. means that for each $\frac{1}{8}$ in. of length shown on the drawing, an actual real-life length of 1 ft − 0 in. is represented.

Example
A room is $4\frac{1}{2}$ in. long on a drawing that has a $\frac{1}{4}$ in. = 1 ft − 0 in. scale. How long is the actual room?

Solution
You must determine how many quarter inches are there in the $4\frac{1}{2}$-in length on the drawing. There are 18 quarter inches ($4\frac{1}{2}$ divided by $\frac{1}{4}$). The actual room length is 18 feet.

The architectural scale shown in Figure 25-50 can simplify this conversion. On its three sides are six different commonly used scales. Using the appropriate scale, the dimension on a drawing

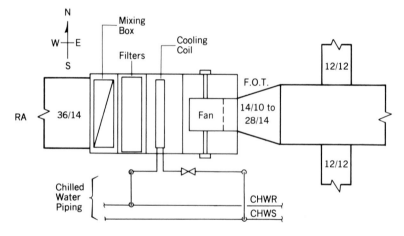

Figure 25-48   Plan view is a top view, as if the building had the roof removed and was viewed from an airplane.

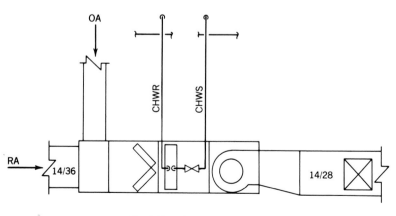

South Elevation

Figure 25-49   An elevation is a side view.

may be measured, and the numbers that you read on the scale are already converted into the actual length represented.

Another type of view that can be represented is the **section drawing.** Figure 25-51 shows how the section arrow on one drawing shows where the section is taken. The section drawing itself then represents the view as if the equipment had been cut with a knife along the section line, and everything behind the section line stripped away. The numbers on the section line tell you the section number (so you will know when you have

found the correct corresponding drawing), the drawing on which the section is taken, and the drawing on which the section is shown. Often, all the sections taken on the various mechanical drawings will be grouped together and shown on one or two drawings titled "Sections."

The ductwork shown on the drawings may be drawn in either **single line** or **double line** (Figure 25-52). Single line drawings leave the routing and fitting up to the installer. Double line drawings have probably been checked by the designer against the structural and drawings of other trades to make sure that there are no interferences. When sizes are shown on a rectangular duct, the dimension facing you is given first. For example, consider a duct that is 16 in. high (vertical dimension), 34 in. wide (horizontal dimension), and is running in an east–west direction. On a north elevation, the duct size might be shown as 16″ × 34″, or 16/34. In a plan view of the same duct, however, its size wil be shown as 34″ × 16″. Round duct size is shown with a single number that represents the duct diameter.

A number of the notations on HVAC drawings

Figure 25-50 Architectural scale has several different scales.

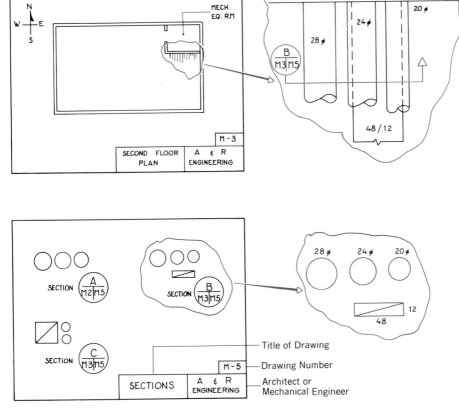

Figure 25-51 A section drawing is a cutaway.

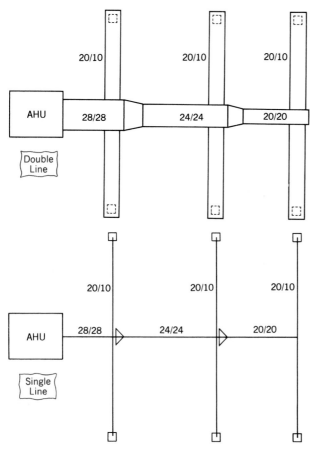

Figure 25-52 Single-line and double-line drawings of the same duct system.

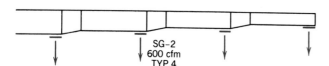

Figure 25-53 Plan view of a duct with sidewall grilles.

are different from those encountered on conventional mechanical drawings. The following listing describes many of these notations.

The sidewall grilles shown in Figure 25-53 are all identical. Instead of labeling and describing each one individually, only one of the four is described. The notation typ. 4 tells you that the description given is typical for all four devices shown. The identification SG-2 is given by the designer. Elsewhere on the drawings or in the specifications, SG-2 is defined to include the manufacturer's name and model number, finish, and other descriptive information too cumbersome to show on the drawing.

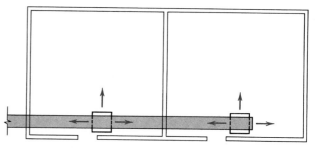

Figure 25-54 Plan view of ceiling diffusers supplying air to two offices.

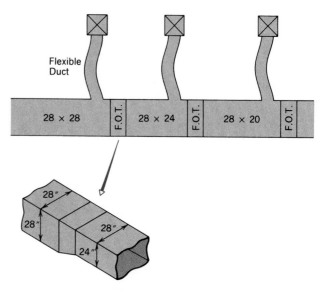

Figure 25-55 Transitions can be flat on top, flat on bottom, or take half the size change on top and bottom.

The arrows on the ceiling diffuser shown in Figure 25-54 describe the direction of the throw. This diffuser is a three-way throw. If not installed as shown, one of the throw directions will direct the air against a wall and cause an objectionable draft.

The transitions shown in Figure 25-55 are not completely described without the FOT or FOB notations (flat on top and flat on bottom). If there is no notation, it may be assumed that the size change is taken equally on the top and on the bottom. On the square to round transition, the lines shown are actually the brake lines that are formed when this transition is fabricated.

The thermostat in Figure 25-56 has an arrow pointing to the unit that it is controlling. The actual routing of the thermostat wire is left to the installer. The note 5 ft AFF means that it is to be located 5 ft above the finished floor.

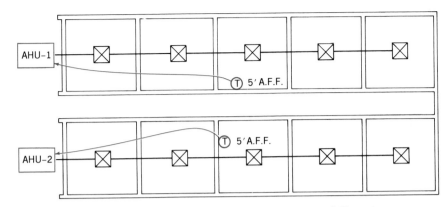

Figure 25-56 Plan view of offices being cooled by two different air-handling units.

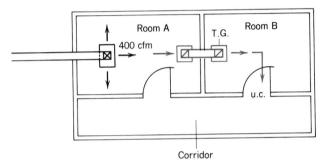

Figure 25-57 Air being supplied to room A moves to room B through a transfer grille, and then through the undercut door.

The rooms in Figure 25-57 have a single supply of 400 cfm to room A. The air leaves room A through a transfer grille (TG), into room B. Finally, the air leaves room B through a door that has been undercut (u.c.) at the bottom to allow space for the air to flow. The same airflow could have been accomplished by a grille in the door of room B.

Duct Symbols

Figure 25-58 shows a number of ductwork symbols used on mechanical drawings. These symbols and abbreviations are not universal. It is common that a drawing set will contain a sheet of the symbols and definitions that are used.

KEY TERMS AND CONCEPTS

Positive, Negative, Neutral Pressure
Fan Curve
Family of Curves
Multirating Table
System Curve
Fan/system Curve
Centrifugal Fan
Axial Fan
FC Fan
BI Fan
AF Fan

Propeller Fan
Tubeaxial Fan
Vaneaxial Fan
Fan-to-Shaft Connection
Pitch Diameter
Variable-Pitch Sheave
V-Belts
Matched Belts
Belt Tension
Fan Laws
Ductwork Sizing

Equivalent Length
Aspect Ratio
Mitered Elbow
Radius Elbow
Turning Vanes
Transitions
Takeoff
Splitter Damper
Parallel and Opposed-Blades
Leaf Damper
Fire Damper

A.F.F.	Above finished floor
AHU	Air-handling unit
BDD	Backdraft damper
B.O.D.	Bottom of duct
—— C ——	Condensate piping
CD	Ceiling diffuser
CFM	Airflow quantity– cubic feet per min.
—— CHWS ——	Chilled water supply piping
—— CHWR ——	Chilled water return piping
CR	Ceiling register
—— CWS ——	Condenser water supply piping
—— CWR ——	Condenser water return piping
E.F.	Exhaust fan
FCU	Fan cool unit
F.D.	Fire damper
F.O.B.	Flat on bottom
—— F.O.S. ——	Fuel oil supply piping
—— F.O.R. ——	Fuel oil return piping
F.O.T.	Flat on top
—— HPS ——	High pressure steam piping
—— HWS ——	Hot water supply piping
—— HWR ——	Hot water return piping
—— LPS ——	Low pressure steam piping
MBH	Thousands of Btu/hr
M.D.	Manual damper
OA	Outside air
OBD	Opposed–blade damper
PRV	Powered roof ventilator or pressure–reducing valve
—— R ——	Refrigerant piping
RA	Return air
REG	Register
RG	Return grille
—— S ——	Steam piping

SA	Supply air
S.D.	Splitter damper
S.G.	Supply grille or sidewall grille
T.G.	Transfer grille
T.O.D.	Top of duct
T.W.	Thermometer well
TYP.4	Typical for four items
U.C.	Undercut
U.H.	Unit heater
U.T.R.	Up through roof
V.D.	Volume damper
V.T.R.	Vent through roof
W.F.	Wall fin
W.H.	Wall heater
	Manual valve
	Automatic valve
	Self-contained regulating valve
	Balancing cock
	Automatic 3-way valve
	Check valve
	Pipe turning down
	Pipe turning up
	Pressure gauge
	Thermometer
	Union
	Strainer
	Pump
	Relief valve
	Size change
	Insulation

Figure 25-58 Ductwork symbols.

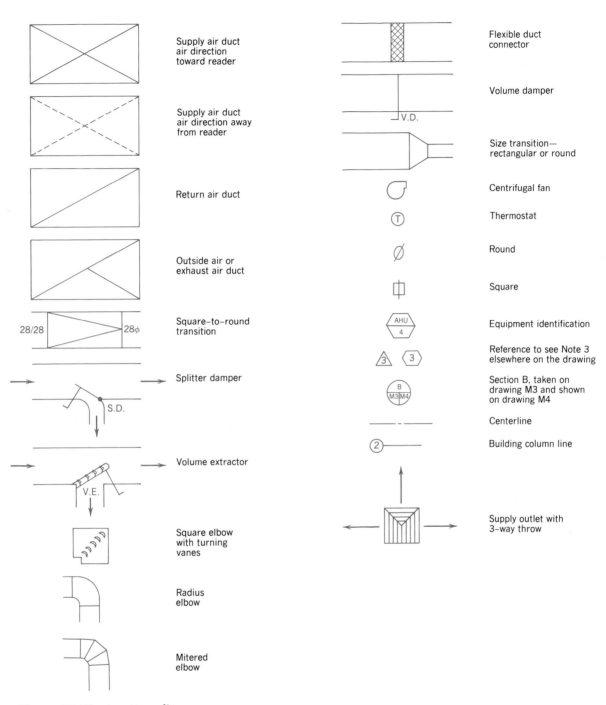

Figure 25-58 (continued)

Ceiling Diffuser Louver Plan View

Linear Diffuser Flexible Ductwork Elevation

Grille Drawing Numbering Isometric Drawing
Conventions
Register Section Drawing

QUESTIONS

1. What two fan performance factors are described by a fan curve?

2. When a duct damper is closed, what happens to the static pressure across the fan? What happens to the cfm being delivered?

3. What determines the friction rate in a duct?

4. If you want to increase the airflow through a duct by 10 percent, how much will the friction through the duct be increased?

5. What are the advantages of using small, high velocity ductwork? The disadvantages?

6. What is the difference between an axial flow fan and a centrifugal fan?

7. Which fan in Question 6 has a flatter fan curve?

8. What is meant by a DWDI fan?

9. Will the sheave on an FC fan generally be smaller or larger than the sheave on a BI fan?

10. Direct-drive fans are generally (higher, lower) rpm than belt-drive fans?

11. A 1750-rpm motor with a 4-in. sheave drives a fan that has an 8-in. sheave. What will be the fan rpm?

12. If you want to increase airflow from a fan, you could increase the size of the (fan, motor) sheave.

13. Name two possible ways in which a worn drive belt can cause airflow to be too low.

14. What are matched belts?

15. A fan is delivering 1500 cfm against a static pressure of 1.2 in. w.c. and the motor is drawing 1.1 hp. If the motor sheave is increased from 3-in. to 4-in. what are the new cfm, static pressure, and hp?

16. Suppose 1000 cfm is moving inside a 10-in. × 14-in. duct. How much pressure drop will be experienced by the air over a run of 200 ft?

17. What is the purpose of a fire damper? How can you tell if it is closed?

18. What is the difference between a grille and a register?

19. What is a plan view on a construction drawing?

20. In a plan view of a duct, the duct runs from the top of the drawing to the bottom. Is it a horizontal duct or a vertical duct?

CHAPTER 26

LOAD CALCULATIONS

In Chapter 23, on psychrometrics, you learned how to determine the properties of air required around the air system, given the heating or cooling load in the room. This chapter explains how to calculate the heating or cooling load within the room. The first part of the chapter is geared toward calculating a winter heating load. In the second portion of the chapter, the additional considerations applicable to summer cooling load are discussed.

The Room Load

Assume we have a residence that is being maintained at 74°F on a cold winter day on which it is 0°F outdoors. Suddenly the furnace stops working, and the temperature of the residence begins to drop. At first, it drops quickly, but the rate then becomes slower and slower (Figure 26-1). Eventually, the residence reaches the same temperature as outdoors. The temperature falls in the residence because there is a heat loss from the room air to the outside air. As long as the furnace is able to add heat at the same rate at which the heat is being lost, the temperature of the room remains constant. In order to know how large the furnace needs to be, it is necessary to calculate the rate of heat loss from the residence. The same applies for calculating the heat gain to a space when figuring the size of the air conditioning system required.

For a heating system, the rate at which heat is being lost from the space depends upon the following factors.

1. The temperature difference desired between the residence and the outside air.
2. The available area of walls, roof, and other surfaces through which heat can flow.
3. The materials of construction.
4. The airtightness of the building, and the amount of outside area that comes into the residence.

Note that the rate of heat loss does not depend upon the number of square feet of floor area or the number of cubic feet of air within the space, except as these factors affect the areas in factor 2. Figure 26-2 demonstrates this point. Both buildings have the same area and volume. But building B will have a higher rate of heat loss because of its greater wall area.

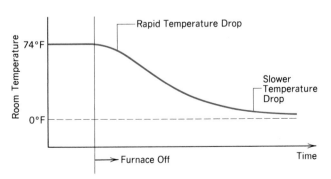

Figure 26-1 The rate of heat loss depends on the difference in temperature between the room and outdoors.

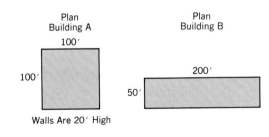

	Building A	Building B
Wall Area	8,000 sq ft	10,000 sq ft
Roof Area	10,000 sq ft	10,000 sq ft
Total Area	18,000 sq ft	20,000 sq ft

Figure 26-2 Both buildings have the same floor area and volume. They will have different heat loss because of the difference in area through which the heat loss takes place.

Sources of Heat Loss

Heat may be lost from a space by two modes, **transmission** and **infiltration.** Transmission is the heat that flows through walls, windows, roofs, and any other surface that separates air at two different temperatures. The rate at which the heat will flow through any of these surfaces is given by the formula

$$Q = U \times A \times \Delta T$$

where

Q = the rate of heat flow, Btu/hr.

U = the U factor of the material of construction.

A = the area of the surface.

ΔT = the temperature difference across the surface.

The method to be used in figuring the total transmission loss is to look at each separate surface through which a transmission loss can occur. The loss is calculated for each surface individually, and then totaled. Examples of this calculation are presented in the sections to follow.

Infiltration refers to the flow of cold air into a building and warm air out of the building (**exfiltration).** While transmission deals with only heat flow, infiltration deals with an actual mass flow of air through a space. The calculation of how much heat must be added to the outside air in order to warm it up to the desired inside temperature is given by

$$Q = 1.08 \times \text{cfm} \times \Delta T$$

where

Q = heat required to offset the loss due to infiltration.

1.08 = a constant that converts cubic feet per minute into pounds per hour, and multiplies by the specific heat of air.

cfm = the amount of cold air that is leaking into the building.

ΔT = the number of degrees that the outside air must be heated for it to reach the indoor temperature.

Note that this formula is the same as the formula given in the chapter on psychrometrics for sensible heating of an air stream.

The cfm depends upon the air-tightness of the building, the presence of any exhaust fans, wind, and frequency of opening of the doors. Note that this cfm is not the same as the cfm that is circulating on the building. They have virtually nothing to do with each other.

Transmission Heat Loss

The first decision to be made in calculating a heating load is what is the temperature difference to be maintained between inside and outside. These are called the **design conditions.** That is, the systems will be designed so that it can maintain the desired indoor temperature when the outside temperature reaches a predetermined level. The recommended outdoor design temperature to be used for your city may be determined by looking up the closest city in Figure 26-3. The outdoor design conditions given do not represent the coldest temperature that can ever be experienced. In fact, the temperature can be expected to drop below the design temperatures for a few hours during the heating season. But for comfort applications it does not make sense to increase the size of the heating system for just a few degrees once or twice a season. When the performance of a system is specified or quoted, it should state what inside temperature the system will be able to maintain, and what is the lowest outdoor temperature at which the system will be able to maintain the desired indoor temperature.

The next factor that must be determined is the **U factor** of the surface through which the heat transmission is taking place. This value must be looked up in a table for the type of construction used. Figures 26-4 through 26-16 give U factors for a number of different types of walls, partitions, roofs, and slabs.

Example

Calculate the heat lost through an exterior wall. The overall wall dimension is 15 ft long and 8 ft high. The wall is constructed of 8-in.-thick hollow concrete block with the inside finished in

State and City	Normal Design Cond.—Summer July at 3:00 P.M.			Avg. Daily Range	Normal Design Cond. Winter		Latitude (deg)
	Dry-Bulb (F)	Wet-Bulb (F)	Moisture Content (gr/lb of dry air)	Dry-Bulb (F)	Dry-Bulb (F)	Annual Degree Days	
Alabama							
Birmingham	95	78	117.5	19	10	2611	34
Mobile	95	80	131	12	15	1566	31
Arizona							
Flagstaff	90	65	81	26	−10	7242	35
Phoenix	105	76	94	30	25	1441	33
Tucson	105	72	77	30	25		32
Arkansas							
Fort Smith	95	76	104.5	16	10	3226	35
Little Rock	95	78	117.5	16	5	3009	35
California							
Bakersfield	105	70	54	25	25		35
El Centro	110	78	94				33
Eureka	90	65	52		30	4758	41
Fresno	105	74	76	35	25	2403	37
Los Angeles	90	70	78	14	35	1391	34
Sacramento	100	72	73	18	30	2680	39
San Diego	85	68	75	10	35	1596	33
San Francisco	85	65	60	17	35	3137	34
Colorado							
Denver	95	64	60	25	−10	5839	40
Connecticut							
Hartford	93	75	102	16	0	6113	40
Delaware							
Wilmington	95	78	117.5	15	0		40
Dist. of Columbia	95	78	117.5	18	0	4561	39
Florida							
Jacksonville	95	78	117.5	17	25	1185	30
Miami	91	79	131	12	35	185	26
Tampa	95	78	117.5	12	30	571	28
Tallahassee	95	78	117.5	14	25	1463	30
Georgia							
Atlanta	95	76	109.5	18	10	2985	34
Columbus	98	76	100				33
Idaho							
Boise	95	65	54.5	31	−10	5678	44
Illinois							
Chicago	95	75	99	19	−10	6282	42

Figure 26-3 Outdoor design conditions, summer and winter. (*Courtesy McGraw-Hill Book Company*)

(continued on next page)

State and City	Normal Design Cond.—Summer July at 3:00 P.M.			Avg. Daily Range	Normal Design Cond. Winter		Latitude (deg)
	Dry-Bulb (F)	Wet-Bulb (F)	Moisture Content (gr/lb of dry air)	Dry-Bulb (F)	Dry-Bulb (F)	Annual Degree Days	
Moline	96	76	103	22	−10		41
Springfield	98	77	106	20	−10	5446	40
Indiana							
Evansville	95	78	117.5	19	0	4410	38
Fort Wayne	95	75	99	20	−10	6232	41
Indianapolis	95	76	104.5	18	−10	5458	40
Iowa							
Davenport	95	78	117.5	18	−15	6252	42
Des Moines	95	78	123	18	−15	6375	42
Kansas							
Topeka	100	78	109.5	19	−10	5075	39
Wichita	100	75	98	21	−10	4644	38
Kentucky							
Louisville	95	78	117.5	22	0	4417	38
Louisiana							
New Orleans	95	80	131	13	20	1203	30
Shreveport	100	78	109.5	15	20	2132	33
Maine							
Augusta	90	73	95	13			45
Eastport	90	70	78	13	−10	8445	45
Portland	90	73	95	13	−5	7377	44
Maryland							
Baltimore	95	78	117.5	18	0	4487	39
Frederick					−5		40
Frostburg					−5		40
Salisbury					10		40
Massachusetts							
Boston	92	75	104	13		5936	42
Nantucket	95	75	99		0		41
Springfield	93	75	102	17	−10		42
Michigan							
Alpena	95	75	99		−10	8278	45
Detroit	95	75	99	19	−10	6560	42
Lansing	95	75	104	20	−10	7149	43
Marquette	93	73	90	20	−10	8745	47
Minnesota							
Duluth	93	73	96	19	−25	9723	47
Minneapolis	95	75	103	17	−20	7966	45
Mississippi							
Meridian	95	79	124	21	10	2330	32

Figure 26.3 (continued)

State and City	Normal Design Cond.—Summer July at 3:00 P.M.			Avg. Daily Range	Normal Design Cond. Winter		Latitude (deg)
	Dry-Bulb (F)	Wet-Bulb (F)	Moisture Content (gr/lb of dry air)	Dry-Bulb (F)	Dry-Bulb (F)	Annual Degree Days	
Missouri							
Columbia	100	78	109.5	19	−10	5070	39
Kansas City	100	76	106.5	19	−10	4962	39
St. Louis	95	78	117.5	20	0	4596	39
Montana							
Billings	90	66	70	20	−25	7213	46
Helena	95	67	71	20	−20	7930	47
Missoula	95	66	49	20	−20	7604	47
Nebraska							
North Platte	95	78	135	26	−20	6384	41
Omaha	95	78	123	20	−10	6095	41
Nevada							
Las Vegas	115	75	76	40	20		36
Reno	95	65	62	41	−5	5621	40
New Hampshire							
Manchester	90	73	95	14			43
New Jersey							
Atlantic City	95	78	117.5	14	5	5015	39
Newark	95	75	99	14	0	5500	41
Trenton	95	78	117.5	14	0	5256	40
New York							
Albany	93	75	102	18	−10	6648	43
Buffalo	93	73	90	18	−5	6925	43
New York City	95	75	99	14	0	5280	41
Rochester	95	75	102	18	−5	6772	43
North Carolina							
Asheville	93	75	114.5	19	0	4236	36
Raleigh	95	78	117.5	15	10	3275	36
North Dakota							
Bismarck	95	73	95.5	19	−30	8937	47
Fargo	95	75	104.5	19	−25		47
Ohio							
Cincinnati	95	78	117.5	22	0	4990	39
Cleveland	95	75	99	19	0	6144	42
Oklahoma							
Oklahoma City	101	77	108	21	0	3670	35
Oregon							
Baker	90	66	71	19	−5	7197	44
Medford	95	70	76	19			42
Portland	90	68	67	19	10	4353	46

State and City	Normal Design Cond.—Summer July at 3:00 P.M.			Avg. Daily Range	Normal Design Cond. Winter		Latitude (deg)
	Dry-Bulb (F)	Wet-Bulb (F)	Moisture Content (gr/lb of dry air)	Dry-Bulb (F)	Dry-Bulb (F)	Annual Degree Days	
Pennsylvania							
Philadelphia	95	78	117.5	14	0	4739	40
Pittsburgh	95	75	105	14	0	5430	40
Scranton	95	75	99	14	−5	6218	41
Rhode Island							
Providence	93	75	102	14	0	5984	42
South Carolina							
Charleston	95	78	117.5	17	15	1866	33
Columbia	95	75	99	17	10	2488	34
South Dakota							
Rapid City	95	70	85	22	−20	7197	44
Sioux Falls	95	75	99	20	−20		43
Tennessee							
Knoxville	95	75	103.5	17	0	3658	36
Memphis	95	78	117.5	18	0	3090	35
Texas							
Abilene	100	74	93		15	2573	32
Corpus Christi	95	80	131		20	965	28
Dallas	100	78	109.5	21	0	2367	33
Houston	95	80	131	14	20	1315	30
Utah							
Salt Lake City	95	65	61	25	−10	5650	41
Vermont							
Burlington	90	73	95	17	−10	8051	44
Virginia							
Richmond	95	78	117.5	16	15	3922	38
Roanoke	95	76	111.5	16	0	4075	38
Washington							
Seattle	85	65	60	17	15	4815	48
Spokane	93	65	54.5	28	−15	6318	48
Walla Walla	95	65	47.5	28	−10	4910	46
West Virginia							
Charleston	95	75	99	16	0		38
Wisconsin							
Green Bay	95	75	99	14	−20	7931	45
La Crosse	95	75	99	17	−25	7421	44
Wyoming							
Cheyenne	95	65	68.5	28	−15	7536	42

Figure 26.3 (continued)

$\frac{3}{8}$-in. gypsum board. There is a 24-in. × 24-in. single-pane window in the wall. Inside design is 73°F, outside design is 5°F.

Solution

Look up the U factors for the wall construction and for single-pane glass. From Figure 26-4, the U factor for the wall is 0.35, and from Figure 26-16 the U factor for single-pane glass is 1.13. The area of the glass is 4 sq ft. The net area of the wall (not including the glass) is 116 sq ft. The ΔT for each surface is 68°F. Calculating the transmission losses for the wall,

$$Q = 0.35 \times 116 \times 68 = 2760 \text{ Btu/hr}$$

and for the glass

$$Q = 1.13 \times 4 \times 68 = 307 \text{ Btu/hr}$$

The total heat loss is 3067 Btu/hr.

Calculating U Factors

For construction that is not listed in Figures 26-4 through 26-16, you may need to calculate a U factor. Or, if the U factor of a surface is known, you may need to calculate what the new U factor will be if insulation is added.

Every material has some resistance to heat transfer. Figure 26-17 lists the resistance values for many different building materials. The relationship between resistance (R) and the heat transmission coefficient (U) is

$$U = \frac{1}{R}$$

and

$$R = \frac{1}{U}$$

When two materials are placed together to form a composite wall, their R values can be added in order to determine the overall resistance of the composite. Their U values cannot be added.

Example

An existing ceiling with a U factor of 0.32 is to have 4 in. of expanded vermiculite insulation added. What will be the new U factor?

Solution

The **R value** for the existing ceiling may be calculated as

$$R = \frac{1}{0.32} = 3.125$$

The resistance of the expanded vermiculite insulation to be added is read from Figure 26-17 as 2.08 per inch. The resistance of 4-in. of this insulation will be $2.08 \times 4 = 8.32$. Adding the resistance of the existing ceiling to the resistance of the insulation, the overall resistance of the composite ceiling will be

$$R = R1 + R2 = 3.125 + 8.32 = 11.45$$

and the U factor of the composite wall is

$$U = \frac{1}{R} = \frac{1}{11.45} = 0.087$$

U factors may also be calculated starting from the component materials. The resistance of the total surface is equal to the total resistance of all the components. In addition to the resistances of the materials you can see, however, there is also a resistance associated with the **air film** on each side of the wall. Figure 26-17 gives the resistance values for air films, as well as for many building materials. In cases where the material of the wall (such as a sheet-metal wall) has a very low resistance to heat transmission, the resistance of the air films is actually greater than the resistance of the material itself.

Example

Calculate the U factor for the walls of a metal shed.

Solution

From Figure 26-17, the resistance of the sheet metal itself is negligable. The air film on the inside (still air) however, is read as 0.68. The resistance of the outside air film should be read assuming a 15 mph wind outside (worst condition). The outside air film resistance is 0.17. Adding all the resistances,

$$\text{total R} = 0.68 + 0 + 0.17 = 0.85$$

and the U factor is

$$U = \frac{1}{\text{total R}} = \frac{1}{0.85} = 1.17$$

TRANSMISSION COEFFICIENT U—MASONRY WALLS

FOR SUMMER AND WINTER

Btu/(hr) (sq ft) (deg F temp diff)

All numbers in parentheses indicate weight per sq ft. Total weight per sq ft is sum of wall and finishes.

EXTERIOR FINISH		THICK-NESS (inches) and WEIGHT (lb per sq ft)	None	⅜" Gypsum Board (Plaster Board) (2)	⅝" Plaster on Wall		Metal Lath Plastered on Furring		⅜" Gypsum or Wood Lath Plastered on Furring		Insulating Board Plain or Plastered on Furring	
					Sand Agg (6)	Lt Wt Agg (3)	¾" Sand Plaster(7)	¾" Lt Wt Plaster(3)	½" Sand Plaster(7)	½" Lt Wt Plaster(2)	½" Board (2)	1" Board (4)
SOLID BRICK	Face & Common	8 (87)	.48	.41	.45	.41	.31	.28	.29	.27	.22	.16
		12 (123)	.35	.31	.33	.30	.25	.23	.23	.22	.19	.14
		16 (173)	.27	.25	.26	.25	.21	.19	.20	.19	.16	.13
	Common Only	8 (80)	.41	.36	.39	.35	.28	.26	.26	.25	.21	.15
		12 (120)	.31	.28	.30	.27	.23	.22	.22	.21	.18	.14
		16 (160)	.25	.23	.24	.23	.19	.18	.18	.18	.16	.12
STONE		8 (100)	.67	.55	.63	.53	.39	.34	.35	.32	.26	.18
		12 (150)	.55	.47	.52	.46	.34	.31	.31	.29	.24	.17
		16 (200)	.47	.41	.45	.40	.31	.28	.28	.27	.22	.16
		24 (300)	.36	.32	.35	.32	.26	.24	.24	.23	.19	.15
ADOBE-BLOCKS OR BRICK		8 (26)	.34	.30	.32	.30	.25	.23	.23	.22	18	.12
		12 (40)	.25	.23	.24	.23	.20	.18	.18	.18	.15	.14
POURED CONCRETE	140 lb/cu ft	6 (70)	.75	.55	.69	.58	.41	.36	.37	.34	.27	.18
		8 (93)	.67	.49	.63	.53	.39	.34	.35	.32	.26	.17
		10 (117)	.61	.44	.57	.49	.36	.32	.33	.31	.25	.17
		12 (140)	.55	.40	.52	.45	.34	.31	.31	.29	.24	.16
	80 lb/cu ft	6 (40)	.31	.28	.30	.27	.23	.21	.22	.21	.18	.14
		8 (53)	.25	.23	.24	.23	.19	.18	.18	.18	.16	.12
		10 (66)	.21	.19	.20	.19	.17	.16	.15	.14	.14	.11
		12 (80)	.18	.17	.17	.15	.15	.14	.14	.14	.12	.10
	30 lb/cu ft	6 (15)	.13	.13	.13	.13	.12	.11	.11	.11	.13	.09
		8 (20)	.10	.10	.10	.10	.09	.09	.09	.09	.10	.07
		10 (25)	.08	.08	.08	.08	.08	.07	.08	.07	.08	.06
		12 (30)	.07	.07	.07	.07	.07	.07	.06	.06	.07	.06
HOLLOW CONCRETE BLOCKS	Sand & Gravel Agg	8 (43)	.52	.44	.48	.43	.33	.29	.30	.28	.23	.17
		12 (63)	.47	.41	.45	.40	.31	.28	.28	.27	.22	.16
	Cinder Agg	8 (37)	.39	.35	.37	.34	.27	.25	.25	.24	.20	.15
		12 (53)	.36	.33	.35	.32	.26	.24	.23	.23	.19	.15
	Lt Wt Agg	8 (32)	.35	.32	.34	.31	.26	.23	.24	.22	.19	.15
		12 (43)	.32	.29	.31	.28	.24	.22	.22	.21	.18	.14
STUCCO ON HOLLOW CLAY TILE		8 (39)	.36	.32	.34	.32	.26	.24	.24	.23	.19	.15
		10 (44)	.32	.29	.31	.28	.23	.22	.22	.21	.18	.14
		12 (49)	.29	.27	.28	.26	.22	.20	.21	.20	.17	.13

1958 ASHAE Guide

Equations: Heat Gain, Btu/hr = (Area, sq ft) × (U value) × (equivalent temp diff).
   Heat Loss, Btu/hr = (Area, sq ft) × (U value) × (outdoor temp − inside temp)

Figure 26-4   Transmission coefficient U—masonry walls.
(Courtesy McGraw-Hill Book Company)

TRANSMISSION COEFFICIENT U—MASONRY VENEER WALLS

FOR SUMMER AND WINTER

Btu/(hr) (sq ft) (deg F temp diff)

All numbers in parentheses indicate weight per sq ft. Total weight per sq ft is sum of wall and finishes.

EXTERIOR FINISH	BACKING	THICK-NESS (inches) and WEIGHT (lb per sq ft)	None	Gypsum Board (Plaster Board) (2)	5/8" Plaster on Wall — Sand Agg (6)	5/8" Plaster on Wall — Lt Wt Agg (3)	Metal Lath Plastered on Furring — 3/4" Sand Plaster(7)	Metal Lath Plastered on Furring — 3/4" Lt Wt Plaster(3)	3/8" Gypsum or Wood Lath Plastered on Furring — 1/2" Sand Plaster(7)	3/8" Gypsum or Wood Lath Plastered on Furring — 1/2" Lt Wt Plaster(2)	Insulating Board Plain or Plastered on Furring — 1/2" Board (2)	Insulating Board Plain or Plastered on Furring — 1" Board (4)
4" Face Brick (43) —or— 4" Stone (50) —or— Precast Concrete (Sand Agg) 4" & 6" (39) (58)	Concrete Block (Cinder Agg)	4 (20)	.41	.37	.39	.35	.28	.26	.26	.25	.21	.16
		8 (37)	.33	.30	.32	.29	.24	.22	.23	.21	.18	.14
		12 (53)	.31	.29	.30	.28	.23	.21	.22	.21	.18	.14
	(Lt Wt Agg)	4 (17)	.35	.32	.34	.31	.25	.23	.24	.22	.19	.15
		8 (32)	.30	.28	.29	.27	.23	.21	.21	.20	.17	.14
		12 (43)	.28	.26	.27	.25	.21	.20	.20	.19	.17	.13
	(Sand & Gravel Agg)	4 (23)	.49	.44	.46	.41	.32	.29	.29	.27	.22	.17
		8 (43)	.41	.37	.39	.35	.28	.26	.26	.25	.21	.16
		12 (63)	.38	.35	.37	.33	.27	.25	.25	.24	.20	.15
	Hollow Clay Tile	4 (16)	.41	.37	.39	.35	.28	.26	.26	.25	.21	.16
		8 (30)	.31	.29	.30	.28	.23	.22	.22	.21	.18	.14
		12 (40)	.26	.25	.25	.24	.20	.19	.19	.18	.16	.13
	Concrete (Lt Wt Agg) 80 lb/cu ft	4 (26)	.35	.31	.34	.31	.25	.23	.24	.22	.19	.15
		6 (40)	.27	.25	.27	.25	.21	.20	.20	.19	.16	.13
		8 (54)	.22	.21	.22	.21	.18	.17	.17	.16	.14	.12
	(Sand & Gravel Agg)	4 (47)	.60	.53	.56	.49	.36	.32	.33	.31	.25	.18
		6 (70)	.55	.49	.52	.45	.34	.31	.32	.29	.24	.17
		8 (95)	.51	.45	.48	.42	.32	.29	.30	.28	.23	.17
	Common Brick	4 (40)	.49	.42	.46	.41	.32	.29	.29	.27	.22	.16
		8 (80)	.35	.31	.34	.31	.25	.23	.24	.22	.19	.15
4" Common Brick (40) —or— Precast Concrete (Sand Agg) 8" & 10" (78) (98) —or— 4" Concrete Block (23) (Sand Agg) —or— 8" Stone (100)	Concrete Block (Cinder Agg)	4 (20)	.36	.33	.35	.32	.26	.24	.24	.23	.19	.15
		8 (37)	.29	.28	.29	.26	.22	.21	.21	.20	.17	.14
		12 (53)	.28	.26	.27	.25	.21	.20	.20	.19	.17	.13
	(Lt Wt Agg)	4 (17)	.32	.29	.30	.28	.23	.22	.22	.21	.18	.14
		8 (32)	.27	.26	.26	.25	.21	.20	.20	.19	.17	.13
		12 (43)	.25	.24	.25	.23	.20	.19	.19	.18	.16	.13
	(Sand & Gravel Agg)	4 (23)	.42	.38	.40	.36	.29	.26	.27	.25	.21	.16
		8 (43)	.36	.33	.35	.32	.26	.24	.24	.23	.19	.15
		12 (63)	.34	.32	.33	.30	.25	.23	.23	.22	.19	.15
	Hollow Clay Tile	4 (16)	.36	.33	.35	.32	.26	.24	.24	.23	.19	.15
		8 (30)	.28	.27	.28	.26	.22	.20	.20	.19	.17	.13
		12 (40)	.24	.23	.23	.22	.19	.18	.18	.17	.15	.12
	Concrete (Lt Wt Agg) 80 lb/cu ft	4 (26)	.32	.29	.30	.28	.23	.22	.22	.21	.18	.14
		6 (40)	.25	.23	.25	.23	.20	.18	.19	.18	.15	.13
		8 (54)	.21	.20	.20	.19	.17	.16	.16	.16	.14	.11
	(Sand & Gravel Agg)	4 (47)	.50	.45	.48	.42	.32	.29	.30	.28	.23	.17
		6 (70)	.47	.42	.44	.39	.31	.28	.29	.27	.22	.17
		8 (95)	.43	.40	.41	.37	.29	.27	.28	.26	.21	.16
	Common Brick	4 (40)	.42	.37	.40	.36	.29	.26	.27	.26	.21	.16
		8 (80)	.32	.29	.30	.28	.23	.22	.22	.21	.18	.14

1958 ASHAE Guide

Equations: Heat Gain, Btu/hr = (Area, sq ft) × (U value) × (equivalent temp diff)

Heat Loss, Btu/hr = (Area, sq ft) × (U value) × (outdoor temp — inside temp)

Figure 26-5 Transmission coefficient U—masonry veneer walls.
(Courtesy McGraw-Hill Book Company)

TRANSMISSION COEFFICIENT U—LIGHT CONSTRUCTION, INDUSTRIAL WALLS

FOR SUMMER AND WINTER

Btu/(hr) (sq ft) (deg F temp diff)

All numbers in parentheses indicate weight per sq ft. Total weight per sq ft is sum of wall and finishes.

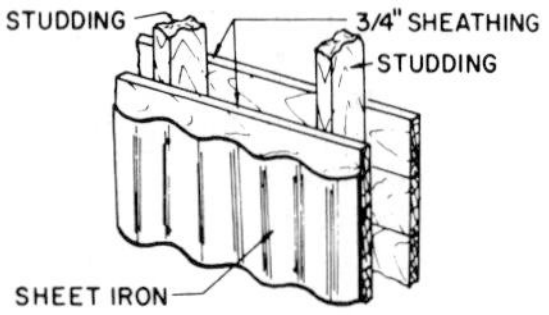

| | | | INTERIOR FINISH | | | | |
| | | | | | Insulating Board | | Wood |
EXTERIOR FINISH	SHEATHING	WEIGHT (lb per sq ft)	None	Flat Iron (1)	½" (2)	25/32" (3)	¾" (2)
3/8" Corrugated Transite	None	(1)	1.16	.55	.32	.26	.36
	½" Ins. Board	(2)	.34	.26	.19	.17	.21
	25/32" Ins. Board	(2)	.27	.21	.17	.15	.18
24 Gauge Corrugated Iron	None	(1)	1.40	.60	.33	.27	.38
	½" Ins. Board	(2)	.36	.27	.20	.17	.21
	25/32" Ins. Board	(2)	.28	.22	.17	.15	.18
	¾" Wood	(3)	.46	.33	.22	.19	.24
¾" Wood Siding	None	(2)	.58	.37	.25	.21	.27

1958 ASHAE Guide

Equations: Heat Gain, Btu/hr = (Area, sq ft) × (U value) × (equivalent temp diff).
　　　　　Heat Loss, Btu/hr = (Area, sq ft) × (U value) × (outdoor temp − inside temp).

Figure 26-6 Transmission coefficient U—light construction, industrial walls.
(Courtesy McGraw-Hill Book Company)

TRANSMISSION COEFFICIENT U—LIGHTWEIGHT, PREFABRICATED CURTAIN TYPE WALLS

FOR SUMMER AND WINTER

Btu/(hr) (sq ft) (deg F temp diff)

All numbers in parentheses indicate weight per sq ft. Total weight per sq ft is sum of wall and finishes.

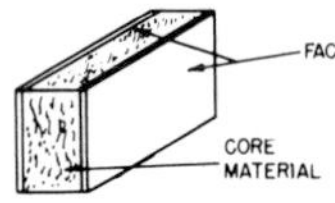

| INSULATING CORE MATERIAL | DENSITY† (lb/cu ft) | METAL FACING (3) | | | | METAL FACING WITH ¼" AIR SPACE (3) | | | |
| | | Core Thickness (in.) | | | | Core Thickness (in.) | | | |
		1	2	3	4	1	2	3	4
Glass, Wood, Cotton Fibers	3	.21	.12	.08	.06	.19	.11	.08	.06
Paper Honeycomb	5	.39	.23	.17	.13	.32	.20	.15	.12
Paper Honeycomb with Perlite Fill, Foamglas	9	.29	.17	.12	.09	.25	.15	.11	.09
Fiberboard	15	.36	.21	.15	.12	.29	.19	.14	.11
Wood Shredded (Cemented in Preformed Slabs)	22	.31	.18	.13	.10	.25	.16	.12	.09
Expanded Vermiculite	7	.34	.20	.14	.11	.28	.18	.13	.10
Vermiculite or Perlite Concrete	20	.44	.27	.19	.15	.35	.23	.18	.14
	30	.51	.32	.24	.19	.39	.27	.21	.17
	40	.58	.38	.29	.23	.43	.31	.25	.20
	60	.69	.49	.38	.31	.49	.38	.31	.26

Equations: Heat Gain, Btu/hr = (Area, sq ft) × (U value) × (equivalent temp diff).
　　　　　Heat Loss, Btu/hr = (Area, sq ft) × (U value) × (outdoor temp − inside temp).

Figure 26-7 Transmission coefficient U—lightweight, prefabricated curtain-type walls.
(Courtesy McGraw-Hill Book Company)

TRANSMISSION COEFFICIENT U—FRAME WALLS AND PARTITIONS

FOR SUMMER AND WINTER

Btu/(hr) (sq ft) (deg F temp diff)

All numbers in parentheses indicate weight per sq ft. Total weight per sq ft is sum of component materials.

		INTERIOR FINISH									
		None		3/8" Gypsum Board (Plaster Board) (2)	Metal Lath Plastered		3/8" Gypsum or Wood Lath Plastered		Insulating Board Plain or Plastered	
			3/4" Wood Panel (2)		3/4" Sand Plaster(7)	3/4" Lt Wt Plaster(3)	1/2" Sand Plaster(7)	1/2" Lt Wt Plaster(2)	1/2" Board (2)	1" Board (4)
EXTERIOR FINISH	**SHEATHING**									
1" Stucco (10) OR Asbestos Cement Siding (1) OR Asphalt Roll Siding (2)	None, Building Paper	.91	.33	.42	.45	.39	.40	.37	.29	.20
	5/16" Plywood (1) or 1/2" Gyp (2)	.68	.30	.37	.40	.35	.36	.33	.26	.19
	25/32" Wood & Bldg Paper (2)	.48	.25	.30	.31	.28	.29	.27	.22	.17
	1/2" Insulating Board (2)	.42	.23	.27	.29	.26	.27	.25	.21	.16
	25/32" Insulating Board (3)	.32	.20	.23	.24	.22	.22	.21	.18	.14
4" Face Brick Veneer (43) OR 3/8" Plywood (1) OR Asphalt Siding (2)	None, Building Paper	.73	.30	.37	.40	.35	.36	.33	.26	.19
	5/16" Plywood (1) or 1/2" Gyp (2)	.57	.28	.33	.36	.32	.32	.30	.24	.18
	25/32" Wood & Bldg Paper (2)	.42	.23	.27	.29	.26	.27	.25	.21	.16
	1/2" Insulating Board (2)	.38	.22	.25	.27	.25	.25	.24	.20	.15
	25/32" Insulating Board (3)	.30	.19	.21	.22	.21	.21	.20	.17	.14
Wood Siding (3) OR Wood Shingles (2) OR 3/4" Wood Panels (3)	None, Building Paper	.57	.27	.33	.35	.31	.32	.30	.24	.18
	5/16" Plywood (1) or 1/2" Gyp (2)	.48	.25	.30	.31	.28	.29	.27	.22	.17
	25/32" Wood & Bldg Paper	.36	.22	.25	.26	.24	.24	.23	.19	.15
	1/2" Insulating Board (2)	.33	.20	.23	.24	.22	.23	.22	.18	.14
	25/32" Insulating Board (3)	.27	.18	.20	.21	.19	.19	.19	.16	.13
Wood Shingles Over 5/16" Insul Backer Board (3) OR Asphalt Insulated Siding (4)	None, Building Paper	.43	.24	.28	.29	.27	.27	.25	.21	.16
	5/16" Plywood (1) or 1/2" Gyp (2)	.38	.22	.25	.27	.24	.25	.23	.19	.15
	25/32" Wood & Bldg Paper	.30	.19	.22	.23	.21	.21	.20	.17	.14
	1/2" Insulating Board (2)	.28	.18	.20	.21	.20	.20	.19	.16	.13
	25/32" Insulating Board (3)	.23	.16	.18	.18	.17	.18	.17	.15	.12
Single Partition (Finish on one side only)			.43	.60	.67	.55	.57	.50	.36	.23
Double Partition (Finish on both sides)			.24	.34	.39	.31	.32	.28	.19	.12

1958 ASHAE Guide

Equations: Walls—Heat Gain, Btu/hr = (Area, sq ft) × (U value) × (equivalent temp diff, *Table 19*).
  —Heat Loss, Btu/hr = (Area, sq ft) × (U value) × (outdoor temp—inside temp).
  Partitions, unconditioned space adjacent—Heat Gain or Loss, Btu/hr = (Area sq ft) × (U value) × (outdoor temp—inside temp—5 F).
  Partitions, kitchen or boiler room adjacent—Heat Gain or Loss, Btu/hr = (Area sq ft) × (U value)
  × (actual temp diff or outdoor temp—inside temp + 15 F to 25 F).

Figure 26-8 Transmission coefficient U—frame walls and partitions.
(Courtesy McGraw-Hill Book Company)

TRANSMISSION COEFFICIENT U—MASONRY PARTITIONS

FOR SUMMER AND WINTER

Btu/(hr) (sq ft) (deg F temp diff)

All numbers in parentheses indicate weight per sq ft. Total weight per sq ft is sum of masonry unit and finish × 1 or 2 (finished one or both sides).

BACKING	THICKNESS (inches) and WEIGHT (per sq ft)	Both Sides Unfinished	No. of Sides Finished	FINISH 3/8" Gypsum Board (Plaster Board) (2)	5/8" Plaster on Wall Sand Agg (6)	5/8" Plaster on Wall Lt Wt Agg (3)	Metal Lath Plastered on Furring 3/4" Sand Plaster (7)	Metal Lath Plastered on Furring 3/4" Lt Wt Plaster (3)	3/8" Gypsum or Wood Lath Plastered on Furring 1/2" Sand Plaster (7)	3/8" Gypsum or Wood Lath Plastered on Furring 1/2" Lt Wt Plaster (2)	Insulating Board Plain or Plastered on Furring 1/2" Board (2)	Insulating Board Plain or Plastered on Furring 1" Board (4)
HOLLOW CONCRETE BLOCK Cinder Agg	3 (17)	.45	One / Both	.39 / .35	.43 / .41	.38 / .33	.30 / .23	.27 / .20	.28 / .20	.26 / .18	.21 / .14	.16 / .10
	4 (20)	.40	One / Both	.36 / .32	.39 / .37	.35 / .31	.28 / .21	.26 / .19	.26 / .19	.25 / .18	.20 / .13	.15 / .11
	8 (37)	.32	One / Both	.29 / .27	.31 / .30	.29 / .26	.24 / .19	.22 / .17	.22 / .17	.21 / .16	.18 / .12	.14 / .09
	12 (53)	.31	One / Both	.28 / .26	.30 / .29	.27 / .25	.23 / .18	.21 / .16	.22 / .17	.21 / .15	.17 / .12	.14 / .09
Lt Wt Agg	3 (15)	.38	One / Both	.34 / .31	.36 / .35	.33 / .30	.27 / .21	.25 / .18	.25 / .19	.24 / .17	.20 / .13	.15 / .09
	4 (17)	.35	One / Both	.31 / .29	.34 / .32	.31 / .27	.25 / .20	.23 / .17	.24 / .17	.22 / .16	.19 / .13	.15 / .09
	8 (32)	.30	One / Both	.27 / .25	.29 / .28	.27 / .24	.22 / .18	.21 / .16	.21 / .16	.20 / .15	.17 / .12	.14 / .09
	12 (43)	.28	One / Both	.25 / .23	.27 / .26	.25 / .23	.21 / .17	.20 / .15	.20 / .16	.19 / .15	.16 / .12	.13 / .08
Sand & Gravel Agg	8 (43)	.40	One / Both	.36 / .32	.39 / .37	.35 / .31	.28 / .21	.26 / .19	.26 / .19	.25 / .18	.20 / .13	.15 / .11
	12 (63)	.38	One / Both	.34 / .30	.36 / .35	.33 / .29	.27 / .21	.25 / .18	.25 / .19	.24 / .17	.19 / .13	.15 / .09
HOLLOW CLAY TILE	3 (15)	.46	One / Both	.40 / .36	.44 / .42	.39 / .34	.31 / .23	.28 / .20	.28 / .20	.27 / .19	.22 / .14	.16 / .10
	4 (16)	.40	One / Both	.36 / .32	.39 / .37	.35 / .31	.28 / .21	.26 / .19	.26 / .19	.25 / .18	.20 / .13	.15 / .11
	6 (25)	.35	One / Both	.31 / .28	.33 / .32	.31 / .27	.25 / .20	.23 / .17	.23 / .18	.22 / .16	.19 / .13	.15 / .09
	8 (30)	.31	One / Both	.28 / .26	.30 / .29	.28 / .25	.23 / .18	.22 / .16	.22 / .17	.21 / .16	.18 / .12	.14 / .09
HOLLOW GYPSUM TILE	3 (9)	.37	One / Both	.33 / .30	.35 / .34	.32 / .29	.26 / .20	.24 / .18	.24 / .18	.23 / .13	.19 / .13	.15 / .09
	4 (13)	.33	One / Both	.30 / .27	.32 / .31	.29 / .26	.24 / .19	.22 / .17	.23 / .17	.22 / .16	.18 / .12	.14 / .09
SOLID GYPSUM PLASTER	1½						.61 (13)	.43 (6)				
	2						.58 (18)	.38 (8)				
	2½						.55 (22)	.34 (9)				

1958 ASHAE Guide

Equations: Partitions, unconditioned space adjacent: Heat Gain or Loss, Btu/hr = (Area, sq ft) × (U value) × (outdoor temp—inside temp—5 F).

Partitions, kitchen or boiler room adjacent: Heat Gain or Loss, Btu/hr = (Area, sq ft) × (U value)

× (actual temp diff or outdoor temp—inside temp + 15 F to 25 F).

Figure 26-9 Transmission coefficient U—masonry partitions.
(Courtesy McGraw-Hill Book Company)

TRANSMISSION COEFFICIENT U—FLAT ROOFS COVERED WITH BUILT-UP ROOFING

FOR HEAT FLOW DOWN—SUMMER. FOR HEAT FLOW UP—WINTER (See Equation at Bottom of Page).

Btu/(hr) (sq ft) (deg F temp diff)

All numbers in parentheses indicate weight per sq ft. Total weight per sq ft is sum of roof, finish and insulation.

TYPE OF DECK	THICK- NESS OF DECK (inches) and WEIGHT (lb per sq ft)	CEILING †	INSULATION ON TOP OF DECK, INCHES						
			No Insu- lation	½ (1)	1 (1)	1½ (2)	2 (3)	2½ (3)	3 (4)
Flat Metal	1 (5)	None or Plaster (6)	.67	.35	.23	.18	.15	.12	.10
		Suspended Plaster (5)	.32	.22	.17	.14	.12	.10	.09
		Suspended Acou Tile (2)	.23	.18	.14	.12	.11	.09	.08
Preformed Slabs—Wood Fiber and Cement Binder	2 (4)	None or Plaster (6)	.20	.16	.13	.11	.10	.09	.08
		Suspended Plaster (5)	.15	.12	.11	.09	.08	.08	.07
		Suspended Acou Tile (2)	.13	.10	.09	.08	.08	.07	.06
	3 (7)	None or Plaster (6)	.14	.11	.10	.09	.08	.08	.07
		Suspended Plaster (5)	.12	.10	.09	.07	.07	.06	.05
		Suspended Acou Tile (2)	.10	.09	.08	.07	.07	.06	.05
Concrete (Sand & Gravel Agg)	4, 6, 8 (47),(70), (93)	None or Plaster (6)	.51	.30	.21	.16	.14	.12	.10
		Suspended Plaster (5)	.28	.20	.16	.13	.12	.10	.09
		Suspended Acou Tile(2)	.21	.16	.13	.11	.10	.09	.08
(Lt Wt Agg on Gypsum Board)	2 (9)	None or Plaster (6)	.27	.20	.15	.13	.11	.10	.08
		Suspended Plaster (5)	.18	.14	.12	.10	.09	.09	.08
		Suspended Acou Tile (2)	.15	.12	.11	.09	.08	.08	.07
	3 (13)	None or Plaster (6)	.21	.16	.13	.11	.10	.09	.08
		Suspended Plaster (5)	.15	.12	.11	.09	.08	.08	.07
		Suspended Acou Tile (2)	.13	.11	.10	.08	.08	.07	.06
	4 (16)	None or Plaster (6)	.17	.14	.11	.10	.09	.08	.07
		Suspended Plaster (5)	.13	.11	.10	.08	.08	.07	.06
		Suspended Acou Tile(2)	.12	.10	.09	.07	.07	.06	.05
Gypsum Slab on ½" Gypsum Board	2 (11)	None or Plaster (6)	.32	.22	.17	.14	.12	.10	.09
		Suspended Plaster (5)	.21	.17	.13	.11	.10	.09	.08
		Suspended Acou Tile (2)	.17	.13	.12	.10	.09	.08	.07
	3 (15)	None or Plaster (6)	.27	.19	.15	.13	.11	.10	.08
		Suspended Plaster (5)	.19	.15	.13	.11	.10	.09	.08
		Suspended Acou Tile (2)	.15	.12	.11	.09	.08	.08	.07
	4 (19)	None or Plaster (6)	.23	.17	.14	.12	.10	.09	.08
		Suspended Plaster (5)	.17	.13	.12	.10	.09	.08	.07
		Suspended Acou Tile (2)	.14	.12	.11	.09	.08	.08	.07
Wood	1 (3)	None or Plaster (6)	.40	.26	.19	.15	.13	.11	.09
		Suspended Plaster (5)	.24	.18	.14	.12	.11	.09	.08
		Suspended Acou Tile (2)	.19	.15	.13	.11	.10	.08	.07
	2 (5)	None or Plaster (6)	.28	.20	.16	.13	.11	.10	.08
		Suspended Plaster (5)	.19	.15	.13	.11	.10	.09	.07
		Suspended Acou Tile (2)	.16	.13	.11	.10	.09	.08	.07
	3 (8)	None or Plaster (6)	.21	.16	.13	.11	.10	.09	.08
		Suspended Plaster (5)	.16	.13	.11	.09	.09	.08	.07
		Suspended Acou Tile (2)	.13	.11	.10	.09	.08	.07	.06

1958 ASHAE Guide

Equations: Summer—(Heat Flow Down) Heat Gain, Btu/hr = (Area, sq ft) × (U value) × (equivalent temp diff, *Table 20*).

Winter—(Heat Flow Up) Heat Loss, Btu/hr = (Area, sq ft) × (U value × 1.1) × (outdoor temp—inside temp).

†For suspended ½" insulation board, plain (.6) or with ½" sand aggregate plaster (5). use values of suspended acou tile.

Figure 26-10 Transmission coefficient U—flat roofs covered with built-up roofing. *(Courtesy McGraw-Hill Book Company)*

TRANSMISSION COEFFICIENT U—PITCHED ROOFS

FOR HEAT FLOW DOWN—SUMMER. FOR HEAT FLOW UP—WINTER (See Equation at Bottom of Page)

Btu/(hr) (sq ft projected area) (deg F temp diff)

All numbers in parentheses indicate weight per sq ft. Total weight per sq ft is sum of component materials.

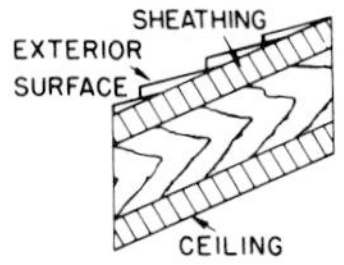

PITCHED ROOFS		CEILING										
		None	3/4" Wood Panel (2)	3/8" Gypsum Board (Plaster Board) (2)	Metal Lath Plastered		3/8" Gypsum or Wood Lath Plastered		Insulating Board Plain or 1/2" Sand Agg Plastered		Acoustical Tile on Furring or 3/8" Gypsum	
					3/4" Sand Plaster (7)	3/4" Lt Wt Plaster (3)	1/2" Sand Plaster (5)	1/2" Lt Wt Plaster (2)	1/2" Board (2)	1" Board (4)	1/2" Tile (2)	3/4" Tile (3)
EXTERIOR SURFACE	SHEATHING											
Asphalt Shingles, (2)	Bldg paper on 5/16" plywood (2)	.51	.27	.30	.32	.29	.29	.28	.22	.17	.23	.21
	Bldg paper on 25/32" wood sheathing (3)	.30	.23	.26	.27	.25	.25	.24	.20	.16	.21	.19
Asbestos-Cement Shingles (3) or Asphalt Roll Roofing (1)	Bldg paper on 5/16" plywood (2)	.59	.28	.34	.37	.33	.33	.31	.25	.18	.25	.22
	Bldg paper on 25/32" wood sheathing (3)	.45	.25	.29	.31	.28	.28	.27	.22	.17	.22	.20
Slates (8) Tile (10) or Sheet Metal (1)	Bldg paper on 5/16" plywood (2)	.64	.29	.36	.38	.34	.35	.47	.26	.19	.26	.23
	Bldg paper on 25/32" wood sheathing (3)	.48	.25	.29	.31	.28	.28	.27	.22	.17	.23	.20
Wood Shingles (2)	Bldg paper on 1" x 4" strips (1)	.53	.26	.31	.33	.30	.30	.28	.23	.17	.24	.21
	Bldg paper on 5/16" plywood (2)	.41	.23	.27	.29	.26	.27	.25	.21	.16	.21	.19
	Bldg paper on 25/32" wood sheathing (3)	.34	.21	.24	.25	.23	.23	.22	.19	.15	.19	.17

1958 ASHAE Guide

Equations: Summer (Heat Flow Down) Heat Gain, Btu/hr = (horizontal projected area, sq ft) × (U value) × (equivalent temp diff).

Winter (Heat Flow Up) Heat Loss, Btu/hr = (horizontal projected area, sq ft) × (U value × 1.1) × (outdoor temp — inside temp).

Figure 26-11 Transmission coefficient U—pitched roofs.
(Courtesy McGraw-Hill Book Company)

TRANSMISSION COEFFICIENT U—CEILING AND FLOOR, (Heat Flow Up)

Based on Still Air Both Sides, Btu/(hr) (sq ft) (deg F temp diff)

All numbers in parentheses indicate weight per sq ft. Total weight per sq ft is sum of ceiling and floor.

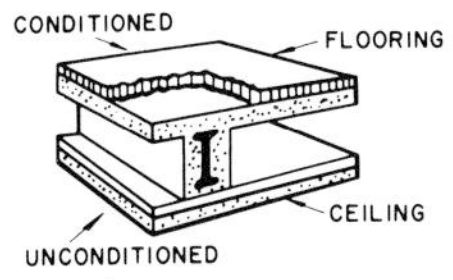

FLOOR	CONCRETE SUBFLOOR	THICKNESS (inches) and WEIGHT (lb per sq ft)	MASONRY CEILING: Not Furred: None or ½″ Sand Plaster (5)	Not Furred: ½″ Lt Wt Plaster (3)	Acoustical Tile Glued: ½″ Tile (1)	Acoustical Tile Glued: ¾″ Tile (1)	Suspended or Furred: Metal Lath Plastered: ¾″ Sand Plaster (7)	Metal Lath Plastered: ¾″ Lt Wt Plaster (3)	⅜″ Gypsum or Wood Lath Plastered: ½″ Sand Plaster (5)	⅜″ Gypsum or Wood Lath Plastered: ½″ Lt Wt Plaster (2)	Insulating Board Plain or ½″ Sand Agg Plastered: ½″ Board (2)	Insulating Board: 1″ Board (4)	Acoustical Tile on Furring or ⅜″ Gypsum: ½″ Tile (1)	Acoustical Tile: ¾″ Tile (1)
None or ⅛″ Linoleum or Floor Tile	Sand Agg	2 (19)	.70	.53	.38	.31	.43	.38	.44	.41	.26	.19	.28	.24
		4 (39)	.63	.49	.36	.30	.41	.36	.41	.38	.25	.18	.26	.23
		6 (59)	.57	.45	.34	.28	.38	.34	.39	.36	.24	.18	25	.22
		8 (79)	.52	.42	.32	.27	.36	.32	.37	.34	.23	.17	.24	.21
		10 (99)	.48	.39	.31	.26	.34	.31	.35	.32	.23	.17	.23	.21
	Lt Wt Agg 80 lb/ft³	2 (15)	.48	.39	.31	.26	.34	.31	.35	.32	.23	.17	.23	.21
		4 (28)	.35	.30	.25	.22	.27	.25	.27	.26	.19	.15	.20	.18
		6 (41)	.27	.24	.21	.18	.22	.21	.22	.21	.17	.13	.17	.15
13/16″ Wood Block on Slab	Sand Agg	2 (20)	.47	.39	.30	.26	.33	.30	.33	.40	.22	.17	.23	.20
		4 (40)	.44	.36	.29	.25	.31	.28	.32	.38	.22	.16	.22	.20
		6 (60)	.41	.34	.28	.24	.30	.27	.30	.36	.21	.16	.22	.19
		8 (80)	.38	.33	.26	.23	.28	.26	.29	.34	.20	.15	.21	.19
		10 (100)	.36	.31	.25	.22	.27	.25	.27	.32	.19	.15	.20	.18
	Lt Wt Agg 80 lb/ft³	2 (16)	.36	.31	.25	.22	.27	.25	.27	.32	.19	.15	.20	.18
		4 (29)	.28	.25	.21	.19	.22	.21	.23	.26	.17	.13	.17	.16
		6 (42)	.23	.21	.18	.16	.19	.18	.19	.21	.15	.12	.15	.14
Floor Tile or ⅛″ Linoleum on ⅝″ Plywood on 2″ x 2″ Sleepers	Sand Agg	2 (22)	.32	.28	.23	.21	.31	.28	.32	.30	.18	.14	.18	.17
		4 (42)	.31	.27	.23	.20	.30	.27	.30	.28	.18	.14	.18	.17
		6 (62)	.29	.26	.22	.19	.28	.26	.29	.27	.17	.14	.18	.16
		8 (82)	.28	.25	.21	.19	.27	.25	.27	.26	.17	.13	.17	.16
		10 (102)	.27	.24	.20	.18	.26	.24	.26	.25	.16	.13	.17	.15
	Lt Wt Agg 80 lb/ft³	2 (19)	.27	.24	.20	.18	.26	.24	.26	.25	.16	.13	.17	.15
		4 (31)	.22	.20	.17	.16	.22	.20	.22	.21	.14	.12	.15	.14
		6 (44)	.19	.17	.15	.14	.18	.17	.19	.18	.13	.11	.13	.12
¾″ Hardwood on 25/32″ Subfloor on 2″ x 2″ Sleepers	Sand Agg	2 (24)	.26	.23	.20	.18	.25	.23	.25	.24	.16	.13	.16	.15
		4 (44)	.25	.22	.19	.17	.24	.22	.24	.23	.16	.13	.16	.15
		6 (64)	.24	.21	.19	.17	.23	.21	.23	.22	.15	.12	.16	.14
		8 (84)	.23	.21	.18	.16	.22	.21	.22	.21	.15	.12	.15	.14
		10 (104)	.22	.20	.17	.16	.21	.20	.22	.21	.14	.12	.15	.14
	Lt Wt Agg 80 lb/ft³	2 (20)	.22	.20	.17	.16	.21	.20	.22	.21	.14	.12	.15	.14
		4 (33)	.19	.17	.15	.14	.18	.17	.18	.18	.13	.11	.13	.12
		6 (46)	.16	.15	.14	.13	.16	.15	.16	.16	.12	.099	.12	.11

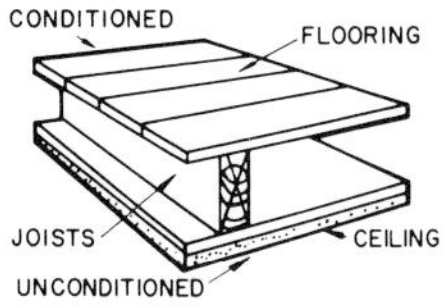

FLOOR	SUBFLOOR	FRAME CONSTRUCTION CEILING: Not Furred: None	Acoustical Tile Glued: ½″ Tile (1)	Acoustical Tile Glued: ¾″ Tile (1)	Suspended or Furred: Metal Lath Plastered: ¾″ Sand Plaster (7)	Metal Lath Plastered: ¾″ Lt Wt Plaster (3)	⅜″ Gypsum or Wood Lath Plastered: ½″ Sand Plaster (5)	⅜″ Gypsum or Wood Lath Plastered: ½″ Lt Wt Plaster (2)	Insulating Board Plain or ½″ Sand Agg Plastered: ½″ Board (2)	Insulating Board: 1″ Board (4)	Acoustical Tile on Furring or ⅜″ Gypsum: ½″ Tile (1)	Acoustical Tile: ¾″ Tile (1)
None	None				.74	.59	.61	.54	.37	.24	.39	.31
	25/32″ Wood (2)	.45	.30	.26	.31	.28	.29	.27	.22	.17	.23	.20
	2″ Wood (5)	.27	.20	.18	.22	.20	.20	.19	.17	.14	.17	.15
½″ Ceramic Tile on 1½″ Cement	25/32″ Wood (21)	.38	.21	.19	.28	.26	.26	.24	.20	.16	.21	.19
	2″ Wood (24)	.24	.19	.17	.20	.19	.19	.18	.16	.13	.16	.15
¾″ Hardwood Floor or Linoleum on ⅝″ Plywood	25/32″ Wood (5)	.33	.24	.21	.25	.23	.23	.22	.18	.15	.19	.17
	2″ Wood (7)	.22	.17	.16	.18	.17	.17	.17	.15	.12	.15	.14
⅛″ Linoleum on ¼″ Hardboard on ⅜″ Insulating Board	25/32″ Wood (5)	.28	.21	.19	.22	.20	.21	.20	.17	.14	.18	.16
	2″ Wood (8)	.20	.16	.15	.17	.16	.16	.16	.14	.12	.14	.13

1958 ASHAE Guide

Equations: Heat flow up, Unconditioned space below: Heat Gain, Btu/hr = (Area, sq ft) × (U value) × (outdoor temp − inside temp − 5 F).
Kitchen or boiler room below: Heat Gain, Btu/hr = (Area, sq ft) × (U value) × (actual temp diff, or outdoor temp − inside temp + 15 F to 25 F).

Figure 26-12 Transmission coefficient U—ceiling and floor, heat flow up. *(Courtesy McGraw-Hill Book Company)*

TRANSMISSION COEFFICIENT U—CEILING AND FLOOR, (Heat Flow Down)

Based on Still Air Both Sides, Btu/(hr) (sq ft) (deg F temp diff)

All numbers in parentheses indicate weight per sq ft. Total weight per sq ft is sum of ceiling and floor.

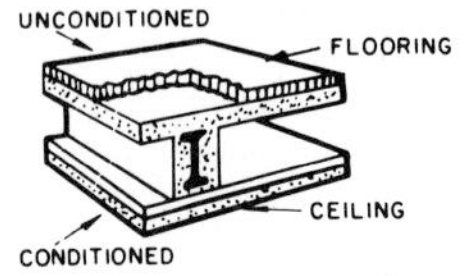

			MASONRY CEILING											
			Not Furred				Suspended or Furred							
		THICK-NESS (inches) and WEIGHT (lb per sq ft)	None or 1/2" Sand Plaster (5)	1/2" Lt Wt Plaster (3)	Acoustical Tile Glued 1/2" Tile (1)	Acoustical Tile Glued 3/4" Tile (1)	Metal Lath Plastered 3/4" Sand Plaster (7)	Metal Lath Plastered 3/4" Lt Wt Plaster (3)	3/8" Gypsum or Wood Lath Plastered 1/2" Sand Plaster (5)	3/8" Gypsum or Wood Lath Plastered 1/2" Lt Wt Plaster (2)	Insulating Board Plain or 1/2" Sand Agg Plastered 1/2" Board (2)	Insulating Board Plain or 1/2" Sand Agg Plastered 1" Board (4)	Acoustical Tile on Furring or 3/8" Gypsum 1/2" Tile (1)	Acoustical Tile on Furring or 3/8" Gypsum 3/4" Tile (1)
FLOOR	CONCRETE SUBFLOOR													
None or 1/8" Linoleum or Floor Tile	Sand Agg	2 (19)	.48	.43	.31	.26	.32	.29	.30	.28	. 23	. 17	.23	.20
		4 (39)	.44	.40	.30	.25	.31	.28	.28	.27	.22	.17	.22	.20
		6 (59)	.41	.37	.28	.24	.29	.27	.27	.26	.21	.16	.22	.19
		8 (79)	.39	.35	.27	.23	.28	.26	.26	.25	.21	.16	.21	.19
		10 (99)	.36	.34	.26	.22	.27	.25	.25	.24	.20	.15	.20	.18
	Lt Wt Agg 80 lb/ft^3	2 (15)	.36	.34	.26	.22	.27	.25	.25	.24	.20	.15	.20	.18
		4 (28)	.29	.26	.21	.19	.22	.21	.21	.20	.17	.14	.17	.16
		6 (41)	.23	.22	.18	.17	.19	.18	.18	.17	.15	.13	.15	.14
13/16" Wood Block on Slab	Sand Agg	2 (20)	.36	.33	.25	.22	.26	.24	.24	.23	.20	.15	.20	.18
		4 (40)	.33	.31	.24	.21	.25	.23	.23	.22	.19	.15	.19	.17
		6 (60)	.32	.29	.23	.21	.24	.22	.22	.21	.18	.15	.18	.17
		8 (80)	.30	.28	.23	.20	.23	.22	.22	.21	.18	.14	.18	.16
		10 (100)	.29	.27	.22	.19	.22	.21	.21	.20	.17	.14	.17	.16
	Lt Wt Agg 80 lb/ft^3	2 (16)	.29	.27	.22	.19	.22	.21	.21	.20	.17	.14	.17	.16
		4 (29)	.23	.22	.19	.17	.19	.18	.18	.17	.15	.13	.15	.14
		6 (42)	.20	.19	.16	.15	.16	.16	.16	.15	.14	.11	.14	.13
Floor Tile or 1/8" Linoleum on 5/8" Plywood on 2" x 2" Sleepers	Sand Agg	2 (22)	.33	.31	.24	.21	.25	.23	.23	.22	.19	.15	.20	.17
		4 (42)	.32	.29	.23	.21	.24	.22	.22	.21	.18	.15	.19	.17
		6 (62)	.30	.28	.23	.20	.23	.21	.22	.21	.18	.14	.18	.16
		8 (82)	.29	.27	.22	.19	.22	.21	.21	.20	.17	.14	.18	.16
		10 (102)	.28	.26	.21	.19	.21	.20	.20	.19	.17	.13	.17	.15
	Lt Wt Agg 80 lb/ft^3	2 (19)	.28	.26	.21	.19	.21	.20	.20	.19	.17	.13	.17	.15
		4 (31)	.22	.21	.18	.16	.18	.17	.17	.17	.15	.12	.15	.14
		6 (44)	. 19	.18	.16	.14	.16	.15	.15	.15	.13	.11	.14	.13
3/4" Hardwood on 25/32" Subfloor on 2" x 2" Sleepers	Sand Agg	2 (24)	.26	.25	.20	.18	.20	.20	.20	.19	.16	.13	.17	.15
		4 (44)	.25	.24	.20	.18	.20	.19	.19	.18	.16	.13	.16	.15
		6 (64)	.24	.23	.19	.17	.19	.18	.19	.18	.15	.13	.16	.14
		8 (84)	.23	.22	.19	.17	.19	.18	.18	.17	.15	.12	.15	.14
		10 (104)	.22	.21	.18	.16	.18	.17	.17	.17	.14	.12	.15	.14
	Lt Wt Agg 80 lb/ft^3	2 (20)	.22	.21	.18	.16	.18	.17	.17	.17	.14	.12	.15	.14
		4 (33)	.19	.18	.16	.14	.16	.15	.15	.15	.13	.11	.13	.12
		6 (46)	.16	.16	.14	.13	.14	.14	.14	.13	.12	.10	.12	. 11

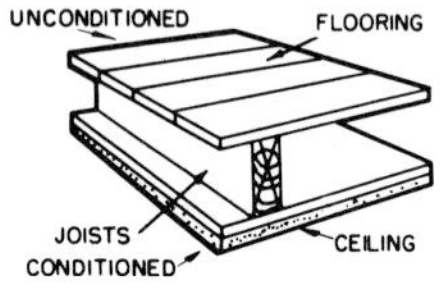

		FRAME CONSTRUCTION CEILING										
		Not Furred			Suspended or Furred							
FLOOR	SUBFLOOR	None	Acoustical Tile Glued 1/2" Tile (1)	Acoustical Tile Glued 3/4" Tile (1)	Metal Lath Plastered 3/4" Sand Plaster (7)	Metal Lath Plastered 3/4" Lt Wt Plaster (3)	3/8" Gypsum or Wood Lath Plastered 1/2" Sand Plaster (5)	3/8" Gypsum or Wood Lath Plastered 1/2" Lt Wt Plaster (2)	Insulating Board Plain or 1/2" Sand Agg Plastered 1/2" Board (2)	Insulating Board Plain or 1/2" Sand Agg Plastered 1" Board (4)	Acoustical Tile on Furring or 3/8" Gypsum 1/2" Tile (1)	Acoustical Tile on Furring or 3/8" Gypsum 3/4" Tile (1)
None	None				.51	.43	.44	.40	.31	.21	.31	.27
	25/32" Wood (2)	.35	.25	.22	.26	.24	.24	.23	.19	.15	.20	.17
	2" Wood (5)	.27	.18	.16	.19	.17	.18	.17	.15	.12	.15	.14
1/2" Ceramic Tile on 1 1/2" Cement	25/32" Wood (21)	.38	.18	.17	.19	.18	.18	.17	.15	.12	.15	.14
	2" Wood (24)	.24	.14	.13	.15	.14	.14	.14	.12	.11	.12	.12
13/16" Hardwood Floor or Linoleum on 5/8" Plywood	25/32" Wood (5)	.33	.17	.16	.18	.17	.17	.16	.14	.12	.14	.13
	2" Wood (7)	.22	.14	.13	.14	.13	.13	.13	.12	.10	.12	.11
1/8" Linoleum on 1/4" Hardboard on 3/8" Insulating Board	25/32" Wood (5)	.29	.16	.15	.16	.15	.16	.15	.13	.11	.14	.13
	2" Wood (8)	.20	.13	.12	.13	.12	.13	.12	.11	.10	.11	.11

1958 ASHAE Guide

Equations: Heat flow down, unconditioned space above: Heat Gain, Btu/hr = (Area, sq ft) × (U value) × (outdoor temp − inside temp − 5 F).
Kitchen above: Heat Gain, Btu/hr = (Area, sq ft) × (U value) × (actual temp diff, or outdoor temp − inside temp + 15 F to 25 F).

Figure 26-13 Transmission coefficient U—ceiling and floor, heat flow down.
(Courtesy McGraw-Hill Book Company)

TRANSMISSION COEFFICIENT U—WITH INSULATION & AIR SPACES

SUMMER AND WINTER

Btu/(hr) (sq ft) (deg F temp diff)

U Value Before Adding Insul. Wall, Ceiling, Roof Floor	Addition of Fibrous Insulation Thickness (Inches)			Add'n of Air Space ¾" or more	Addition of Reflective Sheets to Air Space (Aluminum Foil Average Emissivity = .05) Direction of Heat Flow								
					Winter and Summer Horizontal			Summer Down			Winter Up		
	1	2	3		Added to one or both sides	One sheet in air space	Two sheets in air space	Added to one or both sides	One sheet in air space	Two sheets in air space	Added to one or both sides	One sheet in air space	Two sheets in air space
.60	.19	.11	.08	.38	.34	.18	.11	.12	.06	.05	.36	.20	.14
.58	.19	.11	.08	.37	.33	.18	.11	.12	.06	.05	.36	.20	.14
.56	.18	.11	.08	.36	.32	.18	.11	.11	.06	.05	.35	.20	.14
.54	.18	.11	.08	.36	.31	.17	.11	.11	.06	.05	.34	.19	.14
.52	.18	.11	.08	.35	.30	.17	.10	.11	.06	.05	.33	.19	.14
.50	.18	.11	.08	.34	.29	.17	.10	.11	.06	.05	.32	.19	.13
.48	.17	.11	08	.33	.28	.16	.10	.11	.06	.04	.31	.18	.13
.46	.17	.10	.08	.32	.28	.16	.10	.11	.06	.04	.30	.18	.13
.44	.17	.10	.07	.31	.27	.16	.10	.11	.06	.04	.29	.18	.13
.42	.16	.10	.07	.30	.26	.15	.10	.11	.06	.04	.28	.17	.13
.40	.16	.10	.07	.29	.26	.15	.10	.10	.06	.04	.27	.17	.12
.38	.16	.10	.07	.28	.25	.15	.09	.10	.06	.04	.26	.17	.12
.36	.15	.10	.07	.27	.24	.14	.09	.10	.06	.04	.25	.16	.12
.34	.15	.10	.07	.26	.23	.14	.09	.10	.06	.04	.24	.16	.12
.32	.15	.10	.07	.25	.22	.13	.09	.10	.05	.04	.23	.15	.11
.30	.14	.09	.07	.23	.21	.13	.09	.10	.05	.04	.22	.15	.11
.28	.14	.09	.07	.22	.20	.13	.08	.09	.05	.04	.20	.14	.10
.26	.13	.09	.07	.21	.19	.12	.08	.09	.05	.04	.19	.13	.10
.24	.13	.09	.07	.20	.17	.12	.08	.09	.05	.04	.18	.13	.10
.22	.12	.08	.06	.18	.16	.11	.08	.08	.05	.04	.16	.12	.09
.20	.12	.08	.06	.17	.15	.10	.07	.08	.05	.04	.15	.11	.09
.18	.11	.08	.06	.15	.14	.10	.07	.08	.05	.04	.14	.11	.08
.16	.10	.07	.06	.14	.12	.09	.07	.07	.05	.04	.13	.10	.08
.14	.09	.07	.05	.12	.11	.08	.06	.07	.04	.04	.12	.09	.07
.12	.08	.06	.05	.11	.10	.08	.06	.06	.04	.03	.10	.08	.07
.10	.07	.06	.05	.09	.08	.07	.05	.06	.04	.03	.09	.07	.06

1958 ASHAE Guide

Insulation Added	Air Space Added	Reflective Sheets Added to One or Both Sides	Reflective Sheet in Air Space	Reflective Sheets in Air Space

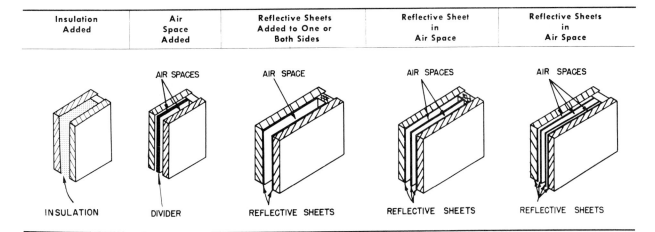

Figure 26-14 Transmission coefficient U—with insulation and air spaces.
(Courtesy McGraw-Hill Book Company)

TRANSMISSION COEFFICIENT U—FLAT ROOFS WITH ROOF-DECK INSULATION

SUMMER AND WINTER

Btu/(hr) (sq ft) (deg F temp diff)

U VALUE OF ROOF BEFORE ADDING ROOF DECK INSULATION	Addition of Roof-Deck Insulation Thickness (in.)					
	½	1	1½	2	2½	3
.60	.33	.22	.17	.14	.12	.10
.50	.29	.21	.16	.14	.12	.10
.40	.26	.19	.15	.13	.11	.09
.35	.24	.18	.14	.12	.10	.09
.30	.21	.16	.13	.12	.10	.09
.25	.19	.15	.12	.11	.09	.08
.20	.16	.13	.11	.10	.09	.08
.15	.12	.11	.09	.08	.08	.07
.10	.09	.08	.07	.07	.06	.05

Figure 26-15 Transmission coefficient U—flat roofs with roof-deck insulation. *(Courtesy McGraw-Hill Book Company)*

TRANSMISSION COEFFICIENT U—WINDOWS, SKYLIGHTS, DOORS & GLASS BLOCK WALLS

Btu/(hr) (sq ft) (deg F temp diff)

GLASS										
	Vertical Glass						Horizontal Glass			
	Single	Double			Triple			Single		Double (¼″)	
Air Space Thickness (in.)		¼	½	¾-4	¼	½	¾-4	Summer	Winter	Summer	Winter
Without Storm Windows	1.13	0.61	0.55	0.53	0.41	0.36	0.34	0.86	1.40	0.50	0.70
With Storm Windows	0.54							0.43	0.64		

DOORS		
Nominal Thickness of Wood (inches)	U Exposed Door	U With Storm Door
1	0.69	0.35
1¼	0.59	0.32
1½	0.52	0.30
1¾	0.51	0.30
2	0.46	0.28
2½	0.38	0.25
3	0.33	0.23
Glass (¾″ Herculite)	1.05	0.43

HOLLOW GLASS BLOCK WALLS	
Description*	U
5¾x5¾x3⅞″ Thick—Nominal Size 6x6x4 *(14)*	0.60
7¾x7¾x3⅞″ Thick—Nominal Size 8x8x4 *(14)*	0.56
11¾x11¾x3⅞″ Thick—Nominal Size 12x12x4 *(16)*	0.52
7¾x7¾x3⅞″ Thick with glass fiber screen dividing the cavity *(14)*	0.48
11¾x11¾x3⅞″ Thick with glass fiber screen dividing the cavity *(16)*	0.44

1958 ASHAE Guide

Equation: Heat Gain or Loss, Btu/hr = (Area, sq ft) × (U value) × (outdoor temp − inside temp)

*Italicized numbers in parentheses indicate weight in lb per sq ft.

Figure 26-16 Transmission coefficient U—windows, skylights, doors, and glass block walls. *(Courtesy McGraw-Hill Book Company)*

THERMAL RESISTANCES R—BUILDING AND INSULATING MATERIALS

(deg F per Btu) / (hr) (sq ft)

MATERIAL	DESCRIPTION	THICK-NESS (in.)	DENSITY (lb per cu ft)	WEIGHT (lb per sq ft)	RESISTANCE R	
					Per Inch Thickness $\frac{1}{k}$	For Listed Thickness $\frac{1}{c}$
BUILDING MATERIALS						
BUILDING BOARD Boards, Panels, Sheathing, etc	Asbestos-Cement Board		120	—	0.25	—
	Asbestos-Cement Board	⅛	120	1.25	—	0.03
	Gypsum or Plaster Board	⅜	50	1.58	—	0.32
	Gypsum or Plaster Board	½	50	2.08	—	0.45
	Plywood		34	—	1.25	—
	Plywood	¼	34	0.71	—	0.31
	Plywood	⅜	34	1.06	—	0.47
	Plywood	½	34	1.42	—	0.63
	Plywood or Wood Panels	¾	34	2.13	—	0.94
	Wood Fiber Board, Laminated or Homogeneous		26	—	2.38	—
			31	—	2.00	—
	Wood Fiber, Hardboard Type		65	—	0.72	—
	Wood Fiber, Hardboard Type	¼	65	1.35	—	0.18
	Wood, Fir or Pine Sheathing	25⁄32	32	2.08	—	0.98
	Wood, Fir or Pine	1⅝	32	4.34	—	2.03
BUILDING PAPER	Vapor Permeable Felt		—	—	—	0.06
	Vapor Seal, 2 Layers of Mopped 15 lb felt		—	—	—	0.12
	Vapor Seal, Plastic Film		—	—	—	Negl
WOODS	Maple, Oak, and Similar Hardwoods		45	—	0.91	—
	Fir, Pine, and Similar Softwoods		32	—	1.25	—
MASONRY UNITS	Brick, Common	4	120	40	—	.80
	Brick, Face	4	130	43	—	.44
	Clay Tile, Hollow:					
	1 Cell Deep	3	60	15	—	0.80
	1 Cell Deep	4	48	16	—	1.11
	2 Cells Deep	6	50	25	—	1.52
	2 Cells Deep	8	45	30	—	1.85
	2 Cells Deep	10	42	35	—	2.22
	3 Cells Deep	12	40	40	—	2.50
	Concrete Blocks, Three Oval Core Sand & Gravel Aggregate	3	76	19	—	0.40
		4	69	23	—	0.71
		6	64	32	—	0.91
		8	64	43	—	1.11
		12	63	63	—	1.28
	Cinder Aggregate	3	68	17	—	0.86
		4	60	20	—	1.11
		6	54	27	—	1.50
		8	56	37	—	1.72
		12	53	53	—	1.89
	Lightweight Aggregate (Expanded Shale, Clay, Slate or Slag; Pumice)	3	60	15	—	1.27
		4	52	17	—	1.50
		8	48	32	—	2.00
		12	43	43	—	2.27
	Gypsum Partition Tile:					
	3"x12"x30" solid	3	45	11	—	1.26
	3"x12"x30" 4-cell	3	35	9	—	1.35
	4"x12"x30" 3-cell	4	38	13	—	1.67
	Stone, Lime or Sand		150	—	0.08	—

Figure 26-17 Thermal resistance R—building and insulating materials. (*Courtesy McGraw-Hill Book Company*)

(continued on next page)

THERMAL RESISTANCES R—BUILDING AND INSULATING MATERIALS (Contd)

(deg F per Btu) / (hr) (sq ft)

MATERIAL	DESCRIPTION	THICKNESS (in.)	DENSITY (lb per cu ft)	WEIGHT (lb per sq ft)	RESISTANCE R: Per Inch Thickness 1/k	RESISTANCE R: For Listed Thickness 1/c
	BUILDING MATERIALS, (CONT.)					
MASONRY MATERIALS Concretes	Cement Mortar		116	—	0.20	—
	Gypsum-Fiber Concrete 87½% gypsum, 12½% wood chips		51	—	0.60	—
	Lightweight Aggregates		120	—	0.19	—
	Including Expanded		100	—	0.28	—
	Shale, Clay or Slate		80	—	0.40	—
	Expanded Slag; Cinders		60	—	0.59	—
	Pumice; Perlite; Vermiculite		40	—	0.86	—
	Also, Cellular Concretes		30	—	1.11	—
			20	—	1.43	—
	Sand & Gravel or Stone Aggregate (Oven Dried)		140	—	0.11	—
	Sand & Gravel or Stone Aggregate (Not Dried)		140	—	0.08	—
	Stucco		116	—	0.20	—
PLASTERING MATERIALS	Cement Plaster, Sand Aggregate		116	—	0.20	—
	Sand Aggregate	½	116	4.8	—	0.10
	Sand Aggregate	¾	116	7.2	—	0.15
	Gypsum Plaster:					
	Lightweight Aggregate	½	45	1.88	—	0.32
	Lightweight Aggregate	⅝	45	2.34	—	0.39
	Lightweight Aggregate on Metal Lath	¾	45	2.80	—	0.47
	Perlite Aggregate		45	—	0.67	—
	Sand Aggregate		105	—	0.18	—
	Sand Aggregate	½	105	4.4	—	0.09
	Sand Aggregate	⅝	105	5.5	—	0.11
	Sand Aggregate on Metal Lath	¾	105	6.6	—	0.13
	Sand Aggregate on Wood Lath		105	—	—	0.40
	Vermiculite Aggregate		45	—	0.59	—
ROOFING	Asbestos-Cement Shingles		120	—	—	0.21
	Asphalt Roll Roofing		70	—	—	0.15
	Asphalt Shingles		70	—	—	0.44
	Built-up Roofing	⅜	70	2.2	—	0.33
	Slate	½	201	8.4	—	0.05
	Sheet Metal		—	—	Negl	—
	Wood Shingles		40	—	—	0.94
SIDING MATERIALS (On Flat Surface)	Shingles					
	Wood, 16″, 7½″ exposure		—	—	—	0.87
	Wood, Double, 16″, 12″ exposure		—	—	—	1.19
	Wood, Plus Insul Backer Board, 5/16″		—	—	—	1.40
	Siding					
	Asbestos-Cement, ¼″ lapped		—	—	—	0.21
	Asphalt Roll Siding		—	—	—	0.15
	Asphalt Insul Siding, ½″ Board		—	—	—	1.45
	Wood, Drop, 1″x8″		—	—	—	0.79
	Wood, Bevel, ½″x8″, lapped		—	—	—	0.81
	Wood, Bevel, ¾″x10″, lapped		—	—	—	1.05
	Wood, Plywood, ⅜″, lapped		—	—	—	0.59
	Structural Glass		—	—	—	0.10
FLOORING MATERIALS	Asphalt Tile	⅛	120	1.25	—	0.04
	Carpet and Fibrous Pad		—	—	—	2.08
	Carpet and Rubber Pad		—	—	—	1.23
	Ceramic Tile	1	—	—	—	0.08
	Cork Tile		25	—	2.22	—
	Cork Tile	⅛	25	0.26	—	0.28
	Felt, Flooring		—	—	—	0.06
	Floor Tile	⅛	—	—	—	0.05
	Linoleum	⅛	80	0.83	—	0.08
	Plywood Subfloor	⅝	34	1.77	—	0.78
	Rubber or Plastic Tile	⅛	110	1.15	—	0.02
	Terrazzo	1	140	11.7	—	0.08
	Wood Subfloor	25/32	32	2.08	—	0.98
	Wood, Hardwood Finish	¾	45	2.81	—	0.68

THERMAL RESISTANCES R—BUILDING AND INSULATING MATERIALS (Contd)

(deg F per Btu) / (hr) (sq ft)

MATERIAL	DESCRIPTION	THICK-NESS (in.)	DENSITY (lb per cu ft)	WEIGHT (lb per sq ft)	RESISTANCE R Per Inch Thickness 1 k	RESISTANCE R For Listed Thickness 1 c
	INSULATING MATERIALS					
BLANKET AND BATT*	Cotton Fiber		0.8 - 2.0	—	3.85	—
	Mineral Wool, Fibrous Form Processed From Rock, Slag, or Glass		1.5 - 4.0	—	3.70	—
	Wood Fiber		3.2 - 3.6	—	4.00	—
	Wood Fiber, Multi-layer Stitched Expanded		1.5 - 2.0	—	3.70	—
BOARD AND SLABS	Glass Fiber		9.5	—	4.00	—
	Wood or Cane Fiber Acoustical Tile	½	22.4	.93	—	1.19
	Acoustical Tile	¾	22.4	1.4	—	1.78
	Interior Finish (Tile, Lath, Plank)		15.0	—	2.86	—
	Interior Finish (Tile, Lath, Plank)	½	15.0	0.62	—	1.43
	Roof Deck Slab Sheathing (Impreg or Coated)		20.0	—	2.63	—
	Sheathing (Impreg or Coated)	½	20.0	0.83	—	1.32
	Sheathing (Impreg or Coated)	25/32	20.0	1.31	—	2.06
	Cellular Glass		9.0	—	2.50	—
	Cork Board (Without Added Binder)		6.5 - 8.0	—	3.70	—
	Hog Hair (With Asphalt Binder)		8.5	—	3.00	—
	Plastic (Foamed)		1.62	—	3.45	—
	Wood Shredded (Cemented in Preformed Slabs)		22.0	—	1.82	—
LOOSE FILL	Macerated Paper or Pulp Products		2.5 - 3.5	—	3.57	—
	Wood Fiber: Redwood, Hemlock, or Fir		2.0 - 3.5	—	3.33	—
	Mineral Wool (Glass, Slag, or Rock)		2.0 - 5.0	—	3.33	—
	Sawdust or Shavings		8.0 - 15.0	—	2.22	—
	Vermiculite (Expanded)		7.0	—	2.08	—
ROOF INSULATION	All Types Preformed, for use above deck Approximately	½	15.6	.7	—	1.39
	Approximately	1	15.6	1.3	—	2.78
	Approximately	1½	15.6	1.9	—	4.17
	Approximately	2	15.6	2.6	—	5.26
	Approximately	2½	15.6	3.2	—	6.67
	Approximatley	3	15.6	3.9	—	8.33
	AIR					
AIR SPACES	POSITION Horizontal	HEAT FLOW Up (Winter) ¾ - 4	—	—	—	0.85
	Horizontal	Up (Summer) ¾ - 4	—	—	—	0.78
	Horizontal	Down (Winter) ¾	—	—	—	1.02
	Horizontal	Down (Winter) 1½	—	—	—	1.15
	Horizontal	Down (Winter) 4	—	—	—	1.23
	Horizontal	Down (Winter) 8	—	—	—	1.25
	Horizontal	Down (Summer) ¾	—	—	—	0.85
	Horizontal	Down (Summer) 1½	—	—	—	0.93
	Horizontal	Down (Summer) 4	—	—	—	0.99
	Sloping 45°	Up (Winter) ¾ - 4	—	—	—	0.90
	Sloping 45°	Down (Summer) ¾ - 4	—	—	—	0.89
	Vertical	Horiz. (Winter) ¾ - 4	—	—	—	0.97
	Vertical	Horiz. (Summer) ¾ - 4	—	—	—	0.86
AIR FILM Still Air	POSITION Horizontal	HEAT FLOW Up	—	—	—	0.61
	Sloping 45°	Up	—	—	—	0.62
	Vertical	Horizontal	—	—	—	0.68
	Sloping 45°	Down	—	—	—	0.76
	Horizontal	Down	—	—	—	0.92
15 Mph Wind	Any Position (For Winter)	Any Direction	—	—	—	0.17
7½ Mph Wind	Any Position (For Summer)	Any Direction	—	—	—	0.25

*Includes paper backing and facing if any.

Transmission Losses through Basements and Slabs

Where basements are unheated, we are not interested in the heat loss of the basement itself. Rather, we are interested in the heat loss through the floor of the heated area above. This may be done by assuming that the basement temperature will be 60°F, and then calculating the heat transmission loss through the floor above in the usual manner.

Where the basement is to be heated to a temperature higher than 60°F, the heat loss through the floor above will be lower, and we must add the heat loss through the basement walls and slab floor.

For portions of the basement wall that are above grade (including windows), calculate heat loss by the methods previously described. For the portions of the basement walls below grade, calculation of the heat transmission loss is inexact, to say the best. The heat loss is difficult to calculate because

1. The temperature of the earth outside the basement surfaces depends upon the depth and upon the outside-air design condition.
2. The heat transfer coefficient of the basement materials below grade depends upon the type of soil, the amount of soil compaction, and the water content of the soil.

For the purposes of approximating this heat loss, use a factor of 18 Btu/hr for each foot of perimeter at or below grade. Apply this factor to basements or to floor slabs to account for the total loss around the perimeter and the floor slab.

Transmission Losses to Attics

The transmission loss from a room, through a ceiling, to an attic is calculated in a similar manner to the transmission through floors to basements. The major difference is that the attic space will be much colder than an unheated basement. As a reasonable approximation, assume that the attic temperature is halfway between the room temperature and the outside air temperature.

Example

How much heat is lost through a 1500 sq ft uninsulated plasterboard ceiling if the inside design is 75°F and the outside design is 5°F?

Solution

From Figure 26-12, read the U value for a $\frac{3}{8}$-in. gypsum board ceiling as 0.61. The temperature of the attic space is halfway between 5°F and 75°F, or 40°F. The temperature difference across the ceiling is (75 − 40) or 35°F.

The heat loss through the ceiling is then calculated as

$$Q = 0.61 \times 1500 \times 35 = 2025 \text{ Btu/hr}$$

Infiltration

In concept, calculating the heat required to offset losses due to infiltration is easy. As stated earlier in this chapter, you need only take the cfm of air infiltrating into the room, mutiply it by 1.08, and multiply it by the number of degrees it must be heated to reach room temperature. The tricky part is figuring how much infiltration takes place.

For systems in which outside air is introduced at the air handler in excess of any exhausts from the room, infiltration may be ignored. The outside air being introduced will pressurize the room to slightly above atmospheric pressure. Any sources of **air leakage** will then have air out-leakage. Very little infiltration will take place, unless there are some unusual wind conditions.

For applications where the room is at a neutral pressure relative to atmosphere, figuring the cfm of infiltration is, at best, a judgment call. Figure 26-18 gives average cfm of infiltration for cold storage rooms, above 32°F, and for average usage conditions. For applications that involve frequent door opening, the cfm values should be doubled.

For occupied buildings, use the **air changes** in Figure 26-19.

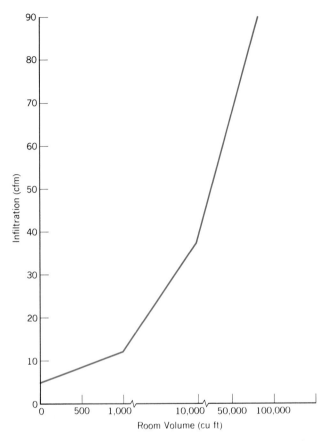

Figure 26-18 Infiltration for cold storage rooms at neutral pressure.

Air Changes Due to Infiltration for Residences with Weather-Stripped Windows and Storm Sash

Building Description	Air Changes per Hour
No operable windows or doors	0.3
Operable windows or doors—1 side	0.6
Operable windows or doors—2 sides	0.9
Operable windows or doors—3 or 4 sides	1.2
Entrances	2.0

Figure 26-19 Air-change factors.

Example

Calculate the heat lost due to infiltration in a residence. Outside design temperature is −5°F, inside design temperature is 75°F. There are windows and doors on four sides. The dimensions of the residence are 25 ft by 80 ft. The ceiling is 8 ft high.

Solution

Calculate the volume of the room

$$\text{Volume} = 25 \times 80 \times 8 = 16,000 \text{ ft}^3$$

From the table in Figure 26-19 estimate the rate of infiltration as 1.2 air changes per hour. Each air change is 16,000 cu ft, therefore the infiltration will be

$$\text{Air changes} = 16,000 \times 1.2 = 19,200 \text{ ft}^3/\text{hr}$$

or

$$19,200/60 = 320 \text{ cfm}$$

The heat loss may then be calculated as

$$Q = 1.08 \times 320 \times 80 = 27,648 \text{ Btu/hr}$$

Vapor Barriers

In a composite wall such as the one shown in Figure 26-20, the temperature at each point

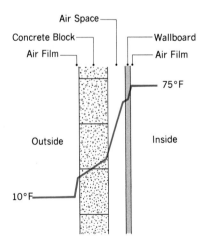

Figure 26-20 Temperature gradient across a composite wall. The higher resistance components will see the most temperature difference.

within the wall will depend upon the distance of that point from the surfaces. The temperature drops across each of the wall components (including the film resistances) are proportional to the resistance of that material. If, for example, the inside film resistance represented 10 percent of the total wall resistance of the wall, then 10 percent of the temperature drop between the inside temperature and the outside temperature would be taken across the inside film.

Example

The wall shown in Figure 26-20 consists of an inside film (R = 0.68), wallboard (R = 0.45), an air space (R = 0.97), concrete block (R = 0.71), and an outside air film (R = 0.17). The outside temperature is 10°F. The inside temperature is 75°F. What is the temperature of the inside surface of the concrete block?

Solution

Calculate the total resistance of the wall as

$$0.68 + 0.45 + 0.97 + 0.71 + 0.17 = 2.98$$

The total resistance of the components between the inside surface of the block and the inside space is

$$0.97 + 0.45 + 0.68 = 2.1$$

The percentage of the total resistance that is between the room and the inside surface of the block wall is

$$\frac{2.1}{2.98} = 0.705$$

or 70.5 percent. So 70.5 percent of the temperature difference will be taken between the inside surface of the block and the room.

$$0.705 \times (75 - 10) = 45.8°F$$

Therefore, of the total 65°F difference between inside and outside, 45.8°F will happen between the inside and the block. The surface temperature of the block wall will be

$$T = 75°F - 45.8°F = 29.2°F$$

The importance of knowing the temperature gradients inside a wall relates to moisture flow through the wall. Typically, during the winter months the moisture content inside the space will be higher than outside. This may be due to humidifiers, cooking, or people. The tendency will be for this moisture to physically pass through the wall, from inside to outside. If this moisture reaches its dew point, it will condense and water will form on (or in) the wall. If, at the point where it condenses, it is at a temperature below 32°F, it will freeze. When this moisture freezes, it expands and can destroy the structure of the wall.

In order to prevent this moisture from traveling far enough through the wall to reach its dew point, a **vapor barrier** is used. The ability of a material to allow flow of moisture is called its **permeance** or its permeability. In general, insulations, plaster, wood, fiberglass, polystyrene (Styrofoam), and asbestos-cement board have relatively high permeability. They will have little effect on stopping the migration of moisture. The following materials are used for their relatively low permeance (high resistance to moisture flow):

Material	Permeance
Aluminum foil	0.0
Polyethelyne (2 mil)	0.16
Polyethelyne (6 mil)	0.06
Roll roofing	0.05
Enamel paint on smooth plaster	1.0
Flat paint on smooth plaster	4.0

The vapor barrier is placed on or close to the warm side of the wall, in order to prevent the migration of moisture to a cold surface. Installation of a vapor barrier (especially polyethelyne sheets) is critical. Tears or other voids can ruin the ability of the entire vapor barrier to work effectively.

Heat Gains

Calculating a cooling or a refrigeration load on a room is more complex than calculating a heat loss. Following are the major considerations that make the cooling load calculation different from the heating load calculation.

1. The sun shining on walls, roofs, and through windows add heat to the space.

2. There are internal loads within the space, such as lights, machines, appliances, and people.

3. There are latent loads, that is, sources of moisture entering the room that will represent a latent load to the cooling coil.

Transmission Heat Gain through Partitions, Walls, and Roofs

Transmission heat gain can occur from adjacent non-air-conditioned indoor spaces, or from the warm outdoors. Generally, we refer to interior walls as partitions. Walls denote exterior walls.

The heat gain from transmission through partitions is handled the same as for heat losses. Look up the U value for the partition, and use the formula

$$Q = U \times A \times \Delta T$$

When figuring the ΔT across partitions, assume that the non-air-conditioned space is at a temperature of 10°F higher than the outside design air temperature.

For transmission heat gain from exterior walls, we have a different problem. The effect of the sun shining on walls is that the temperature of the outside wall surface can become much hotter than the actual outdoor air temperature. You have experienced this effect if you have ever walked barefoot on pavement that has been heated by the sun's radiant energy.

The way we account for this effect is to use an equivalent ΔT for walls and roofs that are exposed to sun. Figures 26-21 and 26-22 give the values for these equivalent ΔT temperatures. Note that the equivalent ΔT to use depends upon the following.

1. The direction in which the surface faces.

2. The color of the surface.

3. The time of day.

4. The type of construction.

The direction is important because the sun's rays will be striking the surfaces from a different angle all the time. The color of the surfaces has an effect because dark-colored surfaces will tend to absorb the radiant energy, while light colors will reflect a higher percentage. Time of day is important because the sun strikes the east walls early and the west walls late in the day. The type of construction affects the heat gain more than just its U factor would indicate. The weight of the wall will determine the time lag. That is, the sun that strikes a very lightweight wall at 2 P.M. will heat the outside surface, and that heat may get to the inside wall by 4 P.M. For a very heavy wall, it may take as long as 12 hr for the effect of the radiant heat to reach the air conditioned space.

Example

Calculate the heat gain through an exterior wall that is 200 sq ft in area, facing south. The wall is lightweight construction, consisting of 1-in. dark-color stucco + an air space + 2 in. of insulation. It has a U factor of 0.106.

Solution

For lightweight construction, find the equivalent ΔT of 33°F for a dark wall at 2 P.M. Calculate the heat gain as

$$Q = U \times A \times \text{equivalent } \Delta T$$
$$= 0.106 \times 200 \times 33$$
$$= 700 \text{ Btu/hr}$$

In choosing what time to use, you need to choose the time at which you think the maximum heat gain will occur. A peak heat gain occurring sometime between 2 P.M. and 6 P.M. is common. But if the space has only one exposure to the outside, and it happens to be an east exposure, the peak could occur well before noon. Or, for a room with west exposure and very heavy construction, the peak could occur well after 6 P.M. You must choose a time, and then use that time in calculating the heat gain for all surfaces. You do not use the time for the peak heat gain for each surface, because they do not occur at the same time.

There are computer programs available to do heat gain calculations. Because they can do so many calculations so quickly, they will do a complete calculation for each time during the day, and then select the time of day for which it found the peak heat gain. For manual calculations, a little common sense and the information provided in this text will be accurate enough for practical purposes.

EQUIVALENT TEMPERATURE DIFFERENCE (DEG F)

FOR DARK COLORED†, SUNLIT AND SHADED WALLS*

Based on Dark Colored Walls; 95 F db Outdoor Design Temp; Constant 80 F db Room Temp;
20 deg F Daily Range; 24-hour Operation; July and 40° N. Lat.†

EXPOSURE	WEIGHT OF WALL (lb/sq ft)	AM							PM												AM				
		6	7	8	9	10	11	12	1	2	3	4	5	6	7	8	9	10	11	12	1	2	3	4	5
Northeast	20	5	15	22	23	24	19	14	13	12	13	14	14	14	12	10	8	6	4	2	0	−2	−3	−4	−2
	60	−1	−2	−2	5	24	22	20	15	10	11	12	13	14	13	12	11	10	8	6	4	2	1	0	−1
	100	4	3	4	4	4	10	16	15	14	12	10	11	12	12	12	11	10	9	8	7	6	6	5	5
	140	5	5	6	6	6	6	6	10	14	16	14	12	10	10	10	10	10	10	10	9	9	8	7	7
East	20	1	17	30	33	36	35	32	20	12	13	14	14	14	12	10	8	6	4	2	0	−1	−2	−3	−3
	60	−1	−1	0	21	30	31	31	19	14	13	12	13	14	13	12	11	10	8	5	4	3	1	1	0
	100	5	5	6	8	14	20	24	25	24	20	18	16	14	14	14	13	12	11	10	9	8	7	6	6
	140	11	10	10	9	8	9	10	15	18	19	18	17	16	14	12	13	14	14	14	13	13	12	12	12
Southeast	20	10	6	13	19	26	27	28	26	24	19	16	15	14	12	10	8	6	4	2	0	−1	−1	−2	−2
	60	1	1	0	13	20	24	28	26	25	21	18	15	14	13	12	11	10	8	6	5	4	3	3	2
	100	7	7	6	6	6	11	16	17	18	19	18	16	14	13	12	11	10	10	10	9	9	8	8	7
	140	9	8	8	8	8	7	6	11	14	15	16	18	16	15	14	13	12	12	12	11	11	10	10	9
South	20	−1	−2	−4	1	4	14	22	27	30	28	26	20	16	12	10	7	6	3	2	1	1	0	0	−1
	60	−1	−3	−4	−3	−2	7	12	20	24	25	26	23	20	15	12	10	8	6	4	2	1	1	0	−1
	100	4	4	2	2	2	3	4	8	12	15	16	18	18	15	14	11	10	9	8	8	7	6	6	5
	140	7	6	6	5	4	4	4	4	4	7	10	13	14	15	16	16	14	12	10	10	9	9	8	7
Southwest	20	−2	−4	−4	−2	0	4	6	19	26	34	40	41	42	30	24	12	6	4	2	1	1	0	−1	−1
	60	2	1	0	0	0	1	2	8	12	24	32	35	36	35	34	20	10	7	6	5	4	4	3	3
	100	7	5	6	5	4	5	6	7	8	12	14	19	22	23	24	23	22	15	10	10	9	9	8	7
	140	8	8	8	8	8	7	6	6	6	7	8	9	10	15	18	19	20	13	8	8	8	8	8	8
West	20	−2	−3	−4	−2	0	3	6	14	20	32	40	45	48	34	22	14	8	5	2	1	0	0	−1	−1
	60	2	1	0	0	0	2	4	7	10	19	26	34	40	41	36	28	16	10	6	5	4	3	3	2
	100	7	7	6	6	6	6	6	7	8	10	12	17	20	25	28	27	26	19	14	12	11	10	9	8
	140	12	11	10	9	8	8	8	9	10	10	10	11	12	14	16	21	22	23	22	20	18	16	15	13
Northwest	20	−3	−4	−4	−2	0	3	6	10	12	19	24	33	40	37	34	18	6	4	2	0	−1	−1	−2	−2
	60	−2	−3	−4	−3	−2	0	2	6	8	10	12	21	30	31	32	21	12	8	6	4	3	1	0	−1
	100	5	4	4	4	4	4	4	4	4	5	6	9	12	17	20	21	22	14	8	7	7	6	6	5
	140	8	7	6	6	6	6	6	6	6	6	6	7	8	9	10	14	18	19	20	16	13	11	10	9
North (Shade)	20	−3	−3	−4	−3	−2	1	4	8	10	12	14	13	12	10	8	6	4	2	0	0	−1	−1	−2	−2
	60	−3	−3	−4	−3	−2	−1	0	3	6	8	10	11	12	12	12	10	8	6	4	2	1	0	−1	−2
	100	1	1	0	0	0	0	0	1	2	3	4	5	5	5	8	7	6	5	4	3	3	2	2	1
	140	1	1	0	0	0	0	0	0	0	1	2	3	4	5	6	7	8	7	6	4	3	2	2	1
		6	7	8	9	10	11	12	1	2	3	4	5	6	7	8	9	10	11	12	1	2	3	4	5
				AM										PM									AM		

SUN TIME

Equation: Heat Gain Thru Walls, Btu/hr = (Area, sq ft) × (equivalent temp diff) × (transmission coefficient U)

*All values are for both insulated and uninsulated walls.

†For other conditions, refer to corrections on Figure 26–22.

Figure 26-21 Equivalent temperature difference for dark-colored, sunlit, and shaded walls. *(Courtesy McGraw-Hill Book Company)*

EQUIVALENT TEMPERATURE DIFFERENCE (DEG F)

FOR DARK COLORED†, SUNLIT AND SHADED ROOFS*

Based on 95 F db Outdoor Design Temp; Constant 80 F db Room Temp; 20 deg F Daily Range;
24-hour Operation; July and 40° N. Lat.†

CONDI-TION	WEIGHT OF ROOF (lb/sq ft)	AM 6	7	8	9	10	11	12	PM 1	2	3	4	5	6	7	8	9	10	11	12	AM 1	2	3	4	5
Exposed to Sun	10	−4	−6	−7	−5	−1	7	15	24	32	38	43	46	45	41	35	28	22	16	10	7	3	1	−1	−3
	20	0	−1	−2	−1	2	9	16	23	30	36	41	43	43	40	35	30	25	20	15	12	8	6	4	2
	40	4	3	2	3	6	10	16	23	28	33	38	40	41	39	35	32	28	24	20	17	13	11	9	6
	60	9	8	6	7	8	11	16	22	27	31	35	38	39	38	36	34	31	28	25	22	18	16	13	11
	80	13	12	11	11	12	13	16	22	26	28	32	35	37	37	35	34	34	32	30	27	23	20	18	14
Covered with Water	20	−5	−2	0	2	4	10	16	19	22	20	18	16	14	12	10	6	2	1	1	−1	−2	−3	−4	−5
	40	−3	−2	−1	−1	0	5	10	13	15	15	16	15	15	14	12	10	7	5	3	1	−1	−2	−3	−3
	60	−1	−2	−2	−2	−2	2	5	7	10	12	14	15	16	15	14	12	10	8	6	4	3	2	1	0
Sprayed	20	−4	−2	0	2	4	8	12	15	18	17	16	15	14	12	10	6	2	1	0	−1	−2	−2	−3	−3
	40	−2	−2	−1	−1	0	2	5	9	13	14	14	14	13	12	9	7	5	3	2	1	0	0	−1	−1
	60	−1	−2	−2	−2	−2	0	2	5	8	10	12	13	14	14	13	12	11	10	8	6	4	2	1	0
Shaded	20	−5	−5	−4	−2	0	2	6	9	12	13	14	13	12	10	8	5	2	1	0	−1	−3	−4	−5	−5
	40	−5	−5	−4	−3	−2	0	2	5	8	10	12	13	12	11	10	8	6	4	2	0	−1	−3	−4	−5
	60	−3	−3	−2	−2	−2	−1	0	2	4	6	8	9	10	10	10	9	8	6	4	2	1	0	−1	−2
		6	7	8	9	10	11	12	1	2	3	4	5	6	7	8	9	10	11	12	1	2	3	4	5
		AM							PM												AM				
									SUN TIME																

Equation: Heat Gain Thru Roofs, Btu/hr = (Area, sq ft) × (equivalent temp diff) × (transmission coefficient U)

*With attic ventilated and ceiling insulated roofs, reduce equivalent temp diff 25%.
 For peaked roofs, use the roof area projected on a horizontal plane.

†For other conditions, refer to corrections below.

CORRECTIONS TO EQUIVALENT TEMPERATURES (DEG F)

OUTDOOR DESIGN FOR MONTH AT 3 P.M. MINUS ROOM TEMP (deg F)	DAILY RANGE (deg F) 8	10	12	14	16	18	20	22	24	26	28	30	32	34	36	38	40
−30	−39	−40	−41	−42	−43	−44	−45	−46	−47	−48	−49	−50	−51	−52	−53	−54	−55
−20	−29	−30	−31	−32	−33	−34	−35	−36	−37	−38	−39	−40	−41	−42	−43	−44	−45
−10	−19	−20	−21	−22	−23	−24	−25	−26	−27	−28	−29	−30	−31	−32	−33	−34	−35
0	−9	−10	−11	−12	−13	−14	−15	−16	−17	−18	−19	−20	−21	−22	−23	−24	−25
5	−4	−5	−6	−7	−8	−9	−10	−11	−12	−13	−14	−15	−16	−17	−18	−19	−20
10	1	0	−1	−2	−3	−4	−5	−6	−7	−8	−9	−10	−11	−12	−13	−14	−15
15	6	5	4	3	2	1	0	−1	−2	−3	−4	−5	−6	−7	−8	−9	−10
20	11	10	9	8	7	6	5	4	3	2	1	0	−1	−2	−3	−4	−5
25	16	15	14	13	12	11	10	9	8	7	6	5	4	3	2	1	0
30	21	20	19	18	17	16	15	14	13	12	11	10	9	8	7	6	5
35	26	25	24	23	22	21	20	19	18	17	16	15	14	13	12	11	10
40	31	30	29	28	27	26	25	24	23	22	21	20	19	18	17	16	15

Figure 26-22 Equivalent temperature difference for dark-colored, sunlit, and shaded roofs. (Courtesy McGraw-Hill Book Company)

Heat Gain through Glass

There are two components of the heat gain through glass, the **transmission** portion and the **solar** portion. When the sun shines on windows, the windows absorb very little of the radiant heat. They allow the radiant heat to pass through and warm whatever inside surfaces it strikes. This solar heat gain is calculated from the Btu/sq ft of window area given in Figures 26-23 through 26-25. Note that the shading factor corrections that are given in Figure 26-26 must be applied to these values. For windows that are shaded from the outside so that no sun strikes the window, the solar portion of the heat gain is zero.

In addition to the solar heat gain through the glass, you must add the transmission through the glass. This is done using the U factor, the window area, and the ΔT, which is the temperature difference between the outdoor and the indoor air design temperatures.

Internal Heat Gains

Internal heat gains are those that occur from people, product, or devices located within the air conditioned or refrigerated space.

Heat gains from people are listed in Figure 26-27. You must choose the activity of the people within the space that most closely matches the activities listed, and read the **sensible** and **latent heat gain** per person. Multiply the table value by the number of people. At the end of the total heat gain calculation, you will need to know the total sensible load and the total latent load, so keep these calculations separated.

Electric motors convert electrical energy into work and heat. Of all the electricity that is supplied to a motor, some is turned to heat directly in the motor. Motor efficiencies for motors from $\frac{1}{6}$ hp to 250 hp range from 60 percent to over 90 percent. The motor efficiency denotes how much of the electricity supplied is converted to shaft horsepower. The rest of the electricity supplied is converted to heat and radiated by the motor casing. The work done by the motor may also eventually turn into heat. There are three different potential locations for the motor and equipment listed for the heat gains in Figure 26-28.

1. Motor in and driven machine in—for example, a motor driving a lathe inside a machine shop. The motor inefficiency as well as the motor work all eventually wind up as heat within the air conditioned space. A motor driving a pedestal fan located within an air conditioned space would produce the same result. The work done by the motor turns a fan that moves air. The air eventually loses its velocity due to friction, and all the work done upon it turns into heat.

2. Motor out, driven machine in—for example, a motor in a warehouse driving a conveyor that is located inside the air conditioned space. The motor inefficiency is heat radiated to the non-air-conditioned warehouse. The work done by the motor is consumed by friction of the rollers, bearings, belts, or other turning devices located within the air conditioned space. The same would be true for the motor on an air handler located in an equipment room. If the motor is not in the air stream, the motor inefficiency heat goes to the mechanical room. The work done by the motor is turned into heat by the work of the fan being done to the air.

3. Motor in, driven machine out—for example, a belt drive exhaust fan where the motor is within the air conditioned space. It drives a fan that is doing work on the air, but the air is leaving the room and therefore does not represent a heat gain.

The heat gain values that are given in Figure 26-28 are for motors that are delivering their nameplate horsepower. For motors that are less than fully loaded, the heat gain will be less. For motors that are only used intermittently, the values in Figure 26-28 are multiplied by the percentage of time that the motor is actually running (the use factor).

Where the motors to be used are actually in operation at the time when you are figuring the heat gain, the most accurate determination of the heat gain may be made by measuring the watts being consumed by the motor, as follows.

1. Measure the amps being drawn by the motor.
2. Measure the voltage of the motor.
3. For single-phase motors, calculate the watts as the volts × amps.
4. For three-phase motors, calculate the watts as the volts × amps × 1.732.

SOLAR HEAT GAIN THRU ORDINARY GLASS (Contd)

30° Btu/(hr) (sq ft sash area) **30°**

30° NORTH LATITUDE		AM						SUN TIME				PM			30° SOUTH LATITUDE	
Time of Year	Exposure	6	7	8	9	10	11	Noon	1	2	3	4	5	6	Exposure	Time of Year
JUNE 21	North	33	29	18	14	14	14	14	14	14	14	18	29	33	South	DEC 22
	Northeast	105	139	130	97	55	19	14	14	14	14	12	10	5	Southeast	
	East	108	156	161	143	98	44	14	14	14	14	12	10	5	East	
	Southeast	42	75	90	90	73	44	17	14	14	14	12	10	5	Northeast	
	South	5	10	12	14	15	19	21	19	15	14	12	10	5	North	
	Southwest	5	10	12	14	14	14	17	44	73	90	90	75	42	Northwest	
	West	5	10	12	14	14	14	14	44	98	143	161	156	108	West	
	Northwest	5	10	12	14	14	14	14	19	55	97	130	139	105	Southwest	
	Horizontal	19	61	131	180	217	240	250	240	217	180	131	61	19	Horizontal	
JULY 23 & MAY 21	North	22	20	14	13	14	14	14	14	14	13	14	20	22	South	JAN 21 & NOV 21
	Northeast	93	131	123	89	46	16	14	14	14	13	12	9	4	Southeast	
	East	100	155	164	145	99	44	14	14	14	13	12	9	4	East	
	Southeast	42	82	100	100	83	53	22	14	14	13	12	9	4	Northeast	
	South	4	9	12	14	20	27	30	27	20	14	12	9	4	North	
	Southwest	4	9	12	13	14	14	14	53	83	100	100	82	42	Northwest	
	West	4	9	12	13	14	14	14	44	99	145	164	155	100	West	
	Northwest	4	9	12	13	14	14	14	16	46	89	123	131	93	Southwest	
	Horizontal	15	66	123	176	214	236	246	236	214	176	123	66	15	Horizontal	
AUG 24 & APR 20	North	6	8	11	13	14	14	14	14	13	13	11	8	6	South	FEB 20 & OCT 23
	Northeast	55	108	100	66	27	14	14	14	13	13	11	8	2	Southeast	
	East	66	147	165	148	102	46	14	14	13	13	11	8	2	East	
	Southeast	37	98	127	129	112	82	39	15	13	13	11	8	2	Northeast	
	South	2	8	13	27	47	58	63	58	47	27	13	8	2	North	
	Southwest	2	8	11	13	13	15	39	82	112	129	127	98	37	Northwest	
	West	2	8	11	13	13	14	14	46	102	148	165	147	66	West	
	Northwest	2	8	11	13	13	14	14	14	27	66	100	108	55	Southwest	
	Horizontal	6	47	107	161	200	225	235	225	200	161	107	47	6	Horizontal	
SEPT 22 & MAR 22	North	0	5	10	12	13	14	14	14	13	12	10	5	0	South	MAR 22 & SEPT 22
	Northeast	0	74	90	40	15	14	14	14	13	12	10	5	0	Southeast	
	East	0	124	158	144	103	48	14	14	13	12	10	5	0	East	
	Southeast	0	98	131	152	141	113	67	25	13	12	10	5	0	Northeast	
	South	0	9	18	60	82	98	105	98	82	60	18	9	0	North	
	Southwest	0	5	10	12	13	25	67	113	141	152	131	98	0	Northwest	
	West	0	5	10	12	13	14	14	48	103	144	158	124	0	West	
	Northwest	0	5	10	12	13	14	14	14	15	40	90	74	0	Southwest	
	Horizontal	0	25	81	135	179	202	212	202	179	135	81	25	0	Horizontal	
OCT 23 & FEB 20	North	0	3	8	11	12	13	14	13	12	11	8	3	0	South	APR 20 & AUG 24
	Northeast	0	33	39	18	12	13	14	13	11	11	8	3	0	Southeast	
	East	0	79	135	132	94	43	14	13	12	11	8	3	0	East	
	Southeast	0	73	142	163	159	136	92	47	15	11	8	3	0	Northeast	
	South	0	18	57	92	121	139	145	139	121	92	57	18	0	North	
	Southwest	0	3	8	11	15	47	92	136	159	163	142	73	0	Northwest	
	West	0	3	8	11	12	13	14	43	94	132	135	79	0	West	
	Northwest	0	3	8	11	11	13	14	13	12	18	39	33	0	Southwest	
	Horizontal	0	6	49	100	143	171	179	171	143	100	49	6	0	Horizontal	
NOV 21 & JAN 21	North	0	1	6	9	11	12	12	12	11	9	6	1	0	South	MAY 21 & JULY 23
	Northeast	0	8	16	9	11	12	12	12	11	9	6	1	0	Southeast	
	East	0	27	109	116	83	35	12	12	11	9	6	1	0	East	
	Southeast	0	28	127	161	162	143	104	64	23	9	6	1	0	Northeast	
	South	0	10	68	109	137	154	159	154	137	109	68	10	0	North	
	Southwest	0	1	6	9	23	64	104	143	162	161	127	28	0	Northwest	
	West	0	1	6	9	11	12	12	35	83	116	109	27	0	West	
	Northwest	0	1	6	9	11	12	12	12	11	9	16	8	0	Southwest	
	Horizontal	0	2	27	71	109	136	145	136	109	71	27	2	0	Horizontal	
DEC 22	North	0	0	4	9	11	12	12	12	11	9	4	0	0	South	JUNE 21
	Northeast	0	0	10	9	11	12	12	12	11	9	4	0	0	Southeast	
	East	0	0	92	105	80	32	12	12	11	9	4	0	0	East	
	Southeast	0	0	114	157	162	143	108	72	28	9	4	0	0	Northeast	
	South	0	0	64	113	142	159	163	159	142	113	64	0	0	North	
	Southwest	0	0	4	9	28	72	108	143	162	157	114	0	0	Northwest	
	West	0	0	4	9	11	12	12	32	80	105	92	0	0	West	
	Northwest	0	0	4	9	11	12	12	12	11	9	10	0	0	Southwest	
	Horizontal	0	0	19	60	97	122	131	122	97	60	19	0	0	Horizontal	
Solar Gain Correction	Steel Sash, or No Sash X 1/.85 or 1.17	Haze −15% (Max.)			Altitude +0.7% per 1000 Ft			Dewpoint Decrease From 67 F +7% per 10 F			Dewpoint Increase From 67 F −7% per 10 F			South Lat. Dec. or Jan. +7%		

Bold Face Values — Monthly Maximums   Boxed Values — Yearly maximums

Figure 26-23 Solar heat gain through ordinary glass, 30 degree latitude.
(Courtesy McGraw-Hill Book Company)

SOLAR HEAT GAIN THRU ORDINARY GLASS (Contd)

40° Btu/(hr) (sq ft sash area) **40°**

40° NORTH LATITUDE		AM				SUN TIME						PM		40° SOUTH LATITUDE	
Time of Year	Exposure	6	7	8	9	10	11	Noon	1	2	3	4	5	6	Exposure	Time of Year
JUNE 21	North	32	20	12	13	14	14	14	14	14	13	12	20	32	South	DEC 22
	Northeast	118	133	112	73	30	14	14	14	14	13	12	10	6	Southeast	
	East	126	161	162	142	95	44	14	14	14	13	12	10	6	East	
	Southeast	51	88	109	111	99	71	34	14	14	13	12	10	6	Northeast	
	South	6	10	12	19	35	44	54	44	35	19	12	10	6	North	
	Southwest	6	10	12	13	14	14	34	71	99	111	109	88	51	Northwest	
	West	6	10	12	13	14	14	14	44	95	142	162	161	126	West	
	Northwest	6	10	12	13	14	14	14	14	30	73	112	133	118	Southwest	
	Horizontal	31	82	134	179	210	232	237	232	210	179	134	82	31	Horizontal	
JULY 23 & MAY 21	North	24	14	12	13	14	14	14	14	14	13	12	14	24	South	JAN 21 & NOV 21
	Northeast	106	127	105	66	26	14	14	14	14	13	12	10	5	Southeast	
	East	118	161	164	144	98	43	14	14	14	13	12	10	5	East	
	Southeast	54	96	119	125	110	82	42	15	14	13	12	10	5	Northeast	
	South	5	10	13	26	44	63	69	63	44	26	13	10	5	North	
	Southwest	5	10	12	13	14	15	42	82	110	125	119	96	54	Northwest	
	West	5	10	12	13	14	14	14	43	98	144	164	161	118	West	
	Northwest	5	10	12	13	14	14	14	14	26	66	105	127	106	Southwest	
	Horizontal	24	73	126	171	203	225	233	225	203	171	126	73	24	Horizontal	
AUG 24 & APR 20	North	7	8	11	13	14	14	14	14	14	13	11	8	7	South	FEB 20 & OCT 23
	Northeast	68	102	82	46	16	14	14	14	14	13	11	8	3	Southeast	
	East	84	147	162	145	101	45	14	14	14	13	11	8	3	East	
	Southeast	48	105	138	146	139	107	66	25	14	13	11	8	3	Northeast	
	South	3	8	24	51	89	97	102	97	89	51	24	8	3	North	
	Southwest	3	8	11	13	14	25	66	107	139	146	138	105	48	Northwest	
	West	3	8	11	13	14	14	14	45	101	145	162	147	84	West	
	Northwest	3	8	11	13	14	14	14	16	46	82	102	68		Southwest	
	Horizontal	9	47	100	150	185	205	214	205	185	150	100	47	9	Horizontal	
SEPT 22 & MAR 22	North	0	5	9	12	13	13	14	13	13	12	9	5	0	South	MAR 22 & SEPT 22
	Northeast	0	51	58	26	13	13	14	13	13	12	9	5	0	Southeast	
	East	0	116	149	139	99	45	14	13	13	12	9	5	0	East	
	Southeast	0	95	144	162	157	133	90	41	14	12	9	5	0	Northeast	
	South	0	12	44	81	110	122	140	122	110	81	44	12	0	North	
	Southwest	0	5	9	12	14	41	90	133	157	162	144	95	0	Northwest	
	West	0	5	9	12	13	13	14	45	99	139	149	116	0	West	
	Northwest	0	5	9	12	13	13	14	13	13	26	58	51	0	Southwest	
	Horizontal	0	21	67	124	153	176	183	176	153	124	67	21	0	Horizontal	
OCT 23 & FEB 20	North	0	2	6	10	11	12	12	12	11	10	6	2	0	South	APR 20 & AUG 24
	Northeast	0	35	33	12	11	12	12	12	11	10	6	2	0	Southeast	
	East	0	85	117	122	88	39	12	12	11	10	6	2	0	East	
	Southeast	0	81	132	161	163	144	107	63	20	10	6	2	0	Northeast	
	South	0	21	59	104	137	154	162	154	137	104	59	21	0	North	
	Southwest	0	2	6	10	20	63	107	144	163	161	132	81	0	Northwest	
	West	0	2	6	10	11	12	12	39	88	122	117	85	0	West	
	Northwest	0	2	6	10	11	12	12	12	11	12	33	35	0	Southwest	
	Horizontal	0	8	29	64	101	123	129	123	101	64	29	8	0	Horizontal	
NOV 21 & JAN 21	North	0	0	3	7	9	10	11	10	9	7	3	0	0	South	MAY 21 & JULY 23
	Northeast	0	0	12	7	9	10	11	10	9	7	3	0	0	Southeast	
	East	0	0	91	100	74	33	11	10	9	7	3	0	0	East	
	Southeast	0	0	109	144	156	144	116	70	27	7	3	0	0	Northeast	
	South	0	0	59	104	139	158	166	158	139	104	59	0	0	North	
	Southwest	0	0	3	7	27	70	116	144	156	144	109	0	0	Northwest	
	West	0	0	3	7	9	10	11	33	74	100	91	0	0	West	
	Northwest	0	0	3	7	9	10	11	10	9	7	12	0	0	Southwest	
	Horizontal	0	0	16	43	73	92	103	92	73	43	16	0	0	Horizontal	
DEC 22	North	0	0	2	6	9	10	10	10	9	6	2	0	0	South	JUNE 21
	Northeast	0	0	7	6	9	10	10	10	9	6	2	0	0	Southeast	
	East	0	0	72	86	68	31	10	10	9	6	2	0	0	East	
	Southeast	0	0	88	134	148	142	115	73	30	7	2	0	0	Northeast	
	South	0	0	51	99	134	158	165	158	134	99	51	0	0	North	
	Southwest	0	0	2	7	30	73	115	142	148	134	88	0	0	Northwest	
	West	0	0	2	6	9	10	10	31	68	86	72	0	0	West	
	Northwest	0	0	2	6	9	10	10	10	9	6	7	0	0	Southwest	
	Horizontal	0	0	8	32	55	76	85	76	55	32	8	0	0	Horizontal	
Solar Gain Correction	Steel Sash, or No Sash × 1/.85 or 1.17	Haze −15% (Max.)		Altitude +0.7% per 1000 Ft		Dewpoint Decrease From 67 F + 7% per 10 F			Dewpoint Increase From 67 F − 7% per 10 F					South Lat. Dec. or Jan. + 7%	

Bold Face Values — Monthly Maximums   Boxed Values — Yearly maximums

Figure 26-24 Solar heat gain through ordinary glass, 40 degree latitude. *(Courtesy McGraw-Hill Book Company)*

SOLAR HEAT GAIN THRU ORDINARY GLASS (Contd)

50°   Btu/(hr) (sq ft sash area)   **50°**

50° NORTH LATITUDE Time of Year	Exposure	6	7	8	9	10	11	Noon	1	2	3	4	5	6	Exposure	50° SOUTH LATITUDE Time of Year
		AM						SUN TIME					PM			
JUNE 21	North	29	12	12	13	14	14	14	14	14	13	12	12	29	South	DEC 22
	Northeast	126	125	94	50	16	14	14	14	14	13	12	10	8	Southeast	
	East	139	164	162	136	94	41	14	14	14	13	12	10	8	East	
	Southeast	64	102	126	135	124	98	61	23	14	13	12	10	8	Northeast	
	South	8	10	16	39	68	87	93	87	68	39	16	10	8	North	
	Southwest	8	10	12	13	14	23	61	98	124	135	126	102	64	Northwest	
	West	8	10	12	13	14	14	14	41	94	136	162	164	139	West	
	Northwest	8	10	12	13	14	14	14	14	16	50	94	125	126	Southwest	
	Horizontal	44	86	133	173	197	214	220	214	197	173	133	86	44	Horizontal	
JULY 23 & MAY 21	North	21	11	12	13	14	14	14	14	14	13	12	11	21	South	JAN 21 & NOV 21
	Northeast	114	117	87	44	15	14	14	14	14	13	12	10	6	Southeast	
	East	131	161	163	141	96	43	14	14	14	13	12	10	6	East	
	Southeast	65	107	134	143	136	109	70	26	14	13	12	10	6	Northeast	
	South	6	10	21	50	80	98	106	98	80	50	21	10	6	North	
	Southwest	6	10	12	13	14	26	70	109	136	143	134	107	65	Northwest	
	West	6	10	12	13	14	14	14	43	96	141	163	161	131	West	
	Northwest	6	10	12	13	14	14	14	14	15	44	87	117	114	Southwest	
	Horizontal	33	75	119	159	188	205	211	205	188	159	119	75	33	Horizontal	
AUG 24 & APR 20	North	8	8	10	12	13	14	14	14	13	12	10	8	8	South	FEB 20 & OCT 23
	Northeast	76	94	70	31	13	14	14	14	13	12	10	8	4	Southeast	
	East	94	145	158	141	98	45	14	14	13	12	10	8	4	East	
	Southeast	53	111	144	157	153	132	89	40	14	13	12	10	4	Northeast	
	South	4	9	36	73	105	130	138	130	105	73	36	9	4	North	
	Southwest	4	8	10	12	13	40	89	132	153	157	144	111	53	Northwest	
	West	4	8	10	12	13	14	14	45	98	141	158	145	94	West	
	Northwest	4	8	10	12	13	14	14	14	14	31	70	94	76	Southwest	
	Horizontal	13	46	89	131	160	179	185	179	160	131	89	46	13	Horizontal	
SEPT 22 & MAR 22	North	0	4	8	10	12	12	12	12	12	10	8	4	0	South	MAR 22 & SEPT 22
	Northeast	0	58	46	16	12	12	12	12	12	10	8	4	0	Southeast	
	East	0	102	138	130	93	43	12	12	12	10	8	4	0	East	
	Southeast	0	86	139	162	163	145	105	56	17	10	8	4	0	Northeast	
	South	0	11	51	93	131	150	158	150	131	93	51	11	0	North	
	Southwest	0	4	8	10	17	56	105	145	163	162	139	86	0	Northwest	
	West	0	4	8	10	12	12	12	43	93	130	138	102	0	West	
	Northwest	0	4	8	10	12	12	12	12	12	16	46	58	0	Southwest	
	Horizontal	0	15	49	88	118	140	148	140	118	88	49	15	0	Horizontal	
OCT 23 & FEB 20	North	0	0	4	7	9	10	11	10	9	7	4	0	0	South	APR 20 & AUG 24
	Northeast	0	29	20	7	9	11	11	10	9	7	4	0	0	Southeast	
	East	0	73	99	105	79	35	11	10	9	7	4	0	0	East	
	Southeast	0	69	111	145	157	144	115	69	24	7	4	0	0	Northeast	
	South	0	17	53	99	137	157	167	157	137	99	53	17	0	North	
	Southwest	0	0	4	7	24	69	115	144	157	145	111	69	0	Northwest	
	West	0	0	4	7	9	10	11	35	79	105	99	73	0	West	
	Northwest	0	0	4	7	9	10	11	10	9	7	20	29	0	Southwest	
	Horizontal	0	2	19	45	72	86	94	86	72	45	19	2	0	Horizontal	
NOV 21 & JAN 21	North	0	0	1	4	6	8	9	8	6	4	1	0	0	South	MAY 21 & JULY 23
	Northeast	0	0	5	4	6	8	9	8	6	4	1	0	0	Southeast	
	East	0	0	51	64	57	28	9	8	6	4	1	0	0	East	
	Southeast	0	0	62	95	127	127	107	67	21	4	1	0	0	Northeast	
	South	0	0	34	70	116	143	153	143	116	70	34	0	0	North	
	Southwest	0	0	1	4	21	67	107	127	127	95	62	0	0	Northwest	
	West	0	0	1	4	6	8	9	28	57	64	51	0	0	West	
	Northwest	0	0	1	4	6	8	9	8	6	4	5	0	0	Southwest	
	Horizontal	0	0	4	13	30	47	53	47	30	13	4	0	0	Horizontal	
DEC 22	North	0	0	0	3	5	6	7	6	5	3	0	0	0	South	JUNE 21
	Northeast	0	0	0	3	5	6	7	6	5	3	0	0	0	Southeast	
	East	0	0	0	27	47	23	7	6	5	3	0	0	0	East	
	Southeast	0	0	0	41	107	116	100	62	25	3	0	0	0	Northeast	
	South	0	0	0	31	99	131	141	131	99	31	0	0	0	North	
	Southwest	0	0	0	3	25	62	100	116	107	41	0	0	0	Northwest	
	West	0	0	0	3	5	6	7	23	47	27	0	0	0	West	
	Northwest	0	0	0	3	5	6	7	6	5	3	0	0	0	Southwest	
	Horizontal	0	0	0	5	19	33	40	33	19	5	0	0	0	Horizontal	

Solar Gain Correction	Steel Sash, or No Sash ×1/.85 or 1.17	Haze −15% (Max.)	Altitude +0.7% per 1000 Ft	Dewpoint Decrease From 67 F +7% per 10 F	Dewpoint Increase From 67 F −7% per 10 F	South Lat. Dec. or Jan. +7%

Bold Face Values — Monthly Maximums   Boxed Values — Yearly maximums

Figure 26-25 Solar heat gain through ordinary glass, 50 degree latitude. *(Courtesy McGraw-Hill Book Company)*

OVER-ALL FACTORS FOR SOLAR HEAT GAIN THRU GLASS

WITH AND WITHOUT SHADING DEVICES

Apply Factors to Figure 26–25

Outdoor wind velocity, 5 mph — Angle of incidence, 30° — Shading devices fully covering window

TYPE OF GLASS	GLASS FACTOR NO SHADE	INSIDE VENETIAN BLIND 45° horiz. or vertical or ROLLER SHADE			OUTSIDE VENETIAN BLIND 45° horiz. slats		OUTSIDE SHADING SCREEN† 17° horiz. slats		OUTSIDE AWNING‡ vent. sides & top	
		Light Color	Medium Color	Dark Color	Light Color	Light on Outside Dark on Inside	Medium Color	Dark § Color	Light Color	Med. or Dark Color
ORDINARY GLASS	1.00	**.56**	.65	.75	.15	.13	.22	.15	.20	.25
REGULAR PLATE (¼ inch)	.94	**.56**	.65	.74	.14	.12	.21	.14	.19	.24
HEAT ABSORBING GLASS††										
40 to 48% Absorbing	.80	**.56**	.62	.72	.12	.11	.18	.12	.16	.20
48 to 56% Absorbing	.73	**.53**	.59	.62	.11	.10	.16	.11	.15	.18
56 to 70% Absorbing	.62	**.51**	.54	.56	.10	.10	.14	.10	.12	.16
DOUBLE PANE										
Ordinary Glass	.90	**.54**	.61	.67	.14	.12	.20	.14	.18	.22
Regular Plate	.80	**.52**	.59	.65	.12	.11	.18	.12	.16	.20
48 to 56% Absorbing outside; Ordinary Glass inside.	.52	**.36**	.39	.43	.10	.10	.11	.10	.10	.13
48 to 56% Absorbing outside; Regular Plate inside.	.50	**.36**	.39	.43	.10	.10	.11	.10	.10	.12
TRIPLE PANE										
Ordinary Glass	.83	**.48**	.56	.64	.12	.11	.18	.12	.16	.20
Regular Plate	.69	**.47**	.52	.57	.10	.10	.15	.10	.14	.17
PAINTED GLASS										
Light Color	.28									
Medium Color	.39									
Dark Color	.50									
STAINED GLASS										
Amber Color	.70									
Dark Red	.56									
Dark Blue	.60									
Dark Green	.32									
Greyed Green	.46									
Light Opalescent	.43									
Dark Opalescent	.37									

Figure 26-26   Overall factors for solar heat gain through glass.
(Courtesy McGraw-Hill Book Company)

HEAT GAIN FROM PEOPLE

DEGREE OF ACTIVITY	TYPICAL APPLICATION	Metabolic Rate (Adult Male) Btu/hr	Average Adjusted Metabolic Rate Btu/hr	ROOM DRY-BULB TEMPERATURE									
				82 F Btu/hr		80 F Btu/hr		78 F Btu/hr		75 F Btu/hr		70 F Btu/hr	
				Sensible	Latent	Sensible	Latent	Sensible	Latent	Sensible	Latent	Sensible	Latent
Seated at rest	Theater, Grade School	390	350	175	175	195	155	210	140	230	120	260	90
Seated, very light work	High School	450	400	180	220	195	205	215	185	240	160	275	125
Office worker	Offices, Hotels, Apts., College	475	450	180	270	200	250	215	235	245	205	285	165
Standing, walking slowly	Dept., Retail, or Variety Store	550											
Walking, seated	Drug Store	550	500	180	320	200	300	220	280	255	245	290	210
Standing, walking slowly	Bank	550											
Sedentary work	Restaurant	500	550	190	360	220	330	240	310	280	270	320	230
Light bench work	Factory, light work	800	750	190	560	220	530	245	505	295	455	365	385
Moderate dancing	Dance Hall	900	850	220	630	245	605	275	575	325	525	400	450
Walking, 3 mph	Factory, fairly heavy work	1000	1000	270	730	300	700	330	670	380	620	460	540
Heavy work	Bowling Alley, Factory	1500	1450	450	1000	465	985	485	965	525	925	605	845

Figure 26-27 Heat gain from people. *(Courtesy McGraw-Hill Book Company)*

HEAT GAIN FROM ELECTRIC MOTORS

CONTINUOUS OPERATION*

NAMEPLATE† OR BRAKE HORSEPOWER	FULL LOAD MOTOR EFFICIENCY PERCENT	LOCATION OF EQUIPMENT WITH RESPECT TO CONDITIONED SPACE OR AIR STREAM‡		
		Motor In - Driven Machine in $\dfrac{HP \times 2545}{\% \text{ Eff}}$	Motor Out - Driven Machine in $HP \times 2545$	Motor In - Driven Machine out $\dfrac{HP \times 2545\,(1 - \% \text{ Eff})}{\% \text{ Eff}}$
		Btu per Hour		
1/20	40	320	130	190
1/12	49	430	210	220
1/8	55	580	320	260
1/6	60	710	430	280
1/4	64	1,000	640	360
1/3	66	1,290	850	440
1/2	70	1,820	1,280	540
3/4	72	2,680	1,930	750
1	79	3,220	2,540	680
1½	80	4,770	3,820	950
2	80	6,380	5,100	1,280
3	81	9,450	7,650	1,800
5	82	15,600	12,800	2,800
7½	85	22,500	19,100	3,400
10	85	30,000	25,500	4,500
15	86	44,500	38,200	6,300
20	87	58,500	51,000	7,500
25	88	72,400	63,600	8,800
30	89	85,800	76,400	9,400
40	89	115,000	102,000	13,000
50	89	143,000	127,000	16,000
60	89	172,000	153,000	19,000
75	90	212,000	191,000	21,000
100	90	284,000	255,000	29,000
125	90	354,000	318,000	36,000
150	91	420,000	382,000	38,000
200	91	560,000	510,000	50,000
250	91	700,000	636,000	64,000

*For intermittent operation, an appropriate usage factor should be used, preferably measured.

†If motors are overloaded and amount of overloading is unknown, multiply the above heat gain factors by the following maximum service factors:

Maximum Service Factors

Horsepower	1/20-1/8	1/6-1/3	1/2-3/4	1	1½-2	3-250
AC Open Type	1.4	1.35	1.25	1.25	1.20	1.15
DC Open Type	—	—	—	1.15	1.15	1.15

No overload is allowable with enclosed motors.

‡For a fan or pump in air conditioned space, exhausting air and pumping fluid to outside of space, use values in last column.

Figure 26-28 Heat gain from electric motors. *(Courtesy McGraw-Hill Book Company)*

5. Calculate the Btu/hr by multiplying the watts by 3.414.

6. Adjust the calculated Btu/hr for applications in which either the motor or its work do not show up as heat within the air conditioned space.

For appliances that are driven by electric motors, obtain the nameplate wattage, and multiply by 3.414. For other appliances, obtain the sensible and latent heat gain values from Figure 26-29.

Lights are handled the same as appliances. Simply multiply the wattage of the bulb by 3.414 to obtain the heat gain in Btu/hr. For fluorescent lights, the normal wattage is 10 W for each foot of tube length. Thus, a 4-ft tube would consume 40 W. There are energy-saving bulbs available that consume less, but it would be conservative to assume that they could, at some time, be re-

HEAT GAIN FROM MISCELLANEOUS APPLIANCES

NOT HOODED*

APPLIANCE	TYPE OF CONTROL	MISCELLANEOUS DATA	MFR MAX RATING Btu/hr	RECOM HEAT GAIN FOR AVG USE		
				Sensible Heat Btu/hr	Latent Heat Btu/hr	Total Heat Btu/hr
ELECTRIC						
Hair Dryer, Blower Type 15 amps, 115 volts AC	Man.	Fan 165 watts, (low 915 watts, high 1580 watts)	5,370	2,300	400	2,700
Hair Dryer, helmet type, 6.5 amps, 115 volts AC	Man.	Fan 80 watts, (low 300 watts, high 710 watts)	2,400	1,870	330	2,200
Permanent Wave Machine	Man.	60 heaters at 25 watts each, 36 in normal use	5,100	850	150	1,000
Pressurized Instrument Washer and Sterilizer		11″ x 11″ x 22″		12,000	23,460	35,460
Neon Sign, per linear ft tube		½″ outside dia ⅜″ outside dia		30 60		30 60
Solution and/or Blanket Warmer		18″ x 30″ x 72″ 18″ x 24″ x 72″		1,200 1,050	3,000 2,400	4,200 3,450
Sterilizer Dressing	Auto. Auto.	16″ x 24″ 20″ x 36″		9,600 23,300	8,700 24,000	18,300 47,300
Sterilizer, Rectangular Bulk	Auto. Auto. Auto. Auto. Auto. Auto. Auto.	24″ x 24″ x 36″ 24″ x 24″ x 48″ 24″ x 36″ x 48″ 24″ x 36″ x 60″ 36″ x 42″ x 84″ 42″ x 48″ x 96″ 48″ x 54″ x 96″		34,800 41,700 56,200 68,500 161,700 184,000 210,000	21,000 27,000 36,000 45,000 97,500 140,000 180,000	55,800 68,700 92,200 113,500 259,200 324,000 390,000
Sterilizer, Water	Auto. Auto.	10 gallon 15 gallon		4,100 6,100	16,500 24,600	20,600 30,700
Sterilizer, Instrument	Auto. Auto. Auto. Auto. Auto.	6″ x 8″ x 17″ 9″ x 10″ x 20″ 10″ x 12″ x 22″ 10″ x 12″ x 36″ 12″ x 16″ x 24″		2,700 5,100 8,100 10,200 9,200	2,400 3,900 5,900 9,400 8,600	5,100 9,000 14,000 19,600 17,800
Sterilizer, Utensil	Auto. Auto.	16″ x 16″ x 24″ 20″ x 20″ x 24″		10,600 12,300	20,400 25,600	31,000 37,900
Sterilizer, Hot Air	Auto. Auto.	Model 120 Amer Sterilizer Co Model 100 Amer Sterilizer Co		2,000 1,200	4,200 2,100	6,200 3,300
Water Still		5 gal/hour		1,700	2,700	4,400
X-ray Machines, for making pictures		Physicians and Dentists office		None	None	None
X-ray Machines, for therapy		Heat load may be appreciable— write mfg for data				
GAS BURNING						
Burners, Laboratory small bunsen	Man.	7⁄16 dia barrel with manufactured gas	1,800	960	240	1,200
small bunsen fishtail burner	Man. Man.	7⁄16 dia with nat gas 7⁄16 dia with nat gas	3,000 3,500	1,680 1,960	420 490	2,100 2,450
fishtail burner large bunsen	Man. Man.	7⁄16 dia bar with nat gas 1½ dia mouth, adj orifice	5,500 6,000	3,080 3,350	770 850	3,850 4,200
Cigar Lighter	Man.	Continuous flame type	2,500	900	100	1,000
Hair Dryer System 5 helmets 10 helmets	Auto. Auto.	Consists of heater & fan which blows hot air thru duct system to helmets	33,000	15,000 21,000	4,000 6,000	19,000 27,000

*If properly designed positive exhaust hood is used, multiply recommended value by .50

Figure 26-29 Heat gain from miscellaneous appliances.
(Courtesy McGraw-Hill Book Company)

placed with conventional bulbs. In addition to the bulb wattage, fluorescent lights use a transformer in the lighting fixture. It is called a ballast, and it will consume an amount of wattage equal to 25 percent of the fixture.

Example

A room has six fixtures, each with two 8-ft fluorescent tubes. How much heat gain will there be due to these fixtures?

Solution

Each fixture has a tube wattage of 160 W from the 16 ft of fluorescent tube. Multiply by 1.25 to account for the ballast, multiply by 3.414 Btu/W, and multiply by 6 for the six fixtures. The heat gain is

$$Q = 160 \times 1.25 \times 6 \times 3.414$$
$$= 4097 \text{ Btu/hr}$$

Figure 26-30 shows a system in which the fluorescent lights are recessed into a ceiling space. For applications such as these, the heat from the ballast should not be added to the room load, but it will be added to the total load seen by the air conditioning system. There are also specialized lighting fixtures that are water cooled. The heat from the lights can thus be removed directly. Manufacturer's data should be consulted for this special application.

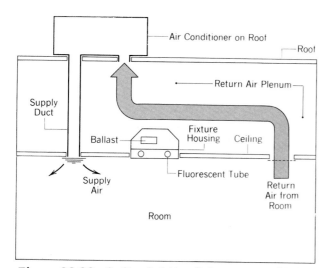

Figure 26-30 Ceiling lighting fixture recessed into a ceiling plenum.

Infiltration Heat Gain

The principles for calculating how much air infiltrates into an air conditioned or refrigerated space are the same as for the heating examples given earlier in the chapter. In cooling applications, however, the infiltration represents both a sensible and a latent load. Once the cfm of infiltration has been determined, the contribution to the room heat gain can be calculated as follows.

$$\text{Sensible heat gain} = 1.08 \times \text{cfm} \times \Delta T$$

and

$$\text{Latent heat gain} = 0.68 \times \text{cfm} \times \Delta H$$

where

ΔT = the temperature difference between indoor and outdoor design.

ΔH = the difference in grains of absolute humidity between outdoor and indoor design.

As with the internal loads from people, the sensible load and the latent load must be calculated separately.

Safety Factor

When all the heat gains or heat losses have been calculated, and the total loads calculated, it is common practice to add 10 percent to the total calculated load as a **safety factor.** The safety factor accounts for losses that may not have been anticipated, U factors that may not be precisely correct, construction that does not exactly correspond to the specified method for which the U factor was based, and reserve capacity for pulldown. The consequences of slightly oversizing the air conditioning or refrigeration system are not usually a problem. If a system is slightly oversized for a job, it will cycle off more frequently than a unit that exactly matches the cooling load or heating load.

A safety factor of more than 10 to 15 percent is not appropriate. Gross oversizing is not good design for the following reasons.

1. The first cost of the additional capacity is not justified by any operating benefits.

2. A grossly oversized unit will cycle on and off much too frequently, resulting in equipment wear and customer dissatisfaction.

3. For air conditioning applications, much closer control of humidity can usually be held when the cooling equipment is run on a more continuous basis.

Zones and Diversity

When there are distinctly different areas within a building where the heating or cooling load cannot be expected to rise and fall together, each area should be served by a separate portion of the air conditioning and heating system. Each of these areas is called a **zone.** Examples of areas that normally require separate zones are as follows.

1. Interior spaces (those with no walls or roofs exposed to outside air) must be zoned differently from exterior zones. There will be times when the interior zones require cooling at the same time that the exterior zones require heating. As a matter of fact, truly interior zones will never require heating, as long as all the surrounding areas are heated.

2. Where different tenants may desire to hold the space at different temperatures, each area requires a separate zone.

3. Different exterior exposures may require different zones, especially if there is a significant amount of glass. On a winter day, it may be normal for an east-facing exposure to require cooling in the morning while the west exposure needs heating, and for their requirements to reverse during the afternoon hours.

4. Where occupancies differ, different zones may be required. For example, a conference room where 20 people can be present for an hour or two requires it own zone.

Where there are different zones, each zone will have its own air-handling system. The heating or cooling system can also be individualized, or there can be a central heating and cooling system. The advantage to the central system over the individualized systems is attributable to diversity.

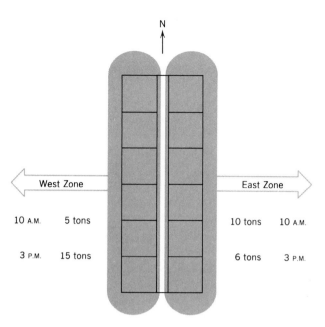

Figure 26-31 Diversity allows the total peak load to be less than the sum of the individual peak loads.

Diversity can be explained by looking at the building in Figure 26-31. During the summer the east half of the building experiences a peak load of 10 tons at 10 A.M. At that time of the morning, the west zone has a load of only 5 tons. During the afternoon, the load in the west zone increases to 15 tons due to the warmer outside air temperature and the location of the sun shining upon the west side of the building. Even though the temperature outside has increased, however, the east zone only requires 6 tons of cooling during the afternoon. If individualized cooling units were used, the owner would need to have 10 tons of capacity for the east zone, and 15 tons of capacity for the west zone for a total of 25 tons of refrigeration capacity. If a central system is used to supply cooling to wherever it is needed, however, it would only need to be 21 tons in capacity. That is the maximum total load that will occur at any time during the day.

Individual zones must have individual heating and cooling load calculations done. These calculations are used to size the individual air systems. If the heating and cooling capacity will be served from a central system, another overall calculation must be done for the entire building in order to determine the maximum simultaneous load.

Methods of Reducing the Load

There are all kinds of helpful hints lists available, each with methods claiming to drastically reduce the heating or cooling load. They may all be valid for some applications, but none are valid for all applications. For example, the addition of 4 in. of insulation may well be a smart investment for a residence with no existing insulation. For another house, however, which is identical in every respect except that it already has 12 in. of insulation, the addition of another 4 in. of insulation may be a very questionable investment.

The way to analyze methods for reducing the heating or cooling load is to do a heat loss or heat gain calculation. With this information, you will quickly see where most of the losses are happening, and steps can be taken to reduce them. If a ceiling represents 40 percent of a total heat loss and windows represent 10 percent, it would be smarter to invest money in insulation to cut the ceiling loss in half, even if the same amount of money could be used to buy window treatments that would virtually eliminate the loss due to the windows.

There are many products available for consideration. They all have good applications and marginal applications.

Window **shading** is very effective in reducing solar heat gain. Exterior shading is the most effective. Overhangs and awnings work well on southern exposure, but do very little for east or west exposures. Outside shading, which also provides wintertime shade, can increase winter heating bills thus offsetting the summer savings. Window tinting, reflective coatings, blinds, and shades all have varying degrees of effectiveness. Storm windows work well to reduce transmission losses, but will do little to reduce solar heat gain through the glass.

Insulation of walls or roofs or ceilings will always provide some measure of load reduction. It is a matter of calculating how much reduction is attained for how much cost, and making a decision of the return on the investment. Aluminum siding sales people have made claims of the insulation value attributable to their product. Only rarely would the marginal increase in insulating value justify the installation of the siding.

Caulking and weatherstripping to reduce infil-

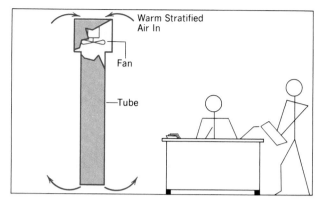

Figure 26-32 Where stratified ceiling tempeatures are 15°F or more higher than near the floor, a vertically ducted system reduces heating costs. Note that 250 cfm covers 2000 square feet, and 400 cfm covers 2500 square feet.

Figure 26-33 Heat recycler. *(Courtesy Environmental Products Company)*

tration is a wise investment whenever any reduction in the infiltration can be attained.

Figures 26-32 and 33 show a device used to reduce stratification in tall buildings. **Stratification** refers to the tendency of warm air to be buoyant and rise to the top of a building. Where the heating is provided for the comfort of the personnel, it is only the occupied area near the floor that is of interest. The tube and fan arrangement shown can be an effective way to capture

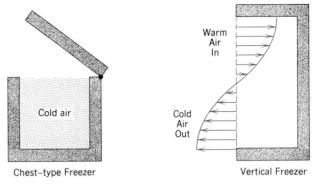

Figure 26-36 Vertical freezer allows the heavier, cold air to fall out each time the door is opened.

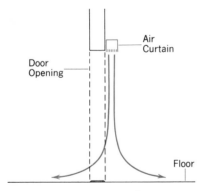

Figure 26-34 Strip door permits free access and reduces heat transfer between two rooms at different temperatures. *(Courtesy Environmental Products Company)*

Figure 26-35 Air curtain provides an air separation between rooms at different temperatures.

the heat that has stratified near the roof and deliver it to the occupied space near the floor.

Where cool air is being lost through traffic openings, there are several products offered. For walk-in boxes, the best method of reducing load is to reduce the amount of time the door stays open. The flexible strips of plastic shown in Figure 26-34 can be added to help contain the cold air. There are also **air curtain** fans (Figure 26-35)

that, if well designed, can form an air barrier that tends to reduce the loss of cool air when doors are open. A poorly designed air curtain can however, actually increase the cooling load by increasing the amount of mixing of cold and warm air. There are fly fans that look similar to air curtains. They are designed to keep flies out. They create a lot of air turbulence, but will not reduce cooling load. They can, in fact, increase the load by increasing the amount of mixing between conditioned and nonconditioned air.

Many grocery chains have discovered that enclosed reach-in boxes require far less refrigeration capacity than open types. Chest freezers, which hold the cold air even with the door open, are far better than vertical cabinets that allow all the cold air to fall out each time the door is opened (Figure 26-36).

Isolation of heat sources is a very practical way to reduce cooling load and cooling operating costs. If there is a large piece of machinery, an oven, a refrigeration unit, or any other large load within an air conditioned space, investigate an enclosure around the equipment that can be separately ventilated. Even if the unit cannot be practically enclosed, it may be possible to provide an exhaust of the warm air created by the machinery. The reduction in room load may well be more than the additional cooling load required to handle the increased requirement for outside air.

Quick Reference Load Estimating

The only accurate and reliable method to determine a heating or cooling requirement is to calculate the heat gain or heat loss. This accounts for the precise materials of construction, internal loads, and usage of the conditioned space. It is useful, however, to have some reference data handy to check against your calculated loads, or to use as rough estimating data. Use of load data for so-called normal applications requires careful judgment from the user. Any unusual factors for a space under consideration could cause the actual load to be quite different from the quick reference material.

Figures 26-37 and 38 present cooling loads for walk-in coolers. Correction factors are given to account for different insulation thicknesses and materials, and for different ambient temperatures. Variations in product load can cause actual loads to be different from these estimated loads.

Figure 26-39 gives compressor requirements for many types of residential refrigerators and freezers. This is particularly useful when it is necessary to replace a compressor that does not have a nameplate, or for which the manufacturer's data are unavailable. Simply find the arrangement that most closely fits your application and select the compressor hp required for the cubic feet of the refrigerator/freezer.

Figure 26-40 gives air conditioning requirements for various types of applications. These values have been based on many assumptions about building construction materials, number and placement of windows, lighting, and occupancy.

OUTSIDE SURFACE AREA QUICK-CHECK

L-ft.	D-ft.	COOLER SIZE (ft. sq.) H - 9 ft.	H - 12 ft.	L-ft.	D-ft.	COOLER SIZE (ft. sq.) H - 9 ft.	H - 12 ft.
8	6	343	432	30	14	1632	1896
10	6	408	504	32	14	1724	2000
12	6	468	576	34	14	1816	2104
14	6	528	648	36	14	1908	2208
16	6	588	720	38	14	2000	2312
18	6	648	792	40	14	2092	2416
20	6	708	864	42	14	2184	2520
22	6	768	936	16	16	1088	1280
24	6	828	1008	18	16	1188	1392
8	8	416	512	20	16	1288	1504
10	8	484	592	22	16	1388	1616
12	8	552	672	24	16	1488	1728
14	8	620	752	26	16	1588	1840
16	8	688	832	28	16	1688	1952
18	8	756	912	30	16	1788	2064
20	8	824	992	32	16	1888	2176
22	8	892	1072	18	18	1296	1512
24	8	960	1152	20	18	1404	1632
26	8	1028	1232	22	18	1512	1752
28	8	1096	1312	24	18	1620	1872
30	8	1164	1392	26	18	1728	1992
32	8	1232	1472	28	18	1836	2112
10	10	560	680	30	18	1944	2232
12	10	636	768	32	18	2052	2352
14	10	712	856	34	18	2160	2472
16	10	788	944	36	18	2268	2592
18	10	864	1032	20	20	1520	1760
20	10	940	1120	22	20	1636	1888
22	10	1016	1208	24	20	1752	2016
24	10	1092	1296	26	20	1868	2144
26	10	1168	1384	28	20	1984	2272
28	10	1244	1472	30	20	2100	2400
30	10	1320	1560	32	20	2216	2528
32	10	1396	1648	34	20	2332	2656
34	10	1472	1736	36	20	2448	2784
36	10	1548	1824	38	20	2564	2912
38	10	1624	1912	40	20	2680	3040
40	10	1700	2000	24	24	2016	2304
12	12	720	864	26	24	2148	2448
14	12	804	960	28	24	2280	2592
16	12	888	1056	30	24	2412	2736
18	12	972	1152	32	24	2544	2880
20	12	1056	1248	34	24	2676	3024
22	12	1140	1344	36	24	2808	3168
24	12	1224	1440	38	24	2940	3312
26	12	1308	1536	40	24	3072	3456
28	12	1392	1632	42	24	3204	3600
30	12	1476	1728	44	24	3336	3744
32	12	1560	1824	46	24	3468	3888
34	12	1644	1920	48	24	3600	4032
36	12	1728	2016	30	30	2880	3240
14	14	896	1064	34	30	3192	3576
16	14	988	1168	38	30	3504	3912
18	14	1080	1272	42	30	3816	4248
20	14	1172	1376	46	30	4128	4584
22	14	1264	1480	50	30	4440	4920
24	14	1356	1584	54	30	4752	5256
26	14	1448	1688	60	30	5220	5760
28	14	1540	1792				

Figure 26-37 Outside surface area quick-check for walk-in coolers.
(Courtesy Russell Coil Company)

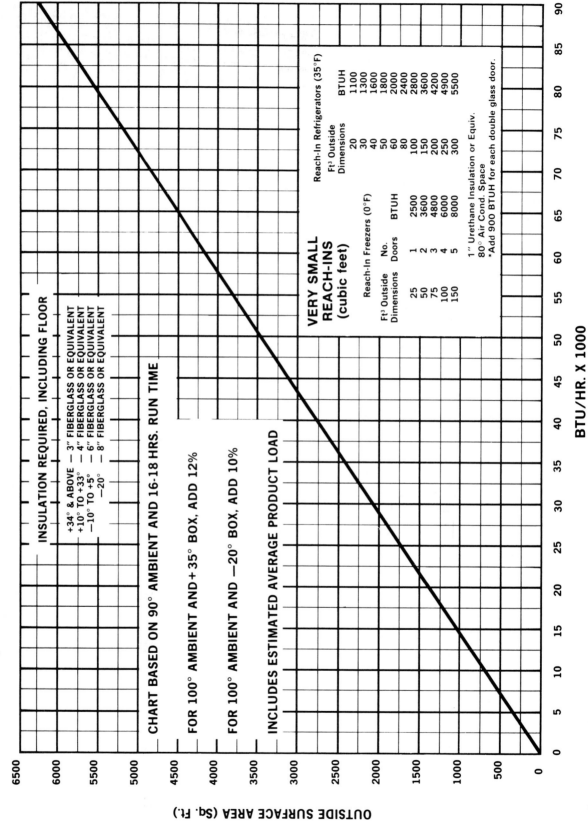

Figure 26-38 Quick reference chart for estimating small-box loads. *(Courtesy Russell Coil Company)*

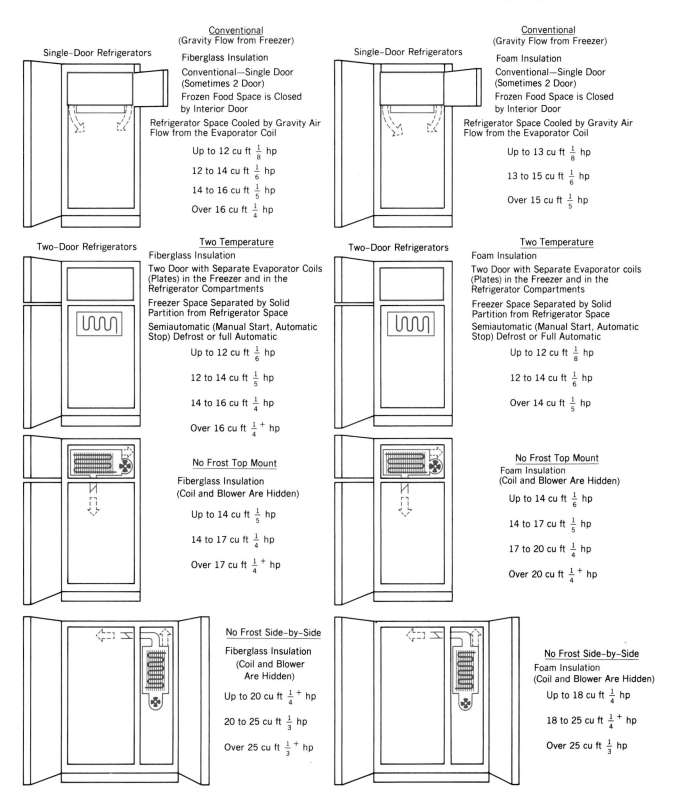

Single-Door Refrigerators

Conventional
(Gravity Flow from Freezer)

Fiberglass Insulation

Conventional—Single Door
(Sometimes 2 Door)

Frozen Food Space is Closed
by Interior Door

Refrigerator Space Cooled by Gravity Air
Flow from the Evaporator Coil

Up to 12 cu ft $\frac{1}{8}$ hp

12 to 14 cu ft $\frac{1}{6}$ hp

14 to 16 cu ft $\frac{1}{5}$ hp

Over 16 cu ft $\frac{1}{4}$ hp

Single-Door Refrigerators

Conventional
(Gravity Flow from Freezer)

Foam Insulation

Conventional—Single Door
(Sometimes 2 Door)

Frozen Food Space is Closed
by Interior Door

Refrigerator Space Cooled by Gravity Air
Flow from the Evaporator Coil

Up to 13 cu ft $\frac{1}{8}$ hp

13 to 15 cu ft $\frac{1}{6}$ hp

Over 15 cu ft $\frac{1}{5}$ hp

Two-Door Refrigerators

Two Temperature

Fiberglass Insulation

Two Door with Separate Evaporator Coils
(Plates) in the Freezer and in the
Refrigerator Compartments

Freezer Space Separated by Solid
Partition from Refrigerator Space

Semiautomatic (Manual Start, Automatic
Stop) Defrost or full Automatic

Up to 12 cu ft $\frac{1}{6}$ hp

12 to 14 cu ft $\frac{1}{5}$ hp

14 to 16 cu ft $\frac{1}{4}$ hp

Over 16 cu ft $\frac{1}{4}^{+}$ hp

Two-Door Refrigerators

Two Temperature

Foam Insulation

Two Door with Separate Evaporator coils
(Plates) in the Freezer and in the
Refrigerator Compartments

Freezer Space Separated by Solid
Partition from Refrigerator Space

Semiautomatic (Manual Start, Automatic
Stop) Defrost or Full Automatic

Up to 12 cu ft $\frac{1}{8}$ hp

12 to 14 cu ft $\frac{1}{6}$ hp

Over 14 cu ft $\frac{1}{5}$ hp

No Frost Top Mount

Fiberglass Insulation
(Coil and Blower Are Hidden)

Up to 14 cu ft $\frac{1}{5}$ hp

14 to 17 cu ft $\frac{1}{4}$ hp

Over 17 cu ft $\frac{1}{4}^{+}$ hp

No Frost Top Mount

Foam Insulation
(Coil and Blower Are Hidden)

Up to 14 cu ft $\frac{1}{6}$ hp

14 to 17 cu ft $\frac{1}{5}$ hp

17 to 20 cu ft $\frac{1}{4}$ hp

Over 20 cu ft $\frac{1}{4}^{+}$ hp

No Frost Side-by-Side

Fiberglass Insulation
(Coil and Blower
Are Hidden)

Up to 20 cu ft $\frac{1}{4}^{+}$ hp

20 to 25 cu ft $\frac{1}{3}$ hp

Over 25 cu ft $\frac{1}{3}^{+}$ hp

No Frost Side-by-Side

Foam Insulation
(Coil and Blower Are Hidden)

Up to 18 cu ft $\frac{1}{4}$ hp

18 to 25 cu ft $\frac{1}{4}^{+}$ hp

Over 25 cu ft $\frac{1}{3}$ hp

Figure 26-39 Compressor requirements for residential refrigerators and freezers.

Figure 26-39   *(continued)*

Freezers

Fiberglass Insulation

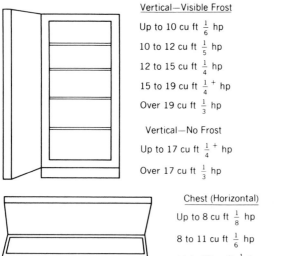

Vertical—Visible Frost

Up to 10 cu ft $\frac{1}{6}$ hp

10 to 12 cu ft $\frac{1}{5}$ hp

12 to 15 cu ft $\frac{1}{4}$ hp

15 to 19 cu ft $\frac{1}{4}^{+}$ hp

Over 19 cu ft $\frac{1}{3}$ hp

Vertical—No Frost

Up to 17 cu ft $\frac{1}{4}^{+}$ hp

Over 17 cu ft $\frac{1}{3}$ hp

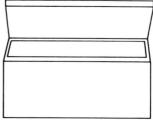

Chest (Horizontal)

Up to 8 cu ft $\frac{1}{8}$ hp

8 to 11 cu ft $\frac{1}{6}$ hp

11 to 13 cu ft $\frac{1}{5}$ hp

13 to 16 cu ft $\frac{1}{4}$ hp

16 to 20 cu ft $\frac{1}{4}^{+}$ hp

Over 20 cu ft $\frac{1}{3}$ hp

Freezers

Foam Insulation

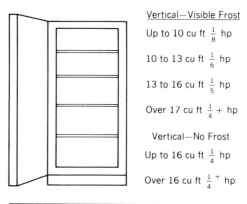

Vertical—Visible Frost

Up to 10 cu ft $\frac{1}{8}$ hp

10 to 13 cu ft $\frac{1}{6}$ hp

13 to 16 cu ft $\frac{1}{5}$ hp

Over 17 cu ft $\frac{1}{4}^{+}$ hp

Vertical—No Frost

Up to 16 cu ft $\frac{1}{4}$ hp

Over 16 cu ft $\frac{1}{4}^{+}$ hp

Chest (Horizontal)

Up to 10 cu ft $\frac{1}{8}$ hp

10 to 13 cu ft $\frac{1}{6}$ hp

13 to 16 cu ft $\frac{1}{5}$ hp

Over 17 cu ft $\frac{1}{4}^{+}$ hp

Apparel Store	200–300 sq ft/ton
Bank	150–300 sq ft/ton
Beauty Shop	100–200 sq ft/ton
Cold Storage Warehouse	350 sq ft/ton
Department	
Store—basement	225 sq ft/ton
first floor	225–250 sq ft/ton
upper floors	
Drug Store	250–275 sq ft/ton
Factory	200–300 sq ft/ton
Fast-Food Restaurant	150–200 sq ft/ton
Food Stores	150–300 sq ft/ton
Hotel Room	400–500 sq ft/ton
Meat Packing Chill	$\frac{3}{4}$ ton/room
Room	4000 lbs meat/24 hrs/ton
Offices—private	180–200 sq ft/ton
general	200–400 sq ft/ton
medical	150–200 sq ft/ton
Printing Shop	200–250 sq ft/ton
Residence—	500–600 sq ft/ton
house	
apartment	1 ton/3-room apartment
building	$1\frac{1}{2}$ ton/5-room apartment
Shoe Store	300–400 sq ft/ton
Theater	17 seats/ton

Figure 26-40   Reference cooling requirements for miscellaneous applications.

KEY TERMS AND CONCEPTS

Room Load

Transmission

Infiltration

Exfiltration

Design Conditions

U Factor

R Factor

Air Film

Calculated U Factors

Heat Loss through Basements

Heat Flow through Attics

Air Leakage

Air Changes

Vapor Barriers

Permeance

Sensible Heat Gain

Latent Heat Gain

Safety Factor

Zones

Load Diversity

Stratification

Air Curtain

Heat Isolation

QUESTIONS

1. What factors determine the amount of heat that will pass through a wall?

2. A wall with a U factor of 0.30 has insulation added with an R factor of 11. What is the new U factor of the wall?

3. In the formula $Q = U \times A \times \Delta T$, what is ΔT?

4. In the formula $Q = 1.08 \times \text{cfm} \times \Delta T$, what is ΔT?

5. There is a ΔT used in calculating heat gains, and a different ΔT used in calculating heat losses. What is the difference between those ΔT's?

6. What is meant by outdoor design conditions?

7. What is meant by indoor design conditions?

8. What ΔT is used to calculate heat loss through the floor over an unheated basement?

9. What ΔT is used to calculate heat loss through the ceiling below an unheated attic?

10. What problems can result from the improper installation of a vapor barrier in a wall?

11. Should the vapor barrier be installed on the warm side or the cold side of a wall?

12. What can happen if an air conditioning supply duct running through an attic is insulated without using a vapor barrier on the outside?

13. What factors must be considered for a heat gain calculation that are unnecessary for a heat loss calculation?

14. How is transmission for heat gain through an outside wall or roof different from the transmission through an interior partition separating the conditioned space from an adjacent non-air-conditioned space?

15. How does the color of a roof affect the amount of heat gain?

16. Why does the peak air conditioning load occur long after the peak outside temperature for the day has been reached?

17. Why is equivalent ΔT different from actual ΔT?

18. Which ΔT (equivalent or actual) will you use to calculate transmission heat gain through glass? Why?

19. A 10 hp motor located within the air conditioned space is driving an exhaust fan. The motor is operating at 80 percent of its full load capability. How much heat gain will be seen by the room?

20. Suppose 1000 cfm of outside air is infiltrating into an air conditioned space. The outside air is at 95°F dry bulb and 76°F wet bulb. The room is maintained at 74°F dry bulb and 50 percent relative humidity. How much are the sensible and latent heat gains from infiltration?

CHAPTER 27

BUILDING SYSTEMS

The objective of this chapter is to describe the common ways in which fans, ducts, heating coils, cooling coils, and dampers are used together in building systems. An understanding of the ways these systems are supposed to work is essential before proceeding to the chapter on pneumatic controls that are used to regulate these building systems.

Air Handlers

The term **air handler** refers to a number of different arrangements of fans, coils, and filters inside a sheet-metal box. Air handlers are commonly located inside mechanical equipment rooms or on roofs. They do not have a clean, finished appearance, as they are not designed to be placed within the occupied areas of a building.

The simplest air handler is simply a **cabinet fan** (Figure 27-1). A centrifugal fan is placed in a sheet-metal box (cabinet) and the cabinet is

placed in the ductwork. Filters may be added to the cabinet fan. Figure 27-2 illustrates a system used to introduce filtered outside air for ventilation only. This ventilation system will probably be operated by a manual switch or automatically by a thermostat whenever the space to be ventilated exceeds a setpoint temperature.

One of the conveniences of cabinet fans (as with all air handlers) is that the location of the air duct openings for supply and return air may be ordered to suit the installation. That is, the return air may be drawn in from the top, the back, or the bottom of the unit, and the fan discharge may also be chosen for direction of discharge. Figure 27-3 shows some of the variations.

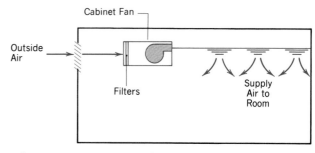

Figure 27-2 A cabinet fan with filters used to provide outside air for ventilation. The air that is supplied will be relieved through open doors or windows or through an exhaust fan.

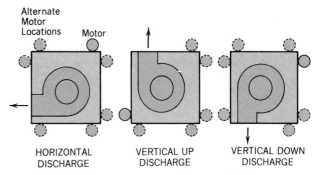

Figure 27-3 Fan sections can be supplied in many different configurations for the fan and motor.

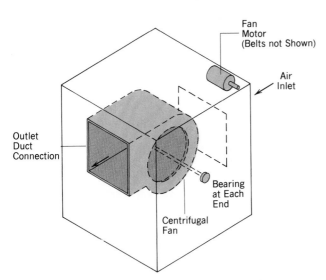

Figure 27-1 A cabinet fan is a centrifugal fan mounted inside a sheet-metal box.

Figure 27-4 shows the same cabinet fan with the addition of a heating coil. The heating coil provides heating for the air from a steam or hot water source. Electric heating coils are also available, but their use is limited due to higher operating costs. This air handler is called a heating and ventilating unit, or simply an **H & V unit.** With this system the fan runs continuously. As the outside air temperature drops, a room thermostat opens a steam or hot water control valve to prevent the room from becoming too cool.

Figure 27-5 expands our H & V unit by adding a cooling coil. The cooling coil may be the evaporator of a refrigeration system (a D-X coil), or for large buildings, the cooling coil will be one of many, in many air handlers that use chilled water produced from a remote water chiller. The air handler includes the drain pan under the

cooling coil to collect condensation. This air handler is called an HVAC unit. It is capable of providing heating, ventilation, air conditioning, and air filtration. The fan will run continuously. The room thermostat will open the steam (or hot water) control valve if the room becomes too cool. If the room becomes too warm, the steam valve is closed, and the chilled water valve is allowed to open.

The assortment of pieces that may be assembled into different air handlers is limitless. The designer of the building system may choose the type of fan wheel, the fan rpm, the motor size, the location of the motor, the physical arrangement and size of the fan, the number of rows deep for both the heating and cooling coils, the end on which the steam, hot water, and chilled water coils will be piped, the type and number of filters, access doors, and dozens of other options. The air handler is truly customized to meet the needs of the building in which it will be installed.

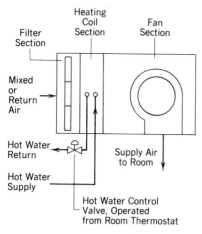

Figure 27-4   Heating and ventilating (H & V) air handler.

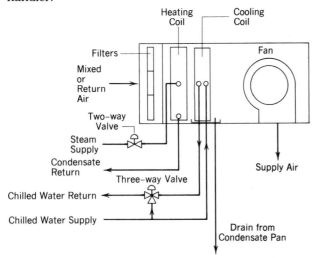

Figure 27-5   An HVAC air handler, using steam for heating and chilled water for cooling.

Fans

The types of fans available for use in air handlers are the centrifugal fans describd in Chapter 25. For most applications, the FC wheel is used, operating at a relatively low fan speed (500 to 1200 rpm). For applications requiring higher static pressures, the BI or AF fan wheels are used at higher rpm (1500 to 2000 rpm is common). All the fans are belt driven. The motors are commonly mounted on the outside of the fan section, with a belt guard covering the area between the fan sheave and the motor sheave (Figure 27-6). It

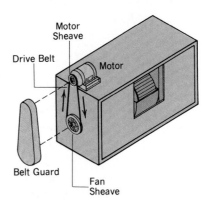

Figure 27-6   Air-handler drive components, including belt guard.

posed-blade damper sections provide better control characteristics.

Face and Bypass Dampers

Figure 27-18 shows an air handler with a method of controlling coil capacity without using a control valve. The **face and bypass damper** is actually two dampers linked together. When full heating (or cooling) is required, the damper section in front of the coil face is full open and the damper section in the bypass is shut. All the air passes through the coil. As the room demand for the coil capacity diminishes, a room thermostat causes a motor to move the face dampers toward the closed position while moving the bypass dampers to a more open position. Face and bypass dampers may be used on either heating or cooling coils. They are not, however, used as extensively as control valves. When a heating or cooling coil is used without a control valve, it is called a wild coil.

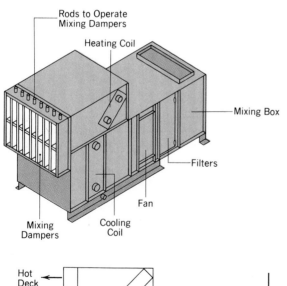

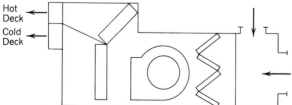

Figure 27-20 Multizone air handler.

Vertical and Horizontal Air Handlers

Air handlers may have their components arranged in either a **vertical** or **horizontal** arrangement. Figure 27-19 shows some examples of each type. Vertical units have the fan section located above or below the coil section, while the fan and coil sections are at the same level on horizontal units. Airflow through the heating and cooling coils is usually horizontal, although heating coils can be used with the face parallel to the ground. Cooling coils used for dehumidification are never installed with the face parallel to the ground, as there would be no way to remove the condensate that forms on the coil.

Note that the direction of air discharge has nothing to do with whether we call a unit horizontal or vertical. Vertical units may have air discharging horizontally, and horizontal units may have air discharging vertically.

Multizone Air Handlers

All the units described to this point are **single-zone** air handlers. That is, at any point in time, the discharge air can be at only one temperature. The discharge air temperature is controlled by a single thermostat located at one point within the occupied space. If there are different areas within a building that will require heating or cooling at different times, more than one single-zone air handler would be required.

The **multizone** air handler shown in Figure 27-20 has the capability of delivering air at several different temperatures through several different ducts, all at the same time. For example, there might be six different supply air ducts from the multizone, each with its own temperature. There would be six different room thermostats in six different locations, each one independently determining the supply air temperature required.

The multizone air handler is a **blow-through** unit as opposed to the **draw-through** single-zone units. Draw-through versus blow-through refers to the position of the fan relative to the coils. When air is drawn through the coil, a very even air distribution across the coil face results; therefore, draw-through units are preferable. Multizone units are blow-through, however. Return air and outside air are mixed together and supplied by the fan to a heating coil and a cooling coil in parallel. The hot and cold sections of the air handler downstream from the fan are called the **hot deck** and the **cold deck.**

The mixing damper section at the discharge of the multizone unit allows air from the hot deck and cold deck to be mixed together in proportions demanded by the room thermostat. For each duct connection to the multizone, there are at least one hot and cold deck damper. They work so that as one damper opens, the other closes. It is possible that one zone may be getting full heating (all hot deck air, no cold deck air) while, at the same time, a different zone may be receiving full cooling.

The advantage to a multizone system is that many different zones may be served from the installation of one air handler. This savings is reduced by high ductwork costs and high operating costs. Also, once installed, the multizone system is not very flexible to change.

In order to understand the high operating costs for multizones, consider the example shown in Figure 27-21. One zone requires heating, one zone requires cooling, and one requires nothing (the 65°F mixed air temperature is exactly right to meet the room needs). The 65°F mixed air splits up, some going through the hot deck where it is heated to 120°F, and the remainder going through the cooling coil where it is cooled to 56°F. The 120°F hot deck air and 56°F cold deck air are then mixed together in the mixing section in order to deliver 65°F air to the zone. While the energy waste may seem hard to believe, there are thousands of multizone systems in use today.

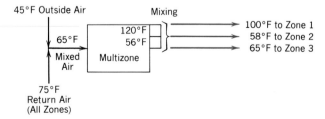

Figure 27-21 Wasted energy with a multizone air handler.

Double-Duct Systems

Figure 27-22 shows a system very similar in concept to the multizone system. The air handler is identical to the multizone, except that there is no mixing damper section. The hot deck discharges into one duct, and the cold deck discharges into a second duct. These two ducts are routed around a building. Every place where a separate zone of control is required, a mixing box is provided (this is not the same as the mixing box used at the inlet of an air handler). The mixing box (Figure 27-23) has an inlet from the cold deck, an inlet from the hot deck, and the internal damper that controls the amount of each that mixes together. The proportion of mixing is controlled by the room thermostat that operates a damper motor on the box.

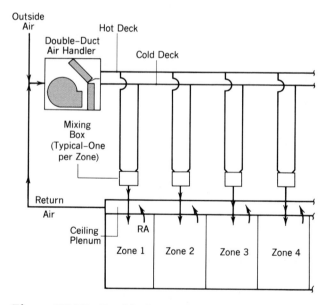

Figure 27-22 Double-duct system.

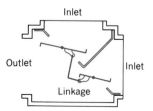

Figure 27-23 Mixing box used at each zone in a double-duct system.

The **double-duct system** has the same high operating costs as the multizone. But it also has the advantage of being very flexible for change and able to handle many zones of control. A double-duct system could handle 50 or more zones, while the multizone systems are usually limited to 8 to 10 zones.

Terminal Reheat

The **terminal reheat** system is shown in Figure 27-24. A single-zone air handler (cooling only) is used to distribute air to the entire occupied area. Each place where a separate zone of control is required, a reheat coil is installed in the duct. The operating costs for this system are high because all the air must be cooled down to the temperature required by the warmest zone. Usually, all the air is cooled to between 54°F and 58°F, and then each zone reheats the air to the desired temperature.

The installed cost of the reheat system is also

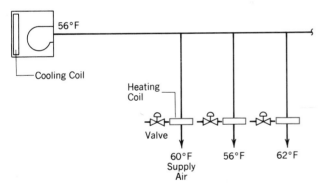

Figure 27-24 Terminal reheat system.

quite high due to all the steam or hot water piping required to the various reheat coils. Electric reheat coils may also be used, with the accompanying high wiring costs and high electrical operating costs. In spite of all the negatives, the terminal reheat system is used because it provides excellent control of room temperatures.

Variable Air Volume

With increased emphasis on building systems that would operate with more energy efficiency, variable air volume (VAV) systems became popular in the 1970s. The original application was for large buildings, in areas that required cooling on a year-round basis. A single-zone air handler supplies air at 55°F to the VAV terminal units. Each VAV terminal, which is controlled by a room thermostat, squeezes down on the supply-air quantity when the cooling load is low. This is not the same as a conventional system that merely has a damper installed in the ductwork to reduce the airflow. Conventional diffusers are designed for a relatively narrow range of airflow. If a duct were to supply a diffuser with less than its normal air quantity, the diffuser would simply dump the cold air instead of causing it to diffuse and mix with the room air (Figure 27-25). Specially designed diffusers and boxes are used in VAV systems to handle the low flow (Figures 27-26 through 28).

Figure 27-29 shows another method of reducing the room airflow. Instead of a squeeze-off box, the VAV terminal handles a constant volume of air. The air is either supplied to the room or it is bypassed into the ceiling plenum where it will

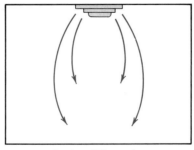

Figure 27-25 A conventional diffuser will dump air into the room without proper mixing if the supply airflow is too low.

Figure 27-26 Shut-off control box for variable air volume. *(Courtesy The Trane Company, LaCrosse, Wis.)*

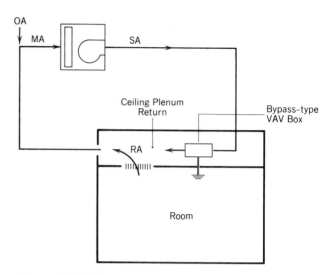

Figure 27-29 Variable air volume control diffuser. *(Courtesy The Trane Company, LaCrosse, Wis.)*

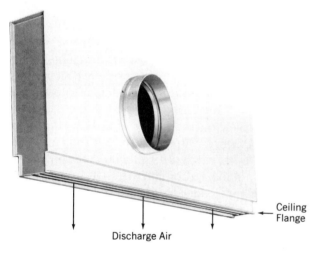

Figure 27-27 Variable air volume slot diffuser. *(Courtesy The Trane Company, LaCrosse, Wis.)*

return to the air handler. This system maintains the basic advantages of the VAV system by preventing the mixing of different temperature airstreams and preventing the cooling and reheating of air.

VAV systems have evolved so that terminal units are now available with reheat coils installed. They can be installed in systems where there will sometimes be a requirement for heating. But when all zones require cooling, they provide the full benefit of a VAV system. A perimeter heating system (Figures 27-30 and 31) can also

Figure 27-28 Bypass VAV box has a constant cfm input. Unneeded air is bypassed directly into the ceiling plenum.

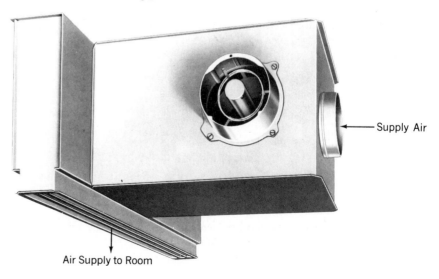

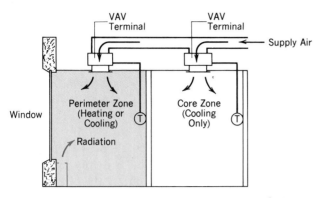

Figure 27-30 VAV system with perimeter radiation.

Figure 27-31 Baseboard heater (perimeter radiation). *(Courtesy Weil-McLain Co.)*

be used instead of reheat coils for the zones that require occasional heating.

Outside Air Economizer Cycle

In the chapter on load calculations, it was noted that the effect of outside air is to drastically increase the total cooling load. This is true for equipment sizing, but is not always true during operation. There are many times when the temperature of the outside air is actually less than the temperature of the room air that is returning to the air handler. The **outside air economizer cycle** is a method of using extra outside air whenever it will reduce the total cooling load.

Figure 27-32 shows a typical outside-air economizer system. It has a return-air fan, extra dampers, and a large outside-air intake. During the warm summer days, the system operates the same as a conventional system. The outside-air damper is close to the closed position, admitting only the minimum required outside-air quantity. Most of the air that is drawn into the air handler is return air. The exhaust-air dampers are almost closed and the return-air damper is open.

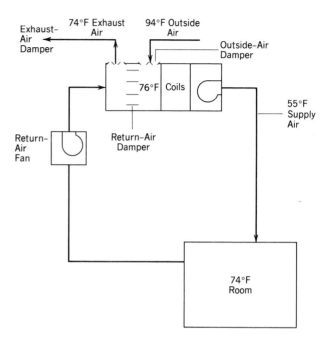

Figure 27-32 Economizer system operating at high outside air temperature, using 10 percent outside air.

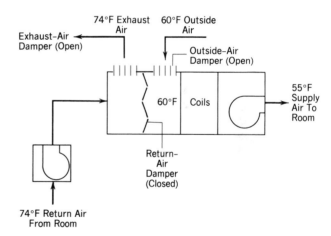

Figure 27-33 Economizer dampers using 100 percent outside air. The cooling coil must only provide enough cooling to drop the air temperature by 5°F.

In Figure 27-33 the outside-air temperature has dropped to 60°F. It is not cool enough to do all the cooling required, but it is cooler than the return air from the room. The dampers have readjusted their positions so that the air handler is now handling 100 percent outside air. The load on the cooling coil has been reduced significantly because it is seeing 60°F air instead of the 72.6°F mixed-air temperature it would be seeing without the economizer cycle (10 percent outside air). You can see in this mode the necessity for the

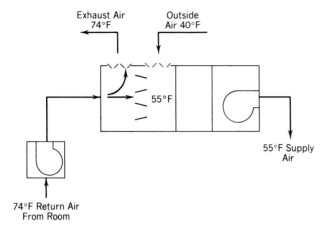

Figure 27-34 When outside air temperature is low, the economizers admit only enough outside air to satisfy the room cooling demand.

return-air fan. In order to allow 100 percent outside air into the building, some means must be provided to let an equal quantity of air to exhaust from the system. Without the return-air fan, the building could become quite pressurized.

As the outside air temperature drops lower than the temperature required by the room, the dampers admit only the amount of cold outside air required by the room (Figure 27-34).

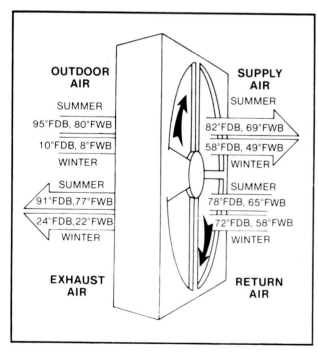

Figure 27-35 Heat wheel preheats outside air during the winter, and precools outside air during summer. *(Courtesy Cargocaire Engineering Corp.)*

Heat Recovery Systems

Some buildings have a requirement for large quantities of outside air to make up for large quantities of exhaust air. This could include hospitals, laboratories, spray booths, factories, or other applications where the room air becomes contaminated and must be thrown away. One method used to recover the heating or cooling that was done to the exhaust air is the **heat transfer wheel** (Figure 27-35). During the winter months the exhaust air heats the media in the rotating wheel. When that portion of the wheel rotates through the outdoor-air intake duct, it gives up its heat to the colder outside air. Other **heat recovery** units have fixed (nonrotating) heat transfer surfaces. They cannot, however, recover the moisture that can be transferred by the rotating wheel.

Water Systems

In offices, motels, and apartments, there are a great number of zones required (possibly as many as one zone for each room). For these applications, **fan-coil** units (Figures 27-36 through 38) are used for each zone. Chilled water or hot water is supplied to the coil. The controls at the cabinet (or mounted remotely on the wall) provide On–Off control to a water valve and speed control on the fan. Fan coils are commmonly mounted on or above the ceiling or under a window. Fan-coil units can be supplied with two coils and two valves (Figure 27-39) for use with four-pipe systems (hot water and chilled water available simultaneously).

Humidification

In cold climates the indoor humidity levels can drop to 5–10% rh. This dryness is uncomfortable for some people to breathe, can cause problems with static electricity, and can cause damage to wood and leather products. At higher levels of relative humidity, people tend to be more comfortable at a lower room temperature. The poten-

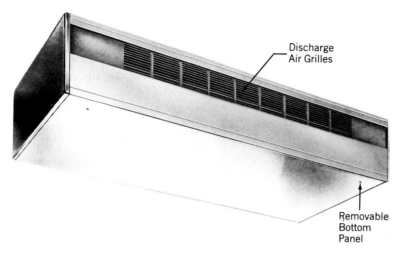

Discharge
Air Grilles

Removable
Bottom
Panel

Figure 27-36 Horizontal cabinet fan-coil unit. *(Courtesy The Trane Company, LaCrosse, Wis.)*

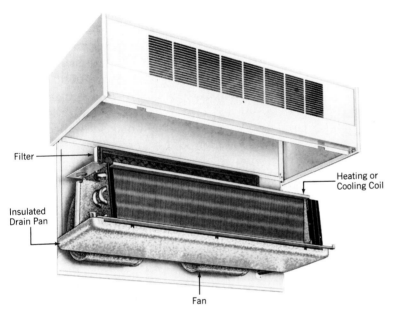

Filter

Insulated
Drain Pan

Heating or
Cooling Coil

Fan

Figure 27-37 Horizontal fan-coil unit, internal view. *(Courtesy The Trane Company, LaCrosse, Wis.)*

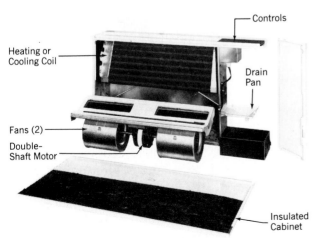

Controls

Heating or
Cooling Coil

Drain
Pan

Fans (2)

Double-
Shaft Motor

Insulated
Cabinet

Figure 27-38 Vertical cabinet fan-coil unit. *(Courtesy The Trane Company, LaCrosse, Wis.)*

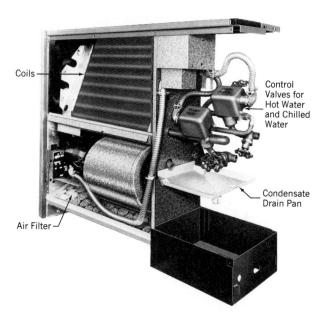

Coils

Control
Valves for
Hot Water
and Chilled
Water

Condensate
Drain Pan

Air Filter

Figure 27-39 Four-pipe fan-coil unit has two coils, two control valves, and can provide heating or cooling at any time.

tial heating cost reduction may more than offset the operating cost of a humidification system.

Two methods are used to add humidity: water evaporation and steam injection. Evaporation can be accomplished by presenting a wetted surface to a warm air stream (Figure 27-40). The wetted surface may be either a brush, pads, or rotating discs (Figure 27-41). As water evaporates into the airstream, solids are left behind. These units should have provision for bleed-off to a drain to prevent accumulation of the solids. The pressure difference between a supply- and return-air duct may be used to push air through the humidifier (Figure 27-42). These humidifiers are capable of adding 10–20 gal of water per day if they run continuously. They only provide humidity when the furnace is operating, however.

Pan humidifiers are self-contained units that use a heat source (electric coil or steam coil) to vaporize water from a pan. These are used where there is no heating system that can be used as the heat source for the evaporation.

Industrial applications requiring very high capacity use **steam-injection humidifiers** (Figure 27-43). Steam that is created in a remote boiler is injected directly into an airstream in a duct or directly into the occupied space.

Figure 27-40 Humidifier installed in a vertical heating duct. *(Courtesy Research Products Corporation)*

Media Drum

Motor Inside
Drives Media
Drum

HERRMIDIFIER
447

Figure 27-41 Humidifier with rotating media for duct installation. *(Courtesy Herrmidifier Company, Inc.)*

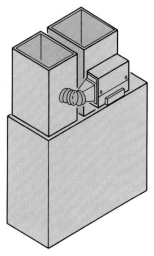

Figure 27-42 Bypass humidifier uses the pressure difference between the supply and return duct to move air across a wetted pad.

Steam Hose Outlet to
Distribution Pipe in Duct

Figure 27-43 Electronic steam-generating humidifier includes a steam cylinder, electric controls, and electrodes. The steam discharge may be to a duct or directly to the room. *(Courtesy Herrmidifier Company, Inc.)*

Humidification can also be accomplished by creating a very fine water mist that readily evaporates into the room air. Industrial applications uses compressed air to help atomize the water (Figures 27-44 and 45).

The capacity of the humidifier is controlled from a humidistat that senses room relative humidity. A relative humidity set point of 30–40% is desirable for many people and products. Humidity inside the occupied space, however, can cause condensation to form on windows. Figure 27-46 shows the maximum relative humidity that can be maintained in a room without allowing condensation on the windows.

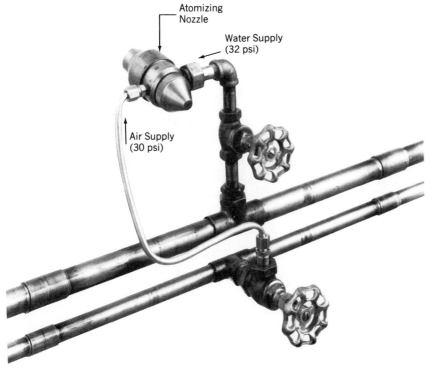

Atomizing
Nozzle

Water Supply
(32 psi)

Air Supply
(30 psi)

Figure 27-45 Air/water atomizing nozzle for humidification. *(Courtesy Herrmidifier Company, Inc.)*

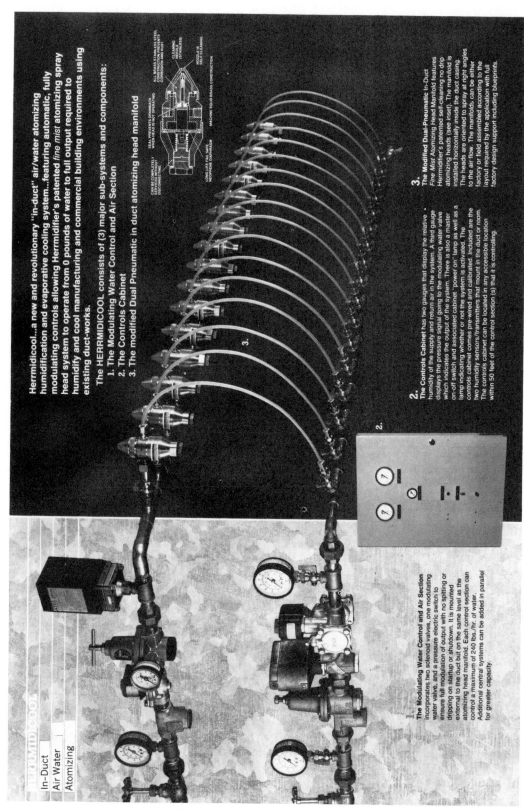

Figure 27-45 Installed bank of air/water atomizing nozzles with controls.
(Courtesy Herrmidifier Company, Inc.)

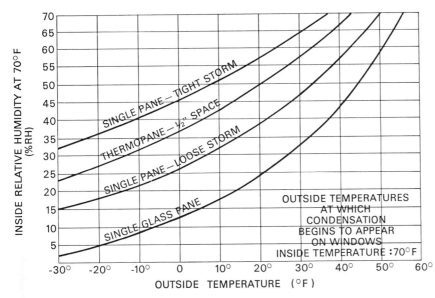

Figure 27-46 Maximum relative humidities without causing condensation on windows. *(Courtesy Research Products Corporation)*

KEY TERMS AND CONCEPTS

Air Handler

Cabinet Fan

H & V Unit

Coil Piping

Flat, Offset, V-Bank

Roll Filter

Prefilter

Bag Filter

HEPA

Mixing Box

Face and Bypass
   Damper

Vertical and
   Horizontal AHUs

Multizone

Draw-through

Blow-through

Hot Deck

Cold Deck

Double-duct System

Terminal Reheat

VAV

Economizer

Heat Recovery

Fan Coil

Humidifiers

QUESTIONS

1. What is a cabinet fan?

2. What types of evaporator coils are used in HVAC units?

3. What types of coils are available for heating in H & V units?

4. On which side (air entering or air leaving) of a cooling coil should chilled water be supplied? Why?

5. What are the common materials of construction for heating and cooling coils?

6. An air handler with a flat filter section measures 48 in. × 96 in. in cross section. Based on allowable face velocities, what airflow would you estimate is being handled by the air handler?

7. How are filters changed when a row of six filters across has access doors on each end?

8. How does a roll filter motor know when to advance the media?

9. What is a HEPA filter?

10. What two air streams are normally mixed in a mixing box?

11. How is a face and bypass damper arrangement used to control coil capacity?

12. Why are cooling coils not installed with the face parallel to the ground?

13. How does a multizone air handler supply different temperatures of supply air to different zones?

14. What is the disadvantage of using fans that blow through a coil versus drawing air through a coil?

15. Can a multizone unit provide heating to one zone at the same time as it is providing cooling to another zone?

16. What is the difference between a multizone system and a double-duct system?

17. What type of air handler is used in a terminal reheat system?

18. How does the variable air volume system save operating costs?

19. What additional components are required in an economizer system?

20. When does a system go into the economizer mode of operation?

CHAPTER 28

PNEUMATIC CONTROLS

Pneumatic (noo-matic) controls deal with the operation of valves and dampers in a system, all of which are powered by compressed air. The purpose of this chapter is to allow you to read pneumatic controls diagrams, and set up and calibrate the common pneumatic controllers.

Proportional Control—An Analogy

When you studied refrigeration sytems, all control was accomplished in what is called On–Off fashion. That is, control of the temperature in a box was controlled by turning the compressor on and off, or loading and unloading compressor cylinders, or opening and closing hot gas bypass valves. The opposite of **On-Off control** is **modulating control**. A chilled water coil may be controlled by modulating a chilled water valve. It may be fully open, fully closed, or at any position in between. The coil may be producing anything between zero and 100 percent of its rated capacity.

Figure 28-1 shows a simple system that uses modulating control. A float valve is maintaining the level in a bucket by controlling the amount of water that is allowed to enter the make-up water line. Let's assume a steady state condition. The water level is at the desired set point, 10 gpm of water is exiting through a valve, and 10 gpm of water is filling the bucket through the make-up water valve. Everything is in equilibrium.

Now let's say that somebody comes along and readjusts the valve on the leaving water line, so that the water flow rate is reduced to 8 gpm. The level of the water in the bucket will begin to rise, and as it rises, it will lift the float. The more the float rises, the more the make-up water flow rate will be reduced, until we reach a new equilibrium position, as shown in Figure 28-2. The valve will now once again maintain a constant level and a constant flow rate, although both the level and the flow rate are not exactly the same as they were. The difference between the new level and the desired level is called **offset.** The originally desired level is called the **set point.** The only way that the flow rate through the valve will be any different than the 10 gpm is if the actual level is different from the set point. And the more the level is different from the set point, the more the flow will be different (higher or lower) than 10 gpm. This is called proportional control. That is, the amount of correction in the flow rate made by the float valve is **proportional** to the amount of offset sensed by the float.

The **sensitivity** of the controller refers to how much correction is made by the controller in re-

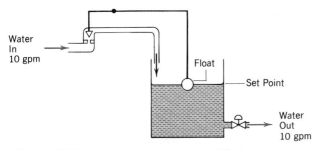

Figure 28-1 A water system in equilibrium.

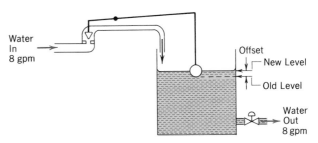

Figure 28-2 With reduced water outflow, a new, higher level is attained, which causes the inflow to match the outflow.

sponse to a given amount of offset. For example, let's say that in our example the original level we are trying to hold in the bucket is 12 in., and when the level is at 12 in., the flow rate is 10 gpm. When the level rises to 11 in. (because of reduced flow out of the bucket), the float adjusts the flow to 8 gpm. The complete range of control will be as follows.

Flow Rate (gpm)	Water Level in.
0	17
2	16
4	15
6	14
8	13
10	12
12	11
14	10
16	9
18	8
20	7

In this example, a total change of 10 in. of level is required in order to make the control go through its full range of output (zero to 20 gpm). The sensitivity for this system is 2 gpm/in. (two gpm per inch). In other words, each additional change in level of 1 in. causes a change in the controlled flow rate of 2 gpm.

If we want to adjust the float valve so that it can control the level at closer to 12-in. as the leaving water flow changes from 0 to 20 gpm, we can adjust the sensitivity. Figure 28-3 shows the same float valve, adjusted so that less change in level is required to stroke the valve. Now a change

of level of 1 in. will cause a change in flow through the valve of 4 gpm. The entire range of potential flow (0 gpm to 20 gpm) will occur as the level in the bucket changes from 9.5-in. to 14.5-in. As we increase the sensitivity, we maintain the level closer to the set point throughout the entire range of control. If we set the sensitivity too high, however, a small change in level will cause an overcorrection in the valve position. This will leave the valve changing from full open to full closed again. This condition is called **hunting.**

The analogy of maintaining a water level in a bucket is very close to the systems used to maintain temperature in a space. The level of water is like the temperature in the room. The flow rate of water leaving the bucket is like the heat gain or heat loss from the room. The inlet water is like the heating or cooling that must be added by the heating and air conditioning system in order to maintain the temperature in the room at close to the set point. The concepts of offset, proportional control, sensitivity, and hunting are the same.

Proportional Control of Temperature

A **pneumatic thermostat** (Figure 28-4) can be used to operate a pneumatic control valve (Figure 28-5). These two devices may be used to create the simplest of pneumatic temperature control systems.

The thermostat has two connections. One connection is supplied with compressed air at a pressure of 20 psi. This is called main air, and is supplied to all the pneumatic controls in the building from an air compressor in the mechanical room. The thermostat is located in the room and senses room temperature. When the room temperature is the same as the setpoint temperature of the thermostat, the **output pressure** is 9 psi. As the temperature of the room rises above the thermostat setpoint, the output pressure will increase. A falling room temperature will cause the output pressure to decrease. A thermostat whose pressure increases as the sensed temperature increases is called **direct acting** (D.A.). Some thermostats are **reverse acting** (R.A.). That is, as the sensed temperature goes down, the output pressure goes up, and as the

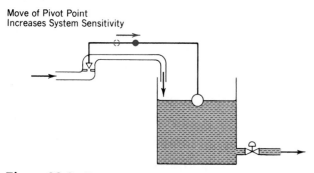

Move of Pivot Point
Increases System Sensitivity

Figure 28-3 By moving the pivot point to the right, the system sensitivity is increased.

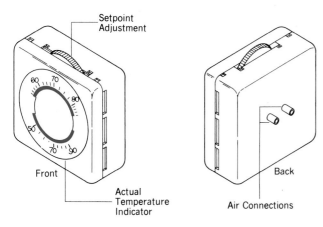

Figure 28-4 Pneumatic room thermostat.

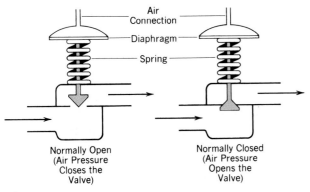

Figure 28-5 Pneumatic valve.

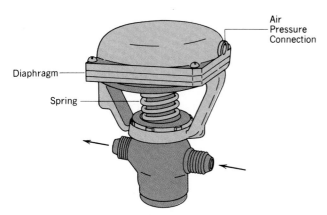

Figure 28-6 Normally open and normally closed pneumatic control valves.

sensed temperature goes up, the output pressure goes down. The useful output range of pressures from the thermostat is 3 psi to 15 psi.

Now let's take a look at the pneumatic control valve. The valve stem is attached to a diaphragm that can move up or down, opening or closing the valve. Air pressure may be applied to the top of

the diaphragm to move the valve stem down. When the air pressure is reduced, the spring pressure takes over and moves the valve stem up. The amount of air pressure required to move the valve stem depends upon the spring that is used. **Spring range** is the term used to describe the charactaristic of the spring. If the valve has a spring range of 4 to 8 psi, this means that the valve stem will just begin to move down when the air pressure reaches 4 psi. As the pressure increases, the valve stem will move downward. If the air pressure reaches 8 psi, the valve stem will reach the end of its travel.

Control valves are classified as **normally open** or **normally closed** (abbreviated n.o. and n.c.). A schematic of the internal construction of each is shown in Figure 28-6. When air pressure is applied to a normally open valve, it will move toward the closed position. A normally closed valve will be opened when sufficient air pressure is applied. The normal position of the valve is determined by its position when there is no air pressure applied. You cannot tell whether a control valve is normally open or normally closed by merely looking at the outside. The stems of each type will move downward as air pressure is increasing. But the normally open valve is closing, and the normally closed valve is opening.

The valves shown in Figure 28-6 are called **two-way valves.** There is one inlet and one outlet. Figure 28-7 shows **three-way valves.** Three-way valves may be either mixing valves or diverting valves. A mixing valve is designed to have two inlets and a common outlet. A diverting valve has a common inlet, and two outlets. The three-way valves have identification on all three ports. One connection is called common. Of the two remaining connections, one is normally open and one is normally closed. Mixing valves are more widely used than diverting valves because they are less costly.

Figure 28-8 is a simple pneumatic control diagram. The room thermostat is direct acting. Its output pressure goes to a normally open (n.o.) steam valve, which controls the flow of steam to a heating coil. The spring range of the valve is 3 to 15 psi. The valve will be fully open at 3 psi, halfway open at 9 psi, and closed at 15 psi. The sensitivity of the thermostat is 3 psi/°F. As the room temperature falls, the output pressure from the thermostat will be reduced, and the steam

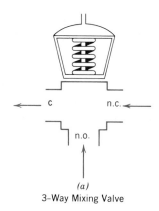

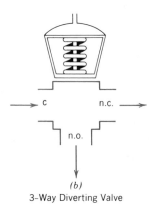

Figure 28-7   Three-way water control valves.

valve will move toward a more open position. The increased steam flow will keep the room temperature from falling further.

Example

In the pneumatic control diagram shown in Figure 28-8, how much change in room temperature will be required in order to cause the steam valve to move from its fully open position to its fully closed position?

Solution

A change of 12 psi (from 3 psi to 15 psi) is required to fully stroke the valve. With a thermostat sensitivity of 3 psi/°F, a change in temperature of 4°F will cause a change in output pressure of 12 psi. The room temperature will be maintained within 2°F above or below the set point.

The graph in Figure 28-9 is another way of looking at the operation of this system. When the room temperature is 2°F below the set point, the output pressure is 3 psi and the steam valve is

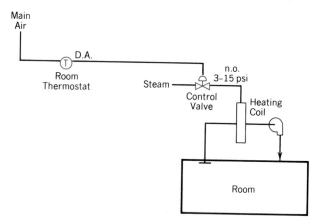

Figure 28-8   As room temperature falls, the thermostat output pressure decreases, causing the steam valve to open.

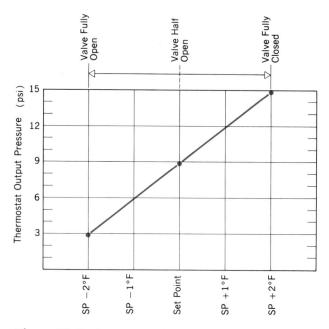

Figure 28-9   Operation of a thermostat (sensitivity of 3 psi/°F) controlling a normally open steam valve (spring range 3–15 psi)

fully open. When the room is at its set point, the thermostat output pressure is 9 psi and the valve is in a partially open position. If the room gets to 2°F higher than the set point, the valve will be fully closed.

One-Pipe and Two-Pipe Systems

The thermostat described in the previous section is called a **two-pipe** thermostat because it has

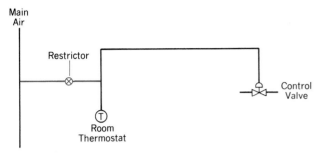

Figure 28-10 One-pipe thermostat piping.

two connections. Some thermostats are **one-pipe.** They are piped into a system as shown in Figure 28-10. The main air is supplied through a **restrictor** into the branch pressure line, which controls the steam valve. The one-pipe thermostat will bleed out the pressure as required in order to accomplish the same result as with a two-pipe thermostat. The one-pipe system has the advantage of requiring only one tube to be

routed to the device. The two-pipe thermostat has the advantage of being faster reacting. If the set point of the thermostat is lowered (creating the same effect as if the room temperature suddenly increased), the two-pipe thermostat will be able to immediately respond by putting out a high pressure signal to the valve. The one-pipe system would take longer for a sufficient quantity of air to bleed through the restricted supply, pressurizing the branch line.

Controlling Valves in Sequence

Figure 28-11 shows a heating and air conditioning system with a chilled water coil and a hot water coil. The spring range on the normally open hot water valve is 3 to 9 psi. The spring range on the chilled water valve is 9 to 15 psi. Both valves receive the same output pressure from the thermostat. The graph in Figure 28-11 shows how, as the room temperature increases

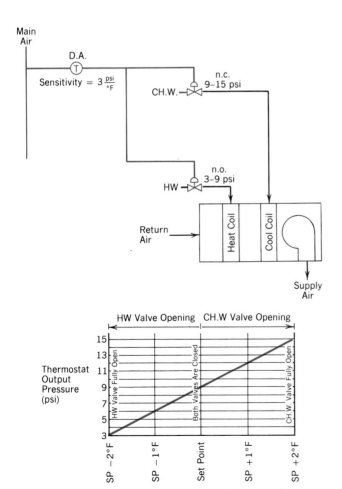

Figure 28-11 Hot water and chilled water valves controlled in sequence.

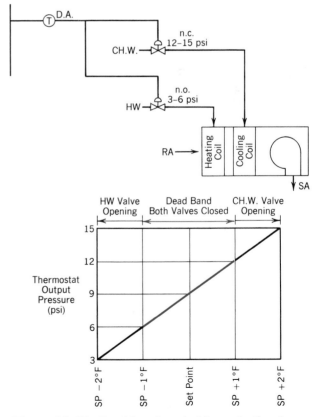

Figure 28-12 Dead-band control keeps both valves closed unless the room temperature gets more than one degree higher or lower than set point.

from a cold start, the hot water valve closes, and then the chilled water valve opens. When the room is exactly at the thermostat set point, the output pressure from the thermostat is 9 psi, and both valves are closed.

Figure 28-12 shows the same control scheme, except that the spring ranges have been changed. The hot water valve will begin to open only if the room temperature drops lower than 1°F below the set point, and will be fully open if the room temperature drops to 2°F below set point. The chilled water valve will only begin to open if the room gets warmer than 1°F above the set point. Any time the room temperature is within the range of 1°F above or below the set point, both valves will be closed. This scheme is said to have a 2°F **dead band.** It assures that no energy will be used to change the room temperature as long as the temperature remains within the dead-band range.

Throttling Range for Thermostats

Many thermostats have adjustable sensitivities that are set in the field. The slider or dial that is adjusted may have numbers that reflect sensitivity or they may have numbers that reflect throttling range. **Throttling range** is the number of degrees of change in sensed temperature that will cause the output pressure of the thermostat to change from 3 psi to 15 psi. The sensitivity and throttling ranges are different ways of expressing the same characteristic of the thermostat. They are related as follows

$$\text{Throttling range} = 12/\text{sensitivity}$$

The thermostat in the example has a sensitivity of 3 psi/°F, therefore it has a throttling range of 4°F.

Temperature Transmitters and Receiver Controllers

A room temperature **transmitter** looks exactly like a thermostat, except that it does not have a setpoint adjustment or a sensitivity adjustment. It is purchased for a specific range of tem-

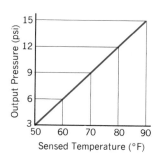

Figure 28-13 Output pressure versus temperature for a 50°F–90°F transmitter (nonadjustable).

perature such as 50°F–90°F or 0°F–100°F. The temperature range indicates what temperature range is required to make the output of the transmitter change from 3 psi to 15 psi. The graph in Figure 28-13 shows the output pressure versus temperature for a 50°F–90°F transmitter. Figure 28-14 is presented as a convenience, to allow you to determine the output pressure from any of several transmitters, at different temperatures.

The sensitivity of transmitters is very low. The 50°F–90°F transmitter has a sensitivity of

$$\begin{aligned} \text{Sensitivity} &= (12 \text{ psi change})/(40°F \text{ change}) \\ &= 0.3 \text{ psi/°F} \end{aligned}$$

The output pressure from the temperature transmitter is piped to a **receiver controller** (Figure 28-15). The small change in output pressure that is produced by the transmitter is amplified by the receiver controller. The amount of amplification that happens in the receiver controller is adjustable, and is called gain.

The control diagram in Figure 28-16 shows the room temperature transmitter, the receiver controller, and a 3–15 psi hot water valve. Note that the transmitter is a one-pipe device.

Let's say that we want to set the gain so that the room temperature is maintained between 70°F and 74°F. We want a **system sensitivity** of

$$\begin{aligned} \text{System sensitivity} &= (12 \text{ psi change})/(4°F \text{ change}) \\ &= 3 \text{ psi/°F} \end{aligned}$$

The receiver controller must therefore be set for a gain of 10. This means that the sensitivity of the transmitter (0.3 psi/°F) will be amplified 10 times, to produce a system sensitivty of 3 psi/°F.

Another way that some receiver controllers will express this amplification is called **throttling range** or **proportional band.** This throttling range is different from the throttling range discussed for thermostats. Throttling range and proportional band both refer to the percentage of the full transmitter range used in order to produce a change in output pressure of 3 to 15 psi from the receiver controller. In the previous example, we want to use a range of only 4°F out of the total transmitter range of 40°F. We would set the throttling range or the proportional band on the receiver controller for 10 percent.

Adjusting the proportional band on the receiver will determine how many degrees of change is required in order to make the branch pressure from the receiver controller change from 3 to 15 psi. But which 4 degrees? The receiver controller also has a set point that allows the technician to choose the appropriate range of control. In the previous example, once the proportional band is set for 10 percent, the set point may be adjusted so that the output varies from 3–15 psi when room temperature varies from 68°F–72°F, or 70°F–74°F, or any other 4°F range of room temperature (see graph in Figure 28-17). The setpoint dial does not have temperature numbers on it because the setting depends upon

Output (psi)	-25 125	-20 80	30 80	50 90	0 100	50 100	-40 120	20 120	25 125	35 135	40 140	50 150	-40 160	0 200	40 240	80 240
3.0	-25	-20	30	50	0	50	-40	20	25	35	40	50	-40	0	40	80
3.24	-22	-18	31	50.8	2	51	-36.8	22	27	37	42	52	-36	4	44	83.2
3.48	-19	-16	32	51.6	4	52	-33.6	24	29	39	44	54	-32	8	48	86.4
3.72	-16	-14	33	52.4	6	53	-30.4	26	31	41	46	56	-28	12	52	89.6
3.96	-13	-12	34	53.2	8	54	-27.2	28	33	43	48	58	-24	16	56	92.8
4.20	-10	-10	35	54	10	55	-24	30	35	45	50	60	-20	20	60	96
4.44	-7	-8	36	54.8	12	56	-20.8	32	37	47	52	62	-16	24	64	99.2
4.68	-4	-6	37	55.6	14	57	-17.6	34	39	49	54	64	-12	28	68	102.4
4.92	-1	-4	38	56.4	16	58	-14.4	36	41	51	56	66	-8	32	72	105.6
5.16	2	-2	39	57.2	18	59	-11.2	38	43	53	58	68	-4	36	76	108.8
5.4	5	0	40	58	20	60	-8	40	45	55	60	70	0	40	80	112
5.64	8	2	41	58.8	22	61	-4.8	42	47	57	62	72	4	44	84	115.2
5.88	11	4	42	59.6	24	62	-1.6	44	49	59	64	74	8	48	88	118.4
6.12	14	6	43	60.4	26	63	1.6	46	51	61	66	76	12	52	92	121.6
6.36	17	8	44	61.2	28	64	4.8	48	53	63	68	78	16	56	96	124.8
6.6	20	10	45	62	30	65	8	50	55	65	70	80	20	60	100	128
6.84	23	12	46	62.8	32	66	11.2	52	57	67	72	82	24	64	104	131.2
7.08	26	14	47	63.6	34	67	14.4	54	59	69	74	84	28	68	108	134.4
7.32	29	16	48	64.4	36	68	17.6	56	61	71	76	86	32	72	112	137.6
7.56	32	18	49	65.2	38	69	20.8	58	63	73	78	88	36	76	116	140.8
7.8	35	20	50	66	40	70	24	60	65	75	80	90	40	80	120	144
8.04	38	22	51	66.8	42	71	27.2	62	67	77	82	92	44	84	124	147.2
8.28	41	24	52	67.6	44	72	30.4	64	69	79	84	94	48	88	128	150.4
8.52	44	26	53	68.4	46	73	33.6	66	71	81	86	96	52	92	132	153.6
8.76	47	28	54	69.2	48	74	36.8	68	73	83	88	98	56	96	136	156.8
9.0	50	30	55	70	50	75	40	70	75	85	90	100	60	100	140	160
9.24	53	32	56	70.8	52	76	43.2	72	77	87	92	102	64	104	144	163.2
9.48	56	34	57	71.6	54	77	46.4	74	79	89	94	104	68	108	148	166.4
9.72	59	36	58	72.4	56	78	49.6	76	81	91	96	106	72	112	152	169.6
9.96	62	38	59	73.2	58	79	52.8	78	83	93	98	108	76	116	156	172.8
10.2	65	40	60	74	60	80	56	80	85	95	100	110	80	120	160	176
10.44	68	42	61	74.8	62	81	59.2	82	87	97	102	112	84	124	164	179.2
10.68	71	44	62	75.6	64	82	62.4	84	89	99	104	114	88	128	168	182.4
10.92	74	46	63	76.4	66	83	65.6	86	91	101	106	116	92	132	172	185.6
11.16	77	48	64	77.2	68	84	68.8	88	93	103	108	118	96	136	176	188.8
11.4	80	50	65	78	70	85	72	90	95	105	110	120	100	140	180	192
11.64	83	52	66	78.8	72	86	75.2	92	97	107	112	122	104	144	184	195.2
11.88	86	54	67	79.6	74	87	78.4	94	99	109	114	124	108	148	188	198.4
12.12	89	56	68	80.4	76	88	81.6	96	101	111	116	126	112	152	192	201.6
12.36	92	58	69	81.2	78	89	84.8	98	103	113	118	128	116	156	196	204.8
12.6	95	60	70	82	80	90	88	100	105	115	120	130	120	160	200	208
12.84	98	62	71	82.8	82	91	91.2	102	107	117	122	132	124	164	204	211.2
13.08	101	64	72	83.6	84	92	94.4	104	109	119	124	134	128	168	208	214.4
13.32	104	66	73	84.4	86	93	97.6	106	111	121	126	136	132	172	212	217.6
13.56	107	68	74	85.2	88	94	100.8	108	113	123	128	138	136	176	216	220.8
13.8	110	70	75	86	90	95	104	110	115	125	130	140	140	180	220	224
14.04	113	72	76	86.8	92	96	107.2	112	117	127	132	142	144	184	224	227.2
14.28	116	74	77	87.6	94	97	110.4	114	119	129	134	144	148	188	228	230.4
14.52	119	76	78	88.4	96	98	113.6	116	121	131	136	146	152	192	232	233.6
14.76	122	78	79	89.2	98	99	116.8	118	123	133	138	148	156	196	236	236.8
15.0	125	80	80	90	100	100	120	120	125	135	140	150	160	200	240	240

TEMPERATURE RANGE (°F)

1 Select column that has the appropriate temperature range.

2 Read down to the sensed temperature.

3 Read to the left the transmitter output pressure.

Figure 28-14 Pressure–temperature relationships for temperature transmitters.

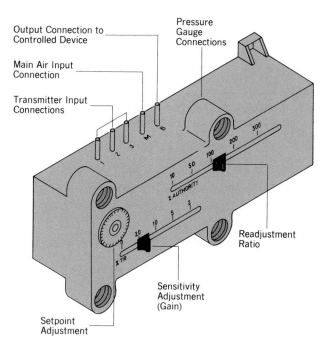

Figure 28-15 Receiver controller with cover removed.

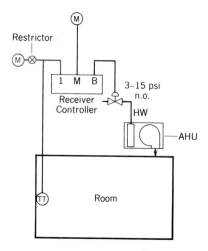

Figure 28-16 Room temperature transmitter controlling a hot water control valve through a receiver controller.

the temperature range of the transmitter that is being used. Once the transmitter and receiver controller are matched, some manufacturers offer stick-on graduated labels to affix to the setpoint dial. Transmitters (as well as thermostats) can be used to sense temperature of air in ducts, water in pipes, or sense at remote points from the device itself (Figure 28-18). The function of these devices is identical to that of the room transmitters and thermostats.

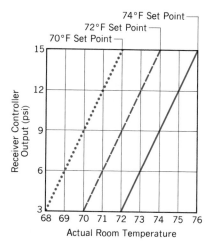

Figure 28-17 Changing the set point of the receiver controller changes which 4°F range of room temperature will cause a 3–15 psi output.

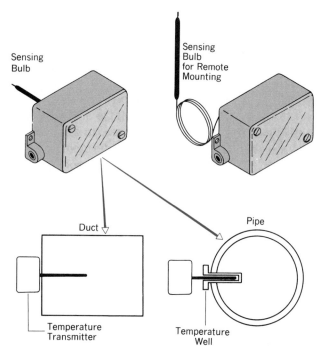

Figure 28-18 Transmitters for sensing temperature in ducts or pipes.

Dual Input Receiver Controllers

Figure 28–19 shows a pneumatic control diagram for hot water temperature control system. The temperature transmitter senses the temperature of the hot water leaving a steam/hot water converter and sends a branch pressure signal to the steam valve in order to maintain the hot water temperature near its set point. It would

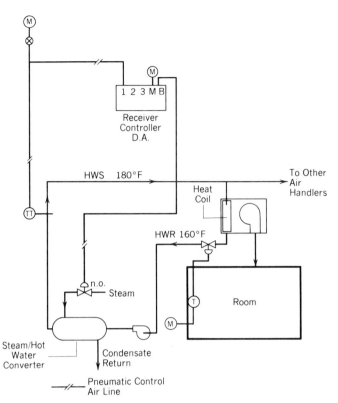

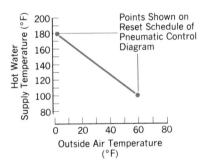

Figure 28-19 The room thermostat controls the hot water valve. The hot water supply temperature transmitter and receiver controller maintain a water supply temperature of 180°F.

Figure 28-20 Hot water temperature set point versus outside air temperature.

be desirable, however, to readjust the set point of the hot water temperature being maintained according to the temperature of the outside air. We might need the hot water to be 180°F when it is 0°F outside, but as the outside temperature rises to 60°F, we would probably be able to accomplish the required heating with 100°F hot water. In fact, the lower hot water temperature is preferable. With hot water temperatures much higher than is actually required, the valves on the heat-

ing coils will be trying to control at very near the shut-off position of the valve. Lower hot water temperature will allow the valves to admit more, lower temperature hot water.

Figure 28-20 shows a graph of how we would like to change the hot water setpoint temperature as outside air temperature varies. It would be very impractical to have an operator measure the outside air temperature each hour, and then readjust the setpoint temperature accordingly. Thankfully, there is a way to accomplish this readjustment on a continuous basis, automatically. Figure 28-21 shows the same hot water temperature control scheme, but with an outside air temperature transmitter added. The outside air transmitter will automatically readjust the set point of the receiver controller. The **reset schedule** is normally shown in the form of a table such as the one shown in the figure. Depending on the manufacturer, the rate of readjustment of the set point may be called either **authority** or **readjustment ratio.** Each is expressed as a percent, and may be set according to another adjustment slider or knob on the receiver controller. The notation DADR on the receiver controller means direct action, direct readjustment. Direct acting means that as the sensed pressure at port 1 increases, the output pressure increases. Direct readjustment means that as the pressure sensed at port 3 increases, the set point is increased.

The formula for calculating the desired authority is

$$\text{Authority } \% = \frac{(\Delta P \text{ at port 1} + \text{psi throttling range})}{\Delta P \text{ at port 3}}$$

Let's assume we have a 40–240°F hot water temperature transmitter, and a −40–160°F outside air temperature transmitter. From the transmitter output pressure table in Figure 28-14, we can calculate the change in pressure at port 1 according to the reset schedule

Pressure at 180°F =	11.4 psi
Pressure at 100°F =	6.6 psi
Δ pressure at port 1 =	4.8 psi

and the change in pressure at port 3 is

Pressure at 60°F =	9.0 psi
Pressure at 0°F =	5.4 psi
Δ pressure at port 3 =	3.6 psi

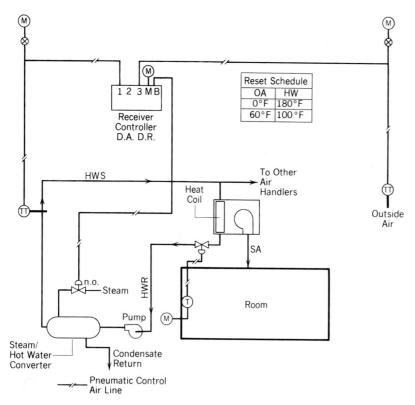

Figure 28-21 Dual-input receiver controller.

The throttling range refers to the range of temperature (pressure) from the hot water transmitter over which the full output range of 3–15 psi will occur, while the set point remains unchanged. If we say that the sensitivity of the system is set up so that the hot water is maintained within 10°F of set point (plus or minus 5°F), then the throttling range corresponds to a 0.75 psi change in transmitter output.

The authority is then calculated as

$$\text{Authority} = \frac{4.8 + 0.75}{3.6}$$
$$= 1.54 = 154 \text{ percent}$$

The receiver controller in this hot water control scheme is said to be direct acting, reverse readjusting. It is direct acting because as the hot water temperature tends to rise, the output pressure of the controller will increase in order to close down the steam valve. It is reverse readjusting because as the temperature sensed by the outside air transmitter increases, the set point of the receiver controller is automatically adjusted downward. Receiver controllers are available as follows.

1. Direct acting, direct readjustment.

2. Direct acting, reverse readjustment.

3. Reverse acting, direct readjustment.

4. Reverse acting, reverse readjustment.

Remote Control Point Adjustment

Thus far, we have discussed four of the five ports on the receiver controller. The last port to be described is the **remote control point adjustment** (port 2). This is the least commonly used port on the receiver controller. It receives a branch pressure signal from a **gradual switch,** which acts to reset the set point of the controller. The gradual switch (Figure 28-22) is actually a small pressure regulator whose set point may be changed manually by turning the dial. The purpose of this arrangement is to allow a person to reset the set point of the system without having to actually go to the location of the receiver controller.

When the remote control point adjustment or

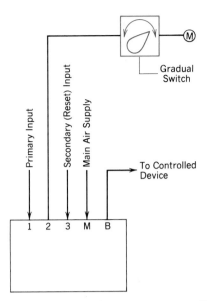

Figure 28-22 Piping to the receiver controller with two transmitter inputs and remote control setpoint adjustment.

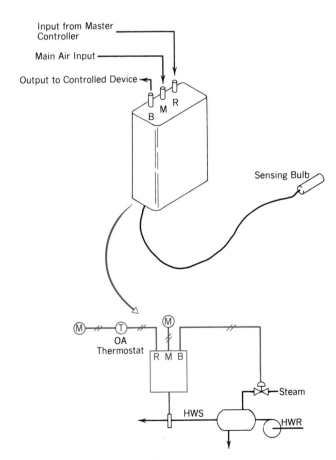

Figure 28-23 Reset of hot water temperature using a submaster controller. The outside air thermostat is the master controller.

any other port is not used, it must be capped or plugged.

Master–Submaster Controls

Figure 28-23 shows a **submaster controller.** A bulb, which is a part of the controller, senses the supply water temperature from the steam/hot water converter. The set point of this controller is determined from another controller, an outdoor air thermostat. The outdoor air thermostat is the **master controller.** The function of this system is the same as the dual-input receiver controller with two temperature transmitters.

Relays

There are many types of relays that are available that make pneumatic controls quite flexible in their application. Some of the relays that are available are as follows.

1. The **averaging relay** takes two different input signals and provides an output signal that is an average of the two inputs. Two thermostats can be used in a large room. The temperatures they sense can be averaged through an averaging relay and passed on to the valve in order to try to maintain a room temperature halfway between the requirements of each of the individual thermostats.

2. The **high pressure selector relay** takes two different input signals and provides an output pressure that is equal to the higher of the two inputs. Figure 28-24 shows two high select relays that are used to determine the single warmest zone in a terminal reheat system. The signal from the one warmest zone is used to control the chilled water valve. Each of the zones also sends the thermostat branch pressure signal to the reheat coil valve to present overcooling.

3. The **low pressure select relay** selects and passes the lower of two input signals.

4. The **reversing relay** provides an output pressure of 3–15 psi as the input pressure varies from 15 to 3 psi. Figure 28-25 shows how a normally open chilled water valve may be used with a

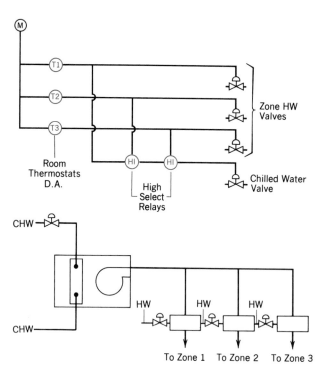

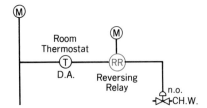

Figure 28-24 Each zone thermostat controls a reheat coil. The zone that is most above its set point controls the chilled water coil.

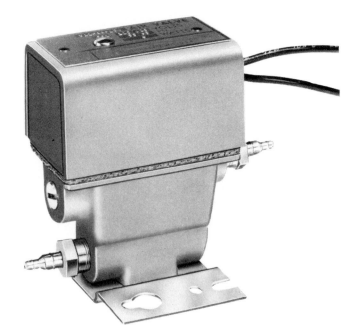

Figure 28-26 Three-way solenoid air valve. (*Courtesy Johnson Controls, Inc.*)

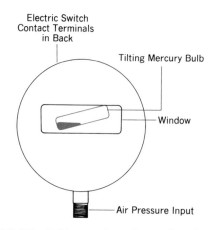

Figure 28-25 Reversing relay increases the output pressure as the input pressure decreases. At 8 psi, input and output pressures are equal.

Figure 28-27 P. E. switch makes or breaks contacts in response to a sensed air pressure.

direct-acting thermostat by the addition of a reversing relay.

5. The **electric–pneumatic relay** (Figure 28-26), sometimes called an E.P. relay or an E.P. switch, is nothing more than a solenoid valve that, when energized, opens to allow the 20 psi main air pressure into the pneumatic control system. A notation is often used on pneumatic control diagrams that says the E.P. relay is wired to L.S.F.S. This means that the solenoid coil is energized whenever the load side of the fan starter is energized. If the fan is not running, main air will not be supplied to the controls. All valves and dampers will return to their normal positions. Note the arrangement of the normally

open, normally closed, and common ports on the valve. The three-way valve is used instead of a two-way valve so that there is a way for all the air pressure in the control system to bleed out when the fan shuts down.

6. The **pressure–electric relay** (Figure 28-27) is commonly referred to as a P.E. switch. It is an electric switch that will make or break a set of contacts when the pressure it receives from a pneumatic controller reaches a predetermined set point. Figure 28-28 shows a room thermostat that is controlling a hot water valve as well as

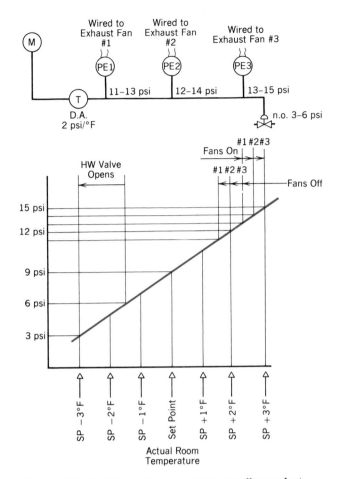

Figure 28-28 Room thermostat controlling a hot water valve and three exhaust fans.

three exhaust fans. When the thermostat output pressure reaches 13 psi, P.E. switch #1 closes, starting exhaust fan #1. If the thermostat output reaches 14 psi, fan #2 will start, and at 15 psi, fan #3 will start. When the output pressure from the thermostat falls to 13 psi, fan #3 goes off. Fan #2 drops out on a fall to 12 psi, and fan #1 stops when the thermostat output drops to 11 psi.

There are many other types of lesser used relays that are not presented here.

Other Pneumatic Functions

Most of the pneumatic controls that you will encounter are temperature controls. There are, however, other variables that are sometimes measured in order to accomplish desired results. There are **humidistats** and humidity transmit-

ters that are used to operate humidifiers and dehumidifying units. There are **enthalpy controllers** that receive a dry-bulb and humidity signal from transmitters, and determine the enthalpy of two different air streams (usually return air and outside air). There are **pressure transmitters** and **differential pressure transmitters** that may be used to activate fans or alarms. There are **velocity transmitters** that indirectly sense cfm by measuring velocity pressure, and sending out a proportional 3–15 psi signal. These and others may be encountered in the field. They will not be explored in depth in this text, but their principles of operation are the same as the principles we have discussed here. Other than the variable that is being measured, there are set points, sensitivities, receiver controllers, offset, and readjustment ratios, all identical to the temperature systems.

Calibration of Thermostats

Pneumatic thermostats have the usual adjustments of temperature set point and sensitivity. These adjustments are not to be confused with calibration. **Calibration** refers to whether or not the thermostat actually controls to the number at which the indicator is set. For a transmitter, calibration refers to whether or not the transmitter pressure output corresponds to what it should be for the temperature being sensed. It is not common for transmitters to need to be field calibrated, because it is done at the factory. For many types of thermostats, however, the calibration is changed every time the sensitivity is adjusted. The following sequence is suggested to completely adjust and calibrate a pneumatic thermostat.

1. Set the desired sensitivity or throttling range according to the numbers on the thermostat.

2. Check the sensitivity adjustment by moving the setpoint adjustment. Note how many degrees you have to change the set point in order to make the thermostat output pressure change from 3 to 15 psi. Compare that number of degrees to the throttling range set on the thermostat.

3. Using a thermometer, measure the temperature being sensed by the thermostat. Move the

set point to match the actual temperature. At this setting, the thermostat should be exactly at its midpoint of control. If the total range of the devices being controlled by this thermostat is 3 to 15 psi, the thermostat output when it is satisfied should be 9 psi. If it is not, adjust the calibration adjustment until the output is 9 psi. If only a portion of the output range is being used for control, the calibration adjustment should be made so that the output is in the middle of that range when the set point and the actual temperature are the same. For example, on a heating-only application, using a hot water valve with a spring range of 4–8 psi, set the output at 6 psi when the thermostat is satisfied.

Other Pneumatic Control Diagrams

Figures 28-29 through 28-32 show additional common pneumatic control schemes in common use. A brief description of each follows. It is left to the student to go through each scheme so that it is thoroughly understood. There are questions about these diagrams in the Questions section at the end of this chapter.

Figure 28-29 incorporates a direct-acting thermostat that is used as a **low limit.** Without the low limit, when the room thermostat is adjusted downward a few degrees, untempered outside air would be supplied to the space until the new

setpoint temperature is reached. The low limit will not pass the higher pressures required to close the hot water valve if the supply air temperature drops below 55°F. It will take longer for the room to reach the new, lower set point, but the blast of untempered outside air will be avoided.

Figure 28-30 uses room air temperature to reset the set point of the supply-air temperature. This is used in large spaces where the response to changes in the valve position will be quite slow. Without the dual input setup, the system would need to be set with a very low sensitivity, allowing large temperature swings in the room. If the throttling range is set lower to hold temperature closer to the set point, hunting will occur. On a call for additional heating, for example, by the time the effect of opening the hot water valve is noticed by the thermostat, too much heat has already been added to the room. The room temperature will overshoot the set point.

Figure 28-31 shows the control scheme for a multizone air handler. Each zone thermostat controls the hot deck and cold deck mixing dampers. The hot deck temperature is maintained at a set point as determined by the outside

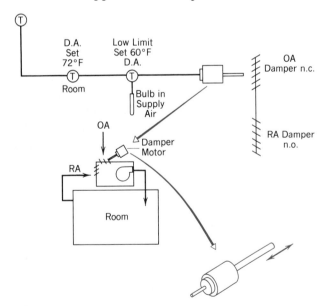

Figure 28-29 Room thermostat controls the outside air quantity, subject to the limitation of the low limit controller.

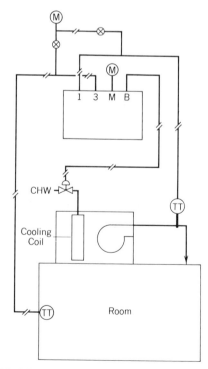

Figure 28-30 When room load increases, the room transmitter resets the set point of the receiver controller to lower the supply-air temperature.

air temperature. The cooling coil is allowed to run wild (no control valve).

Figure 28-32 shows a much more efficient method of controlling the multizone systems. Although most of the multizones in existence use the scheme in Figure 28-31, this scheme repre-

sents an easy retrofit that will significantly reduce energy consumption. The major difference is that the hot water valve is controlled by the coldest zone, and the hot water valve remains closed until the coldest zone has completely closed off its cold deck and opened its hot deck dampers.

Pilot Positioners

A positive positioning relay (**pilot positioner**) is a relay that is attached to a control valve (Figure 28-33). It is actually a simple controller that is widely misunderstood. It is a dual-input controller. The variable that is being controlled is the position of the stem of the control valve. It is sensed by a spring that is attached to the valve stem and to the arm on the pilot positioner. The relay acts to maintain a certain position as the set point. This set point, however, is being readjusted by the branch pressure signal from a thermostat or controller. For example, the reset schedule (if there was one) might look like the following:

Signal from Thermostat (psi)	Valve Position
3	Wide open
4	$\frac{3}{4}$ open
5	$\frac{1}{2}$ open
6	$\frac{1}{4}$ open
7	closed

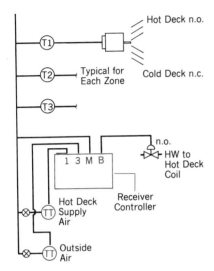

Figure 28-31 Multizone control with reset of the hot deck temperature by outside air temperature.

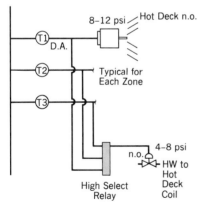

Figure 28-32 Improved method of multizone control. The hot water valve is controlled by whichever zone happens to be the coldest relative to its set point.

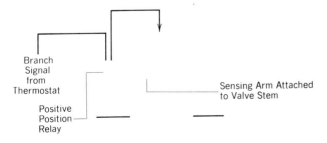

Figure 28-33 Pilot positioner.

When the thermostat signal is 6 psi, the valve position should be almost closed. If high pressure from the fluid is forcing the valve open, or if the stem of the valve is binding, the pilot positioner output presure to the diaphragm will apply whatever pressure is necessary (up to the 20 psi main air pressure) in order to attain the position set point of one-quarter open. With the pilot positioner, the valve will operate to the same position for the same pressure signal from the thermostat, regardless of changing fluid pressures against which the valve may be working. Also, with a pilot positioner, the same result can be accomplished regardless of the spring range of the valve. The controller does not care what output pressure it must apply to the diaphragm. It cares only that the correct valve stem position is attained.

KEY TERMS AND CONCEPTS

On–Off Control

Modulating Control

Offset

Set Point

Sensitivity

Hunting

Pneumatic Thermostat

Output Pressure

Spring Range

N.O. and N.C. Valves

Two-way Valve

Three-way Valve

Direct Acting

Reverse Acting

One-pipe

Two-pipe

Dead Band

Throttling Range

Transmitter

Receiver Controller

System Sensitivity

Proportional Band

Dual Input Receiver
Controller

Reset Schedule

Authority

Readjustment Ratio

Remote Control Point
Adjustment

Gradual Switch

Submaster Controller

Master Controller

Averaging Relay

High or Low Pressure
Selector Relay

Reversing Relay

E.P. Relay

P.E. Switch

Calibration of
Thermostats

Pilot Positioner

QUESTIONS

1. What is meant by the term proportional control?

2. A room thermostat is controlling a normally open steam valve that has a spring range of 3–15 psi.
(a) Is the thermostat D.A. or R.A.?
(b) If the thermostat is set for a throttling range of 6°F and a set point of 72°F, over what range of room temperature will the valve be controlled?

3. Answer Question 2*b* for a valve with a spring range of 3–9 psi.

4. Sketch a pneumatic control diagram using a one-pipe reverse-acting thermostat and a chilled water valve. Label the valve as n.o. or n.c.

5. Sketch a pneumatic control diagram with a direct-acting thermostat controlling a hot water and chilled water valve in sequence. The entire range of control is to take place over a range of room temperature of 68°F to 76°F. Indicate the sensitivity and set point for the thermostat, the normal positions for the valves, and the spring ranges.

6. Repeat Question 5, selecting the spring ranges so that there is a 4°F dead band.

7. What is hunting? What setting would you change (state whether you would increase or decrease) on a thermostat to eliminate hunting?

8. The throttling range on a thermostat is 10°F. It is controlling a 3–15 psi valve. Over what range of room temperature will the valve be controlled?

9. Repeat Question 8 for a valve with a spring range of 12–15 psi.

10. A 0°F–100°F transmitter senses supply-air temperature from a cooling coil. The output from a receiver controller controls a 3–15 psi chilled water valve. Draw the pneumatic control diagram.

11. What is the sensitivity of the transmitter in Question 10?

12. If the chilled water valve in Question 10 is to maintain an air temperature of between 54°F and 58°F, what is the gain setting on the receiver controller?

13. What would the gain setting be in Question 12 if the spring range is 12–15 psi?

14. A dual input controller receives signals from a 40°F–140°F hot deck transmitter and a 0°F-100°F outside air transmitter. Assuming a throttling range of 10°F around any set point, what authority setting is required to maintain the following schedule?

Outside air	20°F	90°F
Hot deck air	140°F	80°F

15. What is the purpose of a remote control point adjustment on a receiver controller?

16. What is a gradual switch?

17. For the control scheme in Figure 28-28, if the thermostat is set at 74°F, how warm must the room become before the first fan will turn on?

18. Referring to Figure 28-29, when the room temperature is 74°F, what is the output pressure to the damper motor? Assume 3 psi/°F on both thermostats and 70°F outside air temperature.

19. What is the answer to Question 18 if the outside air temperature is 59°F?

20. Referring to Figure 28-31, if the outdoor air temperature falls, how will the hot deck supply air temperature set point be affected?

PART FOUR

ELECTRICITY

Part four explains the basics of electricity and the specific components that are most common in the refrigeration, heating, and air conditioning industry. Specific terminal identification information is given wherever such nomenclature is relatively standardized. The circuits described are divided into cooling and heating applications, and the unique aspects of each wiring diagram are explained fully. The chapter on electrical trouble shooting (Chapter 33) enumerates the basic rules of how to use voltmeters and ohmmeters to identify failed switches and defective loads.

OBJECTIVES

After completing Part Four, the student should be able to do the following:

1. Identify the common electrical devices used in the heating, air conditioning, and refrigeration industry.

2. Use a voltmeter and ohmmeter to troubleshoot malfunctioning systems.

3. Read wiring diagrams and explain correct operating sequence.

CHAPTER 29

ELECTRICAL CONCEPTS

The purpose of this chapter is to familiarize you with the basic electrical concepts that are common to air conditioning and all other fields that use electrical circuits.

Electrical Circuits and Water Circuits

Electrical work scares many students. You can't see electricity, and many of us have been frightened by malfunctions of electrical circuits. But thankfully for the student trying to learn the basic electrical concepts, electricity flowing through wires behaves similarly to water flowing through pipes. Let's first discuss a water system. We will frequently refer to this water analogy throughout this chapter.

Figure 29-1 shows a bucket that always contains a constant depth of water. At the bottom of the bucket there is a pipe that allows the water to flow out, through an orifice, through a valve, and out to atmospheric pressure (zero psig). The orifice is a restriction in the line with a small hole in

it to allow water flow. The pressure at the inlet to the orifice remains constant, as long as the depth of the water in the bucket remains constant. Let's assume that the water pressure just ahead of the orifice is 10 psi (never mind that the bucket would have to be quite deep in order for the pressure to actually be this high). When the valve is open, water flows through the orifice. The pressure of the water is 10 psi before the orifice, and 0 psi after the orifice.

Figure 29-2 shows the same setup, but now the valve is closed. There is no flow of water, so there is no pressure drop across the orifice. The pressure remains at 10 psi all the way up to the inlet of the valve, and it is 0 psi at the outlet of the valve.

In Figure 29-1 we can measure the rate of flow (gpm), we can measure the pressure drop of the water flowing through the orifice (psi), and we can measure the resistance of the orifice in some units that would relate to the size of the hole. Note that the available pressure drops to zero across the orifice when there is flow, or it drops to zero across the valve when there is no flow. Let's see what happens as we change the pressure, the size of the hole, and the flow rate of the water through the orifice.

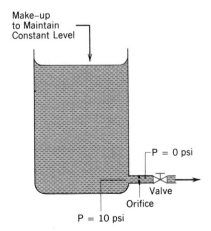

Figure 29-1 Water flow through a pipe and an orifice with the valve open. There is 10 psi of pressure drop caused by friction across the orifice.

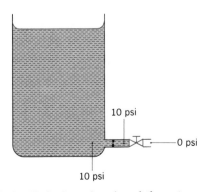

Figure 29-2 With the valve closed there is no flow, no pressure drop across the orifice, and 10 psi of pressure difference across the valve.

511

Using our original setup, if the size of the orifice is reduced the resistance will be increased. Then with the same pressure drop available across the orifice, the rate of water flow (gpm) will be reduced. Similarly, we could use the same size orifice as in the original bucket, and increase the depth of the water in the bucket so that the pressure at the orifice inlet increased to 15 psi instead of 10 psi. The water flow rate would be increased.

So much about water for now. In an electrical circuit, instead of water flowing through pipes, we have electrons moving through a wire (this text will not deal with the atomic theory to describe the physical nature of electrons). In order for the electrons to flow through a device (like a light bulb or a motor), there must be an **electrical pressure** difference available to force the electricity through the device. Electrical pressure is measured in **volts.** The rate of flow of electrons could be in terms of how many electrons flow past a point each second, but this would be quite a large number. Instead, we use a unit called **amperes** (usually shortened to **amps**), which is the equivalent of 6,280,000,000,000,000,000 electrons flowing each second. The resistance of the device to the flow of electrons is measured in **ohms** (Ω). The relationship between volts, ohms, and amps is as follows.

$$\text{Volts} = \text{amps} \times \text{resistance} \qquad E = I \times R$$

This formula can be rearranged by some simple algebra to also say

$$\text{Amps} = \text{volts/resistance} \qquad I = \frac{E}{R}$$

and

$$\text{Resistance} = \text{volts/amps} \qquad R = \frac{E}{I}$$

If we know the resistance of a device (ohms), and we know the pressure (volts) available to push the electrons through, we can calculate how

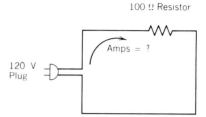

Figure 29-4   Example problem.

much electricity (amps) will flow. Similarly, if we know any two of the circuit factors (volts, ohms, amps), we can calculate the third. Figure 29-3 shows a simple memory aid to help you remember these three formulas. Simply cover the variable you are trying to determine, and the circle will show you what to do with the other two variables in order to calculate the third. This relationship between volts, ohms, and amps is referred to as **Ohm's Law.**

___

Example

The circuit shown in Figure 29-4 has a 100-Ω resistor in a circuit that is supplied with 120-V. How much current will flow in this circuit?

Solution

Using the Ohm's law circle, cover amps with your finger. The part of the circle that is left says volts divided by resistance. Thus, the amps flowing in this circuit will be

$$\text{Amps} = 120 \text{ V/}100 \ \Omega$$
$$\text{Amps} = 1.2$$

___

Example

How much voltage drop (pressure drop) would we have if there is a current of 2 amps flowing through a 20-Ω resistance?

Solution

Covering the volts in the Ohm's law circle, the part of the circle that is left says amps times resistance. The voltage drop across the 20-Ω resistor will be

$$\text{Volts} = 2 \text{ Amps} \times 20 \ \Omega$$
$$\text{Volts} = 40$$

___

E = volts
I = amps
R = ohms

Figure 29-3   Ohm's law circle.

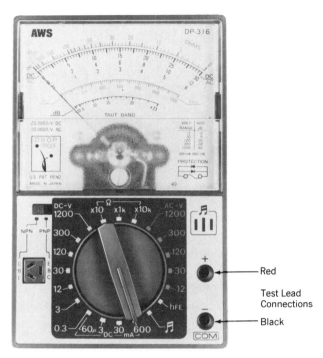

Figure 29-5 Volt-ohmmeter (VOM). *(Courtesy A. W. Sperry Instruments, Inc.)*

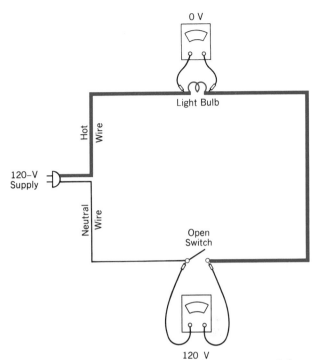

Figure 29-6 Voltage readings with an open switch in the neutral leg.

Measuring Voltage Drops in Circuits

Figure 29-5 shows a voltmeter. Actually it combines the functions of measuring volts and measuring ohms. It is commonly called a VOM (volt-ohmmeter). There are two leads from the voltmeter, and the meter will read the difference in electrical pressure at the two points being touched by the probes. When the probes are touched to the circuit, the meter in no way affects the operation of the circuit. Using a voltmeter to measure voltage differences will be the single most powerful electrical service technique you will have at your disposal.

Consider the circuit shown in Figure 29-6. The plug is inserted into a wall receptacle that supplies 120 V of pressure in one wire and 0 V in the other wire. These two wires are sometimes referred to as **hot** and **neutral.** In household wiring, the hot wire is usually black and the neutral wire is white. If there is a ground wire it will be green. It has the same zero electrical pressure as the neutral wire. When a complete path is provided from the hot wire to the neutral wire through a resistance, current will flow through the circuit. When the switch is open the electrical

pressure of 120 V is available all the way up to the switch. After the switch, the electrical pressure is zero. The voltmeter across the switch will read 120 V because there is a 120-V difference in the voltage sensed by the two probes. *When a voltage is read across a switch, you can be certain that the switch is open.* The open switch functions exactly the same way as the closed water valve on the bucket discussed at the beginning of this chapter. The resistance (light bulb) affects the electrical pressure in the same way as the orifice affected the water pressure—it doesn't affect it at all when there is no flow of water or electricity. If we move the voltmeter to measure the voltage across the resistor, we will find that it reads zero. This does not mean that there is no electrical pressure at that point. In fact, there is 120 V of electrical pressure on both sides of the resistor (just as there was 10 psi of pressure on both sides of the orifice). The voltmeter is merely telling us that there is no *difference* in voltage across the resistor.

Figure 29-7 shows the same circuit, but with the switch located on the hot leg instead of the neutral leg. With the switch in the open position, we will read 120 V when measuring across the

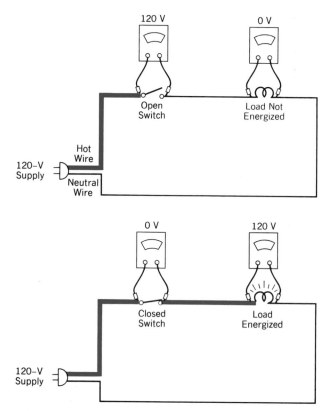

Figure 29-7   Voltage readings with a switch in the hot leg.

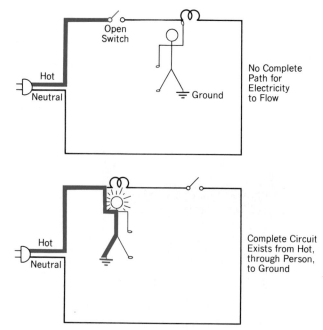

Figure 29-8   Safety difference between switching the hot leg and switching the neutral leg.

switch, and 0 V when measuring across the load. With the switch closed, there is no voltage drop across the switch, and there is a 120-V drop across the load. In this respect, the location of the switch may not seem to matter, whether it is located on the hot leg or the neutral leg. But there is one important safety consideration that is affected by the location of the switch.

In the first case, with the switch on the neutral leg, when the switch is open, there is still voltage available at the load. A person who touches the wiring at the load will receive a shock, even though the switch is in the Off position. Figure 29-8 shows how a person touching the load in this manner will complete a circuit to ground (zero voltage), and will become a part of the circuit. Amps will flow through the person. If the person happens to be wearing rubber-soled shoes, the resistance of the circuit will be very high and the amperage flow will be low.

In the second case, with the switch on the hot leg, when the switch is opened, there is no voltage available at the load. It is therefore safer to have the switch located on the hot leg of the circuit.

In any of the previous circuits, if the switch is closed, the bulb will light. We would be able to read voltage across the load. This is called an **IR drop.** That is, there is a loss in voltage because of the amps passing through a resistance (calculated as amps, I, times the resistance, R). When we measure voltage across an open switch, it is called a potential voltage. That is, there exists a difference in electrical pressure that can cause electricity to flow, given a complete circuit.

Using a voltmeter is the single most important skill that is used in troubleshooting malfunctioning systems. The following list summarizes the conclusions that may be drawn when using a voltmeter across a switch or a load.

a. When you read a voltage across a switch, that switch is open.

b. If no voltage is read across the terminals of a switch, no conclusion can be made. The switch might be closed, or it might be open with no voltage available to either side of the switch.

c. If zero voltage is measured across a load, you cannot tell anything about the load. You must look elsewhere in the system for an open switch that is interrupting the circuit.

d. If you read a voltage across a load, and the

Figure 29-9 Digital multimeter. *(Courtesy A. W. Sperry Instruments, Inc.)*

Figure 29-10 Test light. *(Courtesy Robinair)*

device is not functioning, the device is defective and must be replaced or repaired.

The voltmeter contains a very high resistance. And when the probes are touched to a circuit, the operation of the circuit is unaffected. Do not become confused by thinking that the voltmeter provides a path for electrical flow through the meter.

Voltmeters are manufactured in many configurations. Those with a meter movement are called analog; others are digital display (Figure 29-9). A simple substitute for a voltmeter is a test light (Figure 29-10). When the probes of the test light are placed on two points in a circuit, the light will illuminate if there is a difference in voltage between the two test points. The bulb in the test light must be rated for the expected voltage.

Short Circuit

A **short circuit** is defined as a circuit that provides a path for electrons to travel from the hot leg of a circuit to a neutral leg, without going through a load. Figure 29-11 shows such a cir-

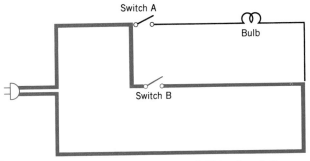

Figure 29-11 The moment switch B is closed there will be a short-circuit condition.

cuit. There only needs to be one such connection from the hot leg to the neutral leg for a short circuit to exist. Using Ohm's law, a circuit with zero resistance (or almost zero resistance) will allow an infinite amount of amperage (or a very high amperage) to flow. The wires used in the circuit, however, are not capable of carrying such high amperage flow without overheating. Circuits are protected from this overcurrent condition by **overcurrent devices,** such as fuses, circuit breakers, or overloads (to be discussed later). A short circuit will cause the overcurrent protection device to trip or to open before the circuit wiring is damaged.

Ohmmeter

An **ohmmeter** is a meter used for measuring resistance. Usually, an ohmmeter and a **voltmeter** are housed together in the same instrument called a volt-ohmmeter **(VOM).** Often, the ohmmeter will be used to check to see if there is **continuity,** or a continuous path for electricity to flow. For example, if the ohmmeter is connected across the two ends of a heating element (Figure 29-12), there should be some resistance. It may be 20, 50, 100 Ω or more, but there is a measurable resistance. If we were to read infinite resistance, that would mean that the wiring inside the heating element had separated, acting as if it were an open switch. In that case, we would say that we have an open circuit, or that there is no continuity through the heater element. Some ohmmeters have a continuity setting, which

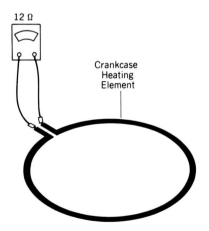

Figure 29-12 Measuring resistance.

makes an audible tone whenever the resistance between the probes is less than 20 Ω.

Several notes are in order for the proper use of an ohmmeter.

1. Do not use the ohmmeter to measure resistance while there is voltage available across a load. It will blow the fuse or a diode inside the meter. Some inexpensive ohmmeters are not furnished with fuse or diode protection. Meters of this type will be rendered useless in a "flash."

2. The ohmmeter uses a small battery inside the meter to create a circuit through the load whose resistance is to be measured. Because the condition of this battery changes as it is used, the meter must be zeroed each time it is used. This is done by touching the probe ends together, and adjusting a dial on the meter until the resistance reads zero.

3. The ohmmeter scale is probably the only one you will ever encounter that starts with zero on the right and increases to infinity on the left. Note that when the ohmmeter is not in use, it is still measuring resistance. It is measuring the resistance of the air space between the probes. This resistance happens to be infinite (air is an excellent insulator), and the meter needle reads at the extreme left end of the scale. Some VOMs have digital readouts, eliminating all confusion over which end of the resistance scale is zero.

4. Resistances to be measured will vary from less than one ohm to hundreds of thousands of ohms. For this reason, there are several scales available on the ohmmeter. They are $R \times 1$, $R \times 10$, $R \times 100$, $R \times 1000$ or 1K, and $R \times 10,000$ or 10K. Depending on the scale being used, the resistance reading shown on the meter must be multiplied by the appropriate number in order to determine the actual resistance being measured.

5. When storing the ohmmeter, make sure that the ends of the probes are not touching each other, or touching a metal surface that provides continuity between the probes. This will complete the circuit through the battery, and the battery will be dead for the next use. If you find that you cannot zero the meter all the way down to zero, your battery is too weak and should be replaced.

DC versus AC

In our discussion thus far, we have used our analogy with a water circuit in which one side of the circuit is maintained at a constant pressure, higher than the neutral side of the circuit. Given a completed circuit, the electrons would always flow in one direction, from the higher pressure side to the lower pressure side. The electrical equivalent to this is called direct current, or **dc.** It is most commonly found in circuits in which a battery is the source of voltage.

Virtually all devices found in the air conditioning industry will not be dc. They will be alternating current, or **ac.** Alternating current is different from direct current, in that the hot leg is sometimes at a higher pressure than the neutral, and sometimes it is at a lower pressure. It alternates between this higher pressure and lower pressure 60 times each second.

Figure 29-13 shows how, with a fluctuating pressure, Ping-Pong balls can be made to flow through a tube. When the piston is pushed to the right, it creates an air pressure higher than the pressure seen by the Ping-Pong balls at the right end of the tube. The Ping-Pong balls will flow toward the right. When the piston is pulled to the left, it creates a vacuum, and the Ping-Pong balls will flow toward the left. So it is with alternating current. The electrical pressure (voltage) in the hot wire is always different from the neutral (zero pressure) leg. The electrons flow first in one direction, and then in the other.

All of the descriptions up to this point of the relationships between volts, amps, and resistance still apply. As the voltage in the ac circuit rises and falls, so too does the amperage.

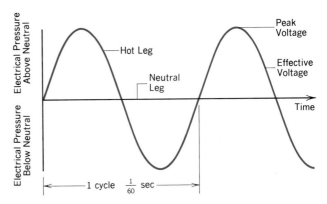

Figure 29-14 Alternating current sine wave.

Figure 29-14 shows the graphical representation of the constantly changing voltage in the hot leg of an ac power supply. The curve is called a **sine wave.** It is constantly repeating after each up and down cycle. Because of some mathematical interpretations used by engineers, we talk about the whole cycle as being represented by the 360 degrees in a circle. Each half-cycle takes 180 degrees. You can think of these degrees in terms of time, with 360 degrees representing one-sixtieth of a second. We will use degrees later to describe when two electrical voltages rise and fall at different times. We say that they are out of phase by a certain number of degrees.

In the United States, all power supplied by public utilities will alternate at the rate of 60 cycles per second. Many other countries have standardized on power that is supplied at an alternating frequency of 50 cycles per second. These frequencies are referred to as 50 Hertz or 60 Hertz, and abbreviated to 50 Hz or 60 Hz (Hertz was a scientist, as were Ohm and Ampere).

In the alternating current power supply, you

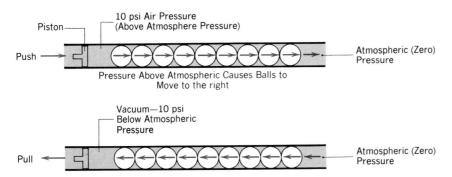

Figure 29-13 Mechanical analogy to alternating current.

will notice that while the voltage in the hot leg is always different from that in the neutral leg (except for two instants each cycle), the amount of difference varies. The greatest difference is called the **peak voltage.** The **effective voltage** is always lower than the peak voltage. For example, when we talk about a 115-V supply, we are saying that on the average, the hot leg is 115 V different from the neutral leg. The effective voltage is 70.7 percent of the peak voltage. The most commonly used voltages are 115 V, 230 V, and 460 V (up to 120 V, 240 V, or 480 V). Household appliances, lights, refrigerators, and small window air conditioners are 115 V. Larger current-consuming appliances, such as electric dryers, electric water heaters, larger window air conditioners, and central air conditioners use 230 V. Commercial and industrial air conditioning applications will be either 230 V or 460 V.

Three-phase Power Supply

In the previous description, there were two wires used as the source of power. One wire is effectively 115 V different from zero, and the neutral wire is at zero potential. This is called **single-phase power.** For large motors, a **three-phase power** supply is used. A three-phase power supply consists of three separate hot wires, each carrying, for example, 460 V. In each wire, the actual voltage is alternating at 60 times per second, but at a different time than each of the other two hot wires. Figure 29-15 shows how the voltage in each of the three wires behaves. They are out of synchronization by one-third of a cycle each. We say that they are 120 degrees out of phase with each other.

Three-phase power is not used when the volt-

age is 115 V, but is only used in the higher voltages. It is not used in residential applications. Motors that use a three-phase power supply are not generally used in sizes smaller than 2 horsepower. Three-phase motors are most commonly used for applications of 5 horsepower (or 5 tons) or larger. Between 2 and 5 horsepower, motors are available as either single-phase or three-phase.

The symbol used to indicate phase is the Greek letter phi (φ). A complete description of the voltage characteristics of a particular motor or other device might be shown as 460 V/60 Hz/3 φ.

230-Volt Single-Phase Power

All residential devices requiring 230 V are supplied with 230-V single-phase power, as are many commercial refrigeration and air conditioning systems of 5 horsepower and smaller. A 230-V single-phase motor has two wires that are connected to the power source. If you were to measure the voltage between the two wires with a voltmeter, you would measure 230 V. But if you were to measure the voltage from either of the wires to ground, you would find that the electrical pressure in each wire is only 115 V!

Figure 29-16 shows what is happening. Even though the electrical pressure in each wire is 115 V different from neutral, they are always different in opposite directions. When the voltage in one wire is 115 V above neutral, the voltage in the other wire is 115 V below neutral. Hence, we read 230-V difference between the two wires. On a

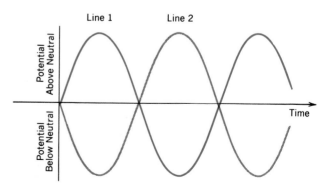

Figure 29-16 Graph showing 230-V single-phase power. Each line is 115-V single-phase relative to neutral.

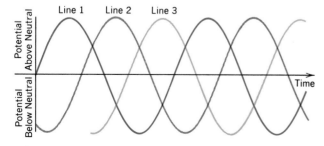

Figure 29-15 Three-phase power supply.

230-V single-phase system, neither of the two wires is neutral unless a third, neutral wire is provided.

Electricity and Magnetism

There is a close relationship between electricity and magnetism. There are several simple experiments that can demonstrate this relationship. Figure 29-17 shows a source of dc voltage (a battery) and a wire that is coiled around a metal bar. When the voltage is applied to the circuit, the bar becomes a magnet.

We talk about a **magnetic field,** and the lines of force in a magnetic field. Figure 29-18 shows another simple experiment that can demonstrate magnetic lines of force. Place a wire through a piece of paper and allow an electric current to flow through the wire. Sprinkle iron filings onto the paper, and then gently shake the paper. The iron filings will align themselves in the directions of the lines of magnetic force. There is a magnetic field created around the wire.

While electricity may be used to produce magnetism, magnetism may also be used to produce electricity. When a wire moves through a magnetic field, we say that it cuts the lines of magnetic force. Figure 29-19 shows another experiment you can do to generate electricity. When the wire is moved through the magnetic lines of force in one direction, a voltage will be measured on the voltmeter. When the wire is moved through the magnetic field in the opposite direction, a reversed voltage is produced. This relationship between electricity and magnetism is the basis for how electricity is produced, and how many electrical devices work.

Resistive and Inductive Loads

There are many devices (loads) that are used on ac systems that behave as if they were resistances in a dc circuit. Lights and heaters of many types fall into this category. You can measure their resistance with an ohmmeter, and calculate how much amperage they will consume in a circuit. These are called **resistive loads.**

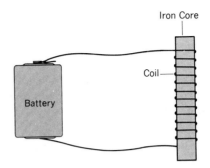

Figure 29-17 When a current is passed through a coil of wire, an electromagnetic field is produced.

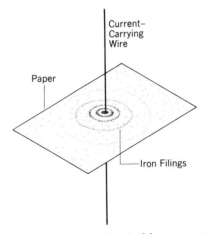

Figure 29-18 Magnetic lines of force created around a wire that is carrying current.

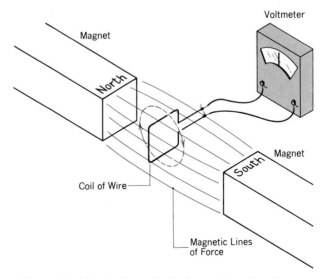

Figure 29-19 As the coil of wire rotates, it cuts through magnetic lines of force, generating a voltage.

There are other loads, however, that behave differently. These are all devices that produce magnetism as part of their operation. This would include any device that uses a coil of wire. Motors are the primary example of this type of load. Loads that use a coil of wire are called **inductive loads.** Their effective resistance in a system is much higher than the resistance you would measure with an ohmmeter. The current that passes through the coil produces magnetic lines of force. Because the current is always alternating in direction, the lines of force are always bulding up and collapsing. The adjacent wires are effectively cutting these lines of force, and they tend to generate a voltage in a direction opposite to the voltage being applied to the coil. This makes it harder for the amps to flow. This property of resistance that only occurs when the coil is energized in a circuit is called inductance.

Ammeters

Figures 29-20 and 29-21 show typical **ammeters** used for measuring the amperage flow in a wire. This ammeter is used in ac circuits only. Its prin-

ciple of operation relies on measuring the magnetic field that is produced around a current-carrying wire. The jaws of the ammeter are placed around the wire. The current flow can be read without even touching the wire. Figure 29-22 shows the correct and incorrect usage of the ammeter. Whenever the current is flowing to the right in one wire, it is flowing to the left in the other wire. The net result is that the magnetic fields produced will be in opposite directions, canceling out the effect of each other. If placed around both wires, the ammeter will read zero.

Sometimes we need to measure very low amperages, much lower than those that can be accurately measured on the scales available on the meter. Figure 29-23 shows how a wire can be run through the jaws of the ammeter a number of times. Each loop is carrying the current in the same direction, and each loop adds to the total magnetic effect. If 10 loops are used, the reading obtained on the ammeter will be 10 times the actual amperage flowing in the wire. Figure 29-24 shows a handy tool you can make called a **multiplying loop.** Simply place the loop in a circuit so that the current that flows through the load is also flowing through the multiplying loop. The amp reading at the multiplying loop may

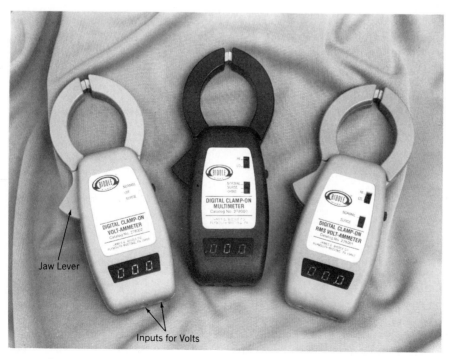

Figure 29-20 Digital volt-ammeter and multimeters.
(*Courtesy Biddle Instruments*)

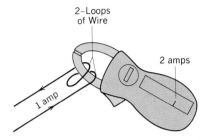

Figure 29-23 Amp reading will be twice the actual amperage flowing in the wire.

Figure 29-21 Analog clamp-on ammeter with ohm and voltmeter. (*Courtesy A. W. Sperry Instruments, Inc.*)

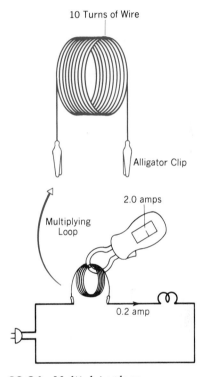

Figure 29-24 Multiplying loop.

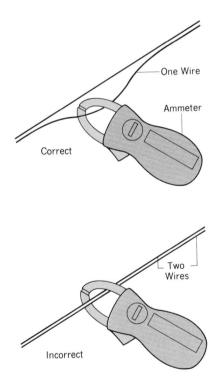

Figure 29-22 Usage of a clamp-on ammeter.

then be divided by 10 in order to determine the amp draw.

Common household circuits can usually carry 15 to 20 amps each. A refrigerator will draw between 3 and 6 amps, a gas valve will draw between 0.2 and 1.0 amps, and a 1-ton window air

conditioner (115 V) will draw between 10 and 14 amps.

Loads in Parallel

Thus far, we have only discussed simple circuits with one load. Air conditioners and refrigeration units will contain more than one load. Usually, these loads will be connected in **parallel,** as shown in Figure 29-25. When loads are connected in parallel, each load has one side connected to the hot leg and one side connnected to the neutral leg. Each load behaves as if it were

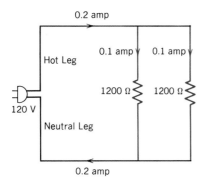

Figure 29-25 Loads in parallel.

the only load in the circuit. The first load has a resistance of 1200 Ω. When it is wired between a 120-V hot leg and a 0-V neutral leg, we have an current flow of 0.1 amp.

When a second resistance of 1200 Ω is added in parallel with the first, it also sees 120 V, and it also has a current flow of 0.1 amp. The wire that is supplying the current to these two loads carries 0.2 amp. As the number of the loads in parallel increases, the overall system resistance is reduced, and the overall system current increases.

Example

What is the overall system resistance of the two 1200-Ω resistances previously described?

Solution

We can think of an equivalent circuit that has only one resistance. The equivalent resistance is one that will allow the same current to flow as our two 1200-Ω resistors. Using Ohm's law, we can calculate this equivalent resistance as

$$R = 120 \text{ V}/0.2 \text{ amp}$$
$$R = 600 \text{ } \Omega$$

Note that the overall system resistance is less than either of the two individual resistances in the circuit. Each additional resistance added in parallel will provide another alternate path for the electron flow from hot to neutral. The system resistance will be decreased even if the resistance of the added load is high. This is reasonable in terms of our earlier example with the bucket of water. If a second outlet is placed at the bottom of the bucket, the total water flow will increase,

even if the new outlet has higher resistance than the original outlet.

When two **equal resistances** are placed in parallel, the overall resistance will be one half of the individual resistances. For three, four, or more equal resistances in parallel, the overall resistance will be one-third, one-fourth, etc., of the individual loads.

For unequal resistances in parallel, the overall system resistance may be calculated from either of the following formulas (algebraically, they are the same formula):

$$\frac{1}{\text{total } R} = \frac{1}{R_1} + \frac{1}{R_2} + \cdots$$

$$\text{total } R = \frac{R_1 \times R_2}{R_1 + R_2}$$

The first formula may be used for any number of resistances in parallel. When there are three or more resistances, the second formula may be used by taking two resistances at a time.

Example

Use each of these formulas to determine the overall resistance of a 2-Ω, a 4-Ω and a 5-Ω resistor in parallel.

Solution A

$$\frac{1}{\text{total } R} = \frac{1}{2} + \frac{1}{4} + \frac{1}{5}$$
$$= 0.50 + 0.25 + 0.20$$
$$= 0.95$$
$$\text{total } R = \frac{1}{0.95} = 1.05 \text{ } \Omega$$

Solution B

The resistance of the 2 Ω and the 4 Ω taken together is calculated as

$$R = \frac{2 \times 4}{2 + 4} = \frac{8}{6} = 1.333 \text{ } \Omega$$

and the resistance of this 1.333 Ω and the 5 Ω taken together is

$$R = \frac{1.333 \times 5}{1.333 + 5}$$
$$= \frac{6.665}{6.333} = 1.05 \text{ } \Omega$$

Loads in Series

Figure 29-26 shows a **series** arrangement of loads. Loads in series will be found far less frequently than loads in parallel. Note that an electron trying to make its way from the hot leg to the neutral leg of the circuit must pass through the first resistance, then the second resistance, and so on. Whereas in the parallel circuit, the electron had a choice of which load to go through, in a series circuit, the electron must go through each of the resistances in sequence. Adding another resistance in series will make it harder for the current to flow (system resistance increases), and the current flow will decrease.

For resistances in series, the overall resistance may be calculated by

$$\text{Total } R = R_1 + R_2 + R_3 + \cdots$$

Using this formula, it is easy to calculate the overall resistance of many loads in series. In air conditioning circuits, however, it would be quite rare to find more than two loads in series. In fact, except for a few special applications, loads will be wired in parallel.

For loads in series (unlike loads in parallel), the same current flow must pass through each load. The voltage drop across each of the loads in parallel will vary, depending on the total resistance.

Example
Two 200-Ω resistances in series are connected between a 120-V source of power. How much voltage drop will be taken across each resistance?

Solution
The total equivalent circuit resistance is the sum of the two individual resistances, or 400 Ω. Using Ohm's law, the amps flowing through the circuit may be calculated as

$$\text{Amps} = \frac{\text{volts}}{\text{resistance}}$$

$$= \frac{120}{440} = 0.3 \text{ amp}$$

and since the same current flow must pass through each of the resistors, the voltage drop across each resistor may be calculated as

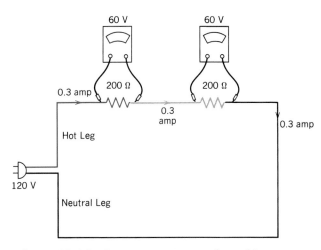

Figure 29-26 Series circuit example problem.

$$\text{Volts} = \text{amps} \times \text{resistance}$$
$$= 0.3 \times 200 = 60 \text{ V}$$

There is a voltage drop across each resistor of 60 V. The total voltage drop across both resistors must equal the total voltage available. In this case, the voltage drop across each load was the same, but only because the resistance of each load was the same. Chances are you will never see two loads of equal resistance in series because there are no practical applications using two devices that operate properly on 60 V each.

For loads of unequal resistance in series, the voltage drop across each load will be different. Those loads with high resistance will see most of the voltage drop. For example, if we were to put a very small heater wire (low resistance) in series with a load of relatively high resistance, we might measure a voltage of 2 V across the heater wire and 118 V across the load. The load will operate properly given any voltage between 110 V and 120 V, and the heater with its small voltage applied will create only a very small amount of heat.

Switches in Series and Parallel

Just as we have described that loads can be wired either in series or in parallel, so too can switches be wired in series or parallel. Figure 29-27 shows one load, controlled by two switches in series. In order to get a completed circuit through the load,

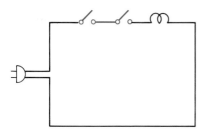

Figure 29-27 Switches in series. Both switches must be closed to complete the circuit.

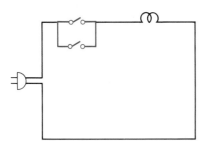

Figure 29-28 Switches in parallel. Either switch can complete the circuit.

both switches must be closed. If either one or both switches are open, there will be no complete circuit.

Switches in series are very common. When we want to have several safety switches all satisfied before allowing a load to be energized, the safety switches will be wired in series with the load.

Figure 29-28 shows a load controlled by two switches in parallel. This load will be energized when at least one of the switches is closed. Each switch provides an independent means of energizing the load. But when any of the switches is opened, there is no guarantee that the load will be deenergized.

Power

Electrical **power** is a measure of the rate of energy consumption. An electric motor delivers mechanical energy and is rated in horsepwoer. An electric heater delivers energy in the form of heat. The rate at which the electrical energy must be supplied to each of these two devices is a function of how many amps and at what pressure (voltage) these amps must be supplied. Power is measured in **watts** (W), and for single-phase applications

for resistive loads it is calculated as

$$\text{Watts} = \text{volts} \times \text{amps}$$

Example

An electric heating element is rated at 2000 watts when operating at 230 V. How many amps will it draw?

Solution

$$2000 = 230 \times \text{amps}$$
$$\text{amps} = \frac{2000}{230}$$
$$\text{amps} = 8.7$$

For three-phase applications, the wattage is calculated as

$$\text{Watts} = \text{volts} \times \text{amps} \times 1.73$$

The 1.73 factor acccounts for the fact that power is being delivered through three individual wires, although not all at the same time.

For circuits with inductive loads (usually motors), neither of the previous methods may accurately calculate the power consumed, due to an effect called **power factor.** When motors are running, the motor windings cut through lines of magnetic force and generate a voltage of their own. This voltage is called **back EMF** (electromotive force), and its direction tends to oppose the voltage being supplied to the motor. This causes the voltage and the amperage cycles to be out of phase with each other. That is, the voltage peak will occur slightly ahead of the amperage peak. This causes an electrical inefficiency, and the power delivered is actually less than you would calculate by either of the previous formulas. Power factor is the ratio of the actual watts delivered compared to the wattage calculated by one of the previous formuals. For circuits with resistive loads only, the power factor is 1.00. Where inductive loads are involved, the power factor will range between 0.85 and 1.00.

Interconnection and Ladder Diagrams

There are two different types of schematic wiring diagrams that are used to describe circuits (Fig-

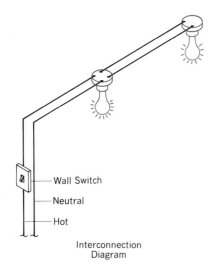

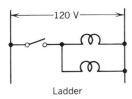

Figure 29-29 Interconnection and ladder diagram.

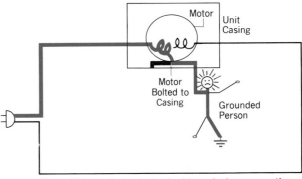

Note: This symbol (⏚) attached to a device means the device casing is attached to the grounded casing or other ground

Figure 29-30 Shock hazard caused by a motor without a casing ground.

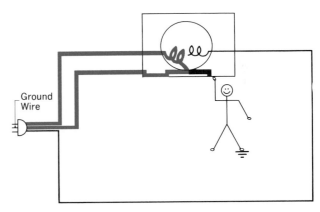

Figure 29-31 In the event of a grounded winding, the circuit will be completed through the casing ground instead of through the person.

ure 29-29). An **interconnection diagram** shows the actual physical relationship of the components. It has the advantage of being able to show you where to look to find any of the individual components in the circuit. A **ladder diagram** is preferred by many people because it makes it easier to figure out how the circuit is supposed to work. We will use both types of diagrams in this text. The student needs to become familiar with both, and to be able to draw a ladder-type diagram from a schematic. Manufacturers will usually put a wiring diagram of the unit on one of the access panels. This could be either an interconnection diagram or a ladder diagram.

Casing Grounding Wires

Figure 29-30 shows a common type of electrical failure that can present a significant safety hazard to people. The motor has grounded to the casing. In this case, if a person were to touch the casing, there is an electrical potential across the person, and there is a shock hazard. The remedy (Figure 29-31) to this problem is to provide a

path for this electrical potential that is far less than the resistance of the person. The electricity will then flow through the **casing ground** wire in preference to the person. Any electrical device may be furnished with a green ground wire. It must be attached to the casing.

Frequently, the current flow through the portion of the motor winding remaining energized will be much higher than normal, and will cause a fuse or circuit breaker in the power wiring to open. In this case, we say that the motor has become grounded, and it must be replaced.

Note that if there is not an electrical failure, the casing ground serves no purpose. Any unit will run satisfactorily without the casing ground, until there is a failure. It is a serious mistake to omit the casing ground where it is required by code or the manufacturer.

There is some misuse of the terms ground,

common, and neutral. A ground wire is at zero electrical potential, and during normal operation, carries no current. A neutral wire is at zero electrical potential, and during normal operation it carries the same amps as the hot wire. Common is a slang term for the neutral wire because many different loads are sometimes wired together into one common neutral wire.

Depending on usage, when we say that a device is grounded, it can mean that it's OK, or it can mean trouble. If used to say that the housing of a device is connected by a green wire to the casing, then all is OK. But the same term can be used to describe a faulty situation in which the wiring that normally carries current is touching part of a housing or casing.

KEY TERMS AND CONCEPTS

Electrical Pressure	VOM	Loads in Parallel
Electrical Flow	ac and dc	Loads in Series
Electrical Resistance	Sine Wave	Switches in Parallel
Volts, Amps, Ohms	Peak Voltage	Equivalent Resistance
Ohm's Law	Effective Voltage	Switches in Series
Hot Leg	Single-phase Power	Power
Neutral Leg	Three-phase Power	Power Factor
Short Circuit	Resistive Loads	Interconnection Diagram
Overcurrent Device	Inductive Loads	
Ohmmeter	Ammeter	Ladder Diagram
Voltmeter	Multiplying Loop	Casing Grounds

QUESTIONS

1. Which electrical measurement is the equivalent of electrical pressure?
2. Which electrical measurement is the equivalent of gallons per minute?
3. Which electrical property can be measured when the circuit is unplugged?
4. How much current will flow in a 24-V circuit that has a total resistance of 48 Ω?
5. What is the resistance in a 120-V circuit that allows a current flow of 0.8 amp?
6. A closed switch is located on the hot leg of a 120-V circuit. Will you get a shock if you touch the side of the switch connected to the hot leg?
7. In Question 6, will you get a shock if you touch the outlet side of the switch?
8. In Question 6, how much voltage would you measure between the inlet side of the switch and the ground?

9. In Question 6, how much voltage would you measure between the outlet side of the switch and the ground?

10. In Question 6, how much voltage would you measure across the terminals of the switch?

11. Answer Questions 6 through 10 again, but this time assume that the switch is open instead of closed.

12. If you find a black wire and a white wire connected to a wall outlet, which one would probably be the hot leg and which the neutral leg?

13. What can you guarantee is true if you measure a voltage across the terminals of a switch?

14. How much voltage would you measure across a lighted household light bulb?

15. How much voltage would you measure across the bulb when the light switch is closed, and the bulb element has burned out?

16. A 120-V bulb is controlled by a switch in the hot leg. With the switch open (off), how much voltage would you measure between the light bulb and ground?

17. A 120-V bulb is controlled by a switch in the neutral leg. With the switch open, how much voltage would you measure between the light bulb and ground?

18. What is a short circuit?

19. What happens to amperage in a circuit where a short circuit occurs?

20. What happens to the amperage flowing in a circuit when a voltmeter is placed across an open switch? A closed switch?

21. When an ohmmeter is in its case and not being used, how many ohms is it reading?

22. Why is it necessary to zero the ohmmeter each time it is used?

23. When reading a resistance on the $R \times 100$ scale, you found that the reading was very close to zero. Which scale would you switch to in order to get an accurate resistance reading?

24. Which voltages would commonly be found in a residence with central air conditioning?

25. What is an electromagnet?

26. Give two examples of resistive loads.

27. Give two examples of inductive loads.

28. What will happen to the amperage flow in a circuit when a second load is added in parallel with an existing load?

29. What will happen to the amerpage in a circuit when a second load is added in series with an existing load?

30. Two 100-Ω loads are placed in series across a 24-V circuit. How many amps will flow?

31. Two 100-Ω loads are placed in parallel across a 24-V circuit. How many amps will flow?

CHAPTER 30

ELECTRICAL DEVICES

The previous chapter explained how switches and loads may be put together in order to make circuits. In this chapter, many of the specific types of switches and loads are described. The next chapter describes the common circuits used in everyday heating, air conditioning, and refrigeration systems.

Types of Switches

The simple switch used in the previous chapter is called a single-pole single-throw switch (**SPST**). It has two wires connected to it, and the switch may be either opened or closed. Figure 30-1 shows other types of switches besides the SPST.

The single-pole double-throw switch (**SPDT**) has one switch that makes a different completed circuit for the two different positions of the switch. In home wiring, this switch is sometimes referred to as a three-way switch. It is meaningless to describe this switch as being open or closed. One circuit is always open when the other is closed.

The double-pole single-throw switch (**DPST**) is actually two separate SPST switches in the same box. The dashed line connecting the two poles indicates that they work together. They are either both open, or they are both closed.

The double-pole double-throw switch (**DPDT**) is a combination of two separate DPST switches, ganged to operate together. When the switch moves from one position to the other, two circuits are opened and two circuits are closed.

Figure 30-2 shows a **momentary switch.** A momentary switch is one that is spring-loaded in either the open or the closed position. When it is pushed manually, its position is changed. When the switch is no longer being pushed, its position returns to the spring-loaded position. These switches are either normally open or normally closed. The "normal" position is the position the switch is in when it is taken out of the box. It does not refer to whether the switch is normally depressed or not when the system is operating normally.

Switches can be either manually or automatically operated. Manual switches rely on an external movement in order to operate. Light switches in the home are manual switches. A person needs to move the pole from one position to a different position. A door switch on a car or a refrigerator is also a manual switch. Its operation depends upon a person opening or closing a door in order to move the pole of the switch.

Automatic switches are devices that sense a variable (usually temperature or pressure). As the temperature or pressure changes, the switch is

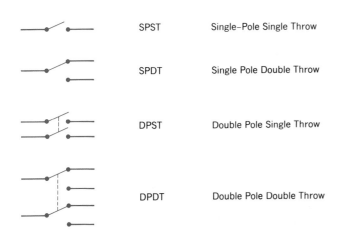

SPST	Single–Pole Single Throw
SPDT	Single Pole Double Throw
DPST	Double Pole Single Throw
DPDT	Double Pole Double Throw

Figure 30-1 Types of switches.

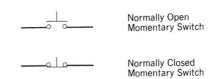

Normally Open Momentary Switch

Normally Closed Momentary Switch

Figure 30-2 Momentary switches.

Symbol	Explanation	Example of Usage
	Opens on a rise in temperature	Heating thermostat
	Closes on a rise in temperature	Cooling thermostat
	Opens on a rise in pressure	High pressure cutout
	Closes on a rise in pressure	Low pressure cutout
	Opens on an increase in level	Safety switch—condensate pumping unit
	Closes on an increase in level	Turns on pump–condensate pumping unit
	Opens on an increase in flow	Alarm circuit to sound if flow stops
	Closes on an increase in flow	Prove airflow before turning on humidifier

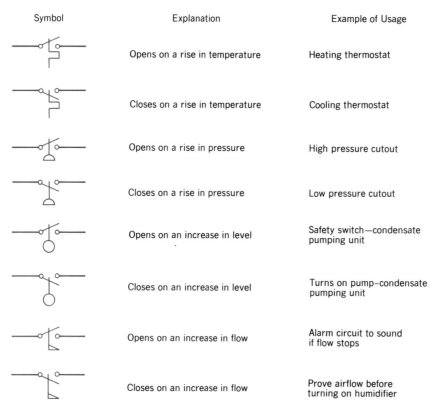

Figure 30-3 Automatically operated switches.

caused to move from one position to the other. Figure 30-3 shows the symbols used in a wiring diagram to describe various types of automatic switches. Note that the symbol gives you a lot of information about the switch in a very concise fashion. It tells you what variable controls the switch, temperature or pressure. There are other symbols for switches operated by a liquid level, flow of air or water, or humidity. The position of the pole shows you whether the switch will open or close as the variable being measured increases. The automatically controlled switch may be either the SPST, SPDT, DPST, or DPDT type. Some automatic switches are shown in Figures 30-4 through 30-8.

A switch that is operated from temperature is called a **thermostat.** A switch that is operated from pressure is called either a **pressure switch,** a **pressurestat,** or a **pressure controller.** A switch that opens or closes in response to humidity is called a **humidistat.** A switch that senses a liquid level in a tank to open or close its contacts is called a **float switch** or a **level switch.** A switch that responds to the flow of air is called a **sail switch,** and a switch that responds to the flow of a liquid (usually water) is called a **flow switch.**

Sensing Bulb for Remote Mounting

Setpoint Adjustment

Figure 30-4 Remote-bulb thermostat. (*Courtesy Johnson Controls, Inc.*)

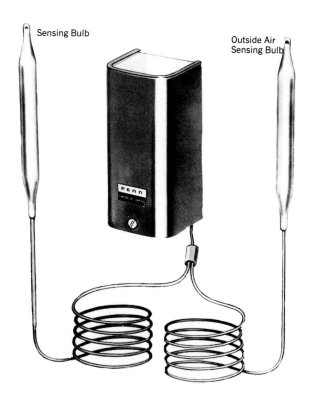

Sensing Bulb

Outside Air Sensing Bulb

Figure 30-5 Electric remote-bulb thermostat with reset. As outside air temperature falls, it resets the set point.

DURATION
SHORTER · LONGER

DEFROST CONTROL

PENN

Figure 30-6 Remote-bulb thermostat used as a defrost termination switch. Shorter–longer adjustment decreases–increases the set point.

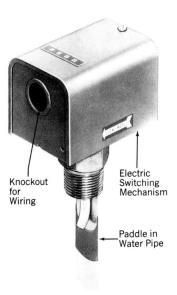

Knockout for Wiring

Electric Switching Mechanism

Paddle in Water Pipe

Figure 30-7 Flow switch. (*Courtesy Johnson Controls, Inc.*)

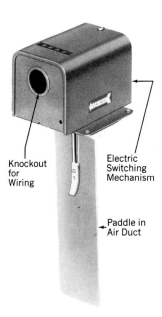

Knockout for Wiring

Electric Switching Mechanism

Paddle in Air Duct

Figure 30-8 Sail switch. (*Courtesy Johnson Controls, Inc.*)

Troubleshooting of a switch is done most easily with a voltmeter. If you measure a voltage across a switch when it is in fact supposed to be closed, then the switch is not completing the circuit. This can be due to a faulty sensing part of an automatic switch, or it may be due to contacts that are pitted and not making good electrical contact when they are brought together. Do not "clean up" the contacts by filing them. New contacts are silver coated to resist corrosion and pitting. Filed contacts have no silver protective coating, and will fail again from corrosion very soon.

Switches can also be tested using an ohmmeter. An open switch will have infinite resistance. A closed switch with clean contacts should have a resistance of zero (less than 0.5 Ω).

Transformers

Transformers are devices that are used to change a source of power from one voltage to another. They can be used to either create a lower voltage (step-down transformer) or a higher voltage (step-up transformer). It is very common to run heavy loads, such as compressors and fans on one voltage, and to use delicate switches, such as thermostats, on a lower voltage. The transformer provides this lower voltage.

Figure 30-9 shows a transformer and its symbol that is used in a schematic diagram. The transformer consists of two coils of wire located near each other. These coils are called the **primary** winding and the **secondary** winding. The same voltage that is used to run the large loads is applied across the primary winding. Because the

primary voltage is alternating current, the magnetic field that is created around the primary coil is constantly changing—building and collapsing. This moving magnetic field cuts through the wires in the secondary winding, inducing a voltage into it. This voltage may then be used to supply electricity to smaller loads, which are designed to run at this lower voltage. The secondary voltage is also alternating current.

The two different sides of the transformer may be referred to as primary and secondary, line side and load side, or **line voltage** and **control voltage.** The most commonly used voltages are either 120 or 240 V on the primary side, and 24 V on the secondary side. The relationship between the voltages is determined by the number of turns of wire on the primary winding compared to the number of turns of wire on the secondary winding. If the number of windings on the secondary is one-fifth of the number of turns on the primary, the secondary voltage will be one-fifth of the primary voltage. A 240/24-V transformer would have ten times as many turns on the primary as it had on the secondary.

Figure 30-9 also shows a transformer with five wires instead of four. This transformer may be used with either of two different primary voltages in order to obtain a 24-V secondary voltage. A common **dual-voltage transformer** is one that can use either 208 V or 240 V on the primary,

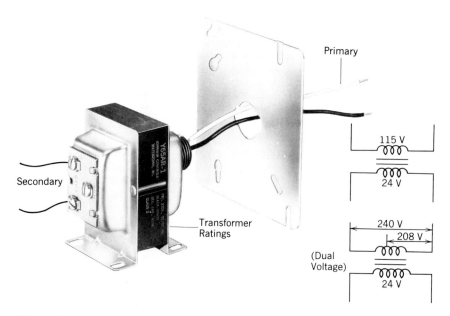

Figure 30-9 Control power transformer. *(Photo Courtesy Johnson Controls, Inc.)*

and produce 24 V on the secondary. Dual-voltage transformers may also be used to produce two different secondary voltages from a single primary voltage.

If one side of the secondary is attached to a ground, then the electrical pressure anyplace in the secondary circuit could be measured relative to the casing ground. In most cases, however, one side of the secondary is not grounded, and the casing cannot be used as an electrical reference when measuring voltage in the secondary circuit.

Transformers are rated according to how much electrical load they can handle. The load is measured in volt-amps, usually abbreviated **VA**. VA refers to the volts times the amps the transformer can produce. For example, if a transformer is being used to supply power to a device that draws 0.5 amp at 24 V, the transformer would need to supply 12 VA. The rating of the transformer would need to be higher than the VA load to be supplied. Residential heating-only systems will usually use a 20-VA transformer, while cooling applications will probably use a 40-VA transformer. The VA input to the transformer is approximately the same as the VA output. The input has much higher volts, but correspondingly lower amps.

Troubleshooting the transformer is easy. Measure the secondary voltage with a voltmeter. If you measure the correct secondary voltage, the transformer is good. If you do not measure any voltage at the secondary, confirm that you have voltage being applied to the transformer primary winding. If you do, and there is no voltage at the secondary, the transformer has failed and must be replaced (after investigating potential causes of the failure).

Contactors

A **contactor** is a device that uses the magnetic field created by a low voltage coil to close a heavy-duty switch. Figure 30-10 shows various arrangements for a contactor, and Figure 30-11 describes its principle of operation. When control voltage is applied across the coil, a plunger is pulled down into the magnetic field. The movement of this plunger causes either one, two, or three independent switches to close. The contactor is described as either one pole, two pole, or three pole, depending upon the number of switches that are closed by the coil. All the poles operate at the same time, making or breaking one, two, or three wires, which are connected to a large motor or resistance heater. There are two important ratings on the contactor. One is the control voltage. That is, the coil is designed to operate on one specific voltage, and it must match the control voltage available. The second important rating is that of the contacts. Common ratings of the contacts will be 20, 25, 30, 40, 50, 60 amps or higher. Sometimes there will be two ratings given on the contactor contacts, the resistive rating in amps and the inductive rating in horsepower. A third rating given on the contactor is the current draw of the coil when it is energized. This amperage, when multiplied by the voltage rating of the coil, tells you how many VA will be drawn from the transformer to operate the contactor.

The simplest contactors consist of only the coil and the load-carrying switches. A variation is sometimes used in which there are auxiliary contacts. Auxiliary contacts are small switches that open or close whenever the contactor switches open or close. There may be one or more of these auxiliary contacts. They will be wired into the low voltage control circuit.

When contactors are shown in control diagrams (Figure 30-12), different portions of the same contactor will show up in different parts of the diagram. The coil will be shown wired into the control voltage part of the circuit, and the contacts will be shown in the line voltage part of the circuit that operates the load.

Control Relay

The **control relay** is quite similar to the contactor, with the following differences.

a. Contactors have switches that are rated for a load-carrying capability of 18 amps or more. Control relay switches are used to switch loads of less than 18 amps.

b. The contactor switches are always normally open. That is, the switches are open except

Single Pole
(25 & 30 amps)

Single Pole with Bus Bar
(25 & 30 amps)

Two Pole
(20, 25, 30 amps)

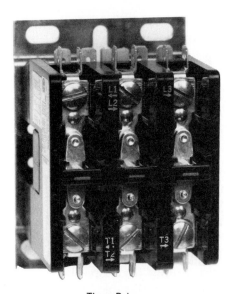

Three Pole
(25, 30, 40 amps)

Three Pole
(50, 60, 75 amps)

Figure 30-10　Contactors. *(Courtesy Joslyn Clark Controls, Inc.)*

when the coil is energized. The contacts on the control relay may be normally open, normally closed, or it may have some normally open and some normally closed contacts.

c. Contactors have from one to three load-carrying switches. Control relays may have as many as three SPDT switches.

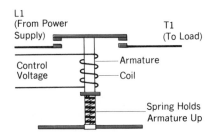

Figure 30-11　Operation of a single-pole contactor.

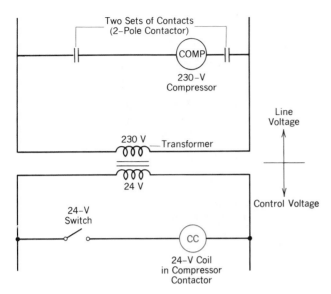

Figure 30-12 Closing the switch allows a small current to pass through the coil. The magnetic field from the coil pulls the switches closed in the 230-V circuit.

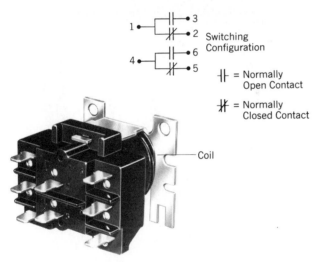

Figure 30-13 General-purpose switching relay or control relay. *(Courtesy Honeywell)*

Figure 30-13 illustrates the control relay. In the schematic control diagram (Figure 30-14), the contacts are shown as either normally open or normally closed. When the manual switch is closed, the coil is energized, turning on bulb A and turning off bulb B. The "normal" position of the switches is the position they assume when the coil is not energized. It does not refer to their position when the unit is operating normally.

Some systems will have more than one control relay. In that case, the coils are all labeled differ-

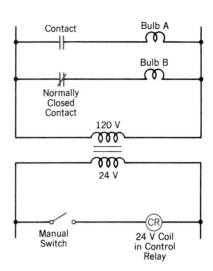

Figure 30-14 Schematic using a control relay.

ently on the wiring diagram. The contacts are labeled with the same nomenclature as is on the coil that operates that set of contacts. The control relays may be labeled CR1, CR2, etc., or they may be labeled to indicate the load that is controlled. For example, if the switches on a control relay are used to control the operation of a fan or blower, the control relay may be identified as FR or BR (fan relay, blower relay).

Thermostats

Thermostats are temperature-actuated switches. Figure 30-15 shows a simple type of thermostat used in a refrigerator. It may also be furnished with a sensing element filled with a liquid or a gas, which expands and contracts as the temperature it senses changes. When **remote-bulb thermostats** (Figure 30-4) are used, the bulb may be placed inside the space whose temperature is to be controlled, while the switching mechanism is located in a more convenient location. The remote element for this type of thermostat may be either a bulb or an averaging element. The bulb senses temperature at a single point. The averaging element may be arranged to sense, for example, the average temperature of air flowing in a duct. These thermostats are called line voltage thermostats. The thermostats themselves are the switches that carry the full amperage of the load that is being controlled.

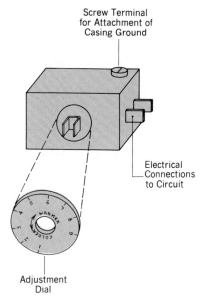

Screw Terminal
for Attachment of
Casing Ground

Electrical
Connections
to Circuit

WARMER
COLDER

Adjustment
Dial

Figure 30-15 Line voltage refrigerator thermostat.

Figure 30-16 Line voltage wall thermostat. *(Courtesy Johnson Controls, Inc.)*

The thermostats that are of the most interest are the room thermostats used to control heating and air conditioning systems. They can be line voltage (Figures 30-16 and 17), but in most cases, they are 24-V thermostats. They may be used to energize 24-V components such as gas valves, compressor contactor coils, and blower relays. Figure 30-18 shows the simplest of these thermostats. It is a heating thermostat that closes on a fall in temperature. When the contacts close, it will cause the heating system to

Figure 30-17 Line voltage thermostat cutaway. *(Courtesy Johnson Controls, Inc.)*

Figure 30-18 Two-wire 24-V room thermostat. *(Courtesy Honeywell)*

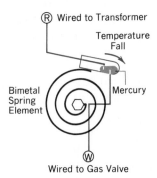

Ⓡ Wired to Transformer

Temperature
Fall

Bimetal
Spring
Element

Mercury

Ⓦ
Wired to Gas Valve

Figure 30-19 Two-wire thermostat. When the room temperature falls, the mercury bulb tilts down to the right, closing the switch between the *R* and *W* terminals.

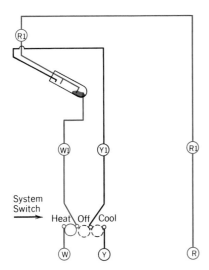

Figure 30-20 Three-wire thermostat with a sub-base. *R1, W1,* and *Y1* are on the thermostat. Terminals *R, W,* and *Y* are on the subbase. In the position shown, the system switch is set for heat and the thermostat is calling for heat, completing the circuit from *R* to *W*. On cooling, the circuit will be complete from *R* to *Y*.

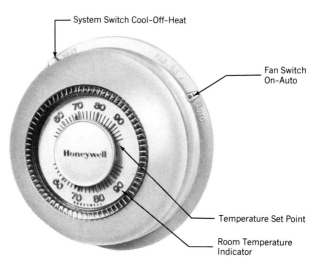

Figure 30-21 Heating and cooling thermostat on a subbase. *(Courtesy Honeywell)*

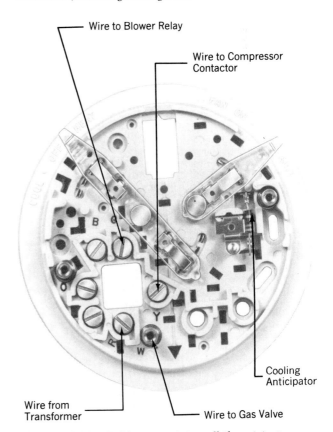

Figure 30-22 Subbase contains all the wiring connections to the thermostat. *(Courtesy Honeywell)*

operate (Figure 30-19). A cooling thermostat that controls a cooling-only system would be similar, closing its contacts on a rise in temperature instead. Figure 30-20 shows a three-wire thermostat schematic that may be used to control either a heating or a cooling system. The bimetal tilts a mercury bulb, which is a SPDT switch. If this switch were to be used to control both heating and cooling, either the heating or the cooling would be on at all times. In order to avoid this, we use a manual switch, which is set either for heating or for cooling, so that only one or the other is available at any time.

Figure 30-21 shows a four-wire thermostat mounted on a subbase. The **subbase** (Figure 30-22) contains all the manual switches. The **system switch** may be set for Heat, Cool, or Off. The **fan switch** may be set for Fan or Auto. Figure 30-23 shows the internal wiring of the thermostat and subbase. In the Heat position, the thermostat will operate the gas valve or other primary heating control. In the Cool position, the bimetal element will cause the compressor and evaporator fan to run. With the fan switch in the Auto position, the fan will run whenever required by the heating or cooling system. With the fan in the On position, the fan will run, regardless of the position of the system switch or the bimetal

temperature sensing switch. Heating plus cooling applications will most commonly use a four-wire thermostat. There are, however, six-wire thermostat systems in use. The extra two terminals are usually energized when the system

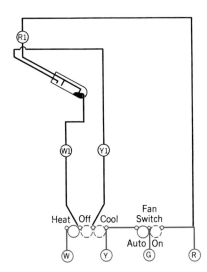

Figure 30-23 Internal wiring of a four-wire thermostat.

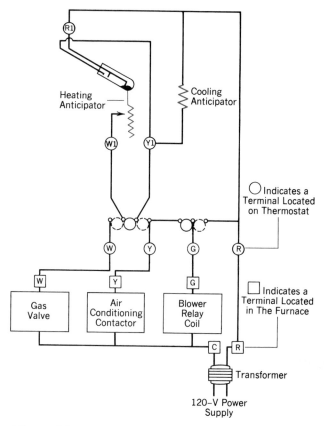

Figure 30-24 Thermostat with heating and cooling anticipators.

switch is set to heating or cooling. The output signal is used to drive electric damper motors that configure duct dampers or valves to correspond to the heating or cooling mode.

Figure 30-24 shows another schematic of a room thermostat, with the addition of two small resistors. These resistors are called **heat anticipator** (Figure 30-25) and a **cooling anticipator**. The heat anticipator is energized whenever the heating system is on. It has a low resistance and therefore takes only a very small voltage drop. It fools the bimetal element into sensing that the room has warmed up before it actually does. The cooling anticipator is energized whenever the air conditioning has cycled off. It also fools the bimetal element, causing the air conditioner to cycle on sooner than the bimetal is able to sense. The anticipators allow closer control of room temperatures by causing more frequent cycling of the heating and air conditioning system. The heating anticipator is located on the thermostat itself, and it is adjustable. It should be set to match the amperage draw of the device that is being controlled (i.e., the gas valve). The cooling anticipator is located on the subbase (Figure 30-22), and is not adjustable.

In commercial applications, thermostats may be enclosed to prevent occupants from making adjustments. Then, a thermostat that automatically changes the system from heating to cooling is required. An **automatic changeover thermostat** (Figure 30-26) has two bimetal elements,

Figure 30-25 Heat anticipator. *(Courtesy Honeywell)*

two mercury switches, and two set points. The construction of the thermostat is such that the heating set point is at least two degrees lower than the cooling set point. This prevents unnecessary cycling between heating and cooling.

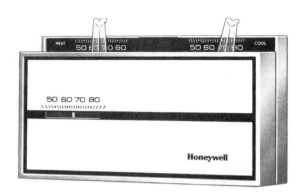

Figure 30-26 Automatic changeover thermostat has a heating set point and a cooling set point, and will automatically change from heating to cooling and back as required.

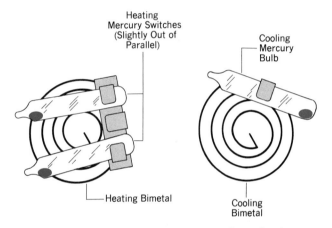

Figure 30-27 Two-stage heating and single-stage cooling thermostat.

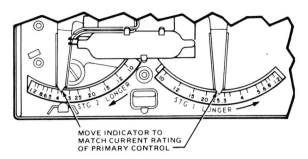

Figure 30-28 Two heat anticipators on a two-stage thermostat. *(Courtesy Honeywell)*

Figure 30-29 Microelectronic thermostat. *(Courtesy Johnson Controls, Inc.)*

Staging thermostats (Figure 30-27) have two mercury switches on the same bimetal element. They may be two-stage heating, two-stage cooling, or both. For a two-stage heating thermostat, as the room temperature drops 1°F below the set point, the first mercury bulb will tilt, turning on the first stage of heating (i.e., heat pump). If the capacity of the first-stage heater is sufficient to match the room heat loss, the system will cycle on and off using the first stage only. But if the room temperature drops to 2°F below the set point, then the second-stage bulb will also tilt, bringing on additonal heating capacity (i.e., auxiliary electric heat). Each stage on the multiple-stage thermostat can have its own heating or cooling anticipator (Figure 30-28).

Microelectronic thermostats (Figures 30-29 through 30-31) are the most sophisticated of all the thermostats. The wiring to the subbase connections are similar to conventional thermostats, but the thermostat itself must be programmed like a computer. They have the capability of choosing different set points during different hours of the day or during different days of the week. Each manufacturer provides detailed instructions on programming their particular thermostat. Frequently, the service technician will be asked to provide programming instructions to the customer.

The **terminal identification** on thermostats is not universal, but the following systems are very common.

1. Power connection from the transformer is R (for red) or V (for voltage). Where separate transformers are used for the heating and cooling systems, they may be labeled R_c and R_h. If both terminals are available but only a single transformer is being used, place a jumper wire between R_c and R_h.

2. Heating: the terminal that leads to the load that provides the heating is labeled W (for white) or H (for heat). If it is a two-stage heating thermostat, the terminals could be labeled W_1 (first stage) and W_2 (second stage).

3. Cooling: the terminal that leads to the cooling contactor will be either Y (for yellow) or C (for cooling). For two stages, Y_1 and Y_2.

4. Fan: the terminal that provides power to the blower relay will be G (for green) or F (for fan).

5. Heating damper (not commonly used): B (for blue). This is simply an auxiliary switch that is energized whenever the system is set for heating.

6. Cooling damper (not commonly used): O (for orange). This auxiliary switch is energized whenever the system switch is set for cooling.

Solenoid Valves

A **solenoid valve** is a valve that is opened or closed electrically (Figure 30-32). The principle of operation is the same as for the contactor, except that instead of the plunger moving a set of switches, the plunger causes a valve to open or close. If the current on the coil causes the valve to open, it is a normally closed valve. If the valve is open when there is no voltage on the coil, then it is a normally open valve. The valves that are operated by the solenoid coil may be two-way, three-way, or four-way valves. The electrical terminals on a solenoid valve are not always easy to reach with the voltmeter probes. An easy way to tell whether the valve is energized or not is to touch a lightweight screwdriver blade to the housing of

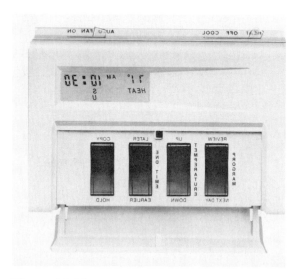

Figure 30-30 Programming the microelectronic thermostat. *(Courtesy Johnson Controls, Inc.)*

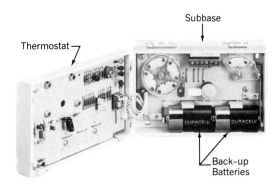

Figure 30-31 Subbase for the microelectronic thermostat. *(Courtesy Johnson Controls, Inc.)*

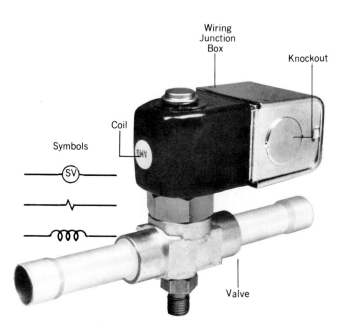

Figure 30-32 Solenoid valve. *(Photo Courtesy Alco Controls Division Emerson Electric Co.)*

the solenoid. If the coil is energized, you will be able to feel the magnetic field that is being created by the coil. Be certain that you do not touch any exposed wiring with the screwdriver.

The two most commonly used types of solenoid valves are the liquid solenoid valve on a refrigeration system and the gas valve on gas-fired heaters.

Heating Elements

Figure 30-33 shows the symbol used to describe an electric heating element or a resistor. The heating element may be a major load such as in an electric furnace, a small heater such as the one in the walls of a refrigerator to eliminate sweating, or the heating element may be only a

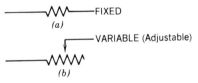

Figure 30-33 Resistor or heating element. (*a*) Fixed. (*b*) Variable (adjustable).

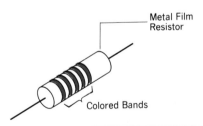

small resistor used to produce heat inside another controller (such as a cooling anticipator). A resistor shown with an arrow (Figure 30-33*b*) is a variable resistor. Its resistance may be adjusted as with a heat anticipator.

Normally, you will not be dealing with repair of components that use resistors, as you will probably only be replacing them. It is, however, sometimes useful to be able to read a resistor. Figure 30-34 shows how to read the resistance value for resistors with colored bands.

Time-Delay Switches

Sometimes it is desirable to cause one device to operate some seconds, minutes, or hours after another event has happened. For example, once a large motor has started, we may not want to allow it to start again for 30 minutes. Or on a home furnace, we want the fan to come on only after the gas valve has been open for one minute. **Time-delay switches** may operate off a timer clock, a **bimetal switch** and heater element, or a solid state timer.

Figure 30-35 shows a clock-type timer. The clock motor is connected directly to a source of voltage. As it turns, a series of cams is actuated to open or close switches. This is the type of timer used to regulate the operation of automatic defrost refrigerators.

Figure 30-36 shows a bimetal-type timer. The box contains two circuits, a resistor (sometimes called the heater), and a bimetal switch. When a

	Color	Number Value
1. First three bands—resistance value	Black	0
	Brown	1
2. Fourth band—number of zeros to add to resistance value, except if fourth band is gold, divide resistance value by 10. If fourth band is silver, divide resistance value by 100	Red	2
	Orange	3
	Yellow	4
	Green	5
3. Fifth band—tolerance (accuracy)	Blue	6
2% = Red	Violet	7
5% = Gold	Gray	8
10% = Silver	White	9

Examples: brown, white, green, gold, gold = 19.5 Ω ± 5%; brown, white, blue, red, silver = 19,600 Ω ± 10%.

Figure 30-34 Reading a metal film resistor.

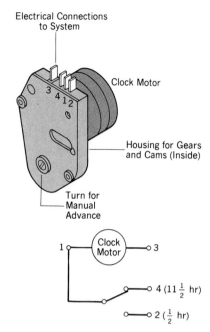

Figure 30-35   Defrost time clock used in domestic refrigerator.

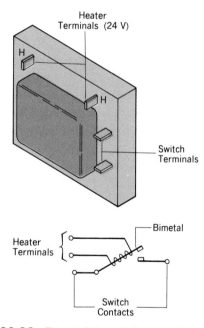

Figure 30-36   Bimetal time-delay switch.

voltage is applied to the resistor, it heats up. The heat from the resistor causes the bimetal switch to warp and make (or break) the switch. The resistor may be on the low voltage side of the circuit, while the switch is on the line voltage side of the circuit. The time delay with this type of switch is usually limited to less than 2 min-

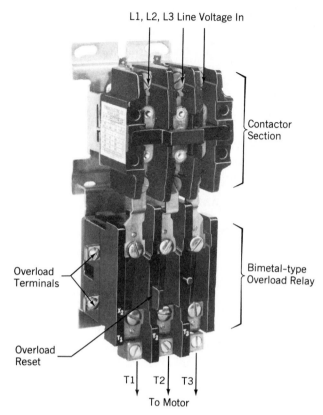

Figure 30-37   Magnetic starter. *(Courtesy Furnas Electric Company)*

utes. When the voltage is removed from the heater, it begins to cool. There will also be a time delay for the switch to return to its normal position.

Magnetic Starter

A **magnetic starter** (Figure 30-37) performs the same function as a contactor—it uses a control voltage to create a magnetic field that causes a switch or switches to close, starting a motor or other load. The magnetic starter, however, has an additional feature in that it provides **overload** protection as well. The magnetic starter has heater elements that create heat as they carry current. The more current they carry, the more heat they produce. The heaters are selected to match the maximum amperage rating of the load being started. If this rating is exceeded, the heaters will cause a solder pot to melt, and the control circuit to the coil will open. The device must then be manually reset before it will operate again.

Some magnetic starters have built-in heaters that are adjustable with a screwdriver. This eliminates the need for physically changing out heater elements to match the controlled load.

Note that the overload protection provided is to protect the motor. This is not the same as **overcurrent** protection that is provided by a fuse (see Chapter 31, on power wiring). The wiring that supplies power to a motor is capable of carrying more amps than the motor can safely use. A fuse protects the wiring.

Capacitors

A **capacitor** (Figure 30-38) consists of two conducting plates separated by a dielectric (insulating) material. When a voltage is applied to a capacitor, electrons build up on one plate, charging the capacitor. When the charge builds up on one plate, electrons are moved from the other plate. The mechanical analogy of a capacitor is an air bladder inside a water pipe. As pressure builds up on one side of the bladder, the bladder is compressed. Then, after some time delay, the pressure builds up downstream from the bladder.

When a capacitor is used in an alternating-current circuit, the build-up of charge can be used to amplify the voltage as it builds in the opposite direction. Capacitors are of two types, and used for two different purposes.

1. **Start capacitors** are commonly round in cross section, and designed to amplify a voltage to increase the starting torque of a motor. Start capacitors are designed to be used for only a few seconds at a time (during the start-up of a motor). After this time, a switch wired in series must open, taking the start capacitor out of the circuit.

2. **Run capacitors** are commonly oval or rectangular in cross-section shape, and designed to align the voltage and amperage cycles that have been separated by the back EMF generated by a motor winding. This improves the power factor and reduces the operating cost for a motor.

The unit of capacitance is the **farad.** A farad, however, is a very large unit of capacitance. In order to avoid using very small numbers, capaci-

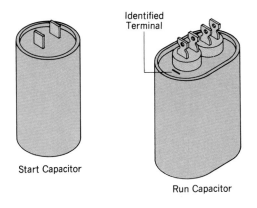

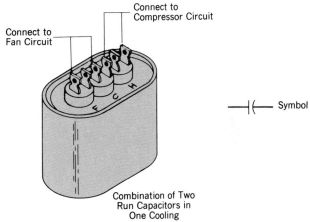

Figure 30-38 Capacitors.

tors are rated in **microfarads** (μF). Run capacitors are usually lower microfarad rating (2–40 μF) than start capacitors (several hundred μF).

Capacitors are rated according to their capacitance and voltage capability. Whenever possible, exact replacement capacitors should be used to replace a failed capacitor. It is possible, however, to combine capacitors just as it is possible to combine resistors. Figure 30-39 shows how capacitors add, when wired in series and in parallel. Figure 30-40 shows the consequences that will result if capacitors are installed that do not match the original ratings.

There are testers available that can test the capacitance rating of capacitors. Field checks can be performed using only an ohmmeter. First, discharge any stored charge on the capacitor by shorting across the terminals with a high resistance (15,000 to 20,000 Ω). Technicians commonly discharge the capacitor with an insulated screwdriver, but this method is not recommended by capacitor manufacturers, as it may cause failure of the capacitor.

If there is already a **bleed resistor** installed

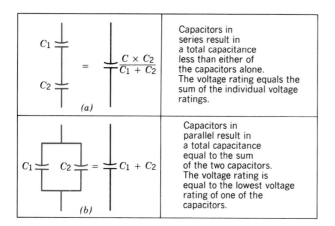

Figure 30-39 Combining capacitors (*a*) in series and (*b*) in parallel.

Replacement Capacitor Compared to Original	Possible Results
mfd rating too low	Capacitor fails
mfd rating too high	Stalled motor, lower running amps, low motor rpm
Voltage rating too low	Capacitor fails
Voltage rating too high	No problem

Figure 30-40 Potential problems associated with replacing a capacitor with other than an exact replacement.

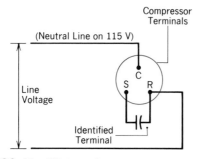

Figure 30-41 Wiring of a run capacitor in a compressor circuit.

across the capacitor terminals, it will need to be disconnected to test the capacitor. Then, with the ohmmeter set on a high resistance scale (R × 10,000), measure the resistance across the terminals. If the capacitor is good, the needle will swing toward zero, and then slowly return to a high resistance reading. The capacitor resistance is increasing as it stores the charge being fur-

nished by the battery in the ohmmeter. A second check is required to determine whether the capacitor is shorted to the metal casing. Still using the R × 10,000 scale, measure resistance from each terminal to the casing. The needle should not move (infinite resistance). Run capacitors have one terminal identified with either a dot, a dash, an arrow, or a red dot. This is the terminal that is more likely to ground to the casing.

When run capacitors are wired to compressors, they are wired between the start and run terminals. The identified terminal should be the one connected to the run winding (Figure 30-41). In this way, if the capacitor shorts to ground, a fuse will blow without the high current passing through the motor windings. If the capacitor is installed backwards, the likelihood of failure is higher, and the chance of compressor damage

Figure 30-42 Turn-past timer. (*Courtesy Paragon Electric Co., Inc.*)

is also higher. The chance of failure is increased because at the start terminal a voltage higher than the line voltage exists, due ot the EMF generated in the start winding. This adds to the line voltage.

Time Clocks

Time clocks are popular in the air conditioning industry in order to reduce operating costs by turning off air conditioners during unoccupied times. Figures 30-42 through 30-45 show a number of different types of timers. A turn-past clock can be located in a room such as a conference room or a classroom. When entering the room, the occupants will turn the timer past the number of minutes they expect to occupy the room. A spring-operated clock will then open the switch that number of minutes later. A seven-day timer has tripppers around the perimeter that

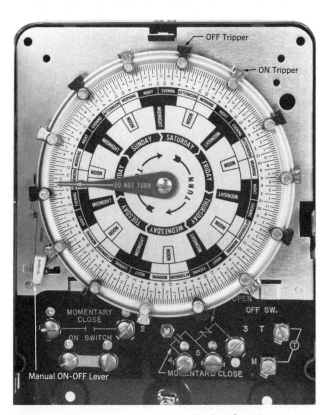

Figure 30-43 Seven-day time clock. (*Courtesy Paragon Electric Co., Inc.*)

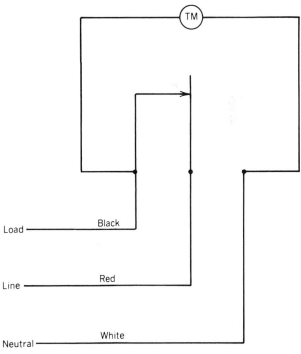

Figure 30-44 Twenty-four-hour miniature time control. One On and one Off operation each day. (*Photo Courtesy Paragon Electric Co., Inc.*)

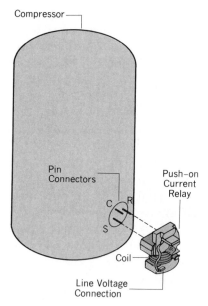

Figure 30-46 Current relay. Inrush current through the coil is strong enough to momentarily lift the contacts to close the switch.

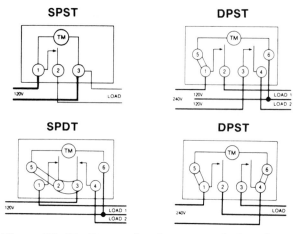

Figure 30-45 Twenty-four-hour time clock with several switching options. *(Courtesy Paragon Electric Co., Inc.)*

can be positioned to turn a unit on and off once (or more) at a specific time each day. Typically, the Monday to Friday schedule will be the same, and the weekend schedule will be different. Defrost time clocks may have two time settings. One is the frequency of the defrost (two, three, or four times per day). The other is the duration of the defrost mode of operation (10 min to 1 hr).

Starting Relays

Starting relays are devices that are used only during the first few seconds of starting up a single phase compressor. There are many types of start relays (Figures 30-46 through 30-48) using different principles of operation. The most popular types in use today are current relay, potential relay, and solid state relay.

The purpose of each of the starting relays is the same. That is, to provide a completed circuit through the start winding for just long enough for the compressor to get started. Once running, a switch in the start relay opens, taking the start winding out of the circuit. The differences between the relay types is how each senses that it is time to open the switch to the start winding.

Figure 30-49 shows the schematic wiring diagram for a **current-type starting relay.** The coil of the relay is wired in series with the run winding. When line voltage is applied, there is a high inrush current through the run winding. This high inrush current passing through the coil of the current relay creates a strong magnetic field, and lifts a switch that completes the circuit through the start winding. With both windings energized, the compressor starts and comes up to full oper-

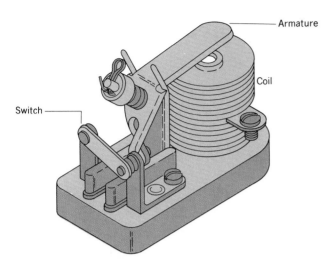

Figure 30-47 Potential relay. As the magnetic field builds from the coil, the armature is pulled down and the switch is pushed open.

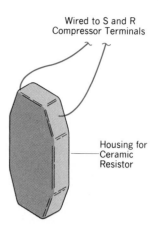

Figure 30-48 Solid state relay.

ating speed. As the compressor speeds up, the inrush current diminishes down to the normal operating current. With this current reduction, the magnetic field of the coil in the current relay is reduced. When the compressor is almost up to full speed, the magnetic field around the current relay coil is no longer strong enough to hold the switch closed, and it falls open due to gravity. This drops the start winding out of the circuit, and the compressor continues to run with only the run winding energized.

There are two precautions in using current relays. First, it must be installed absolutely level and in an upright position. This is because the switch relies upon gravity of the motive force to open the switch. Second, the current relay must be matched exactly to the compressor on which

it is to be installed. The starting current characteristics of each compressor are different, so there is no universal-type current relay that will work for all the different compressors.

The current relay should not be used on motors that use both a start capacitor and a run capacitor, as there is the possibility that the switch contacts will be ruined due to the arcing created by the stored electricity in the capacitors. Where both start and run capacitors are required, either a potential or a solid state starting relay should be used.

A **potential-type starting relay** is wired with a capacitor start compressor as in the schematic in Figure 30-50. Unlike the current relay, the switch in the potential relay is closed while the compressor is deenergized. When line voltage is applied,

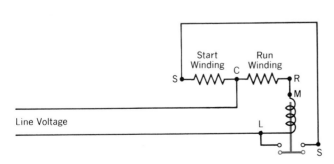

Figure 30-49 Current relay wiring schematic.

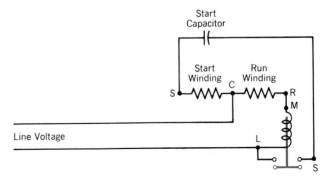

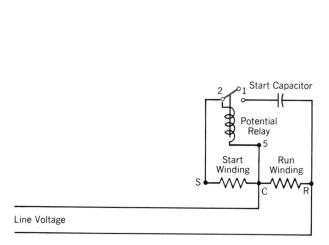

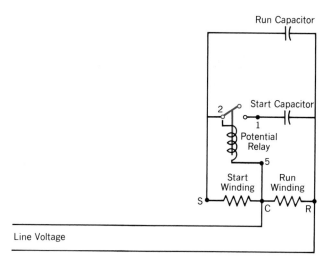

Figure 30-50 Potential relay wiring schematic.

there are immediately two complete parallel circuits. One is through the start winding, and the other is through the run winding. As the compressor comes up to speed, the start winding begins acting as a generator, and generates a back EMF. It is this voltage that is applied across the coil of the potential relay. When the motor is almost up to speed, the back EMF generated by the start winding is strong enough that the magnetic field around the coil can pull open the switch to the start winding. With the start winding out of the circuit, it continues to generate sufficient voltage for the coil to hold open the switch. The potential relay can be recognized by its terminal numbers, usually 5, 2, and 1. The rhyme 5-2-1, common-start-run may help you remember how it is wired into the circuit.

The **solid state relay** does not actually use a switch to take out the start winding. This relay is actually a ceramic device that has the characteristic of a resistance that varies with its temperature. Figure 30-51 shows the installation wiring diagram for the solid state relay. When the compressor is idle, the relay is cool and its resistance is very low (like a closed switch). When voltage is applied, the start and run windings are energized simultaneously, and the compressor starts. Within a few seconds, the start winding current that is passing through the relay causes the relay to get quite hot. This causes its resistance to increase dramatically, and it then acts almost like an open switch. The only disadvan-

tage to the solid state relay is that its operation is not closely related to the starting of the compressor. It is more closely related to time only. Therefore, the solid state relay may leave the start winding in the circuit longer than either of the

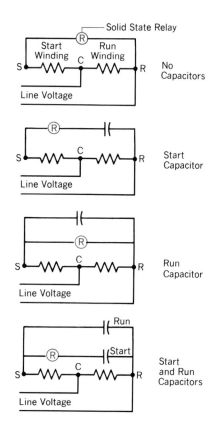

Figure 30-51 Solid state relay wiring schematic.

other two types of relay. The advantage of the solid state relay is its universal application. A single solid state relay can be used to replace all current- and potential-type relays on compressors from $\frac{1}{12}$ hp to $\frac{1}{3}$ hp. This characteristic has made it quite popular with service technicians.

The relays described in this section are usually called potential relays, current relays, and solid state relays, rather than the more correct starting relays. Their use is for the specific application described in this section. Do not confuse their use with that of the control relay that finds general application in many switching applications.

KEY TERMS AND CONCEPTS

SPST	VA Rating	Solenoid Valve
SPDT	Contactor	Time-delay Switches
DPST	Control Relay	Bimetal Timer Switch
DPDT	Subbase	Magnetic Starter
Momentary Switch	System Switch	Overload
Automatic Switches	Fan Switch	Overcurrent
Transformer Primary	Heating Anticipator	Capacitor
Transformer Secondary	Cooling Anticipator	Microfarads
Line Voltage	Staging Thermostats	Bleed Resistor
Control Voltage	Microelectronic Thermostats	Time Clocks
Dual-voltage Transformer	Thermostat Terminal Identification	Starting Relays

QUESTIONS

1. Draw and label an SPDT switch that is temperature operated. On a temperature rise the switch breaks between terminals 1 and 2, and makes between terminals 2 and 3.

2. Draw a momentary switch used as a door switch that would be used to turn on a refrigerator light when the door is opened.

3. Draw a momentary switch used as a door switch that would be used to turn off an evaporator fan when the door is opened.

4. What is the difference between a thermostat and a thermometer?

5. If you measure infinite resistance across a switch, is the switch open or closed?

6. You measure 120 V across a transformer primary and 0 V across the secondary. Which winding (primary, secondary, either one) has failed?

7. Will a transformer work on either ac voltage or dc voltage? Explain your answer.

8. Why do some transformers have five wires?

9. A 120/24-V transformer provides control voltage to a gas valve that has a current rating of 0.4 amp, and a resistor that has a current rating of 0.2 amp wired in parallel with the gas valve. What is the minimum VA rating required on the transformer?

10. How many wire terminals (line voltage and control voltage) would you find on a three-pole contactor?

11. What is the difference between a contactor and a control relay?

12. A 24-V thermostat is to be used with a control relay to turn off a 120-V heater element and turn on a 120-V motor that will provide cooling. Draw the schematic showing the thermostat, control relay coil and contacts, heater, motor, and transformer.

13. What device would be connected to the thermostat with a red wire? A green wire? A yellow wire? A white wire?

14. What is the function of a heating anticipator? When does the heating anticipator heat?

15. What is the function of a cooling anticipator? When does the cooling anticipator heat?

16. Why are there two set points on an automatic-changeover thermostat?

17. A liquid solenoid valve is energized (opens) when the compressor runs. Is this an n.o. or an n.c. valve?

18. What two devices would you find inside a bimetal-type time-delay switch?

19. What is the difference between a contactor and a starter?

20. What is the difference between overcurrent protection and overload protection? What device is used to provide each?

21. What is the common shape for a run capacitor? A start capacitor?

22. What are the three common types of start relay?

23. When the compressor is idle, which types of start relay(s) has (have) its (their) switch contacts open?

CHAPTER 31

POWER WIRING AND ELECTRIC MOTORS

This chapter describes the heavy-duty wiring that supplies electricity to motors, and the motors that consume this power.

Power Distribution

The power company distributes electricity from the power-generating station at very high voltage, 120,000 V or higher. The reason for such a high transmission voltage is to reduce power loss, and to minimize the required transmission line sizes. Both the line size and the power loss ($I^2 \times R$) vary with the amperage. Therefore, the power company distributes power at high voltage and low amperage.

This high voltage power is supplied to a **substation** that contains transformers that reduce the voltage down to 4800 V, and increase the available amperage. For residential use, additional transformers are supplied by the power company to provide **single-phase** power to residences. For commercial or industrial applications, the customer may provide their own transformers to further break down the 4800-V supply. The following sections describe the various types of transformers that can be used to break down the 4800 V for use by the customer. Figure 31-1 shows a transformer used to deliver 115-V and 230-V single-phase power. This is the type of transformer usually provided by the utility company to supply residences. Three wires are supplied to the residence. In the circuit breaker box, the individual circuits may take off from either hot leg and a ground, providing a 115-V circuit. Or two circuit breakers are ganged together to supply a circuit from the two hot legs, providing 230-V single-phase power.

Figure 31-2 shows the primary and secondary windings of a **three-phase transformer** called a **delta transformer.** It is called delta because of the shape formed by the three secondary windings. For customers who require three-phase power, a three-leg transformer is used. The legs of the sec-

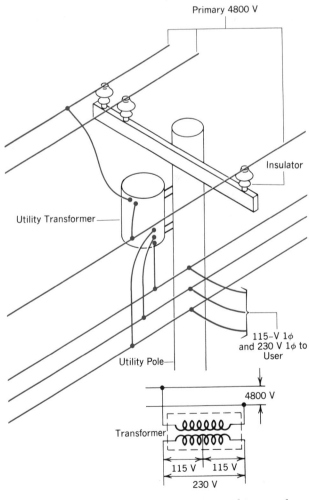

Figure 31-1 Two primary wires are used to supply a single transformer providing 115-V 1φ or 230-V 1φ power.

551

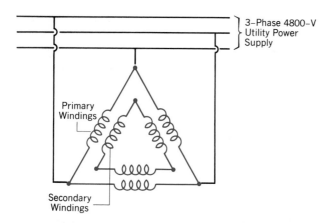

Figure 31-2 Delta transformer.

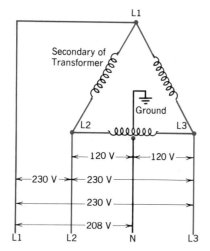

Figure 31-3 Delta transformer providing 230-V 3φ power and 120-V 1φ power.

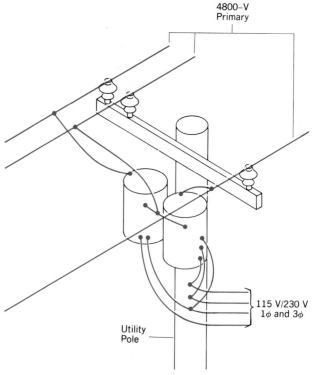

Figure 31-4 Three primary 4800-V wires are used to supply 115-V and 230-V single- or three-phase power.

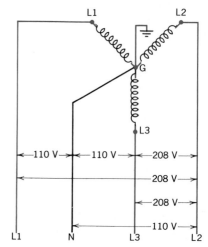

Figure 31-5 Wye transformer secondary providing 208-V 3φ and 110-V 1φ power.

ondary windings of the transformer may be connected to form either a delta or a **wye.** The delta transformer in Figures 31-3 and 31-4 supplies four wires to the consumer. The consumer may select the appropriate wires in order to obtain 230-V single-phase power (between any of the two hot legs, L1, L2, or L3). The consumer may also obtain 120-V single-phase power by wiring a circuit between either the L2 or L3 hot legs and ground. Note that the L1 terminal opposite the center tapped leg is at a voltage of 208 V when compared to ground. This leg is referred to as the high leg, the wild leg, or the stinger leg. Of course, all three hot legs may be taken together to supply 230-V three-phase power. This type of transformer is used when the primary requirement of the consumer is for 230-V three-phase power, with only a minor requirement for the 120-V single-phase power.

If the primary requirement for power is 110-V

single-phase, the consumer may choose the wye-type transformer shown in Figure 31-5. Of the four wires supplied, 110-V single-phase power may be taken from any of the hot legs and ground. Between the hot legs, 208 V is available. Although 208 V is suitable for most motors rated at 230 V, it is not as desirable. It is near the low end of the acceptable motor supply voltage, and will tend to cause somewhat higher amp draws

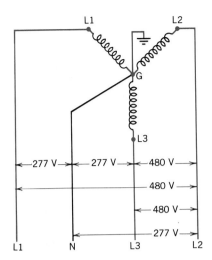

Figure 31-6 Wye transformer providing 4800-V 3φ power and 277-V 1φ for the lighting circuits.

on motors. During periods of peak utility usage, the voltage supply from the utility may drop slightly. If this happens, the voltage available from the wye transformer may drop below 208 V, and cause motors to trip out on overload. Motors are commonly rated to operate satisfactorily at plus or minus 10 percent of the rated voltage. A 230-V motor requires a minimum of 207 V.

Figure 31-6 shows a wye connection transformer that is popular in many industrial applications. It provides 480-V three-phase power, or 277-V single-phase. Because virtually all wiring insulation that is used on 208-V or 230-V systems is rated for 600 V anyway, 480 V is a popular voltage. Therefore, the distribution wiring for a 480-V system may be sized smaller (lower amps for the same power requirement), and it is far less expensive. The 480-V motors are approximately the same cost as comparable horsepower 230-V motors.

The 277-V single-phase circuits available with this transformer configuration are used for fluorescent lighting. There is no way to obtain either 115-V or 230-V single-phase power with this system. Separate transformers must be provided if either of these two voltages is required.

Motors

This text will not describe the theory of operation of each of the different types of motors. In most cases, the service technician will not repair

motors. If they don't run when the proper voltage is applied, they are usually replaced. In the case of large motors where repair would cost significantly less than replacement, motor rewinding shops are available.

Motor Types

The following motor descriptions are to familiarize you with the types of single-phase induction motors that are available.

Shaded-pole motors (Figure 31-7) are low-cost motors used in direct-drive fan and blower applications. They are not used for belt-drive applications because of their limited starting torque capability. They are used in applications from $\frac{1}{100}$ of a horsepower to $\frac{1}{4}$ hp.

Permanent split capacitor **(PSC)** motors (Figure 31-8) are more efficient then shaded-pole motors, and are used in similar applications. They are also commonly used in air conditioning hermetic compressors up to 5 hp.

Split-phase motors (Figure 31-9) have moderate starting torque. They would be suitable for use on belt-drive fans of from $\frac{1}{20}$ hp to $\frac{1}{2}$ hp. They may also be used on centrifugal pumps. A centrifugal switch is mounted so that it rotates with the motor shaft. As the motor nears its rated

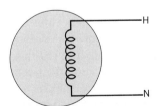

Figure 31-7 Shaded-pole motor has only a single winding.

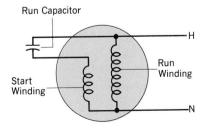

Figure 31-8 Permanent split capacitor (PSC) motor has a start and run winding that both remain energized as long as the motor is running.

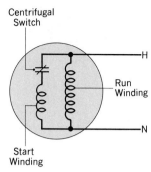

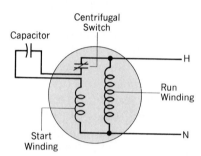

Figure 31-9 Split-phase motor has a start winding that is switched out of the circuit as the motor comes up to operating speed.

Figure 31-10 Capacitor-start motor uses a start capacitor in series with the start winding. Both are switched out of the circuit as the motor reaches operating speed.

speed the centrifugal switch will open, taking the start winding out of the circuit.

Capacitor-start motors (Figure 31-10) have high starting torque, and are therefore suitable for domestic refrigeration compressors and small commercial refrigeration compressors of up to $\frac{3}{4}$ or 1 hp. In refrigeration applications, a start relay is used instead of the centrifugal switch.

Capacitor-start capacitor-run motors have similar starting torque but higher running efficiencies than the capacitor start motor. They would be used in most compressor applications of from 1 to 5 hp.

Aside from the different types of motors, there are a number of other terms that are applicable when describing motors.

Motor Enclosures

Most motor enclosures are rated as **drip proof** (Figure 31-11). Drip proof means that the open-

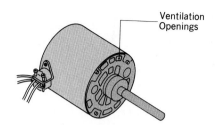

Figure 31-11 Open drip-proof motor enclosure.

ings in the housing that allow for ventilation air to pass through are located so that droplets of liquid falling from above will not affect the motor performance. Open drip-proof motors are used indoors, or outdoors where the motor will not be subjected to rain.

Where the air available for cooling the motor windings is dirty or corrosive, a motor will be selected with an enclosure that is totally enclosed. This may be totally enclosed fan cooled **(TEFC)** or totally enclosed nonventilated **(TENV)**. The TEFC motor has a shrouded fan that blows air over the enclosure, but there is no air circulation through the enclosure. The TENV motor has no integral source of ventilation. It relies on convection of air over the warm motor housing for its cooling. Some TENV motors are further rated as **air over.** These motors are used to drive fans and blowers. The motor must be placed in the moving air stream in order to provide the required amount of motor cooling.

Explosion-proof motors are rated for service in atmospheres that may be explosive. The arcing from switches and the like may produce an explosion. Explosion-proof motors are of very heavy-duty construction. They are totally enclosed and the housing is sufficiently heavy to withstand an internal explosion. There are classifications of explosion-proof design, depending upon which explosive materials may be present. Some of these classifications and materials are as follows.

Class I

Group A: Acetylene.

Group B: Hydrogen and some manufactured gasses.

Group C: Acetaldehyde, diethyl ether, ethylene.

Group D: Acetone, ammonia, benzene, butane, ethanol, gasoline, natural gas.

Class II

Group E: Aluminum and magnesium dusts.

Group F: Carbon or coke dusts.

Group G: Dusts from flour or grain.

Class III

Any easily ignited fibers, such as cotton or rayon.

Explosion-proof motors are used only infrequently due to their high cost. Where potentially explosive atmospheres do exist, the National Electric Code (**NEC**) should be consulted for correct motor selection.

Motor Mounting

There are five basic ways in which motors are mounted in equipment. Figure 31-12 shows a **rigid mounting.** The motor is bolted directly to the equipment by means of a bracket that is welded or bolted to the motor shell.

Figure 31-13 is a more commonly used type of mounting that uses **resilient rubber rings** around the end bells of the motor. A bracket or a cradle type of base can be fastened to the rings. This mounting arrangement reduces the amount of motor vibration that may be transmitted to the equipment, and therefore reduces noise transmission.

The **flange-mount** motor shown in Figure 31-14 has a flat mounting surface on the shaft end. Holes in this surface allow easy bolt-on mounting to equipment. This type of mounting is used on oil burners and only infrequently on fans.

The **stud-mount** motor in Figure 31-15 has bolts extending from the end of the motor. This type of mounting is found on small fans, usually direct drive.

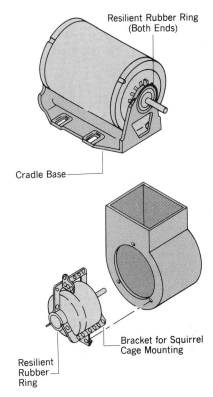

Figure 31-13 Split-phase motors with resilient rings.

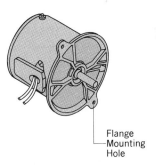

Figure 31-14 Flange-mounted motor.

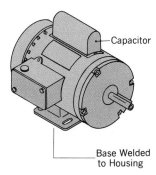

Figure 31-12 Rigid welded base mounting.

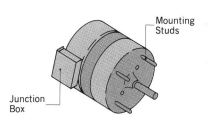

Figure 31-15 Stud-mounted motor.

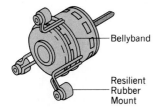

Figure 31-16 Belly-mount motor.

Where no mounting provisions are supplied with the motor, a **belly-mount** bracket (Figure 31-16) is frequently used.

Motor Rotation

When replacing a motor, it is imperative that the new motor rotates in the same direction as the original. For three-phase motors (Figure 31-17), the motor may be wired for rotation in either direction. As a matter of fact, when three-phase motors are installed, the installer usually has no

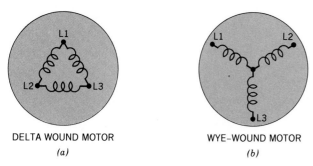

Figure 31-17 Three-phase motor windings.

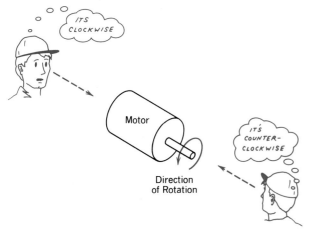

Figure 31-18 Clockwise or counterclockwise? It all depends on the manufacturer's convention.

idea which way it will rotate when it is started for the first time. If it rotates in the incorrect direction, switching any two of the three power leads will change the direction of rotation.

Single-phase motors may be designated as **clockwise, counter-clockwise,** or **reversible.** Usually, the direction of motor rotation is designated as the direction you would see when viewing the motor from the shaft end (Figure 31-18). This is not true for all manufacturers, however. You must confirm that the motor is the correct rotation. Reversible motors have two ways in which the wiring can be connected. The wiring options will be described in the motor directions.

Motor Bearings

The rotating portion of the motor is supported at each end by a **bearing.** The bearing at the end where the shaft emerges from the motor is called the inboard bearing. The other is the outboard bearing. Ball bearings and sleeve bearings are both used. Some bearings require no lubrication. Where lubrication ports are provided, the motor should be serviced according to the manufacturer's directions. In the absence of any directions, a few drops of 30W oil once a year will suffice for most applications. When replacing a motor, if the new motor has oil ports, make sure that the motor is installed so that the oil ports are on top (Figure 31-19). Otherwise, it will be very difficult to make the oil defy the laws of gravity and flow uphill to the bearing.

The primary causes of bearing failure in motors is lack of lubrication (where required), overloading, and overheating. Overloading of a bearing is most commonly caused on belt-drive applications where the belt has been tensioned too tightly. The sideways pull on the bearing will result in premature failure. On direct-drive fans, an out of balance wheel can cause the same re-

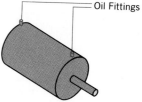

Figure 31-19 Position the oil fittings on top so that oil will flow by gravity into the motor bearings.

sult. Overheating can be prevented by maintaining proper motor ventilation by keeping the ventilation ports clean.

Motor Speed

The speed of a motor is determined by the power supply (60 cycles per second) and the wiring configuration of the motor. The stator is the stationary (nonrotating) part of the motor. The number of poles on the stator determines the motor speed (Figure 31- 20).

Multispeed Motors

Residential and commercial central heating and air conditioning applications represent a common application for single-phase, **multispeed**

blower motors. The system usually operates so that on air conditioning, the blower operates on high speed, and on heating, the blower operates on low speed. The motor may be a three- or four-speed motor, with the installer able to choose any two of the available speeds to be used. Both the shaded pole and the PSC motors may be tap-wound (Figure 31-21). Other multispeed motors may have separate speed windings. A three-speed motor has three different windings (Figure 31-22). All three windings come together at one end, and the wire that emerges from this junction is referred to as the common wire (it is usually white). The power supply to the motor is applied between the common wire and either the high, medium, or low speed wire. Any leads that connect to windings that are not being used must not be left bare. When the motor is rotating, the unused windings will act as if they were generators, and will create a voltage. Wires to the unused speeds should be taped to avoid contact with the metal casing.

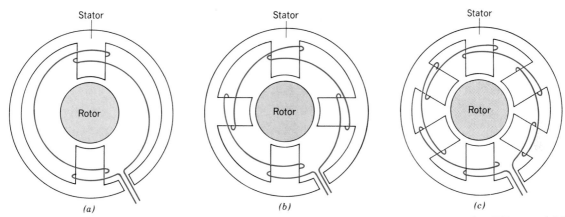

Figure 31-20 (*a*) Two-pole motor. Synchronous speed is 3600 rpm. Actual speed will be 3450–3500 rpm. (*b*) Four-pole motor. Synchronous speed is 1800 rpm. Actual speed will be 1725–1750 rpm. (*c*) Six-pole motor. Synchronous speed is 1200 rpm. Actual speed will be approximately 1075 rpm.

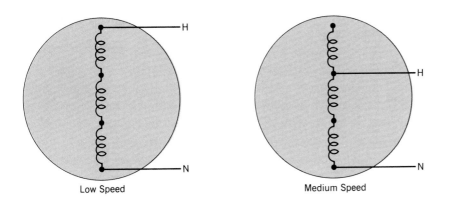
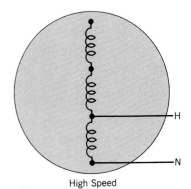

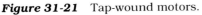

Figure 31-21 Tap-wound motors.

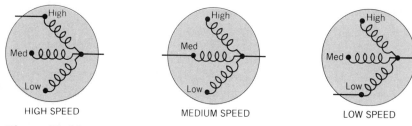

Figure 31-22 Multispeed motor with three separate speed windings.

The high, medium, and low speed windings must be identified in some fashion in order for you to know which is which. The wires may be numbered (for example, 1 = high, 2 = medium, 3 = low). There may be a key that identifies three wires as center, ribbed, and plain. You may need to inspect the wiring carefully to identify which is a ribbed wire. The ribbing may only be a slight line on the insulation of that wire that is not found on the other two wires.

For direct-drive motors, a commonly used speed is 1075 rpm on high speed. Belt-drive motors will most commonly be 1750 rpm at high speed, with 3600 rpm and several other speeds also being used.

Dual-voltage Motors

Three-phase motors are commonly available with **dual-voltage** ratings. That is, the same motor can be used on either a 230-V or a 460-V power supply, depending on how the motor is wired. There are actually six windings inside the motor. Nine wires emerge from the motor into the junction box. The nine wires are attached to the coils as shown in Figure 31-23. Depending upon which voltage is being used, the installer will connect the wiring as required. For example, a 230/460-V motor is to be used on a 230-V power supply. The installer would wire the wires numbered 4, 5, and 6 together into one wire nut. Wires 3 and 9 will be wired together with L1 from the power supply. Similarly, L2 will be connected to 2 plus 8, and L3 will be wired with 1 and 7.

Motor Current Requirements

Motor nameplates will have the following current characteristics.

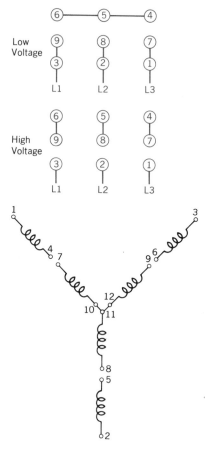

Figure 31-23 Connections for a three-phase dual-voltage motor.

1. **Full load amps (FLA).** This is the amount of current that the motor will draw when it is operating at its rated horsepower output. The chart in Figure 31-24 shows what some average values are for the full load amperage of various compressor motors. At part load, the motor amps will be roughly proportional to the load.

2. **Locked rotor amps (LRA).** When the motor starts up, its current draw is very high, and diminishes rapidly as the motor begins to rotate. The high current draw for the first few seconds is the locked rotor amps. the locked rotor amp rat-

		Motor HP	150	125	100	75	60	50	40	30	25	20	15	10	7½	5	3	2	1½	1	¾	½
Single Phase	115 V	Full Load Current												100	80	56	34	24	20	16	13.8	9.8
		Power Factor %												89	88.5	87.5	86	84	83	80	77	73
		Starting Current												575	460	322	195	138	115	92	80	56
	230 V	Full Load Current												50	40	28	17	12	10	8	6.9	4.9
		Power Factor %												89	88.5	87.5	86	84	83	80	77	73
		Starting Current												288	230	161	98	69	58	46	40	28
Three Phase	220 V	Full Load Current	353	293	223	180	144	120	103	75	64	52	40	27	22	15	9	6.5	5	3.5	2.8	2
		Power Factor %	91.5	91.4	91.2	91	90.8	90.6	90.4	90.2	90.1	90	89.5	89	88.5	87.5	86	84	83	80	77	73
		Starting Current	2118	1758	1338	1080	864	720	618	450	384	312	240	162	132	90	54	39	30	21	16.8	12
	440 V	Full Load Current	172	144	117	90	72	60	52	38	32	26	20	14	11	7.5	4.5	3.3	2.5	1.8	1.4	1
		Power Factor %	91.5	91.4	91.2	91	90.8	90.6	90.4	90.2	90.1	90	89.5	89	88.5	87.5	86	84	83	80	77	73
		Starting Current	1032	864	702	540	432	360	312	228	192	156	120	84	66	45	27	19.8	15	10.8	8.4	6
	550 V	Full Load Current	138	117	94	72	58	48	41	30	26	21	16	11	9	6	4	2.6	2	1.4	1.1	.8
		Power Factor %	91.5	91.4	91.2	91	90.8	90.6	90.4	90.2	90.1	90	89.5	89	88.5	87.5	86	84	83	80	77	73
		Starting Current	828	702	564	432	348	288	246	180	156	126	96	66	54	36	24	15.6	12	8.4	6.6	4.8

Figure 31-24 Average motor current requirements.

ing will be from four to six times greater than the full load amp rating.

3. **Service factor (SF).** Some industrial service motors have a service factor of 1.10, 1.15, 1.25, or 1.35. This means that the motor may be operated at 10, 15, 25, or 35 percent higher than the full load amps for short periods of time. Most motors will have a service factor of 1.0. They should not be allowed to draw more than the FLA.

Frame Sizes

The **frame size** for a motor refers to its outside dimensions that have been standardized by the National Electrical Manufacturers Association **(NEMA).** If you replace a motor with another motor of identical outside dimensions, that does not guarantee that you have matched the horsepower rating. Figure 31-25 shows the range of motor sizes that may be packaged in each different NEMA frame size.

Compressor Motor Wiring

Single-phase compressor motors have a **start winding** and a **run winding,** arranged as shown in Figure 31-26. On window and central air conditioners, a permanent split capacitor motor is used. The line voltage is applied across the *R* and *C* terminals, and a run capacitor is placed between the R and the S terminals. Both the run and start windings remain a part of the circuit whenever the compressor is energized.

Small refrigeration systems and domestic refrigeration compressors are designed so that the

NEMA Frame	Shaft Diameter	hp Range
42	⅜	1/20 1/15 1/12 1/10 ⅛ ⅙ ⅕
48	½	⅙ ¼ ⅓ ½ ¾
56	⅝, ⅞	1/12 ⅛ ⅙ ¼ ⅓ ½ ¾ 1 1½ 2
66	¾	1 1½
143T	⅞	1 1½
145T	⅞	1 1½ 2 3

Figure 31-25 NEMA frame sizes used on motors to 3 hp.

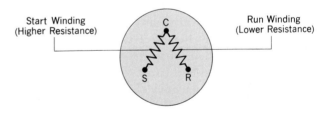

Figure 31-26 Single-phase compressor motor windings.

start winding only stays in the circuit for a few seconds. Once the compressor gets started, a switch opens, causing the start winding to drop out of the circuit. A start relay is the device that senses that the motor is running and that contains the switch that drops out the start winding. There are many types of starting relays, each operating on a different principle to accomplish the purpose just stated. The most commonly used types of starting relays are

a. Current relay.

b. Potential relay.

c. Solid state relay.

The start relay replaces the function of the centrifugal switch in Figure 32-3. The centrifugal switch cannot be used where the motor is located in the refrigerant circuit. The spark created each time the switch opens would react with the refrigerant to form an acid, and eventually cause the motor windings to fail.

Motor Overloads

A **motor overload** protector (called overload) is designed to detect when a motor is drawing too many amps, and stop the motor operation. The overload must therefore sense motor heat or current and open a switch. The switch may open the power circuit or a control circuit that will in turn shut down the compressor.

Figure 31-27 shows a **line-break overload** that is used in virtually all household-refrigeration-type compressors. It consists of a bimetal disc and a resistor that carries the current being carried by the compressor. When too much current is being drawn, the resistor heats the bimetal and opens the switch (with an audible pop sound). It will reset within one or two minutes, and will try to start the compressor again. This overload protector is often called a Klixon overload, although there are many different manufacturers. It comes either as a two-wire device as previously described, or a three-wire device that includes a resistance heater to help sense current. When a three-wire device is used, terminal 1 is connected to line voltage, terminal 2 is connected to the run pin, and terminal 3 is connected to the start pin. These types of overload protectors are installed against the side of the compressor shell. The heat from the shell is an important factor in determining at what point the bimetal switch will open.

In systems larger than 3 to 5 tons, overload protection is usually the type that breaks the control circuit. Figure 31-28 shows a **current overload** or **current relay** (not to be confused with a current-type starting relay). Two terminals are located in the power wiring to the compressor that cause the compressor current to flow through a coil. When the current flow exceeds a predetermined limit for any period of time more than a few seconds, a set of contacts in the control circuit will open, deenergizing the compressor contactor coil. Three phase compressors can use one of these overloads in each of the three legs.

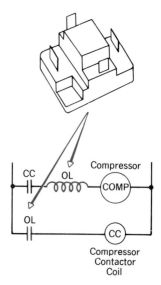

Figure 31-28 Current overload. If the compressor draws too much current, the overload coil will produce sufficient magnetic strength to pull open the overload contacts.

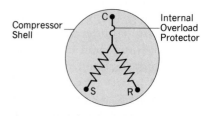

Figure 31-29 Internal overload protection. The overload is imbedded inside the compressor motor windings to sense overtemperature.

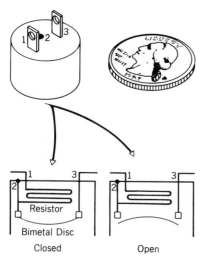

Figure 31-27 Line-break overload carries all the current carried by the compressor.

Some motors have an **internal overload protector** (Figure 31-29). This is a temperature-sensing element (a bimetal or a thermistor) that is imbedded in the motor windings. When high temperature limits are exceeded, the contacts open. Some internal overloads are not wired into the motor windings inside the shell. Instead, the wires from this switch emerge through the compressor shell to be wired in series with the compressor contactor.

Power Wiring

Power wiring is the wiring that supplies voltage to the major users of electricity, such as motors or electric heating elements. It is distinctly different from control wiring, which supplies voltage to low-amperage-consuming coils, lights, or resistors. Power wiring is the business of the qualified electrician. Heating, refrigeration, and air conditioning service technicians will only do power wiring that is incidental to the installation of equipment. The HVAC & R technician must also recognize when a problem rests with the power wiring and not with the HVAC & R equipment. Because power wiring can carry such large amounts of current, the laws that govern its installation are primarily concerned with two potential hazards:

a. The size of the conductor must be a large enough cross section to carry the required amount of amps.

b. The physical installation of the power wiring must be mechanically sound to prevent power wiring from shorting together.

Power Wiring Sizing

The ability of a wire to carry amperage is analogous to the ability of a pipe to carry water. If you try to push too much water through a pipe, its velocity will be very high, there will be a high friction rate between the water and the inside of the pipe wall, and there will be a high rate of pressure drop in the pipe. When wiring is too small to carry the required load, there is a voltage drop in the wire. The voltage drop times the amperage flow is a power loss, and the loss in power

AWG	Ohms per 1000 ft	Ampacity (copper)	Ampacity (aluminum)
0000	0.05	195	155
0	0.10	125	100
4	0.25	70	55
6	0.40	55	40
8	0.64	40	30
10	1.02	30	25
12	1.62	20	15
14	2.58	15	—

Figure 31-30 Amp carrying capacities of wire based on 140°F temperature rating.

is converted into heat. This heat will melt the insulation on the wiring and will result in a short circuit.

Wire size is defined according to its **AWG** (American Wire Gauge). The largest wire available is 0000 (4/0). As the numbers get larger, the wire diameter gets smaller. The smallest standard wire is number 50. Figure 31-30 gives information for some of the more commonly used wire sizes. The wiring used to supply power to a unit must have an ampacity (amp carrying capacity) greater than the current draw of the equipment being served. Where exceptionally long runs of power wiring are required, larger wire sizes may be required. The resistance values of the wire can be used to determine how much voltage drop will occur in a wire.

Example

Select a copper wire suitable to handle a 35-amp load at 440 V. The length of the power wiring is 60 ft.

Solution

A number 10 wire can only carry 30 amps, therefore we would choose a number 8 wire with an ampacity of 40 amps. The resistance of this wire is

$$0.64 \ \Omega/1000 \ \text{ft} \times 60 \ \text{ft} = 0.0384\Omega$$

The voltage drop in the wire is

$$\text{volts} = \text{amps} \times \text{resistance}$$
$$\text{volts} = 35 \times 0.0384$$
$$\text{volts} = 1.34 \ \text{V}$$

HP	Up to 30 ft		20 to 50 ft		50 to 100 ft		100 to 200 ft		200 to 400 ft	
	115V	230V	115V	230V	115V	230V	115V	230V	115V	230V
1/6	18	18	18	18	18	18	16	18	14	16
1/4	18	18	18	18	18	18	16	18	12	16
1/3	18	18	18	18	16	18	14	18	10	14
1/2	16	18	16	18	16	18	14	16	10	14
3/4	14	18	14	18	14	16	12	14	8	10
1	14	16	14	16	12	16	10	14	8	10
1½	12	16	12	16	12	14	8	12	6	8
2	12	14	12	14	10	14	8	12	6	8

Figure 31-31   Minimum required wire sizes for electric motors.

Copper is the most popularly used conductor. Aluminum is sometimes used, however, due to its low cost. Aluminum wiring cannot carry as much amperage as copper for a given AWG size. Note also that the ampacity values in Figure 31-30 are based on a 140°F rating on the wire insulation. For higher temperature ratings, the ampacity would be higher.

Figure 31-31 gives minimum wire sizes to be used for motors up to 2 hp. Oversizing wire causes no operational problems but undersizing wire causes overheating, voltage drop, and motor burnout.

Insulation for Power Wiring

In the previous section, we saw that wire size was determined by the current carrying requirement for the wiring. It had nothing to do with the voltage in the conductor. Selection of the **insulation** requirements for wiring depends only upon the voltage inside and the potential physical abuse the insulation will likely receive that will make it break down. Some of the wire insulation designations are as follows.

1. Type RH is a heat-resistant rubber insulation used in dry applications where the temperature will not exceed 167°F.
2. Type RHH is similar, except that it is rated up to 194°F.
3. Type RHW is a heat-and-moisture-resistant rubber used in dry and wet locations at temperatures not to exceed 167°F.

4. Type TW insulation is a moisture-resistant thermoplastic for wet and dry locations up to 140°F.
5. Type THHN is a heat-resistant thermoplastic for dry locations up to 194°F. It has an outer covering consisting of a nylon jacket.
6. Type THW is a moisture-and-heat-resistant insulation for wet and dry applications up to 167°F.

Conduit

To protect power wiring from mechanical damage, it is run inside a protective pipe called a **conduit.** A conduit can be threaded like a water pipe or it may be thin-wall tubing using special connectors. The thin-wall conduit is the most popular, and is referred to as **EMT (electromechanical tubing).** Wires are pulled through the conduit from the power source to the load. Any wire connections that are required are made with wire nuts inside junction boxes. A typical junction box is shown in Figure 31-32. A conduit may be rated for indoor or for outdoor service. Outdoor conduit connections are gasketed to keep rain from getting inside to the wiring. Conduits are available in sizes starting at ½-in. diameter. The size of the conduit required is determined by the number and size of the wire it will carry.

Figures 31-33 and 31-34 show **flexible conduit** as opposed to **rigid conduit.** This may also be rated for indoor or outdoor service. Codes

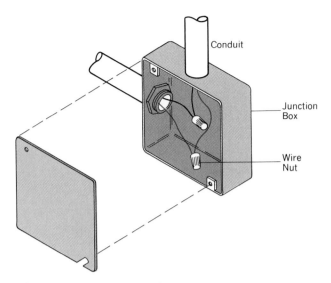

Figure 31-32 Junction box.

Figure 31-33 Weatherproof Sealtite flexible conduit. (*Courtesy Anamet Inc.*)

Figure 31-34 Installation of flexible conduit. (*Courtesy Anamet Inc.*)

govern the maximum allowable lengths of flexible conduit that may be used. Rigid conduit is used for the long, straight runs. At the end near the air conditioning or refrigeration unit, flexible conduit is used to simplify the installation. Indoor flexible conduit is commonly called armored cable or BX cable. Outdoor flexible conduit is commonly called Sealtite, although this alludes to only one of many brands that are available.

Disconnect Switch

Figure 31-35 shows a **disconnect switch.** It is usually located on or near the equipment that it controls. It is basically a knife switch that opens all the legs (two or three legs) of the power supply to a motor when the handle is in the down (Off) position. The cover is sometimes locked in the closed position when the handle is up (On) to prevent tampering. It is a switch that is not designed to be operated frequently, such as an operating control, which controls the normal On–Off operation of the unit. It is to be used to shut the unit down for servicing or for an emergency shutdown. Many local codes require the installation of a disconnect switch within a line

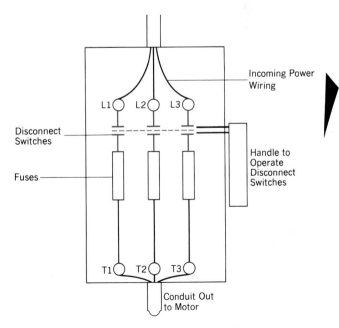

Figure 31-35 Fused disconnect switch.

Figure 31-36 Fuse pullers. *(Courtesy Bussmann Division Cooper Industries)*

of sight from the controlled unit. This is a safety precaution, so that if the service technician shuts down the unit, there is some measure of assurance that the unit will not be turned back on by another person. As an added precaution, the disconnect switch may be locked in the closed position with a padlock. **Locking out** the unit is common on construction jobs where there are many crafts working at the same time. It is also common in industrial sites where there is no local disconnect, and the power is turned off in an MCC room (motor control center) where all the motor starters are grouped together. There are some important safety considerations around the use of lockouts.

1. IF YOU LOCK OUT A UNIT, HAVE A TAG ON YOUR LOCK THAT CLEARLY IDENTIFIES YOU AS THE OWNER OF THE LOCK. THE TAG MUST BEAR SUFFICIENT IDENTIFICATION THAT ANYONE WHO SEES THE TAG WILL BE ABLE TO LOCATE YOU.

2. IF THERE IS ALREADY ANOTHER PERSON'S LOCK ATTACHED TO THE DISCONNECT OR STARTER, DO NOT ASSUME THAT THIS IS SUFFICIENT PROTECTION FOR YOU. ADD YOUR OWN LOCK.

3. NEVER TAKE IT UPON YOURSELF TO CUT OFF ANOTHER PERSON'S LOCK, EVEN IF YOU ARE POSITIVE THAT THEY MAY HAVE GONE HOME FOR THE DAY. ADVISE THE OWNER'S REPRESENTATIVE, FOREMAN, OR OTHER RESPONSIBLE PERSON IN CHARGE. THERE MAY BE A REASON FOR THE SWITCH TO REMAIN LOCKED.

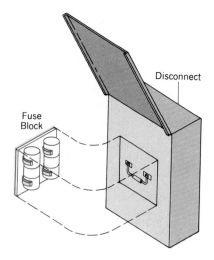

Figure 31-37 Disconnect switch with fuse block.

The disconnect switch can be either fused or nonfused. Cartridge-type fuses are used in the fused disconnect switch. Usually, you cannot tell whether this type of fuse is good or has blown by merely looking at it. You can measure voltage across it while it is in the circuit (a voltage reading other than zero indicates a blown fuse), or you can remove the fuse and measure continuity from one end to the other (a good fuse will show a resistance of zero on the R × 1 scale). When removing the fuse, be sure that you first switch the disconnect to the Off position, and use a fuse puller to remove the fuse (Figure 31-36).

Figure 31-37 shows another type of fused disconnect. It is probably a safer design, because the fuses can be removed by simply pulling on the handle and removing the entire fuse block. The fuse block may be returned upside down to the receptacle, and this will provide the same protection as having turned a handle to the Off position.

Fuses

Figure 31-38 shows various types of **fuses.** Fuses are rated according to the amperage they can carry. Some fuses come in unique physical sizes to prevent replacement with the wrong fuse size. Never attempt to replace a fuse with a larger fuse

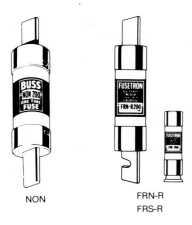

Figure 31-38 Fuses. *(Courtesy Bussmann Division Cooper Industries)*

that will not provide adequate protection for the power wiring.

A fuse functions by providing a conductor that is only capable of carrying the rated amperage. Higher amperage will cause the conductor to melt from excess heat, and thus open the circuit before damage is done to the wiring. The fuse is an excellent protection device against short circuits. It is not unusual for a fuse to be able to handle a short-circuit amperage of 100,000 amps and open the circuit quickly enough to protect the wiring.

Fuses that protect motor circuits have a problem that is not faced by fuses that protect circuits with resistive loads. The motor will draw locked rotor amps on start-up, but will drop down to its normal operating voltage within a few seconds. For this application, **dual-element** or **slow blow fuses** are used. This time-delay type of fuse will allow the momentary surge of power, which is 4 to 6 times the normal operating current. If this current persists for more than a few seconds, the fuse will blow. In the event of a short-circuit condition, there will be no time delay in the operation of the fuse.

Circuit Breakers

In modern residences and in some commercial and industrial applications, **circuit breakers** are used instead of fuses. If an overcurrent condition has existed, the circuit breaker will trip. A tripped circuit breaker can be identified because the switch will not be all the way in the On position, but will be slightly out of line with the rest of the circuit breaker switches. The circuit breaker has an advantage over fuses in that the switch may be reset, eliminating the need for replacing the fuse. To reset a circuit breaker, move the switch all the way to the Off position, and then move it all the way to the On position.

Circuit breakers may be found either individually or ganged together. The single circuit breaker opens the hot leg of a 120-V circuit. Where two circuit breakers are ganged together, they will open both legs of a 230-V single-phase power supply. A gang of three circuit breakers is used for three-phase applications. The circuit breaker may be easily removed from the circuit breaker panel. Most are attached with clips, and may be easily pulled out. Others are attached to the main breaker by screws.

If circuit breakers have advantages of convenience over fuses, they also have the following operational disadvantages.

1. When a circuit breaker switch has not been operated for a long period of time (one year or more), the contacts tend to become somewhat fused together. It will still provide adequate short-circuit protection, but for a gradually added overload, the circuit breaker will carry far more than its rated amperage.

2. Where the fuse can handle a short-circuit condition of 100,000 amps, the interrupting capacity of a common circuit breaker is limited to around 10,000 amps.

KEY TERMS AND CONCEPTS

Substation	Explosion Proof	Frame Size
Single-phase Transformer	NEC	NEMA
Three-phase Transformer	Rigid Mount	Internal Overload
Delta Transformer	Resilient Mount	Line-break Overload
Wye	Flange Mount	Power Wiring Sizing
Shaded Pole	Stud Mount	AWG
PSC	Belly Mount	Wiring Insulation
Split-phase	Motor Rotation Direction	Conduit
Capacitor Start	Motor Rpm	Disconnect Switch
Drip Proof	Multispeed Motor	Lockouts
TEFC	Dual-voltage Motor	Fuses
TENV	FLA	Time-delay Fuses
Air Over	LRA	Circuit Breakers
	SF	

QUESTIONS

1. What type of compressor motor would you expect to find in a 2-ton window air conditoner?

2. Compressors with low starting torque are usually limited to use on capillary tube systems. Why?

3. What type of motor would you expect to find driving the evaporator fan in a domestic refrigerator?

4. What are the limitations for the use of an open drip-proof motor enclosure?

5. Name one application where you might expect to find a TENV air-over motor.

6. What type of motor mounting is used to isolate any motor vibration from being transmitted to the driven equipment?

7. How can the direction of rotation of a three-phase motor be reversed?

8. When would you expect a motor to draw FLA?

9. When would you expect a motor to draw LRA?

10. Explain what is meant by a 1.15 service factor.

11. What is the difference between a line break type of overload and a pilot duty type of overload?

12. What is the difference between power wiring and control wiring?

13. What is the minimum size of wire you would expect to find serving a compressor that draws 42 amps at full load?

14. What are the two factors that affect the selection of wire insulation?

15. Why are residential heating/air conditioning evaporator fans supplied with multispeed motors?

16. What are the advantages of fuses over circuit breakers?

17. What are the advantages of circuit breakers over fuses?

18. How can the service technician protect against accidental starting of a unit that is being serviced?

19. What is the difference in fuses that would be used on an electric heater versus a compressor motor?

20. How many wires emerge from a dual-voltage three-phase motor? How many power leads are brought to the motor from the power source?

CHAPTER 32

ELECTRICAL CIRCUITS

The purpose of this chapter is to present many different electrical circuits, with a description of only the unique features of each. By understanding many different ways in which control functions are accomplished, you will learn to be able to read schematic diagrams that you have not previously seen.

Cooling Circuits

Household Refrigerator

Figure 32-1 shows a simple refrigeration schematic. It would apply to a household refrigerator or small commercial refrigerator. The line voltage thermostat (cold control) senses the box temperature, and closes to turn on the refrigeration system. The compressor, condenser fan motor, and the evaporator fan motor are all wired in parallel, so they will all start and stop together. Each of the motors is rated for 115-V service. The start relay used with the compressor is a current relay, although any other type of start relay might also be used. The overload is in the compressor

circuit only. It is wired in series with the common terminal of the compressor, and it will detect too much current through either the start winding or the run winding. If an overload condition is sensed and the overload switch contacts open, only the compressor will be shut down. The condenser fan and evaporator fan will continue to function.

The cabinet light is operated by a door switch. It is a normally closed momentary switch that will energize the cabinet light whenever the door is opened. The operation of the cabinet light portion of the circuit is totally independent from the refrigeration portion of the circuit.

Frequently, schematics will show a single-phase compressor without showing the internal windings or the starting relay. It is understood that this includes some type of relay and overload where used.

Figure 32-2 adds some features to the circuit. The cabinet lights are wired in parallel, and are

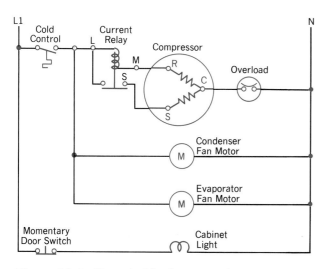

Figure 32-1 Household refrigerator/freezer wiring.

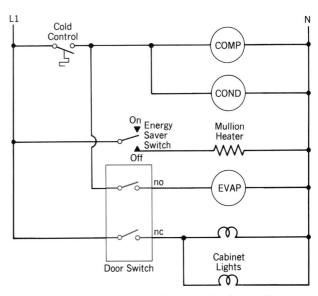

Figure 32-2 Switching of evaporator fan and cabinet lights with a DPST door switch.

569

controlled through a DPST switch that also operates the evaporator fan. If the door is opened while the refrigeration system is operating, the evaporator fan will be shut down. This feature reduces the loss of cold air from the box while the door is open.

The heater shown may be called a mullion heater, a stile heater, a door heater, depending upon where they are installed. It is actually a nickel–chromium wire installed in places around the cabinet that may be sufficiently cold to condense moisture out of the air. The heat will be sufficient to eliminate condensation of moisture on the cabinet. Some systems have the heater energized continuously, while others use an energy saver switch. During periods of low humidity, the owner can turn the energy saver to On, which actually opens the switch, turning the heaters off.

Compressor Controlled by a Low Pressure Cutout

Figure 32-3 shows a refrigeration system, using a low pressure cutout (**LPC**) as the operating control. This wiring diagram might be found on a commercial-type walk-in box. The compressor is cycled on and off with an LPC. Because this system is somewhat larger and more costly, additional safety controls are provided. The HPC and the OPC are wired in series with the LPC. The wiring shown for the oil pressure cutout is actually a simplification, showing only the safety switch portion of the device. More detailed wiring is shown in a later section.

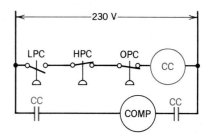

Figure 32-4 Compressor operated by a two-pole compressor contactor.

In systems with multiple **safety** controls, all the safeties are wired in series. In order for the compressor to run, all of the safety switches must be closed. In this circuit, all the safety controls carry the full current draw of the compressor.

For a somewhat larger compressor, Figure 32-4 shows the LPC energizing a compressor contactor instead of the compressor itself. In this case, the control voltage and the line voltage are identical, but the current carrying requirements of the safety and operating controls are significantly reduced.

Defrost Systems

The refrigeration system in Figure 32-5 includes a four-wire **defrost timer,** a defrost heater, and a defrost termination switch. The clock motor in the timer is wired between terminals 1 and 3, and is energized all the time. For $7\frac{1}{2}$ hr the switch from 1 to 4 is closed, and the switch from 3 to 2 is

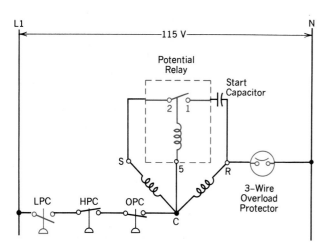

Figure 32-3 Compressor operated by a LPC using a three-wire overload and a potential relay.

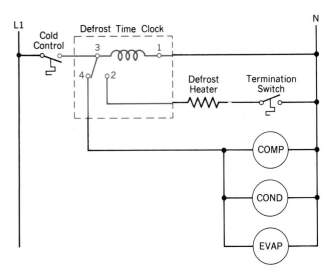

Figure 32-5 Electric defrost controlled by a four-wire defrost time clock.

open. The system runs normally, as if the defrost heater is not even there. But after $7\frac{1}{2}$ hr, the cam in the timer mechanism causes 3 to 4 to open, and 3 to 2 to close. The refrigeration system is disabled, and the defrost heater is energized for one-half hour. After that half hour, the system returns to normal operation. If the evaporator becomes defrosted in less than the half hour, the rise in evaporator temperature will be sensed by the defrost termination switch. When defrost is accomplished, the termination switch will open, turning off the defrost heater. Then nothing is energized until the clock motor returns the system to the normal operating mode. With some systems, when the time clock ends the defrost period, only the compressor and condenser fans are restored to normal operation. The evaporator fan is wired ahead of the defrost termination thermostat, keeping it off until the evaporator coil has cooled. Another variation on this scheme is a defrost time clock that has a fifth terminal

that turns the evaporator fan on shortly after the compressor has been energized.

The terminal numbers on the defrost timers are not universal. Figure 32-6 shows the numbering conventions used by several different manufacturers. On some commercial defrost time clocks, an X terminal is provided for a solenoid coil inside the timer. The solenoid is energized when the defrost termination switch warms, and returns the system to normal cooling.

There are several alternative ways that the defrost timer may be wired. Figure 32-7 shows a defrost that is accomplished by the use of hot gas from the compressor instead of using an electric heater. With this scheme, the compressor remains running during the defrost cycle in order to provide the hot gas. The evaporator fan remains off, however, as we do not wish to heat up the box, only the evaporator coil. Note that this defrost system does not operate a fixed number of

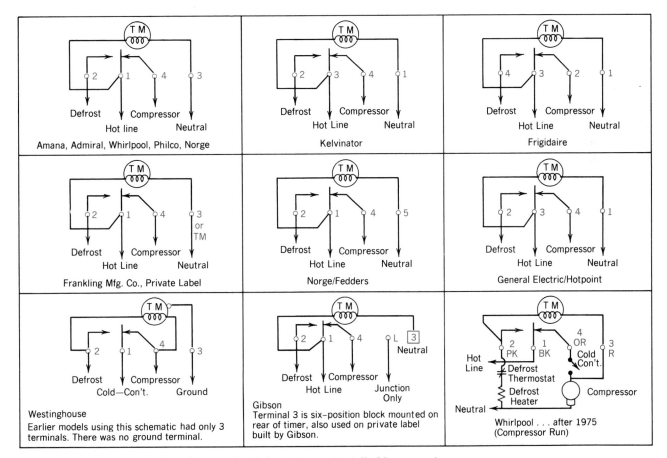

Figure 32-6 Terminal identification for defrost timers installed by several manufacturers. Switches are shown in the refrigeration mode.

times each day. When the cold control is open, the clock motor is not energized. This system might be set to go into defrost each 6 hr, but it will be 6 hr of running time.

Figure 32-8 shows a three-wire defrost timer. During normal operation, the switch is in a position that closes switch 1–2. The compressor is operated directly from the voltage between L1 and L2. The clock motor is energized from L1 to terminal 1 to terminal 2, through the clock motor, then through the defrost heater, and to L2. Even though the clock motor and the defrost heater are in series, only the clock motor operates. The resistance of the clock motor is very high (several thousand ohms) compared to the heater, so most of the voltage drop takes place across the motor. Only a very small current flows through the heater, and it does a negligible amount of heating. During the defrost cycle, the switch changes position. The heater gets line

voltage from L1 to 1 to 3, through the heater to L2. The clock motor receives voltage from L1 to 3, through the motor, and then through the compressor. The small current that passes through the compressor windings does not have any effect on the compressor.

Ice Maker, Household

The **household ice maker** (Figures 32-9 and 32-10) starts with room temperature water in an ice cube mold. The thermostat (set at 25°F) is open, and no part of the ice maker is energized. Then the water freezes, the thermostat closes, and energizes a motor and a mold heater. The motor serves the following three purposes.

1. It drives an arm that will push the ice cubes out of the mold.

2. It drives a cam that will operate the electrical switches.

3. It drives an arm that will move through the ice storage bin (ice-sensing switch).

Soon after the motor starts, it stalls (stops moving) because the harvest arm cannot push

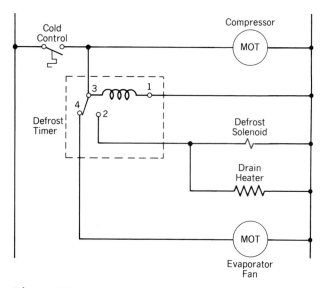

Figure 32-7 Hot gas defrost on a domestic refrigerator/freezer. The condenser is natural draft.

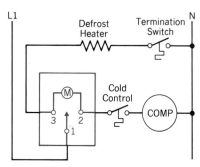

Figure 32-8 Three-wire defrost time clock.

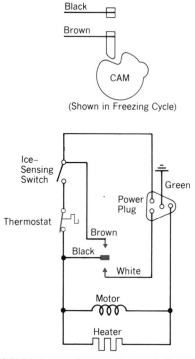

Figure 32-9 Ice maker used in a G.E. household refrigerator/freezer. When the thermostat initiates the harvest cycle, the motor drives the cam.

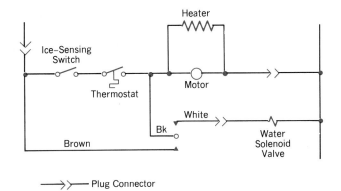

Figure 32-10 Ladder diagram for G.E. ice maker.

the ice out of the mold. After some time, the mold heater releases the ice from the mold, and the motor resumes its rotation. After the ice has been pushed out of the mold into the storage bin, one switch (black to brown) is closed by the cam in order to bypass the thermostat and keep the motor running. It also bypasses the ice-sensing switch. When the second switch (black to white) is also closed by the cam, a circuit is completed through a plug connection to the water solenoid valve. As the cam continues to move, the water solenoid becomes deenergized, and then the by-pass switch opens. The motor stops because the thermostat contacts have opened due to the in-troduction of the new room temperature water. While the cam is rotating, the ice-sensing switch opens. If the bin is not full of ice, it recloses, and the cycle begins again. If the bin is full, the ice-sensing switch remains open, stopping the cycle.

Pumpdown

A **pumpdown** circuit (Figure 32-11) is one in which the thermostat or low pressure control does not directly operate the compressor. The thermostat senses space temperature, and when cooling is required, energizes the liquid solenoid valve. Refrigerant flows through the LSV into the low pressure side of the system, causing the low

pressure cut-out switch to close. This energizes the contactor for the compressor.

One problem with the pumpdown control is the potential that short cycling will result if the liquid solenoid valve has a slight leak. The low side pressure will slowly rise, until the LPC switch closes. The compressor will run for a few seconds, and then shut down. The solution to this problem is the nonrecycling pumpdown (Figure 32-12). Once the system has pumped down, the system will not restart on the pressure switch before the thermostat has once again closed. During normal operation, all the control relays are energized, and all the relay contacts are closed. The LPC is an SPDT switch. When the thermostat opens, CR2 is deenergized, causing the liquid solenoid valve to close (deenergized). Additional CR2 contacts open, but CR1 and the compressor contactor remain energized through CR3 contacts and CR1 contacts. When the sys-tem has pumped down, the LPC switches to the lower position, deenergizing CR3. CR3 contacts open, deenergizing CR1 and the compressor con-tactor. CR4 remains energized through CR4 con-tacts. If the LPC were now to switch to the upper position before the thermostat closed, the com-pressor would remain off. Even though CR3 ener-gizes, CR2 will remain deenergized.

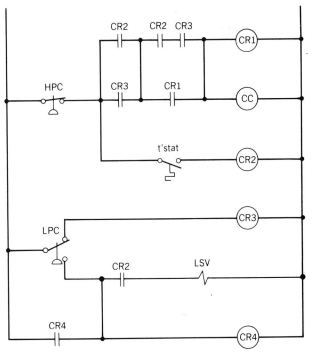

Figure 32-12 Nonrecycling pumpdown.

Figure 32-11 Pumpdown control.

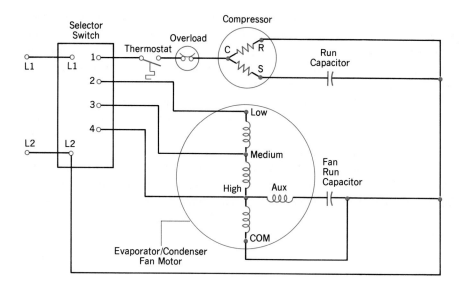

SELECTOR SWITCH POSITION	CONTINUITY BETWEEN
High cool	L1,1,4
Medium cool	L1,1,3
Low cool	L1,1,2
High fan	L1,4
Medium fan	L1,3
Low fan	L1,2

Figure 32-13 Window air conditioner.

In the case of a power outage, CR4 will be deenergized. When power is restored, the compressor will restart because CR2 and CR3 are energized. The LSV is deenergized because the LPC is in the upper position. The compressor will pump down the low side until the LPC contacts cause the LSV and CR4 to once again become energized.

Window Air Conditioner

Most window and residential central air conditioning units use compressors with PSC motors. The start winding remains in the circuit, and there is no relay. There may or may not be a run capacitor between the start and run windings. There is usually no start capacitor.

The **window air conditioner** in Figure 32-13 can energize the double-shafted fan motor (runs both the condenser and the evaporator fan) on any of three speeds (High Fan, Med Fan, Low Fan). It can also energize the compressor with any of these fan speeds (High Cool, Med Cool, Low Cool). The compressor runs the same, regardless of the High, Med, or Low setting. It will cycle on and off in response to the line voltage thermostat.

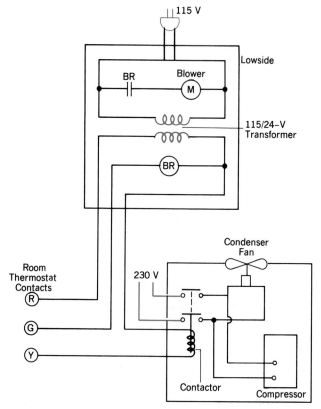

Figure 32-14 Central split-system air conditioner.

Residential Central Air Conditioner

The **residential central air system** shown in Figure 32-14 has a condensing unit located outside, and a lowside (fan, filter, and evaporator coil) located indoors. The indoor unit runs on 115 V, the outdoor condensing unit runs on 230 V, and the control system runs on 24 V. When the cooling thermostat is closed (closing R to Y), it energizes two devices—the compressor contactor and the blower relay. The compressor contactor switch closes, providing 230 V to the compressor and the condenser fan (they operate off the same contactor). The compressor contactor may be either two pole, breaking both hot wires to the compressor as shown, or, where allowed by local codes, single-pole, breaking only one hot lead.

At the same time as the compressor starts, the blower relay contacts complete a 115-V circuit through the evaporator fan motor, and the complete system is running.

Figure 32-15 shows the same central air system, but it is used in conjunction with a gas-fired forced-air furnace. The furnace fan is used as the

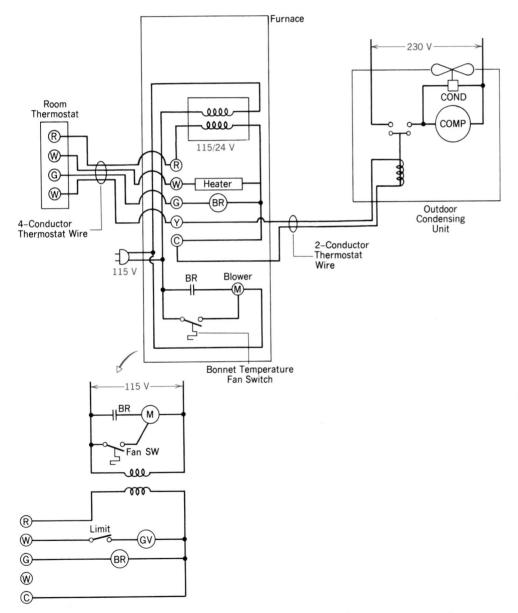

Figure 32-15 Residential heating and air conditioning. The compressor and condenser fan motors operate on 230 V, the blower motor operates on 115 V, and the control voltage is 24 V.

blower motor, and the evaporator coil sits atop the furnace in the ductwork. Note the terminals that are available in the furnace (R, W, G, Y, C). On modern furnaces, they are arranged on a terminal strip. The thermostat terminals are wired to the terminal in the furnace with the corresponding terminal identification. Note also, with this system, that the blower motor is two speed. When the blower is energized by the bonnet fan switch, it operates on low speed. When the blower is energized by the blower relay for air conditioning or for ventilation, the high speed winding is energized.

Motor Starter Circuit—Push-Button Control

Figure 32-16 illustrates the use of a remote starter used to energize a three-phase motor. This type of **motor push-button control** will be found on large field assembled systems. Packaged refrigeration systems will use contactors rather than motor starters.

The starter contains the holding coil, the overload heaters, the overload contacts (in the control circuit), and an auxiliary switch. The Start–Stop switch is located on the face of the starter cabinet. When the start button is pushed, the holding coil is energized from L1-1-Stop switch-Start switch-coil-overload switches-L2. When the Start push button is released, the holding coil remains energized from L1-1-Stop switch-2-3 (through the auxiliary switch)-coil-overload switches-L2. The system shown is a three-pole starter with two legs of overload protection. Three legs of overload protection may be provided with the addition of a third overload heater and switch.

In this example, the auxiliary switch in the starter was used for the seal-in circuit for the Start push button. Auxiliary switches may also be used to energize additional loads. A common sequence might be to have a push-button station energize an air-handler fan. The auxiliary switch on the fan starter is then used to energize the control circuit for a compressor.

The magnetic starters described in this and the following sections are all called **across-the-line starters.** They are the least expensive and

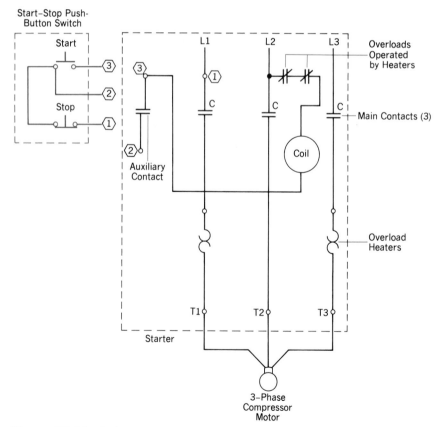

Figure 32-16 A three-phase compressor started by a push-button switch and a motor starter.

most common starter used for three-phase compressors. Some power companies, however, require the use of starters that reduce the inrush current, especially on larger motors. Starters used to accomplish this are of five types

1. Part winding.

2. Star Delta.

3. Autotransformer.

4. Primary resistor.

5. Reduced voltage starting.

Each functions in a different manner and provides different amounts of reduction in starting torque. Many times a specially wound motor is required to accommodate the use of these specialty starters.

Condenser Fan Cycling

Figure 32-17 presents two features that have not been included in previous control schemes:

1. A 230-V single-phase control voltage is taken from a three-phase power supply.

2. The condenser fans operate from their own contactors. They may be energized individually, and are controlled by outside air temperature.

The power wiring and the control wiring are both shown in this diagram. All of the contactors are three pole, although some units may use two-pole contactors on a three-phase power supply. The control relay contacts noted as the operating control are energized by a room thermostat not shown as part of this unit. The temperature switches sense outside air temperature. As the outside air temperature falls, first one, and then the second temperature switch is opened. This allows the unit to operate at low outside air temperatures, without the head pressure becoming too low.

Motor Starter with Low Voltage Thermostat

The starter may also be used to start a compressor through the operation of a thermostat (Figure 32-18). The control relay contacts are closed when the thermostat energizes the relay coil in the low voltage portion of the circuit. This completes the circuit through the coil in the magnetic starter through the normally closed contacts L-M in the oil pressure cutout. The

compressor starts, and the heater in the OPC is energized. The system has 60 sec in which to establish oil pressure and open switch 1-2, deenergizing the heater. Otherwise, after 60 sec, the heater will cause switch L-M to open, deenergizing the magnetic starter and shutting down the compressor. Note that the heater in the oil pressure cutout is rated at 115 V. When using 230 V, the dropping resistor consumes half of the voltage, leaving 115 V for the heater.

Antirecycle Timer

The **antirecycle timer** is a device that prevents a compressor from starting too frequently, or short cycling. This is a desirable feature because each time the compressor starts, the inrush current heats up the motor windings. If the compressor is allowed to short cycle, the cumulative effect of

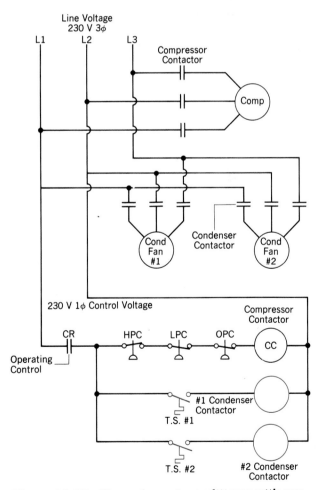

Figure 32-17 Three-phase air conditioner with condenser fan cycling for head pressure control.

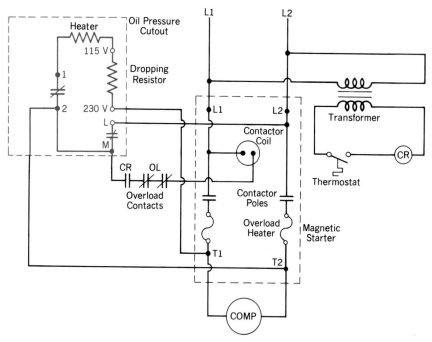

Figure 32-18 A 230-V single phase compressor controlled by a thermostat, magnetic starter, and oil pressure cutout.

the inrush current with each start will cause the motor to burn out.

The wiring of the antirecycle time is shown in a four-sequence diagram in Figure 32-19. During the off cycle (Figure 32-19a), all switches are in the position shown. When the thermostat calls for cooling (Figure 32-19b), CR1 is energized. The timer motor becomes energized from L1 through CR1 contacts and timer switch B2-B. A second control relay, CR2, is simultaneously energized from L1 through CR1 contacts and switch A2-A. The timer motor runs for 15 s with the switches in the position shown. Then the timer switches change position (Figure 32-19c). The timer motor becomes deenergized because the normally closed CR2 contacts are open. CR2 remains energized from L1 through CR1 contacts and the normally open CR2 contacts. The compressor contactor CC is energized from L1 through CR1 contacts, CR2 normally open contacts, and switch A-A1.

When the thermostat is satisfied (Figure 32-19d), CR1 contacts open, and the compressor stops. Even if the thermostat were to immediately close again, the compressor will not restart. The timer must now remain energized for 10 min

before it will return the timer switches to the original positions in Figure 32-19a.

Reset Relay

Figure 32-20 shows the addition of a standard control relay with one set of contacts to a standard compressor circuit. With this scheme, automatic reset safety controls may be made to perform like manual reset controls. With all of the safety controls satisfied (closed), the control relay RR **(reset relay)** does not see any voltage difference and remains deenergized. When any of the safety controls open, the voltage drop across that safety control is imposed upon the reset relay, and it is energized. The normally closed contacts open, assuring that the reset relay remains energized and the compressor locked out of the circuit even though the safety control may reset itself. The system may only be reset by deenergizing the reset relay. This may be done by either switching the unit disconnect off and back on again, or by turning the thermostat up (opening its contacts), and then returning it to its normal setting. The reset relay scheme may also use a normally closed momentary switch in series with the control relay coil. It will act as a reset button

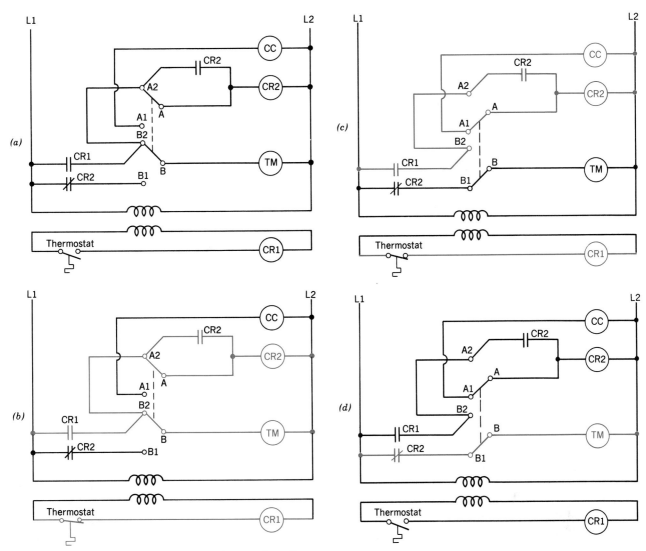

Figure 32-19 (*a*) Antirecycle timer during the off cycle. (*b*) Zero to fifteen seconds following the closing of the room thermostat. (*c*) Compressor running. (*d*) Timer switch position for the 10 min following the opening of the room thermostat.

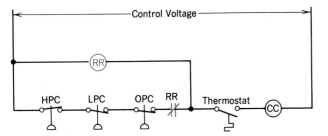

Figure 32-20 A control relay used as a reset relay.

that, when pressed, will deenergize the control relay.

Ice Maker — Auger Delay Switch

Figure 32-21 is the schematic for a small ice maker that uses an auger inside a cylindrical evaporator to make flaked ice. The low pressure switch and the **auger delay switch** both sense refrigerant suction pressure. The auger delay switch is set to switch positions at 20 psi, and the low pressure switch will open at 10 psi.

When the bin thermostat closes (calling for the ice maker to start), the auger motor becomes energized through terminals 3-2 on the auger delay switch. When the auger motor starts, the centrifugal switch on top of the auger motor closes, allowing the compressor and condenser fan to start. The purpose of the centrifugal switch is to prevent the refrigeration system from operating if the auger motor has failed for any reason. Within a few seconds, the suction pressure drops to below 20 psi, causing the position of the auger delay switch to change. The switch from 3 to 2 opens, and the switch from 1 to 2 closes.

After sufficient ice has been made and delivered into the bin by the auger, the ice touches the sensing element of the bin thermostat, and the

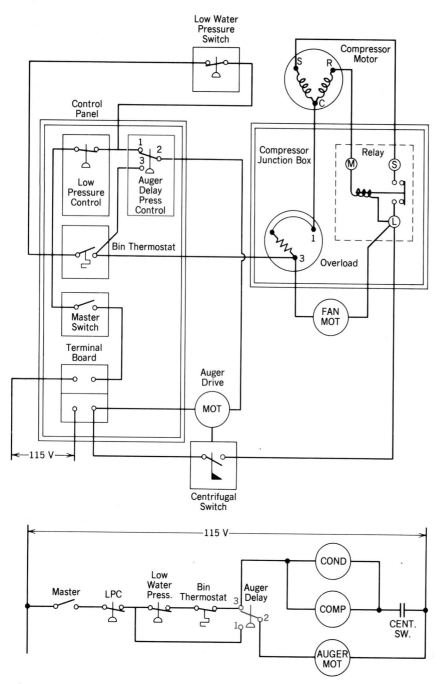

Figure 32-21 Interconnection and ladder diagrams for a 115-V ice flaker with auger delay.

thermostat opens. The compressor and condenser fans turn off immediately, but the auger continues to run. The suction pressure rises because the compressor is off. When it rises to 30 psi, the auger delay switch returns to its original position. By allowing the auger to run for a few seconds after the compressor has stopped, all the ice that has been formed is delivered to the bin. Without the auger delay, the ice may prevent the auger motor from starting on the next On cycle.

Ice Cuber with Evaporator Thermostat

Figure 32-22 illustrates the schematic of a simple ice maker that uses a pump to recirculate water over a tilted, cold evaporator plate to build up a slab of ice. When a sufficient thickness of ice has been formed, a **harvest cycle** is initiated, which puts hot gas through the evaporator. This releases the ice slab to slide off onto a **cutting grid.** The weight of the slab pushes it through the grid, forming cubes that fall into the bin.

With the system switch in the down position, ice will be made. With the bin thermostat closed, the compressor, the condenser fan, and the recirculating pump are energized. The water inlet solenoid valve and the hot gas defrost solenoid

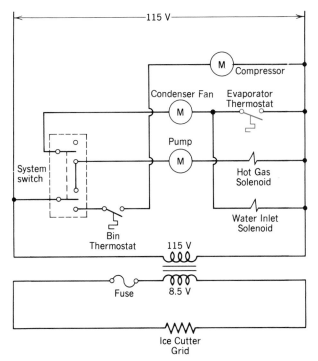

Figure 32-22 Ice cuber with harvest initiated by an evaporator thermostat.

are not energized, because the evaporator thermostat is closed, shorting out both of these solenoids. As the ice thickness builds, the evaporator becomes insulated, and the evaporator pressure and temperature drop. When the slab reaches the correct thickness, the evaporator temperature drops sufficiently low to open the evaporator thermostat (usually between 0°F and 10°F). At that point, the harvest cycle begins. The hot gas solenoid and water inlet solenoid are energized. It would appear that the condenser fan and recirculating pump are also energized, as they are in series with the solenoids. This is not the case, however. The solenoids have a much higher resistance than the pump motor and condenser fan, therefore virtually all the voltage drop occurs across the solenoids. The condenser fan motor and pump motor serve only to complete the cycle during the harvest cycle. During the harvest cycle, the hot gas warms the evaporator, releasing the ice slab, and the water inlet solenoid allows a fresh supply of water into the pump reservoir.

When the slab has been released to the ice-cutting grid (which is always energized), the evaporator warms, the evaporator thermostat closes, and the freezing cycle begins again.

The up position on the system selector switch is for the Clean or Wash mode. In this position, only the recirculating pump is in operation. Cleaning chemicals may be added to the sump and recirculated throughout the water system to dissolve away any build up of lime and scale.

Ice Cuber with Conductivity Switch

Another system that is very similar to the scheme previously described uses a slightly different evaporator and a different method of initiating the harvest cycle. The evaporator is in the shape of an ice cube tray standing vertically on end. The water falls down the tray, gradually building up the ice thickness. Instead of using the evaporator temperature to initiate the harvest, a pair of probes **(conductivity switch)** are located so that when the ice cubes are formed, the water flowing over them will touch both probes, creating the same effect as a closed circuit. The water itself serves as the conductor to complete the circuit. Note that the designs utilizing an ice mold that forms cubes instead of slabs eliminate the need for a cutting grid.

Ice Cuber with Thickness Switch

Another method of initiating a harvest cycle uses a **thickness switch.** It consists of a thickness motor that rotates a probe over the slab once each 4 min. When the correct ice slab thickness has been formed, the probe causes a SPDT switch to operate. The condenser fan turns off, the hot gas solenoid turns on, and the slab is released to the cutter grid.

Ice Maker—Water Level Switch Control

The ice maker in Figure 32-23 has an evaporator shaped like an ice cube mold, a water plate to distribute water over the mold, and a circulation tank fastened to the water plate. The amount of water in the circulating tank is sensed by a switch that is operated by the weight of water.

The cycle begins with the water solenoid open and the water plate and the defrost and pump toggle switch all in the up position. The pump operates, and the defrost valve is closed because the cold water control is sensing the warm inlet water temperature. When the correct amount of water has entered the system, the weight of water causes the water level control switch to open, closing the water valve. No additional water will enter the system until the cycle is complete. The freezing continues, with the ice level building on the evaporator mold, and the level of remaining water decreasing. When the water quantity is sufficiently lowered (indicating the ice is completed), the actuator toggle switch snaps up, completing a circuit through the actuator motor. The actuator motor moves the water plate away from the evaporator, and rotates cams. The cam snaps the toggle switch open, causing the actuator motor to stop.

As the water plate moved away from the evaporator, the defrost and pump toggle switch opened, causing the pump to stop. This switch also completed a circuit to the defrost valve to keep it open until the end of the defrost. As the defrost continues, the ice falls out into the storage bin. After all the ice has been harvested, the actuator thermostat moves to the warm position. This completes the circuit from the weight switch to the reversing side of the actuator toggle switch and the actuator motor. The motor runs in the reverse direction, lifting the water plate toward the evaporator and pushing up the defrost and pump toggle switch. The pump starts again to begin another freezing cycle.

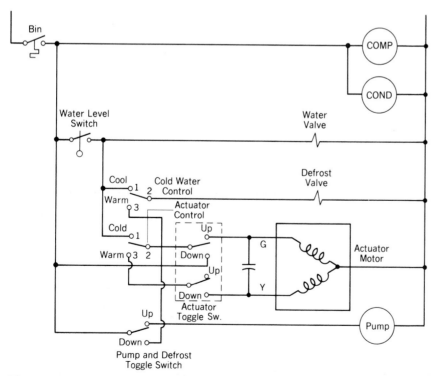

Figure 32-23 Ice cuber using water weight to initiate harvest.

Heating Circuits

Furnaces—Standing Pilot

The furnace is plugged into the wall, providing 115 V to be used by the fan motor and the primary winding of the control transformer (Figure 32-24). The transformer supplies 24 V to the red (or V) terminal of the thermostat. When the thermostat closes, the 24 V passes to the combination gas valve, causing it to open (provided that the thermocouple has proved the flame). With the main flame burning, the bonnet temperature rises, and is sensed by the fan switch. When the bonnet temperature reaches the cut-in setting of the fan switch, the fan motor will be energized through the 115 V fan switch contacts. The limit switch also senses bonnet temperature, and will open if the bonnet temperature rises to the setting of the limit switch. The limit switch will cut off the gas valve, but it will continue to allow the fan to run, removing the heat from the furnace.

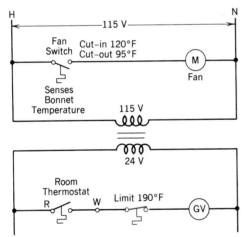

Figure 32-24 Furnace with high limit switch in the gas valve circuit.

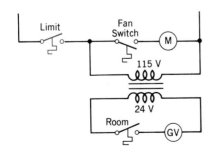

Figure 32-25 Furnace with a line voltage limit switch.

Figure 32-25 shows a limit switch installed in the 115 V circuit. When this limit switch opens, the entire system, including the fan, will shut down.

Figure 32-26 shows the wiring for a downflow or horizontal furnace. It uses two limit switches and a timed-start fan control. When the thermostat makes red to white, it energizes both the gas valve and a resistance heater in the fan control. After the gas valve has been opened for one minute, the resistance heater has warmed sufficiently to close the 115 V bimetal contact, starting the fan motor. One of the limit switches is located downstream from the heat exchanger to detect overheating caused by low airflow or by overfiring. The other limit switch is located high upstream of the heat exchanger to detect overheating caused by no airflow.

Figure 32-27 shows a furnace with a bimetal-type pilot safety. The gas valve in this case is just a simple solenoid valve, with a position for manually lighting the pilot. Once the pilot flame is lit, the pilot safety switch will remain closed, and the gas solenoid valve will energize and open each time the thermostat closes red to white. Note another potential location for the limit switch. This time it is in the transformer circuit. This system also uses a two-speed fan motor, with a High–Low SPDT switch to choose the desired speed. The **summer switch** is wired in parallel with the fan switch. It may be turned on for

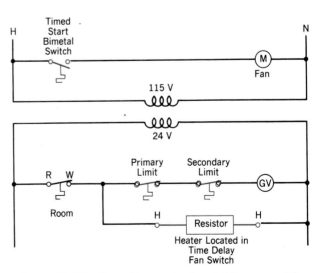

Figure 32-26 Downflow or horizontal furnace with two limit switches and a timed start fan switch.

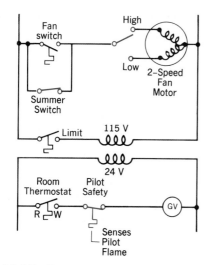

Figure 32·27 Furnace with bimetal pilot safety switch, summer switch, and fan-speed selector switch.

ventilation during the summer, or for continuous air circulation during winter operation.

Automatic Pilot Relighting

Figure 32-28 illustrates an **automatic pilot relighting** system used on commercial rooftop units. The pilot safety is a three-wire device, with a single-pole double-throw switch. The gas valve is a simple solenoid valve (no thermocouple). When pilot flame is lost, the pilot safety switch returns to its normal position. A second transformer is energized through the normally closed pilot safety contacts. This transformer provides 2.4 V to a small glow coil that is located at the pilot burner. The glow coil will reignite the pilot flame, and will operate the bimetal in the pilot safety switch. The pressure switch senses gas pressure. If, for any reason, gas is not available to the furnace, the pressure switch will prevent the glow coil from burning out.

Figure 32-29 shows the same automatic reignition system, except that a 24-V sparker module is used to relight the pilot instead of the glow coil. The pilot burner assembly is manufactured so that the same assembly is capable of accepting either the glow coil or the igniter.

Intermittent Ignition

Figure 32-30 shows an **intermittent ignition** system, using Honeywell equipment. The S86 module is a "black box" that is not serviceable. It must be used with a matching gas valve and pilot

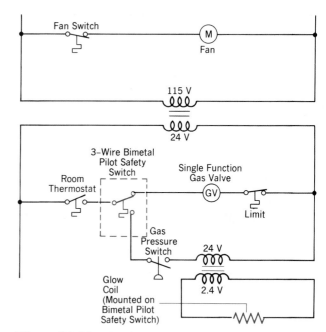

Figure 32-28 Rooftop furnace with automatic pilot relighting using a glow coil.

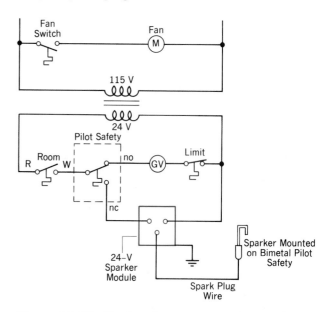

Figure 32-29 Rooftop furnace using a spark generator for pilot relighting.

ignitor-sensor. When the thermostat closes, it provides 24-V to the S86 control module. The module will immediately provide 24 V to the PV (pilot valve) terminal, and pilot gas will be admitted to the pilot burner. At the same time, high voltage will be supplied to the ignitor-sensor assembly through the spark plug wire, and the ignitor will begin sparking.

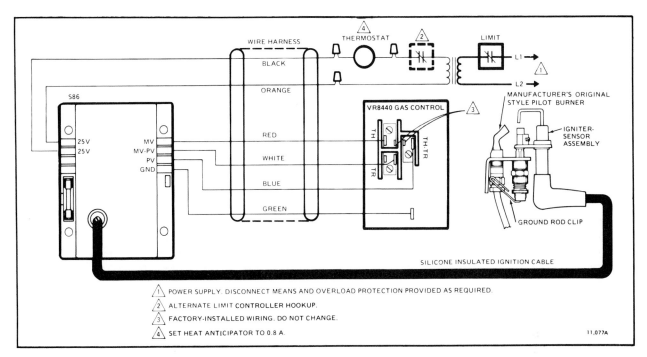

Figure 32-30 Intermittant ignition system.
(Courtesy Honeywell)

When the pilot flame has been established, it will be sensed electronically by the same wire that provided the spark. This signal is sent back to the control module, which will then stop the sparking, and energize the MV (main valve) terminal. Main gas will then be admitted to the furnace.

Figure 32-31 shows a Carrier or BDP type of electronic ignition system. The gas valve contains three coils, called **Pick, Hold,** and **Main.** When the pick coil is energized, it will pick up and open the pilot gas valve. When the hold coil is energized, it is able to hold the pilot valve open once the pick coil has opened it. The hold coil cannot lift the pilot valve to the open position. When the main coil is energized, the main gas valve will open. The sequence begins when the thermostat closes, energizing the pick coil and the hold coil, and providing 24 V to the sparking module. When the pilot flame is established, it is sensed by the three-wire bimetal pilot safety switch. As the bimetal switch moves, the normally closed contacts open, and then the normally open contacts close. When the normally closed contacts open, the sparker stops, and the pick coil is deenergized. The pilot valve remains open, however, held open by the hold coil. When

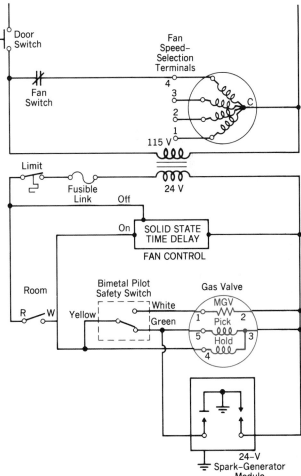

Figure 32-31 Furnace using a Pick-Hold gas valve.

the normally open contact of the bimetal switch closes, the main gas valve is opened.

Fan control on this system is accomplished by a solid state timing device located on a circuit board in the furnace. It is not adjustable or serviceable. The fan switch is shown as normally closed, but when 24 V is available to the solid state time-delay module, the fan switch immediately opens. It will require another 24-V input from the thermostat portion of the circuit to the fan module to reclose the contacts. When the customer complains that the fan won't shut off, check for an open fusible link (most likely), or limit switch, or defective transformer. Note the door switch on this furnace. It is located on the panel covering the fan compartment. It will only complete the circuit if the fan access panel is properly in place. Many states have legislated the use of this switch on all furnaces now being manufactured. For the owner, it is a safety feature. For the service technician, it is an easy service call when the owner complains that the furnace hasn't worked since the filters were changed.

Another way of accomplishing an intermittent ignition control scheme uses a liquid-filled bulb to sense the pilot flame. The bulb operates a switch located on the top of the gas valve. Figure 32-32 illustrates a White Rodgers-type of valve.

Only the 24-V portion of the system is shown, as the fan control is handled in a conventional fashion. When the thermostat calls for heat, the pilot gas valve is energized through the flame sensor contacts that are closed when the mercury bulb is closed (3-1). At the same time, the sparker is energized, providing the ignition source for the pilot gas. When the mercury bulb senses the pilot flame, the SPDT switch moves to the terminal labeled Hot, energizing the main gas valve. The pilot gas valve continues to be energized through the pressure switch that is located on the valve body, and closes whenever the pilot gas pressure is sensed downstream from the pilot valve. The sparker turns itself off when the sparking rod detects that the pilot flame has been established.

The pressure switch in this system, and the Hold coil in the Carrier-type system serve a similar function called **redundancy.** Simply stated, this means that there must be two independent means of proving the pilot before opening the main gas valve.

Vent Damper

The vent damper is a relatively new device introduced to conserve energy. It is a damper installed in the flue gas duct. It opens when the furnace is operating, and closes when the furnace shuts down. The wiring is shown in Figure 32-33. The damper is closed by a small motor, and opened by a spring. When the thermostat is open, the motor is kept running continuously through the normally closed control relay contacts. When the thermostat closes, the control relay coil is ener-

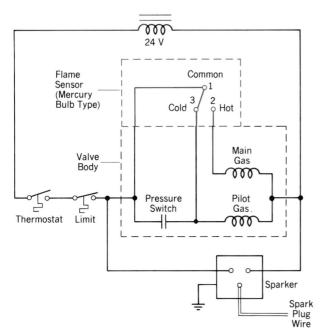

Figure 32-32 Spark ignition system using a mercury-filled bulb to sense pilot flame.

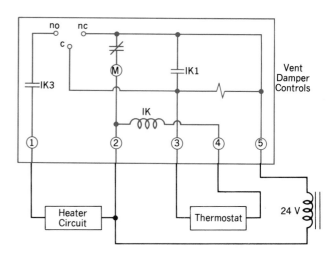

Figure 32-33 Vent damper.

gized through the single-pole double-throw (SPDT) switch, operating three sets of contacts. The control relay coil remains energized, sealed in by one set of its contacts. The motor becomes deenergized, allowing the spring to return the SPDT switch to its normally open position. When the end switch reaches the normally open terminal, the ignition system is energized from terminal 5, through the control relay contacts to Common on the end switch, through the second set of normally open control relay contacts.

The purpose of the resistor between terminals 3 and 5 is to allow proper operation of some clock thermostats. Without this resistor, clocks that operate off the secondary transformer voltage would lose a few seconds each time the end switch was traveling between the normally open and normally closed contacts.

Humidifier

The humidifier (Figure 32-34) is energized whenever the humidity is too low and the furnace fan is running. When the fan switch energizes the blower motor, it also energizes the transformer furnished with the humidifier. The 24-V humidistat completes a circuit through a relay that energizes the humidifier motor and the water valve. Figure 32-35 shows an alternate method in which the transformer and 24-V humidistat are replaced by a 115-V sail switch and a 115-V humidistat. This eliminates the need to wire the humidifier into the furnace circuit. The humidistat and sail switch is sometimes packaged to-

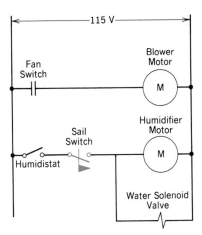

Figure 32-35 Humidifier operated by a sail switch and humidistat.

gether in a single switch that is mounted on the cold air return duct.

Residential Hot Water Boiler

The residential hot water boiler (Figure 32-36) controls the water inside the boiler at a constant temperature (between 160°F and 200°F) using a water-temperature-sensing thermostat to open and close the gas valve (or oil burner). The room thermostat controls the pump, drawing the heat from the boiler. The sequence might be initiated by the room thermostat energizing the pump. Water leaves the boiler at 180°F, and returns at 160°F, cooling the boiler water. The boiler fires up, and remains firing as long as the pump is drawing heat out. When the room thermostat is satisfied, the pump turns off. The boiler con-

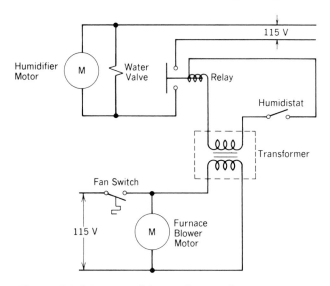

Figure 32-34 Humidifier with transformer.

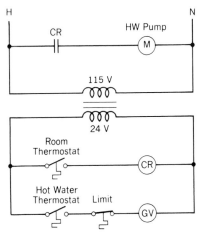

Figure 32-36 Small hot water boiler serving a single zone.

tinues to fire until the water temperature returns to 180°F.

The hot water system is also adaptable for use with several different zones. Figure 32-37 shows a two-zone system where each room thermostat controls a **motorized zone valve.** The zone valve is normally closed. Upon receiving a 24-V input, the valve opens slowly (60 sec). At the end of the travel of the valve stem, an auxiliary switch (end switch) is made. The auxiliary switch is wired in the 115-V circuit, so that the pump will run if either (or both) of the zones requires heat. If no zones are calling for heat, the pump will not operate.

Zone control may also be accomplished by providing a separate pump for each zone (Figure 32-38). This system uses the room thermostats to energize control relays. The control relay contacts are wired in parallel to control the gas valve. In this way, during the summer months when the boiler does not need to be maintained at 180°F, it will be allowed to cool to room temperature.

It may be that the boiler normally operates at 180°F room temperature, but during the summer months it must be maintained at 160°F for heating domestic hot water. In that case, a DPST relay is used, and a second boiler water thermostat set at 160°F is wired in parallel with the second set of control relay contacts (Figure 32-39).

Oil Burner Controller

Figure 32-40 shows the internal schematic wiring of a primary controller. This wiring diagram describes the function of the control. When a malfunction occurs, however, the controller itself is replaced rather than repaired. When the thermostat calls for heating, a circuit is completed through the 1K relay coil, safety switch

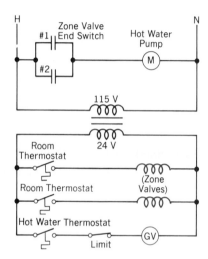

Figure 32-37 Small hot water boiler serving two zones, each controlled by a motorized zone valve.

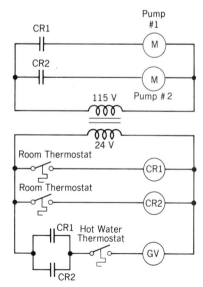

Figure 32-38 Hot water boiler using multiple pumps and hot water temperature that is not maintained unless one of the zones requires heating.

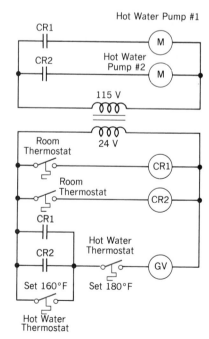

Figure 32-39 Hot water boiler. The water temperature is maintained at 180°F while heating. When there is no demand for heating, the hot water temperature is allowed to drop, but no lower than 160°F.

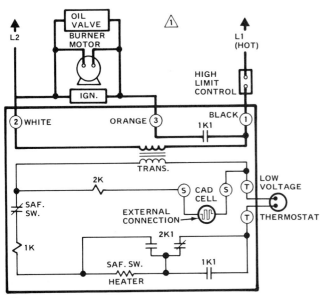

PROVIDE OVERLOAD PROTECTION AND DISCONNECT MEANS AS REQUIRED.

Figure 32-40 Oil burner primary controller with cad cell. *(Courtesy Honeywell)*

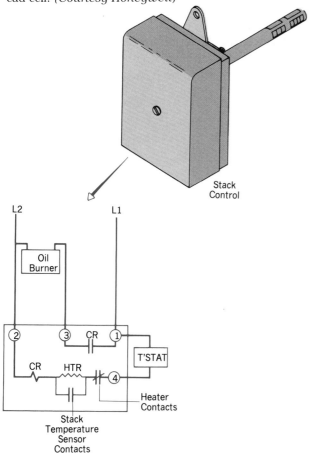

Stack Control

Heater Contacts

Stack Temperature Sensor Contacts

Figure 32-41 A stack-mounted primary control. For the oil burner to start, the stack temperature sensor contacts must close, energizing the control relay and shorting out the heater.

heater, and the normally closed contacts of 2K1. The 1K contacts complete the line voltage circuit through the ignitor transformer, the burner motor, and the oil solenoid valve. If a flame is established and sensed by the cad cell, the 2K relay coil will become energized. This will create a short around the safety switch heater, causing it to stop heating. If the cad cell circuit is not completed within 15 to 45 sec (depending on the model of the controller), the safety switch heater will open the 1K coil circuit.

Figure 32-41 shows an internal wiring diagram for the **oil burner primary control** that is mounted on the stack. The temperature-sensing probe in the stack senses when the oil flame has been established, shorting out the heater as in the previous example.

Electric Heat

The simplest electric furnace (Figure 32-42) operates a 230-V heating element through a heat relay that is controlled by a low voltage thermostat. Fuses are provided on both sides of the blower motor circuit and heater element circuit because on the 230-V supply, each leg carries a hot 115 V. The limit switch will open on high air temperature if the airflow is restricted. The thermal fuse in the heater circuit will open on high temperature caused by any reason. The electronic blower control senses current passing through the heating element, and immediately energizes the blower motor (no time delay).

Multiple heating elements are common, staged to come on in sequence. A two-stage thermostat may be used to energize one or two elements as required by the room (Figure 32-43). Even if the thermostat is turned up, the elements will be sequenced. When both switches in the thermostat close, only #1 heat relay is energized. This energizes #1 heat element, and also heat relay 2. Only after heat relay 2 contacts close will the #2 heat element be energized through heat relay 2a. Figure 32-44 shows a three-element heater, with all three elements sequenced by a single-stage low voltage room thermostat. Figure 32-45 adds an outdoor thermostat that locks out one heating element if the remaining heating element will eventually be able to satisfy the thermostat. This effectively accomplishes the same end as the two-stage thermostat, avoiding fast cycling of the furnace during periods of light load.

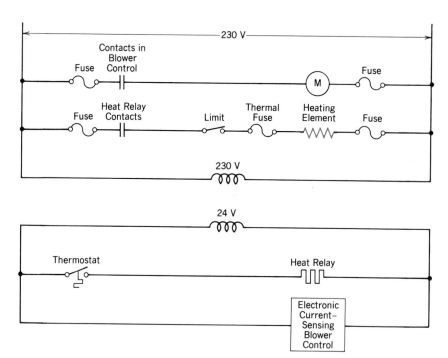

Figure 32-42   Electric heat using a single heating element.

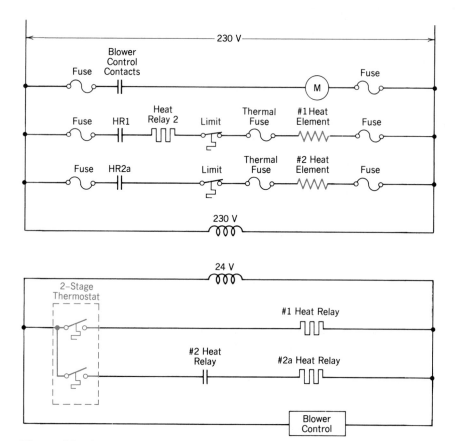

Figure 32-43   Electric heat using two elements operated form
a two-stage room thermostat.

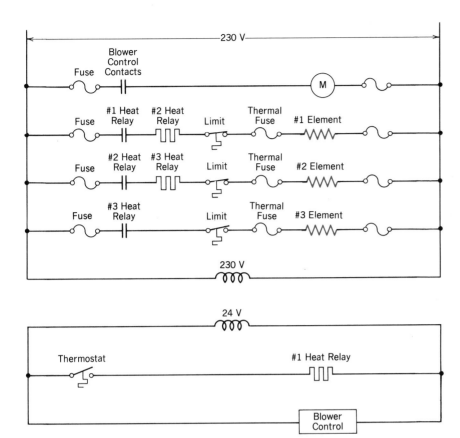

Figure 32-44 Electric heat using three sequenced elements.

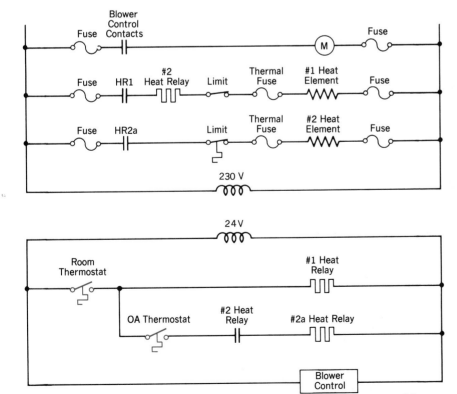

Figure 32-45 Electric heat using two elements that are sequenced from one room thermostat to an outdoor air thermostat.

KEY TERMS AND CONCEPTS

LPC Control

Safeties in Series

Defrost Timers

Household Ice Maker

Pumpdown

Window Air
Conditioner

Split-system Air
Conditioner

Motor Push-button
Control

Across-the-line Starter

Condenser Fan
Cycling

Antirecycle Timer

Reset Relay

Auger Delay Switch

Harvest Cycle

Cutting Grid

Conductivity Switch

Thickness Switch

Summer Switch

Automatic Pilot
Relighting

Intermittent Ignition

Pick-Hold Valve

Redundancy

Motorized Zone Valve

Oil Burner Controller

Sequenced Heating
Elements

QUESTIONS

1. If the door in Figure 32-2 is opened while the compressor is running, which load(s) are energized? Deenergized?

2. What is the minimum number of switches in Figure 32-4 that must be closed in order to keep the compressor running?

3. What is the function of the defrost termination switch in Figure 32-5? When will it open?

4. What is the purpose of the drain heater in Figure 32-7? When does it operate?

5. Referring to Figure 32-8, how many volts would you measure across the cold control during the defrost cycle? Volts across the compressor?

6. What is the purpose of the heater in Figure 32-9?

7. If the liquid solenoid valve in Figure 32-11 leaks, what symptom would you observe?

8. The unit in Figure 32-12 has pumped down, and the pressures subsequently equalized due to a leaking LSV. Which control relay(s) are energized?

9. The unit in Figure 32-13 is running on High Cool. What will happen to the evaporator fan when the thermostat is satisfied?

10. What will happen if you remove the thermostat from its subbase in Figure 32-14 and you jumper the R terminal to the G terminal?

11. The system in Figure 32-14 is idle. How would you operate the thermostat to cause the indoor fan (and nothing else) to operate?

12. What will happen if you jumper R to Y in the furnace terminal strip shown in Figure 32-15? R to C? (*Caution!* Please do not try it.)

13. Referring to Figure 32-16, what is the first load that will be deenergized when the stop button is depressed?

14. The compressor in Figure 32-17 runs without either of the condenser fans until it cuts out on high head pressure. Could the problem be that the HPC is faulty? Explain.

15. Referring to Figure 32-18, what is the purpose of the switch between terminals 1 and 2 in the OPC?

16. What is the purpose of a reset relay?

17. What type of starting relay is used in Figure 32-21?

18. Referring to Figure 32-22, how many volts would you measure across the hot gas solenoid during the freezing cycle?

19. Referring to Figure 32-24, how will the operation of the furnace be changed if the fan switch cut-in is changed to 115°F?

20. If the blower in Figure 32-27 is not operable due to an open winding, what symptoms will you be able to observe?

21. Referring to Figure 32-28, the tenants have left for the day and turned the thermostat down to 50°F. During the early evening the pilot flame blows out. Will it attempt to relight? Explain.

22. In Figure 32-31, the pilot flame lights, and several seconds later goes out. This repeats itself every 30 seconds. What is the problem?

23. Referring to Figure 32-32, when will the pressure switch in the gas valve close?

24. Redraw the wiring diagram in Figure 32-37, adding whatever is necessary for a third zone.

25. What is the purpose of the normally closed 2K1 contact in Figure 32-40?

26. Referring to Figure 32-43, what is the physical difference between heat relay 2 and heat relay 2a?

CHAPTER 33

ELECTRICAL TROUBLESHOOTING

Although this is the last chapter in the text, it is probably the most important in terms of describing a required skill of the service technician. Of all the times that a unit stops working properly, over 80 percent of those times on the average are attributable to electrical malfunctions. Of all the time that a service technician spends repairing air conditioning and refrigeration units, 50 percent of that time will be spent on electrical troubleshooting and repair.

The Elements of Troubleshooting

When an electrical malfunction is suspected, it is the job of the technician to

1. Locate the failed part or parts quickly.
2. Determine if the part failed for a reason, or if it was a random failure.
3. Replace the defective parts and test for normal operation.

The following sections describe many techniques that can be used for troubleshooting systems of various voltages. Some are techniques that are not commonly described in textbooks due to potential safety implications if used improperly. But they are used by trained personnel every day in the field because they are good techniques. **THE STUDENT IS CAUTIONED TO STRICTLY OBSERVE THE SAFETY REMINDERS IN THIS CHAPTER.**

The Troubleshooting Approach

Most electrical problems within a refrigeration, heating, or air conditioning unit can be attributed to one of two causes.

1. A switch is open, not allowing a device (load) to become energized.
2. The control system is working properly, but a load (motor, solenoid, etc.) has failed.

Note that the open switch may be a defective switch, or it may be a good switch that has correctly sensed an out of limit condition (high pressure, low pressure, etc.), and it is properly doing its job.

The experienced service technician will be able to draw conclusions based upon observations of which parts of the system are not operating, and devise a logical troubleshooting sequence. The symptoms observed and the accessibility of each of the parts of the system influence the troubleshooting sequence. Blindly following prescribed troubleshooting procedures for anything more complex than a single device will not be effective. Above all, the effective electrical troubleshooter must be a good puzzle solver.

A Voltmeter vs. An Ohmmeter

Troubleshooting of the same devices may be done with either a voltmeter or an ohmmeter. For example, if a switch is not making good contact, a voltmeter will indicate a voltage across the switch, or an ohmmeter will indicate infinite resistance. The voltmeter method, however, is preferable for the following reasons.

1. When using the ohmmeter, external wires have to be removed. This takes additional time.
2. When removing wiring connections, there is a chance of causing more problems if terminals are broken or the wires are otherwise damaged.
3. In the case of a poor connection, wiggling of

595

the wires may temporarily fix the problem. You will no longer be able to troubleshoot the problem, and you can expect a callback.

Cutting the Problem in Half

The most common situation encountered by the service technician is that the unit does not run. Units like residential condensing units have a single contactor that controls both the compressor motor and the condenser fan motor. If neither is running, you can probably rule out a failed compressor, because if that was the only problem the condenser fan motor would still run. You can probably rule out a failed condenser fan motor for the same reason. It is theoretically possible that both motors have failed at the same time. Statistically, this is so remote that our first assumption will be that it is not the case.

With both motors not running, we can surmise that there exists one of three problems.

1. There is no control voltage at the contactor, therefore it is not closing the switch in the power circuit.

2. The control voltage has caused the contactor switches to close, but there is no line voltage available to cause electrical flow through the switch to the motors (blown circuit breaker).

3. The control voltage has asked the contactor to close, but the contactor remains open because it is not working properly.

Troubleshooting this system may be done as described in Figure 33-1. As with all troubleshooting charts, it is a guideline only. For example, if the contactor is difficult to access, and a fused disconnect is staring you in the face, by all means check the power through the fuses first.

The nontextbook way of troubleshooting the contactor is as follows.

1. Remove the cover on the contactor that protects the contacts (Figure 33-2).

2. With the power On, using an insulated screwdriver, manually press down on the plastic parts that will cause the contacts to move to the contact position. BE CERTAIN THAT YOU DO NOT TOUCH ANY METAL PARTS WITH THE SCREWDRIVER.

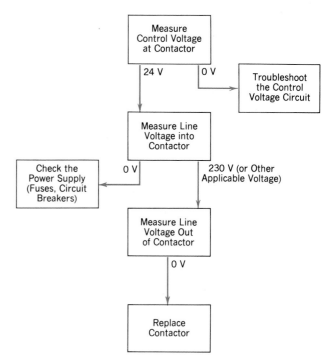

Figure 33-1 Troubleshooting a condensing unit that has a compressor and condenser fan motor, both not running.

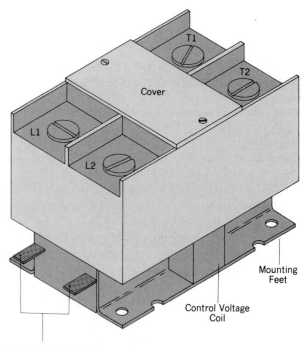

Figure 33-2 Check incoming voltage by measuring between L1 and L2, leaving voltage between T1 and T2.

3. If the unit then runs, you have a control wiring or contactor problem. Check for voltage at the low voltage contacts to determine which it is.

4. If the unit does not run, you have a power wiring problem or a contactor problem. Check the incoming line voltage between L1 and L2.

This technique is useful, quick, and must be done very carefully to avoid slipping with the screwdriver. Some technicians will use this technique on voltages of 230 or lower, and the meter technique for 460-V systems.

Check All the Switches

Where there is no wiring diagram available, and where the wiring of the unit is complex, it may be quickest to go to each switch and determine whether there is a voltage across it. Without having to figure out the entire circuit, an open switch may be identified by a non-zero voltage reading. In order to confirm that replacement of a switch will, in fact, cure the problem, a jumper wire may be used as follows.

1. Turn the unit off at the disconnect.

2. Attach the jumper wires across the terminals of the switch. THE JUMPER WIRE GAUGE MUST BE SUFFICIENTLY LARGE TO CARRY THE LOAD.

3. Close the disconnect switch. If the unit then runs properly, proceed with replacement of the switch. IF THE UNIT DOES NOT RUN IMMEDIATELY, OPEN THE DISCONNECT SWITCH.

Finding an open switch does not always mean that you have found the cause of a problem. In Figure 33-3, the compressor contactor coil has shorted. The rush of current caused the contacts of the thermostat to burn, and now both are de fective. If you simply replace the thermostat, the new one will fail as soon as you energize the circuit. In that case, using the jumper wire across the switch would result in one of two outcomes.

1. If the contactor coil is still shorted, when power is applied, the fuse protecting the control circuit will open. If no fuse is provided (very common), the transformer may be ru-

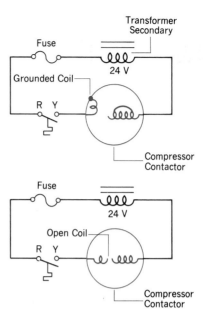

Figure 33-3 Thermostat–contactor circuit.

ined if the jumper is left in place for more than a second or two.

2. The contactor coil is open, following the short. When power is applied, the compressor will still not run.

Measuring Circuit Resistance

An ohmmeter can be used to determine whether a potential short-circuit condition exists, which would damage the replacement switch. Turn off the power. Attach a jumper wire across the open switch. Measure the resistance across the plug (for a 115-V system) or across the transformer (for a 24-V system) as in Figures 33-4 and 33-5. If there is a short anywhere in the system, it will be indicated by a zero resistance reading. The advantage of this method is that it avoids the potential of blowing a fuse or taking a chance on damaging a transformer.

Hopscotch Method

Where you have a wiring diagram available, or where wiring is easily followed, the **hopscotch method** will pinpoint an open switch, an open wire, or a poor connection. Consider the wiring

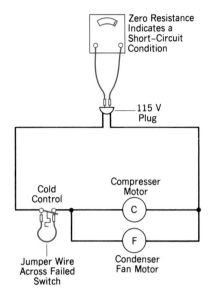

Figure 33-4 Investigating a potential short-circuit condition.

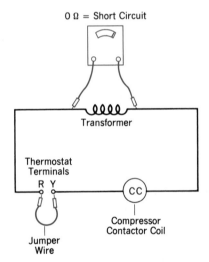

Figure 33-5 Investigating a potential short-circuit condition in the secondary.

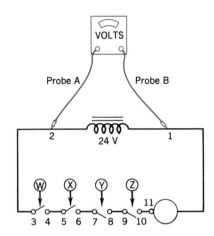

Figure 33-6 Hopscotch method—move probe A to points 3, 4, 5, etc., until the voltage reading becomes zero.

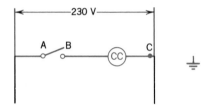

Figure 33-7 Voltage between points A, B, or C to ground will be 115 V, regardless of the switch position.

diagram in Figure 33-6. Start with the voltmeter probes at points 1 and 2 to measure secondary voltage (if there is no secondary voltage, you will move backwards into the primary voltage circuit). If secondary voltage is read, move probe A to point 3, then 4, 5, and so on, until you no longer read a voltage. If you read voltage at 7, but no voltage at 8, then switch Y is open. You can move your voltmeter probes to points 7 and 8, and you will measure a voltage directly across the switch.

If you measure a voltage at point 8 and no voltage at point 9, you have discovered a broken wire or a poor connection connecting points 8 and 9.

The hopscotch method may also be used on 115-V systems. You may start with your probes on a hot and a neutral, or you may start with your probes on a hot and a casing ground (screw, sheet metal, etc.). Using the casing ground will not work, however, on most 24-V systems, as neither side of the 24-V transformer is usually grounded. The casing ground should also not be used on systems with a control voltage of 230 V. You will read 115 V at all points in the circuit (Figure 33-7). This will be of no value in helping you locate the source of the problem.

Any time you measure a voltage across a load (motor, control relay coil, contactor coil, heater, etc.) and the load is not doing what it is supposed to do, the load is defective. There is no need to look any further. The device must be repaired or replaced. In order to confirm your diagnosis, the load may be removed from the circuit and tested with an ohmmeter. Reading infinite ohms across the load confirms that it has an open coil or wire. Even if the ohm reading is

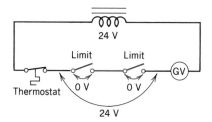

Figure 33-8 If two switches are open, the voltmeter will read zero across each.

correct, the device may have still failed mechanically.

Two Open Switches

The circuit in Figure 33-8 has two limit switches, and they are both open. In checking for voltage across the switches, you will not read 24 V. This is because for each switch measurement, one of the voltmeter probes is attached to a section of wire that is not connected to anything else in the circuit! If the voltmeter is placed across the section of the circuit that includes both switches, a 24-V reading will result.

Common Problems

When a device (load) does not work, either there is no voltage being applied to the device (a switch is open, a fuse has blown, etc.) or the device is defective. Troubleshooting this type of problem with the voltmeter is relatively simple, as previously described. The descriptions that follow are less clear-cut.

Example
You discovered a small, single-phase hermetic compressor that has been operating for years. It is cycling on the overload protector.

Solution
Cycling on the overload protector indicates that either the compressor is drawing too much current, or it is getting too hot, or else the compressor is fine and the overload is defective. When it is operating, observe the condenser fan (if applicable) to make sure it is running while you measure the amperage through the compressor. If it draws locked rotor amps until the overload opens, check the starting relay. It is not taking the start winding out of the circuit. Other possibilities are a shorted run capacitor, start capacitor, or a defective compressor motor. Check the motor winding resistances first, check for grounded windings, test the capacitors with an ohmmeter, and if the compressor will still not run properly, replace the relay. If all this is to no avail, the compressor may be mechanically stuck, or there may be a restriction in the capillary tube that is not allowing the high-side and low-side pressures to equalize.

If your amperage reading on the compressor shows a normal starting current, and then a running current that is above nameplate, the compressor might be beginning to seize. If it is a hermetic compressor, it will need to be replaced.

If your amperage reading on the compressor shows normal starting and running current, change the overload.

Example
A new central air conditioning system installed for a residence. On start-up, the condenser fan runs but the compressor does not run. Then the circuit breaker trips.

Solution
On a new installation, first check the line voltage to make sure that it is within 10 percent of the rated voltage. The compressor is probably a PSC motor. Suspect that the cylinder-to-piston clearance may be tight, and the compressor has insufficient starting torque. Install (with alligator clips) a large start capacitor between the S and R terminals. Energize the unit for only a couple of seconds. The capacitor may provide sufficient additional starting torque to break loose the tight fit. If so, remove the start capacitor and try the unit again with its normal configuration. If it still does not start, permanently install a hard-start kit. This kit provides the start capacitor and the starting relay to take the start winding out of the circuit.

For stubborn cases of a mechanically tight compressor, there are two additional steps that may be taken if the start capacitor does not work. If it is a 115-V unit, try it on 230-V for *just a couple of seconds*. If this does not do the trick, as a last resort you can try to run the compressor motor in the opposite direction. Rewire the compressor as if the run terminal is actually the start terminal, and the start terminal is actually the run terminal. Again, energize the circuit for just a couple of seconds, just to intially break the compressor loose.

Customer Relations

The service technician frequently enjoys the freedom of being able to work in customers' homes or places of business without direct supervision. But this priviledge also carries many responsibilities.

1. Never surprise the customer (or your boss for that matter). When you determine what needs to be done, advise your customer of at least the approximate cost before you proceed. It is a courtesy to allow the customer to decide whether to proceed before you spend a lot of time or install expensive parts.

2. Respect the customer's premises. Take care not to track in mud (flat soled shoes are good) or get finger prints on walls or ceiling tiles.

3. Be a diplomat. You might have to remove ten cartons of junk in order to gain access to a unit to be serviced. If the homeowner apologizes that "I've been meaning to clean this up for quite some time," it would not be appropriate to respond "Yeah, this is one of the worst messes I've seen in a month!." A nonjudgmental "I have trouble finding enough time for my projects around the house too," would be far less offensive.

4. Be a good ambassador from your company. Keep yourself neat in appearance, listen to the customer, and be aware of potential sales of additional equipment.

5. Keep everybody informed. If you become delayed either on a job or due to unexpected circumstances, let other people know. The boss has a right to know, and the customer has a right to expect you at the time that had been estimated for your arrival.

6. Honesty and integrity above all. For the well-trained technician, there will always be work. Companies grow based upon satisfied customers. It is a good practice to repair systems the best way you know how. Recommend replacement of marginally functioning parts. But *never* intentionally replace a part (and charge a customer) you know to be good. All parts that are removed from a system should be left with the customer so it will be obvious that they have no value to you. As a courtesy, you might ask the customer, "Would you like me to dispose of the old parts?"

KEY TERMS AND CONCEPTS

Cutting the Problem in Half

Manual Contactor Operation

Finding the Open Switch

Finding the Defective Load Device

Hopscotch Method

230-V Troubleshooting

Customer Relations

QUESTIONS

1. On a residential split-system central air conditioner, you observe that the compressor runs and the condenser fan does not. Name at least two potential problems you would investigate first.

2. You have rewired a domestic refrigerator. What measurement can you take to assure that there is not a short circuit somewhere in the system?

3. Why is it preferable to troubleshoot with a voltmeter instead of an ohmmeter?

4. Describe two ways to test for a blown fuse. Which is better? Why?

5. What do you accomplish by manually depressing the switches on a contactor?

6. A residential split-system air conditioner uses the furnace blower as the evaporator fan. The furnace has been unplugged. What voltage will you measure between L1 and L2 at the condensing unit? Between T1 and T2?

7. In Figure 32-1 on page 569, the compressor will not run but the condenser and evaporator fan motors will. Name three possible causes.

8. In Figure 32-2 on page 569, if one of the cabinet lights does not work, could the problem be a defective door switch? Explain.

9. In Figure 32-3 on page 570, if the overload and the OPC are both open, what voltage would you measure across the OPC?

10. In Figure 32-4 on page 570, with the compressor running, what voltage would you measure from one of the HPC terminals to the casing ground?

11. In Figure 32-5 on page 570, the compressor, condenser, and evaporator all won't run. Across which terminals would you place a jumper wire to determine whether the defrost time clock was at fault?

12. The cooling system in Figure 32-11 (see page 573) does not run. You measured zero volts across the thermostat, 24 V across the LSV, zero volts across the LPC, and 24 V across the compressor contactor coil. What will you do to repair the system?

13. In Figure 32-11, if you measured 24 V across the LSV and 24 V across the LPC, name three possible problems.

14. The window air conditioner in Figure 32-13 (see page 574) runs on Low, Cool, or Medium Cool, but not high cool. Across which terminals in the selector switch could you use a jumper wire to determine whether the switch is defective?

15. You push on the contactor in Figure 32-14 on page 574. The compressor runs and the condenser fan doesn't. What is the problem?

16. In Figure 32-20 on page 579, what voltage would you expect to find across the reset relay coil when the unit is operating normally?

17. In Figure 32-21 on page 580, the auger motor is the only device that runs. What problem would you suspect?

18. In Figure 32-24 on page 583, where could you place a jumper wire to make the fan come on without the flame?

19. In Figure 32-27 on page 584, you measure 115 V across the motor and zero volts across the fan switch. What does that tell you about the fan switch?

20. In Figure 32-30 on page 585, you measured 24 V between MV and GND. You observe a good pilot flame but no main flame. What will you do next?

21. In Figure 32-32 on page 586, you observe the pilot flame coming on intermittently, but the main flame does not come on. What do you suspect is the problem? Explain your answer.

APPENDIXES

APPENDIX A

MECHANIC'S DATA

Tap Drill Sizes

NATIONAL COARSE OR U.S.S.

Screw & Tap Size	Threads Per Inch	Use Drill Number
No. 5	40	39
No. 6	32	36
No. 8	32	29
No. 10	24	25
No. 12	24	17
1/4	20	8
5/16	18	F
3/8	16	5/16
7/16	14	U
1/2	13	27/64
9/16	12	31/64
5/8	11	17/32
3/4	10	21/32
7/8	9	49/64
1	8	7/8
1 1/8	7	63/64
1 1/4	7	1 7/64
1 1/2	6	1 11/32

NATIONAL FINE OR S.A.E.

Screw & Tap Size	Threads Per Inch	Use Drill Number
No. 5	44	37
No. 6	40	33
No. 8	36	29
No. 10	32	21
No. 12	28	15
1/4	28	3
5/16	24	1
3/8	24	Q
7/16	20	W
1/2	20	29/64
9/16	18	33/64
5/8	18	37/64
3/4	16	11/16
7/8	14	13/16
1 1/8	12	1 3/64
1 1/4	12	1 11/64
1 1/2	12	1 27/64

Decimal Equivalent Size of the Number Drills

Drill No.	Decimal Equivalent	Drill No.	Decimal Equivalent	Drill No.	Decimal Equivalent
80	.0135	53	.0595	26	.1470
79	.0145	52	.0635	25	.1495
78	.0160	51	.0670	24	.1520
77	.0180	50	.0700	23	.1540
76	.0200	49	.0730	22	.1570
75	.0210	48	.0760	21	.1590
74	.0225	47	.0785	20	.1610
73	.0240	46	.0810	19	.1660
72	.0250	45	.0820	18	.1695
71	.0260	44	.0860	17	.1730
70	.0280	43	.0890	16	.1770
69	.0292	42	.0935	15	.1800
68	.0310	41	.0960	14	.1820
67	.0320	40	.0980	13	.1850
66	.0330	39	.0995	12	.1890
65	.0350	38	.1015	11	.1910
64	.0360	37	.1040	10	.1935
63	.0370	36	.1065	9	.1960
62	.0380	35	.1100	8	.1990
61	.0390	34	.1110	7	.2010
60	.0400	33	.1130	6	.2040
59	.0410	32	.1160	5	.2055
58	.0420	31	.1200	4	.2090
57	.0430	30	.1285	3	.2130
56	.0465	29	.1360	2	.2210
55	.0520	28	.1405	1	.2280
54	.0550	27	.1440		

Decimal Equivalent Size of the Letter Drills

Letter Drill	Decimal Equivalent	Letter Drill	Decimal Equivalent	Letter Drill	Decimal Equivalent
A	.234	J	.277	S	.348
B	.238	K	.281	T	.358
C	.242	L	.290	U	.368
D	.246	M	.295	V	.377
E	.250	N	.302	W	.386
F	.257	O	.316	X	.397
G	.261	P	.323	Y	.404
H	.266	Q	.332	Z	.413
I	.272	R	.339		

APPENDIX A *(Continued)*

Decimal Equivalents of the Common Fractions

1/64 .... = .0156	21/64 .... = .3281	43/64 .... = .6719
1/32 .... = .0313	11/32 ...... = .3438	11/16 = .6875
3/64 .... = .0469	23/64 .... = .3594	45/64 .... = .7031
1/16 = .0625	3/8 = .3750	23/32 ...... = .7188
5/64 .... = .0781	25/64 .... = .3906	47/64 .... = .7344
3/32 .... = .0938	13/32 ...... = .4063	3/4 = .7500
7/64 .... = .1094	27/64 .... = .4219	49/64 .... = .7656
1/8 = .1250	7/16 = .4375	25/32 ...... = .7813
9/64 .... = .1406	29/64 .... = .4531	51/64 .... = .7969
5/32 .... = .1563	15/32 ...... = .4688	13/16 = .8125
11/64 .... = .1719	31/64 .... = .4844	53/64 .... = .8281
3/16 = .1875	1/2 = .5000	27/32 ...... = .8438
13/64 .... = .2031	33/64 .... = .5156	55/64 .... = .8594
7/32 .... = .2188	17/32 ...... = .5313	7/8 = .8750
15/64 .... = .2344	35/64 .... = .5469	57/64 .... = .8906
1/4 = .2500	9/16 = .5625	29/32 ...... = .9063
17/64 .... = .2656	37/64 .... = .5781	59/64 .... = .9219
9/32 .... = .2813	19/32 ...... = .5938	15/16 = .9375
19/64 .... = .2969	39/64 .... = .6094	61/64 .... = .9531
5/16 = .3125	5/8 = .6250	31/32 ...... = .9688
	41/64 .... = .6406	63/64 .... = .9844
	21/32 ...... = .6563	

APPENDIX B

FOOD STORAGE CHARACTERISTICS

(Courtesy Russell Coil Company)

General Product Information

Commodity	Storage Temp, F	Relative Humidity, %	Approximate Storage Life*	Water Content, %	Highest Freezing Point, F	Specific Heat Above Freezing Btu/lb/F	Specific Heat Below Freezing Btu/lb/F	Latent Heat (Calculated) Btu/lb
Alfalfa meal	0 or below	70-75	1 year, plus	—	—	—	—	—
Apples	30-40	90	3-8 months	84.1	29.3	0.87	0.45	121
Apricots	31-32	90	1-2 weeks	85.4	30.1	0.88	0.46	122
Artichokes (Globe)	31-32	95	2 weeks	83.7	29.9	0.87	0.45	120
Jerusalem	31-32	90-95	5 months	79.5	27.5	0.83	0.44	114
Asparagus	32-36	95	2-3 weeks	93.0	30.9	0.94	0.48	134
Avocados	45-55	85-90	2-4 weeks	65.4	31.5	0.72	0.40	94
Bananas	—	85-95	—	74.8	30.6	0.80	0.42	108
Beans (Green or snap)	40-45	90-95	7-10 days	88.9	30.7	0.91	0.47	128
Lima	32-40	90	1 week	66.5	31.0	0.73	0.40	94
Beer, keg	35-40	—	3-8 weeks	90.2	28.0	0.92	—	129
bottles, cans	35-40	65 or below	3-6 months	90.2	—	—	—	—
Beets								
Bunch	32	95	10-14 days	—	31.3	—	—	—
Topped	32	95-100	4-6 months	87.6	30.1	0.90	0.46	126
Blackberries	31-32	95	3 days	84.8	30.5	0.88	0.46	122
Blueberries	31-32	90-95	2 weeks	82.3	29.7	0.86	0.45	118
Bread	0	—	3 weeks to 3 months	32-37	—	0.70	0.34	46-53
Broccoli, sprouting	32	95	10-14 days	89.9	30.9	0.92	0.47	130
Brussels sprouts	32	95	3-5 weeks	84.9	30.5	0.88	0.46	122
Cabbage, late	32	95-100	3-4 months	92.4	30.4	0.94	0.47	132
Candy	0-34	40-65	—	—	—	—	—	—
Canned foods	32-60	70 or lower	1 year	—	—	—	—	—
Carrots, Topped, immature	32	98-100	4-6 weeks	88.2	29.5	0.90	0.46	126
Topped, mature	32	98-100	5-9 months	88.2	29.5	0.90	0.46	126
Cauliflower	32	95	2-4 weeks	91.7	30.6	0.93	0.47	132
Celery	32	95	1-2 months	93.7	31.1	0.95	0.48	135
Cherries, sour	31-32	90-95	3-7 days	83.7	29.0	0.87	—	120
Frozen	0 to -10	—	1 year	—	—	—	0.45	—
Sweet	30-31	90-95	2-3 weeks	80.4	28.8	0.84	—	—
Cocoa	32-40	50-70	1 year, plus	—	—	—	—	—
Coconuts	32-35	80-85	1-2 months	46.9	30.4	0.58	0.34	67
Coffee (green)	35-37	80-85	2-4 months	10-15	—	0.30	0.24	14-21
Collards	32	95	10-14 days	86.9	30.6	0.90	—	—
Corn, sweet	32	95	4-8 days	73.9	30.9	0.79	0.42	106
Cranberries	36-40	90-95	2-4 months	87.4	30.4	0.90	0.46	124
Cucumbers	50-55	90-95	10-14 days	96.1	31.1	0.97	0.49	137
Currants	31-32	90-95	10-14 days	84.7	30.2	0.88	0.45	120
Cheese, Cheddar	30-34	65-70	18 months	37.5	8.0	0.50	0.31	53
Cheddar	40	65-70	6 months	37.5	8.0	0.50	0.31	53
Processed	40	65-70	12 months	39.0	19.0	0.50	0.31	56
Grated	40	60-70	12 months	31.0	—	0.45	0.29	44
Butter	40	75-85	1 month	16.0	-4 to 31	0.50	—	23
Butter	-10	70-85	12 months	16.0	-4 to 31	—	0.25	23
Cream	-10 to -20	—	6-12 months	55-75	31.0	0.66-0.80	0.36-0.42	79-107
Ice cream	-20 to -15	—	3-12 months	58-63	21.0	0.66-0.70	0.37-0.39	86
Milk, fluid whole								
Pasteurized, Grade A	32-34	—	2-4 months	87.0	31.0	0.93	—	125
Pasteurized, Grade A	-15	—	3-4 months	87.0	31.0	—	0.46	125
Condensed, sweetened	40	—	15 months	28.0	5.0	0.42	0.28	40
Evaporated	70	—	12 months	74.0	29.5	0.79	0.42	106
Evaporated	40	—	24 months	74.0	29.5	0.79	0.42	106
Milk, dried Whole	70	low	6-9 months	2-3	—	—	—	4
Non-fat	45-70	low	16 months	2-4.5	—	0.36	—	4
Whey, dried	70	low	12 months	3-4	—	0.36	—	4
Dates	0 or 32	75 or less	6-12 months	20.0	3.7	0.36	0.26	29
Dewberries	31-32	90-95	3 days	84.5	29.7	0.88	—	—
Dried foods	32-70	low	6 months to 1 year, plus	—	—	—	—	—
Dried fruits	32	50-60	9-12 months	14.0-26.0	—	0.31-0.41	0.26	20-37
Eggplant	45-50	90-95	7-10 days	92.7	30.6	0.94	0.48	132
Eggs, Shell	29-31	80-85	5-6 months	66.0	28.0	0.73	0.40	96
Shell, farm cooler	50-55	70-75	2-3 weeks	66.0	28.0	0.73	0.40	96

APPENDIX B *(Continued)*

General Product Information

Commodity	Storage Temp, F	Relative Humidity, %	Approximate Storage Life*	Water Content, %	Highest Freezing Point, F	Specific Heat Above Freezing Btu/lb/F	Specific Heat Below Freezing Btu/lb/F	Latent Heat (Calculated) Btu/lb
Frozen, whole	0 or below	—	1 year, plus	74.0	—	—	0.42	106
Frozen, yoke	0 or below	—	1 year, plus	55.0	—	—	0.36	79
Frozen, white	0 or below	—	1 year, plus	88.0	—	—	0.46	126
Whole egg solids	35-40	low	6-12 months	2-4	—	0.22	0.21	4
Yolk solids	35-40	low	6-12 months	3-5	—	0.23	0.21	6
Flake albumen solids	Room Temp	low	1 year, plus	12-16	—	0.31	0.24	20
Figs, Dried	32-40	50-60	9-12 months	24.0	—	0.39	0.27	34
Fresh	31-32	85-90	7-10 days	78.0	27.6	0.82	0.43	112
Fish, Fresh	33-35	90-95	5-15 days	62-85	28.0	0.70-0.86	—	89-122
Frozen	-20- 0	90-95	6-12 months	62-85	—	—	0.38-0.45	89-122
Smoked	40-50	50-60	6-8 months	—	—	0.70	0.39	92
Brine salted	40-50	90-95	10-12 months	—	—	0.76	0.41	100
Mild cured	28-35	75-90	4-8 months	—	—	0.76	0.41	100
Shellfish, Fresh	30-33	85-95	3-7 days	80-87	28.0	0.83-0.90	—	113-125
Frozen	-20- 0	90-95	3-8 months	—	—	—	0.44-0.46	113-125
Frozen-pack fruits	-10- 0	—	6-12 months	—	—	—	—	—
Frozen-pack vegetables	-10- 0	—	6-12 months	—	—	—	—	—
Furs and Fabrics	34-40	45-55	several years	—	—	—	—	—
Garlic, dry	32	65-70	6-7 months	61.3	30.5	0.69	0.40	89
Gooseberries	31-32	90-95	2-4 weeks	88.9	30.0	0.90	0.46	126
Grapefruit	50-60	85-90	4-6 weeks	88.8	30.0	0.91	0.46	126
Grapes, American type	31-32	85-90	2-8 weeks	81.9	29.7	0.86	0.44	116
European type	30-31	90-95	3-6 months	81.6	28.1	0.86	0.44	116
Greens, leafy	32	95	10-14 days	—	—	—	—	—
Guavas	45-50	90	2-3 weeks	83.0	—	0.86	—	—
Honey	—	—	1 year, plus	18.0	—	0.35	0.26	26
Hops	29-32	50-60	several months	—	—	—	—	—
Horseradish	30-32	95-100	10-12 months	73.4	28.7	0.78	0.42	104
Kale	32	95	3-4 months	86.6	31.1	0.89	0.46	124
Lard (without antioxidant)	45	90-95	4-8 months	0	—	—	—	—
Lard (without antioxidant)	0	90-95	12-14 months	0	—	—	—	—
Lemons	32 or 50-58	85-90	1-6 months	89.3	29.4	0.91	0.46	127
Lettuce, head	32-34	95-100	2-3 weeks	94.8	31.7	0.96	0.48	136
Limes	48-50	85-90	6-8 weeks	86.0	29.1	0.89	0.46	122
Mangoes	55	85-90	2-3 weeks	81.4	30.3	0.85	0.44	117
Maple syrup	—	—	—	35.5	—	0.48	0.31	51
Meat								
Bacon, frozen	-10- 0	90-95	4-6 months	—	—	—	—	—
cured (Farm style)	60-65	85	4-6 months	13-29	—	0.30-0.43	0.24-0.29	18-41
cured (Packer style)	34-40	85	2-6 weeks	—	—	—	—	—
Beef, fresh	32-34	88-92	1-6 weeks	62-77	28-29	0.70-0.84	—	89-110
frozen	-10- 0	90-95	9-12 months	—	—	—	0-38-0.43	—
Fat backs	34-36	85-90	0-3 months	6-12	—	0-25-0.30	0.22-0.24	9-17
Hams and shoulders, Fresh	32-34	85-90	7-12 days	47-54	28-29	0.58-0.63	—	67-77
Frozen	-10- 0	90-95	6-8 months	—	—	—	0.34-0.36	—
Cured	60-65	50-60	0-3 years	40-45	—	0.52-0.56	0.32-0.33	57-64
Lamb, Fresh	32-34	85-90	5-12 days	60-70	28-29	0.68-0.76	—	86-100
Frozen	-10- 0	90-95	8-10 months	—	—	—	0 38-0.51	—
Livers, Frozen	-10- 0	90-95	3-4 months	70.0	—	—	0.41	100
Pork, Fresh	32-34	85-90	3-7 days	32-44	28-29	0.46-0.55	—	46-63
Frozen	-10- 0	90-95	4-6 months	—	—	—	0.30-0.33	—
Smoked Sausage	40-45	85-90	6 months	60.0	—	0.68	0.38	86
Sausage Casings	40-45	90-95	—	—	—	—	—	—
Veal, Fresh	32-34	90-95	5-10 days	64-70	28-29	0.71-0.76	—	92-100
Frozen	-10- 0	90-95	8-10 months	—	—	—	0.39-0.41	—
Melons, Cantaloupe	36-40	90-95	5-15 days	92.0	29.9	0.93	0.48	132
Casaba	45-50	85-95	4-6 weeks	92.7	30.1	0.94	0.48	132
Honeydew and Honey Ball	45-50	90-95	3-4 weeks	92.6	30.3	0.94	0.48	132
Persian	45-50	90-95	2 weeks	92.7	30.5	0.94	0.48	132
Watermelons	40-50	80-90	2-3 weeks	92.1	31.3	0.97	0.48	132
Mushrooms	32	90	3-4 days	91.1	30.4	0.93	0.47	130
Mushroom, Manure spawn	34	75-80	8 months	—	—	—	—	—
Grain spawn	32-40	75-80	2 weeks	—	—	—	—	—
Nectarines	31-32	90	2-4 weeks	81.8	30.4	—	—	—

General Product Information

Commodity	Storage Temp, F	Relative Humidity, %	Approximate Storage Life*	Water Content, %	Highest Freez-ing Point, F	Specific Heat Above Freezing Btu/lb/F	Specific Heat Below Freezing Btu/lb/F	Latent Heat (Calcu-lated) Btu/lb
Nursery stock	32-35	85-90	3-6 months	—	—	—	—	—
Nuts	32-50	65-75	8-12 months	3-6	—	0.22-0.25	0.21-0.22	4-8
Oil (vegetable salad)	70	—	1 year, plus	0	—	—	—	—
Okra	45-50	90-95	7-10 days	89.8	28.7	0.92	0.46	128
Oleomargarine	35	60-70	1 year, plus	15.5	—	0.32	0.25	22
Olives, fresh	45-50	85-90	4-6 weeks	75.2	29.4	0.80	0.42	108
Onions (dry) and onion sets	32	65-70	1-8 months	87.5	30.6	0.90	0.46	124
green	32	95	3-4 weeks	89.4	30.4	0.91	—	—
Oranges	32-48	85-90	3-12 weeks	87.2	30.6	0.90	0.46	124
Orange juice, chilled	30-35	—	3-6 weeks	89.0	—	0.91	0.47	128
Papayas	45	85-90	1-3 weeks	90.8	30.4	0.82	0.47	130
Parsley	32	95	1-2 months	85.1	30.0	0.88	0.45	122
Parsnips	32	98-100	4-6 months	78.6	30.4	0.84	0.44	112
Peaches and nectarines	31-32	90	2-4 weeks	89.1	30.3	0.90	0.46	124
Pears	29-31	90-95	2-7 months	82.7	29.2	0.86	0.45	118
Peas, green	32	95	1-3 weeks	74.3	30.9	0.79	0.42	106
Peppers, Sweet	45-50	90-95	2-3 weeks	92.4	30.7	0.94	0.47	132
Peppers, Chili (dry)	32-50	60-70	6 months	12.0	—	0.30	0.24	17
Persimmons	30	90	3-4 months	78.2	28.1	0.84	0.43	112
Pineapples, Mature green	50-55	85-90	3-4 weeks	—	30.2	—	—	—
Ripe	45	85-90	2-4 weeks	85.3	30.0	0.88	0.45	122
Plums, including fresh prunes	31-32	90-95	2-4 weeks	82.3	30.5	0.88	0.45	118
Pomegranates	32	90	2-4 weeks	—	26.6	—	—	—
Popcorn, unpopped	32-40	85	4-6 months	13.5	—	0.31	0.24	19
Potatoes, Early crop	50-55	90	—	81.2	30.9	0.85	0.44	116
Late crop	38-50	90	—	77.8	30.9	0.82	0.43	111
Poultry, Fresh	32	85-90	1 week	74.0	27.0	0.79	—	106
Frozen, eviscerated	0 or below	90-95	8-12 months	—	—	—	0.42	—
Pumpkins	50-55	70-75	2-3 months	90.5	30.5	0.92	0.47	130
Quinces	31-32	90	2-3 months	85.3	28.4	0.88	0.45	122
Radishes—Spring, prepackaged	32	95	3-4 weeks	93.6	30.7	0.95	0.48	134
Winter	32	95-100	2-4 months	93.6	—	0.95	0.48	134
Rabbits, Fresh	32-34	90-95	1-5 days	68.0	—	0.74	0.40	98
Frozen	-10- 0	90-95	0-6 months	—	—	—	—	—
Raspberries, Black	31-32	90-95	2-3 days	80.6	30.0	0.84	0.44	122
Red	31-32	90-95	2-3 days	84.1	30.9	0.87	0.45	121
Frozen (red or black)	-10- 0	—	1 year	—	—	—	—	—
Rhubarb	32	95	2-4 weeks	94.9	30.3	0.96	0.48	134
Rutabagas	32	98-100	4-6 months	89.1	30.1	0.91	0.47	127
Seed, vegetable	32-50	50-65	—	7.0-15.0	—	0.29	0.23	16
Spinach	32	95	10-14 days	92.7	31.5	0.94	0.48	132
Squash, Acorn	45-50	70-75	5-8 weeks	—	30.5	—	—	—
Summer	32-50	85-95	5-14 days	94.0	31.1	0.95	—	135
Winter	50-55	70-75	4-6 months	88.6	30.3	0.91	—	127
Strawberries, Fresh	31-32	90-95	5-7 days	89.9	30.6	0.92	—	129
Frozen	-10- 0	—	1 year	72.0	—	—	0.42	103
Sweet Potatoes	55-60	85-90	4-7 months	68.5	29.7	0.75	0.40	97
Tangerines	32-38	85-90	2-4 weeks	87.3	30.1	0.90	0.46	125
Tobacco, hogsheads	50-65	50-55	1 year	—	—	—	—	—
Bales	35-40	70-85	1-2 years	—	—	—	—	—
Cigarettes	35-46	50-55	6 months	—	—	—	—	—
Cigars	35-50	60-65	2 months	—	—	—	—	—
Tomatoes, Mature green	55-70	85-90	1-3 weeks	93.0	31.0	0.95	0.48	4
Firm ripe	45-50	85-90	4-7 days	94.1	31.1	0.94	0.48	134
Turnips, roots	32	95	4-5 months	91.5	30.1	0.93	0.47	130
Vegetable seed	32-50	50-65	—	7.0-15.0	—	0.29	0.23	16
Yams	60	85-90	3-6 months	73.5	—	0.79	—	105
Yeast, compressed baker's	31-32	—	—	70.9	—	0.77	0.41	102

WEIGHT OF REFRIGERANT IN TUBING

(Courtesy Russell Coil Company)

WEIGHT OF REFRIGERANT IN COPPER LINES
Pounds per 100 feet of Type L Tubing

O.D. Line Size	I.D. Line Size	Volume per 100 Ft. in Cu. Ft.	Weight of Refrigerant, Pounds					
			Liquid @ 100° F.	Hot Gas @ 120° F. Condensing	Suction Gas (Superheated to 65°)			
					-40° F.	-20° F.	20° F.	40° F.
R-12								
3/8	.315	.054	4.25	.171	.011	.018	.044	.065
1/2	.430	.100	7.88	.317	.021	.033	.081	.120
5/8	.545	.162	12.72	.514	.033	.054	.131	.195
7/8	.785	.336	26.4	1.065	.069	.112	.262	.405
1 1/8	1.025	.573	45.0	1.82	.118	.191	.464	.690
1 3/8	1.265	.872	68.6	2.76	.179	.291	.708	1.05
1 5/8	1.505	1.237	97.0	3.92	.254	.412	1.01	1.49
2 1/8	1.985	2.147	169.0	6.80	.441	.715	1.74	2.58
2 5/8	2.465	3.312	260.0	10.5	.680	1.10	2.68	3.98
3 1/8	2.945	4.728	371.0	15.0	.97	1.57	3.82	5.69
3 5/8	3.425	6.398	503.0	20.3	1.32	2.13	5.18	7.70
4 1/8	3.905	8.313	652.0	26.4	1.71	2.77	6.73	10.0
R-22								
3/8	.315	.054	3.84	.202	.013	.021	.052	.077
1/2	.430	.100	7.12	.374	.024	.04	.096	.143
5/8	.545	.162	11.52	.605	.038	.064	.156	.232
7/8	.785	.336	24.0	1.26	.079	.134	.323	.480
1 1/8	1.025	.573	40.8	2.14	.136	.228	.550	.820
1 3/8	1.265	.872	62.1	3.26	.207	.348	.839	1.25
1 5/8	1.505	1.237	88.0	4.62	.294	.493	1.19	1.77
2 1/8	1.985	2.147	153.0	8.04	.51	.858	2.06	3.06
2 5/8	2.465	3.312	236.0	12.4	.78	1.32	3.18	4.72
3 1/8	2.945	4.728	336.0	17.7	1.12	1.88	4.55	6.75
3 5/8	3.425	6.398	456.0	24.0	1.51	2.55	6.15	9.14
4 1/8	3.905	8.313	592.0	31.1	1.97	3.31	8.0	11.19
R-502								
3/8	.315	.054	3.98	.284	.020	.033	.077	.112
1/2	.430	.100	7.38	.525	.037	.061	.143	.208
5/8	.545	.162	11.95	.852	.061	.098	.232	.337
7/8	.785	.336	24.8	1.77	.126	.204	.481	.700
1 1/8	1.025	.573	42.3	3.01	.215	.347	.820	1.19
1 3/8	1.265	.872	64.4	4.60	.327	.527	1.25	1.81
1 5/8	1.505	1.237	91.2	6.5	.465	.750	1.77	2.57
2 1/8	1.985	2.147	159.0	11.3	.806	1.30	3.08	4.48
2 5/8	2.465	3.312	244.0	17.4	1.24	2.0	4.74	6.90
3 1/8	2.945	4.728	349.0	24.8	1.77	2.87	6.76	9.84
3 5/8	3.425	6.398	471.0	33.6	2.40	3.87	9.15	13.32
4 1/2	3.905	8.313	612.0	43.8	3.12	5.03	11.90	17.30

PRESSURE-ENTHALPY DIAGRAMS

(Courtesy E. I. duPont de Nemours and Company)

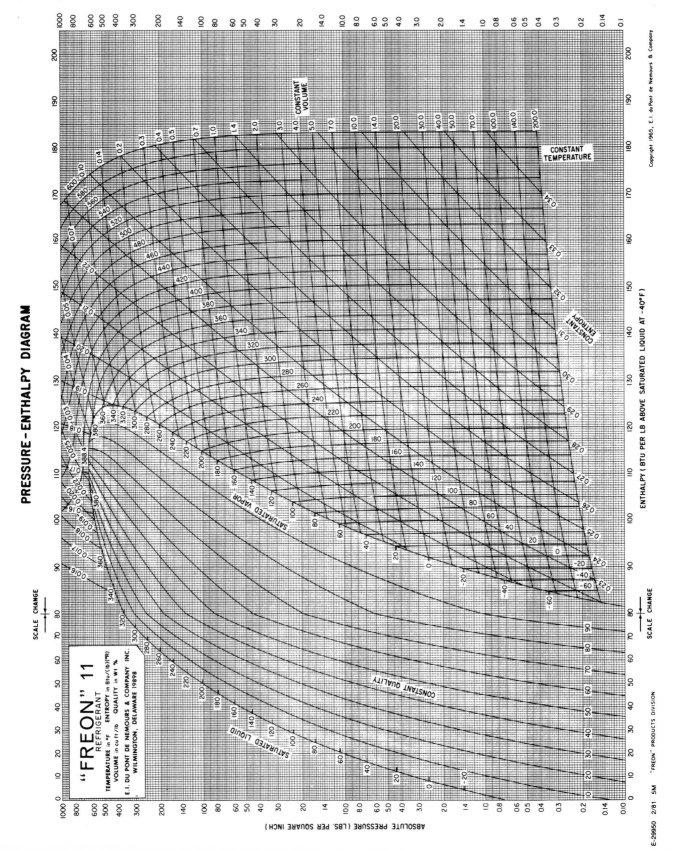

PRESSURE-ENTHALPY DIAGRAM

"FREON" 11
REFRIGERANT
TEMPERATURE in °F ENTROPY in Btu/(lb)(°R)
VOLUME in cu ft/lb QUALITY in Wt %
E. I. DU PONT DE NEMOURS & COMPANY INC.
WILMINGTON, DELAWARE 19898

ENTHALPY (BTU PER LB ABOVE SATURATED LIQUID AT -40°F)

ABSOLUTE PRESSURE (LBS. PER SQUARE INCH)

CONSTANT VOLUME

CONSTANT TEMPERATURE

CONSTANT ENTROPY

SATURATED VAPOR

SATURATED LIQUID

CONSTANT QUALITY

SCALE CHANGE

Copyright 1965, E. I. duPont de Nemours & Company

E-29950 2/81 5M "FREON" PRODUCTS DIVISION

APPENDIX D *(Continued)*

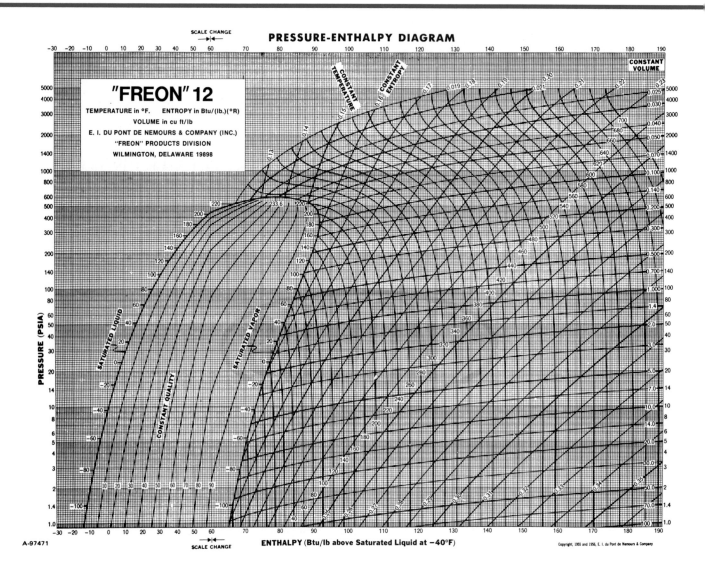

PRESSURE-ENTHALPY DIAGRAM

"FREON" 12

TEMPERATURE in °F.   ENTROPY in Btu/(lb.)(°R)

VOLUME in cu ft/lb

E. I. DU PONT DE NEMOURS & COMPANY (INC.)

"FREON" PRODUCTS DIVISION

WILMINGTON, DELAWARE 19898

ENTHALPY (Btu/lb above Saturated Liquid at −40°F)

A-97471

Copyright, 1955 and 1956, E. I. du Pont de Nemours & Company

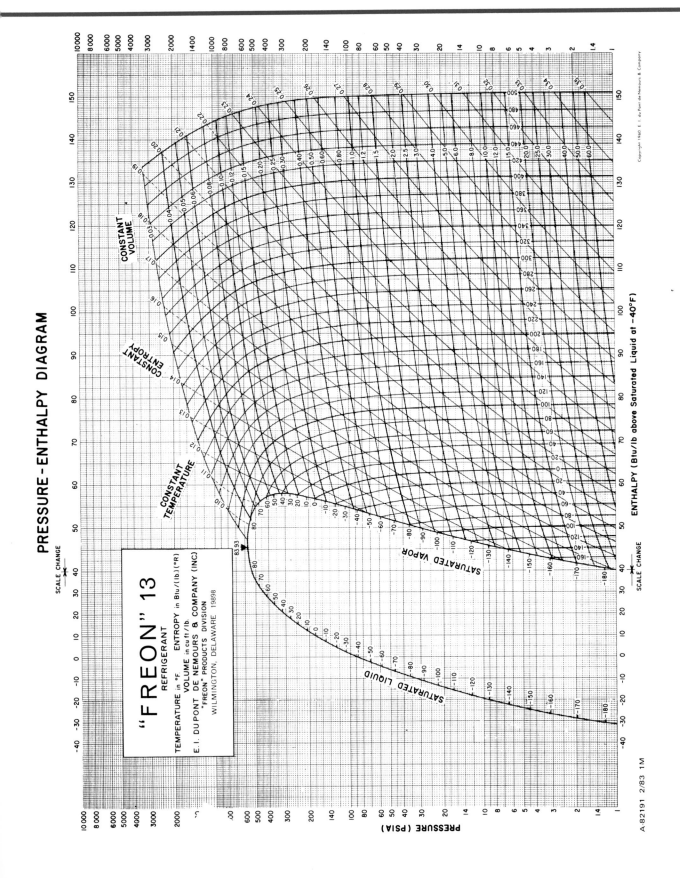

PRESSURE-ENTHALPY DIAGRAM

"FREON" 13
REFRIGERANT

TEMPERATURE in °F. ENTROPY in Btu/(lb.)(°R)
VOLUME in cu.ft./lb.
E. I. DU PONT DE NEMOURS & COMPANY (INC)
"FREON" PRODUCTS DIVISION
WILMINGTON, DELAWARE 19898

CONSTANT VOLUME

CONSTANT ENTROPY

CONSTANT TEMPERATURE

SATURATED LIQUID

SATURATED VAPOR

PRESSURE (PSIA)

ENTHALPY (Btu/lb above Saturated Liquid at −40°F)

SCALE CHANGE

Copyright 1960 E. I. du Pont de Nemours & Company

A-82191 2/83 1M

APPENDIX D (Continued)

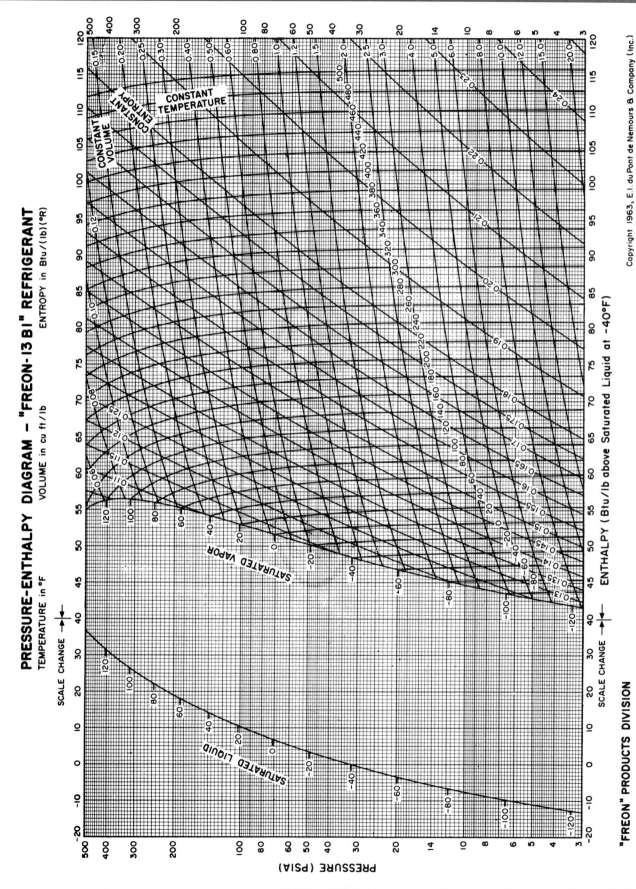

PRESSURE-ENTHALPY DIAGRAM – "FREON-13 B1" REFRIGERANT

TEMPERATURE in °F VOLUME in cu ft/lb ENTROPY in Btu/(lb)(°R)

ENTHALPY (Btu/lb above Saturated Liquid at –40°F)

PRESSURE (PSIA)

CONSTANT TEMPERATURE

CONSTANT VOLUME

CONSTANT ENTROPY

SATURATED VAPOR

SATURATED LIQUID

SCALE CHANGE

Copyright 1963, E.I. duPont de Nemours & Company (Inc.)

"FREON" PRODUCTS DIVISION   E.I. DU PONT DE NEMOURS & CO. (INC.) WILMINGTON 98, DELAWARE

"FREON" 14 FLUOROCARBON   PRESSURE-ENTHALPY DIAGRAM

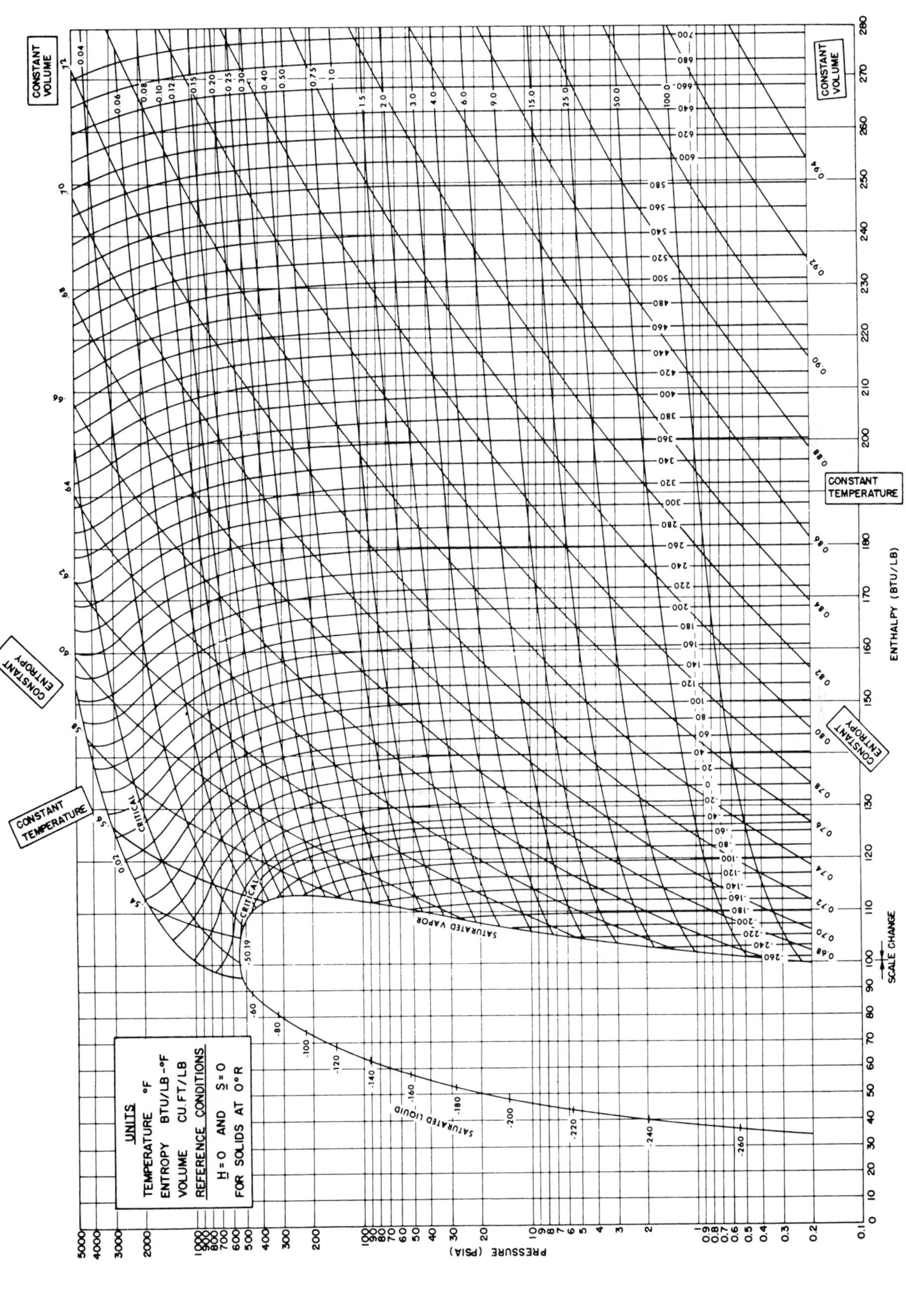

UNITS

TEMPERATURE °F
ENTROPY BTU/LB-°F
VOLUME CU FT/LB

REFERENCE CONDITIONS

H = 0 AND S = 0
FOR SOLIDS AT 0°R

CONSTANT VOLUME

CONSTANT ENTROPY

CONSTANT TEMPERATURE

CRITICAL

SATURATED VAPOR

SATURATED LIQUID

PRESSURE (PSIA)

ENTHALPY (BTU/LB)

SCALE CHANGE

CONSTANT TEMPERATURE

CONSTANT ENTROPY

CONSTANT VOLUME

Reprinted by permission from a dissertation by N. C. S. Chari, University of Michigan, based on Du Pont-supported research

PRINTED IN U.S.A.

6/78 1M

APPENDIX D *(Continued)*

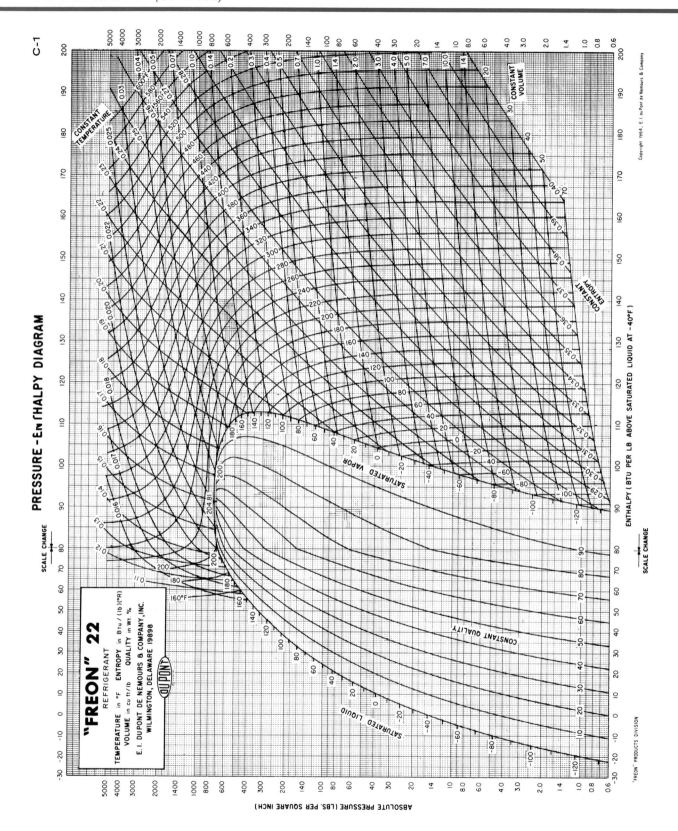

PRESSURE - EnTHALPY DIAGRAM

"FREON" 22
REFRIGERANT

TEMPERATURE in °F   ENTROPY in Btu / (lb)(°R)
VOLUME in cu ft/lb   QUALITY in Wt. %

E. I. DUPONT DE NEMOURS & COMPANY, INC.
WILMINGTON, DELAWARE 19898

DUPONT

ABSOLUTE PRESSURE (LBS. PER SQUARE INCH)

ENTHALPY (BTU PER LB ABOVE SATURATED LIQUID AT -40°F)

Copyright 1964, E. I. du Pont de Nemours & Company

"FREON" PRODUCTS DIVISION

A-85951 3/83 5M

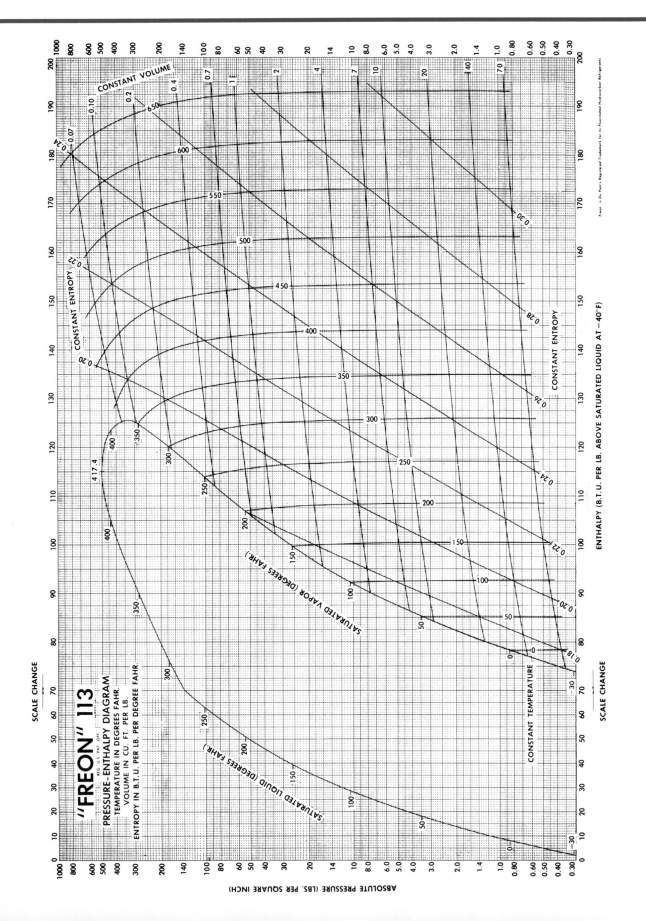

"FREON" 113
REG. U.S. PAT. OFF.
PRESSURE-ENTHALPY DIAGRAM
TEMPERATURE IN DEGREES FAHR.
VOLUME IN CU. FT. PER LB.
ENTROPY IN B.T.U. PER LB. PER DEGREE FAHR.

ENTHALPY (B.T.U. PER LB. ABOVE SATURATED LIQUID AT −40°F)

ABSOLUTE PRESSURE (LBS. PER SQUARE INCH)

A-95291 7/80 2M

APPENDIX D (Continued)

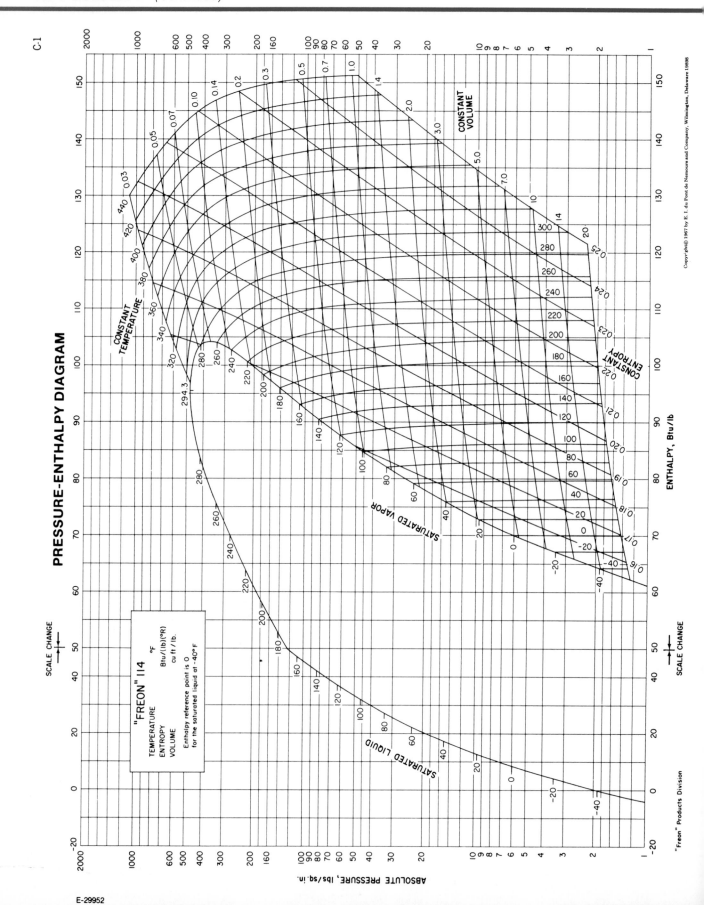

PRESSURE-ENTHALPY DIAGRAM

C-1

"FREON" 114

TEMPERATURE °F
ENTROPY Btu/(lb)(°R)
VOLUME cu ft / lb.

Enthalpy reference point is 0
for the saturated liquid at -40° F

CONSTANT TEMPERATURE

CONSTANT VOLUME

CONSTANT ENTROPY

SATURATED VAPOR

SATURATED LIQUID

ENTHALPY, Btu/lb

ABSOLUTE PRESSURE, lbs/sq.in.

SCALE CHANGE

Copyright© 1967 by E. I. du Pont de Nemours and Company, Wilmington, Delaware 19898

"Freon" Products Division

E-29952

"FREON" 502 REFRIGERANT PRESSURE—ENTHALPY DIAGRAM

TEMPERATURE in °F, ENTROPY Btu/(lb)(°R), VOLUME in cu ft/lb

ENTHALPY, Btu/lb.

PRESSURE, PSIA

Copyright, 1968, E. I. dupont de Nemours & Company

"FREON" PRODUCTS DIVISION

A-82310 4/82 5M

APPENDIX D *(Continued)*

PRESSURE - ENTHALPY DIAGRAM

"FREON-503"
REFRIGERANT
TEMPERATURE in °F   ENTROPY in Btu/(lb)(°R)
VOLUME in cu ft/lb

E. I. DU PONT DE NEMOURS & COMPANY, INC.
WILMINGTON, DELAWARE 19898

Copyright 1966, E.I. du Pont de Nemours & Company, Inc.

CONSTANT VOLUME

CONSTANT TEMPERATURE

CONSTANT ENTROPY

SATURATED VAPOR

SATURATED LIQUID

SCALE CHANGE

ENTHALPY (BTU PER LB ABOVE SATURATED LIQUID AT −40°F)

ABSOLUTE PRESSURE (LBS. PER SQUARE INCH)

"FREON" PRODUCTS DIVISION

ABBREVIATIONS OF HVAC TERMS

A	Amps, area.	MA	Mixed air.
Btu	British thermal unit.	MAT	Mixed air temperature.
C	Capacitance.	μF	Microfarad.
cfh	Cubic feet per hour.	mfd	Microfarad.
cfm	Cubic feet per minute.	Ω	Ohms of electrical resistance.
d	Diameter	OA	Outside air.
D	Displacement.	OAT	Outside air temperature.
db	Dry bulb.	p.f.	Power factor.
Δ	(Delta.) Difference.	pf	Power factor.
E	Volts.	psi	Pounds per square inch.
EAT	Entering air temperature.	R	Resistance.
EDB	Entering dry-bulb temperature.	RA	Return air.
gph	Gallons per hour.	RAT	Return air temperature.
gpm	Gallons per minute.	rh	Relative humidity.
H	Absolute humidity.	rpm	Revolutions per minute.
h	Enthalpy.	SA	Supply air.
Hg	Mercury.	SAT	Supply air temperature.
I	Amperes.	SHR	Sensible heat ratio.
kW	Kilowatts.	T	Temperature, time.
kWh	Kilowatt-hours.	U	Heat transmission coefficient.
L	Inductance.	V	Velocity, volume, or volts.
LAT	Leaving air temperature.	VA	Volt-amps.
LDB	Leaving dry-bulb temperature.	wb	Wet bulb.

APPENDIX F

CONVERSION FACTORS AND FORMULAS

CONVERSION FACTORS

Weight

1 gallon of water = 8.33 pounds

1 cubic foot of water = 62.4 pounds

1 cubic foot of air = 0.075 pound

1 pound = 7000 grains

Volume

1 gallon = 231 cubic inches
$\qquad$ = 0.137 cubic foot

1 cubic foot = 1728 cubic inches
$\qquad$ = 7.48 gallons

1 barrel (oil) = 42 gallons

Head

1 psi = 2.31 feet of water
$\qquad$ = 2.04 inches of mercury

1 atmosphere = 29.92 inches of mercury
$\qquad$ = 33.94 feet of water
$\qquad$ = 14.7 psi

Energy

1 hp = 746 watts
$\qquad$ = 2546 Btu/hr

1 kW = 1000 watts
$\qquad$ = 1.34 hp

1 kWh = 3414 Btu

1 therm = 100,000 Btu

FORMULAS

Heat

Sensible heating or cooling of air

$$Q = 1.08 \times \text{cfm} \times \Delta T$$

Latent cooling or humidification of air

$$Q = 0.68 \times \text{cfm} \times \Delta H$$

Total change in heat content of air

$$Q = 4.45 \times \text{cfm} \times \Delta h$$

Heat transmission

$$Q = U \times A \times \Delta T$$

Motors

$$\text{Synchronous speed} = \frac{120 \times \text{frequency}}{\text{number of poles}}$$

$$\text{hp (single phase)} = \frac{V \times A \times \text{efficiency}}{746}$$

$$\text{hp (three phase)} = \frac{V \times A \times \text{efficiency} \times \text{p.f.} \times 1.73}{746}$$

Electrical

To Find	AC System	
	3 Phase	**1 Phase**
hp output	$\dfrac{I \times E \times 1.73 \times \%Eff \times pf}{746}$	$\dfrac{I \times E \times \%Eff \times pf}{746}$
Amperes when hp is known	$\dfrac{HP \times 746}{1.73 \times E \times \%Eff \times pf}$	$\dfrac{HP \times 746}{E \times \%Eff \times pf}$
kVA	$\dfrac{I \times E \times 1.73}{1000}$	$\dfrac{I \times E}{1000}$
Amperes when kVA is known	$\dfrac{kVA \times 1000}{1.73 \times E}$	$\dfrac{kVA \times 1000}{E}$
kW	$\dfrac{I \times E \times 1.73 \times pf}{1000}$	$\dfrac{I \times E \times pf}{1000}$
Amperes when kW is known	$\dfrac{kW \times 1000}{1.73 \times E \times pf}$	$\dfrac{kW \times 1000}{E \times pf}$

I = amperes; E = volts; kVA = kilo-volt-amperes;
kW = kilowatts; %Eff = percent efficiency; pf = power factor.

APPENDIX G

MECHANICAL AND ELECTRICAL DATA

Properties of Saturated Steam

Temperature °F	Pressure psia	Enthalpy Btu/lb	
		Saturated Liquid	Saturated Vapor
40	0.12	8.1	1079
60	0.26	28.1	1087
80	0.51	48.0	1096
212	14.7	180	1150
250	29.8	218	1164
300	67.0	270	1180
350	135	322	1192
400	247	375	1201
450	423	430	1205

Recommended Duct Gauge

Maximum Side, (in.)	Gauge
Up to 12	26
13 to 30	24
31 to 60	22
61 to 90	20

Full-Load Currents in Amperes
Single-Phase Alternating-Current Motors

hp	115 V	230 V
1/6	4.4	2.2
1/4	5.8	2.9
1/3	7.2	3.6
1/2	9.8	4.9
3/4	13.8	6.9
1	16	8
1 1/2	20	10
2	24	12
3	34	17
5	56	28
7 1/2	80	40
10	100	50

Full-Load Current
Three-Phase Alternating-Current Motors

hp	Induction Type Squirrel-Cage and Wound-Roter Amperes		
	115 V	230 V	460 V
1/2	4	2	1
3/4	5.6	2.8	1.4
1	7.2	3.6	1.8
1 1/2	10.4	5.2	2.6
2	13.6	6.8	3.4
3		9.6	4.8
5		15.2	7.6
7 1/2		22	11
10		28	14
15		42	21
20		54	27
25		68	34
30		80	40
40		104	52
50		130	65
60		154	77
75		192	96
100		248	124

Maximum Number of Conductors in Trade Sizes of Conduit or Tubing

Type Letters	Conductor Size AWG, MCM	½	¾	1	1¼	1½	2
	Conduit Trade Size (Inches)						
TW, T, RUH,	14	9	15	25	44	60	99
RUW.	12	7	12	19	35	47	78
XHHW (14 thru 8)	10	5	9	15	26	36	60
	8	2	4	7	12	17	28
RHW and RHH	14	6	10	16	29	40	65
(without outer	12	4	8	13	24	32	53
covering),	10	4	6	11	19	26	43
THW	8	1	3	5	10	13	22
TW,	6	1	2	4	7	10	16
T,	4	1	1	3	5	7	12
THW,	3	1	1	2	4	6	10
RUH (6 thru 2),	2	1	1	2	4	5	9
RUW (6 thru 2),	1		1	1	3	4	6
FEPB (6 thru 2),							
RHW and RHH							
(without outer							
covering)							
	14	13	24	39	69	94	154
	12	10	18	29	51	70	114
THWN,	10	6	11	18	32	44	73
	8	3	5	9	16	22	36
THHN,	6	1	4	6	11	15	26
FEP (14 thru 2),	4	1	2	4	7	9	16
FEPB (14 thru 8),	3	1	1	3	6	8	13
PFA (14 thru 4/0)	2	1	1	3	5	7	11
PFAH (14 thru 4/0)	1		1	1	3	5	8
Z (14 thru 4/0)							
XHHW (4 thru 500MCM)							
	14	3	6	10	18	25	41
	12	3	5	9	15	21	35
RHW,	10	2	4	7	13	18	29
	8	1	2	4	7	9	16
RHH	6	1	1	2	5	6	11
	4	1	1	1	3	5	8
(with	3	1	1	1	3	4	7
outer	2		1	1	3	4	6
covering)	1		1	1	1	3	5

APPENDIX G *(Continued)*

Loss of Air Pressure Due to Friction

In psi per 1000 ft of pipe at 100 psig

Cu Ft Free Air per Min	Pipe Size					
	$\frac{1}{2}''$	$\frac{3}{4}''$	1"	$1\frac{1}{4}''$	$1\frac{1}{2}''$	2"
10	6.50	0.99	0.28	—	—	—
20	25.90	3.90	1.11	0.25	0.11	—
30	58.5	9.01	2.51	0.57	0.26	—
40	—	16.0	4.45	1.03	0.46	—
50	—	25.1	6.96	1.61	0.71	0.19
60	—	36.2	10.0	2.32	1.02	0.28
70	—	49.3	13.7	3.16	1.40	0.37
80	—	64.5	17.8	4.14	1.83	0.49
90	—	82.8	22.6	5.23	2.32	0.62
100	—	—	27.9	6.47	2.86	0.77
125	—	—	48.6	10.2	4.49	1.19
150	—	—	62.8	14.2	6.43	1.72
175	—	—	—	19.8	8.72	2.36
200	—	—	—	25.9	11.4	3.06
250	—	—	—	40.4	17.9	4.78
300	—	—	—	58.2	25.8	6.85
350	—	—	—	—	35.1	9.36
400	—	—	—	—	45.8	12.1
450	—	—	—	—	58.0	15.4
500	—	—	—	—	71.6	19.2

Loss of Water Pressure Due to Friction

In psi per 100 ft of Type "K" copper tube at 60 psig

Flow Gal. Per Min	Size					
	$\frac{1}{2}''$	$\frac{3}{4}''$	1"	$1\frac{1}{4}''$	$1\frac{1}{2}''$	2"
1	1.3	0.25	0.06	—	—	—
2	4.5	0.85	0.23	0.07	—	—
3	9.0	1.70	0.45	0.15	0.063	—
4	15.0	2.80	0.75	0.25	0.10	—
5	22.0	4.20	1.10	0.36	0.16	0.042
6	31.0	5.80	1.50	0.50	0.22	0.058
7	40.0	7.50	2.00	0.65	0.28	0.073

Loss of Water Pressure Due to Friction

In psi per 100 ft of clean, general-use pipe

Flow Gal. per Min	Pipe Size					
	$\frac{1}{2}''$	$\frac{3}{4}''$	1"	$1\frac{1}{4}''$	$1\frac{1}{2}''$	2"
2	3.3	—	—	—	—	—
3	6.9	1.8	—	—	—	—
4	11.8	3.0	—	—	—	—
5	17.8	4.6	1.4	—	—	—
10	—	16.5	5.1	1.3	—	—
15	—	—	10.8	2.8	1.3	—
20	—	—	18.4	4.8	2.3	—
25	—	—	—	7.2	3.4	1.0
35	—	—	—	13.6	6.4	1.9
50	—	—	—	—	12.3	3.7

APPENDIX H

INDUSTRY ASSOCIATIONS

AABC
Associated Air Balance Council
1518 K St. NW, Suite 503
Washington, DC 20005

ABMA
American Boiler Manufacturers Association
1500 Wilson Blvd.
Arlington, VA 22209

ACCA
Air Conditioning Contractors of America
1228 17th St. NW
Washington, DC 20036

ADC
Air Diffusion Council
230 N. Michigan Ave.
Chicago, IL 60601

ADI
Air Distribution Institute
4415 W. Harrison St., #242-C
Hillside, IL 60162

AFFI
American Frozen Food Institute
1700 Old Meadow Rd., Suite 100
McLean, VA 22209

AGA
American Gas Association
1515 Wilson Blvd.
Arlington, VA 22209

AHAM
Association of Home Appliance Manufacturers
20 N. Wacker Dr.
Chicago, IL 60606

AIPE
American Institute of Plant Engineers
3975 Erie Ave.
Cincinnati, OH 45208

AMCA
Air-Movement and Control Association Inc.
30 W. University Dr.
Arlington Heights, IL 60004

ANSI
American National Standards Institute
1430 Broadway
New York, NY 10018

ARI
Air Conditioning and Refrigeration Institute
1501 Wilson Blvd., 6th Floor
Arlington, VA 22209

ASHRAE
American Society of Heating, Refrigeration and
   Air Conditioning Engineers
1791 Tullie Circle N.E.
Atlanta, GA 30329

CGA
Compressed Gas Association Inc.
1235 Jefferson Davis Hwy.
Arlington, VA 22202

CRMA
Commercial Refrigerator Manufacturers
   Association
1101 Connecticut Ave.
Washington, DC 20036

GAMA
Gas Appliance Manufacturers Association
1901 N. Fort Myer Dr.
P.O. Box 9245
Arlington, VA 22209

GVI
Gas Vent Institute
6531 Drake Ave.
Lincolnwood, IL 60645

HI
The Hydronics Institute
35 Russo Pl.
Berkeley Heights, NJ 07922

HVI
Home Ventilating Institute
Division of AMCA
30 West University Dr.
Arlington Heights, IL 60004

IIAR
International Institute of Ammonia
    Refrigeration
111 E. Wacker Dr.
Chicago, IL 60601

NAFEM
National Association of Food Equipment
    Manufacturers
111 E. Wacker Dr.
Chicago, IL 60601

NATTS
National Association of Trade & Technical
    Schools
2251 Wisconsin Ave. NW
Washington, DC 20007

NEMA
National Electrical Manufacturers Association
Suite 300
2101 L St. NW
Washington, DC 20037

NHAW
Northamerican Heating and Airconditioning
    Wholesalers Association
1389 Dublin Rd.
P.O. Box 16790
Columbus, OH 43216

PMAA
Petroleum Marketers Association of America
1701 H St., 11th Floor
Washington, DC 20006

RCRA
Refrigeration Compressor Rebuilders
    Association
P.O. Box 19047
Kansas City, MO 64141

RETA
Refrigeration Engineers & Technicians
    Association
230 N. Michigan Ave.
Chicago, IL 60601

RRF
Refrigeration Research Foundation
7315 Wisconsin Ave.
Bethesda, MD 20814

RSES
Refrigeration Service Engineers Society
1666 Rand Rd.
Des Plaines, IL 60016

SEIA
Solar Energy Industries Association
Suite 520
1156 15th St. NW
Washington, DC 20005

SMACNA
Sheet Metal & Air Conditioning Contractors
    National Association
P.O. Box 70
Merrifield, VA 22116

VMA
The Valve Manufacturers Association
6845 Elm St., Suite 711
McLean, VA 22101

GLOSSARY

ABSOLUTE HUMIDITY The amount of moisture contained in one cubic foot of dry air, expressed in grains per cu ft or lb per cu ft.

ABSOLUTE PRESSURE Pressure expressed on a scale that starts at zero psia for a perfect vacuum. For pressures above atmospheric, absolute pressure equals gauge pressure plus atmospheric pressure (14.7 psi).

ABSOLUTE TEMPERATURE Temperature expressed on a scale that starts at zero degrees corresponding to absolute zero.

ABSOLUTE ZERO The lowest temperature that can exist. It is equivalent to $-460°F$ or $-273°C$.

ACCESSIBLE HERMETIC A type of compressor in which the motor and compressor are housed together in a casing that is bolted together. Also referred to as serviceable hermetic or semihermetic compressor.

ACCUMULATOR A tank located in the suction line between the evaporator and compressor. It is designed to collect small amounts of liquid refrigerant that may pass through the evaporator for short periods of time. The liquid vaporizes before passing on to the compressor.

AIR-COOLED CONDENSER A heat exchanger used to condense refrigerant vapor from the compressor, using air as the condensing medium.

AIR HANDLER Cabinet containing a fan that supplies air to a room. It may also contain a cooling coil, a heating coil, and/or filters.

AIR VENT Plumbing accessory located at the high point in a water or steam piping system, used to release air from inside the system.

AIR WASHER Bank of water sprays through which air is passed in order to be cooled, cleaned, or humidified.

AMBIENT TEMPERATURE The temperature of the air that surrounds an object. May also refer to room air temperature or outside air temperature.

AMMETER A meter used to measure the flow of current in a wire.

AMPERE The unit of electrical measurement that describes the flow rate of electrons. Amps.

ANEMOMETER An instrument that measures air velocity.

ANNEAL A heat treatment process consisting of heating a metal and letting it cool slowly. Annealed metals become soft and are easily formed or shaped.

ASME American Society of Mechanical Engineers

ASPECT RATIO Long dimension of a duct cross section divided by the short dimension.

ATMOSPHERIC PRESSURE The pressure that exists due to the existence of air in the atmosphere.

ATOMIZE Breaking down a liquid into small droplets, usually by spraying through a nozzle. Liquids are generally atomized in order to increase the amount of surface area.

AUGER The screwlike center part of an ice flaker that scrapes ice from a cylindrical evaporator.

AUTOMATIC EXPANSION VALVE A type of metering device that senses low-side pressure, and modulates in order to maintain the low-side pressure constant.

BACK EMF Impedence.

BACK PRESSURE The low-side pressure entering the compressor. Also suction pressure.

BACK PRESSURE REGULATOR An automatic valve in the suction line that prevents the suction pressure from dropping below the set point.

BACK SEAT Screwing a valve to the full counter-clockwise position. Valves that have a backseat position are commonly found on refrigeration service valves and oxygen tank valves.

BEARING Support device for shaft that must be allowed to rotate within the support.

BELLOWS A cylindrical container with accordion sides that moves in response to changes in pressure.

BENDING SPRING A spring that is placed on the inside or outside of a tube, allowing it to be bent into shape without kinking.

BIMETAL Temperature-sensing device consisting of two dissimilar metals bonded together into a strip. A temperature change causes unequal expansion of the two metals, causing the combination strip to bend.

BLOWER A fan.

BOILER Tank containing water that is heated by the burning of a fuel in order to produce hot water or steam.

BOILER HORSEPOWER A means of describing the heating capacity of a boiler. One boiler horsepower is equivalent to 33,475 Btu/hr.

BONNET The part of a furnace around the heat exchanger.

BORE The diameter of a compressor cylinder.

BOURDON TUBE Pressure-sensing device shaped like a question mark. An increase of pressure applied to the inside of the tube causes it to straighten out, causing a mechanical motion.

BRAZING Joining metals together by melting a non-ferrous filler material with a melting temperature of higher than 800°F.

BREAKER STRIP The plastic closure strip between the plastic interior of a refrigerator and the outside sheet-metal casing.

BRINE Saltwater.

BTU British Thermal Unit. A measure of heat, equal to the amount of heat required to raise the temperature of one pound of water by a temperature of one degree Fahrenheit.

CAD CELL Cadmium sulfide cell. A device that senses a flame in an oil-fired system.

CALORIE A measure of heat, equal to the amount of heat required to raise the temperature of one gram of water by a temperature of one degree Celsius.

CAPACITOR Electrical device consisting of two plates separated by a dielectric. Usually installed between the start and run terminals of a motor.

CAPILLARY TUBE A long, small-diameter tube that can be used either as a metering device or to transmit a pressure to a controller.

CARBON MONOXIDE A colorless, odorless product of combustion that can be formed when there is insufficient combustion air. Very toxic.

CASCADE SYSTEM A refrigeration system in which a very low temperature refrigeration cycle uses a standard-type refrigeration cycle to provide the condensing medium for the low temperature system.

CAVITATION The result of a pump operating with insufficient suction pressure. Vapor bubbles are formed locally at the impeller inlet.

CELSIUS The temperature scale in the metric system in which freezing water is defined as zero degrees, and boiling water is 100 degrees.

CENTIGRADE Celsius.

CENTRIFUGAL COMPRESSOR A compressor that uses the centrifugal force produced by a rotating wheel to compress refrigerant. Commonly used in systems over 100 tons capacity.

cfm Cubic feet per minute.

CHANGE OF STATE The changing of a substance from one state (solid, liquid, or vapor) to a different state. May be caused by a change in heat or a change in pressure.

CHARGING CYLINDER A calibrated container in which a predetermined weight of refrigerant may be charged or dispensed into a refrigeration system. Commonly used on small systems.

CHECK VALVE A valve that permits flow in one direction only.

CHIMNEY EFFECT The tendency of a gas to rise when heated, as in a fireplace chimney.

CIRCUIT BREAKER An electrical device that senses current flow, and opens the circuit if the current exceeds the rating of the breaker. Used to protect the circuit wiring.

CLEARANCE SPACE The space that is left inside the cylinder of a compressor when the piston is at the top of its stroke. Manufacturers try to design compressors that minimize this volume.

CLUTCH Electrically operated device used on automobile air conditioning systems. When energized, it couples the compressor to the pulley, which is being driven by the engine.

COEFFICIENT OF EXPANSION The change of length of a material per degree change in temperature.

COEFFICIENT OF PERFORMANCE Measure of energy efficiency for a refrigeration cycle. Equal to the Btus of refrigeration effect divided by the Btus of energy consumed.

COLD A subjective measure of temperature.

COLD DECK The portion of a multizone air handler between the cooling coil and the discharge from the unit. Also on a double-duct air handler.

COMBUSTION AIR The quantity of air that must be provided to achieve complete combustion.

COMPOUND GAUGE A gauge that indicates pressures above and below atmospheric pressure. Commonly used on the low side of manifold gauges.

COMPOUND SYSTEM A refrigeration system in which two compressors are used in series in order to produce the required pressure difference between the low-side and the high-side pressures.

COMPRESSION RATIO The compressor absolute discharge pressure divided by the absolute suction pressure.

COMPRESSION TANK A closed tank containing air that can be compressed to provide additional room for expanding water.

COMPRESSOR Used to increase the pressure of a vapor by using mechanical energy to squeeze the vapor into a smaller space.

COMPRESSOR DISPLACEMENT The volume rate of flow that is produced by a compressor.

CONDENSATE The liquid that is formed when a vapor loses enough heat to form a liquid. Condensate is commonly formed by refrigerant vapor in a condenser, by air on the outside of an evaporator, or by steam inside a heating coil.

CONDENSING PRESSURE The pressure inside a condenser at which the condensation is taking place.

CONDENSING TEMPERATURE The temperature inside the portion of a condenser where liquid and vapor exist together. Must correspond to the condensing pressure.

CONDENSING UNIT A single unit that packages together the compressor, condenser, and controls.

CONDUCTIVITY A measure of the ease or difficulty with which heat can flow through a substance. Also, electrical conductivity.

CONDUCTOR A wire.

CONTACTOR An electrical switch that provides voltage to a large electric load. The switch is closed by the magnetic field created by a coil of wire.

CONTAMINANT Any substance inside a refrigeration system other than refrigerant and oil. May be moisture, air, scale, or acid.

CONTINUITY Having a continuous path for electrical flow.

CONTROL VALVE An automatic valve that is operated by a sensing mechanism in order to control the flow of a fluid.

CONTROL RELAY A switch or switches together with a magnetic coil that controls the switch operation.

CONTROL VOLTAGE The voltage used in the electrical circuit that contains the low current devices such as thermostats.

CONVECTION The movement of heat by means of a flow of liquid or vapor.

COOLING TOWER Used in conjunction with water-cooled condensers. Receives the warmed water from the condenser, and cools it for reuse by the condenser.

COUNTER EMF Electromotive force (voltage) produced by an electrical coil, which tends to oppose the voltage being applied.

COUNTERFLOW The flow of two fluids in opposite directions.

CONNECTING ROD The part of a compressor that connects the crankshaft to the piston.

CRANKCASE PRESSURE REGULATOR An automatic valve installed in the suction line that prevents the pressure in the crankcase from exceeding the set point.

CRANKSHAFT The shaft in a compressor that is driven by a motor, and rotates to provide movement to the pistons.

CRANKSHAFT SEAL The seal that must be provided on open-type compressors where the crankshaft passes through the body of the compressor.

CROSS CHARGED A sealed container (TXV bulb) that contains two different refrigerants.

CRYOGENICS Science dealing with the production and use of ultralow temperatures.

CURRENT The flow of electricity through a circuit.

CURRENT RELAY Type of motor starting relay in which the current through the run winding also passes through the coil of the relay.

CUT-IN The pressure or temperature setting of a control device at which the contacts of the device will close.

CUT-OUT The pressure or temperature setting of a control device at which the contacts of the device will open.

CYLINDER The space inside a compressor in which the vapor is compressed.

CYLINDER HEAD The top plate on a compressor cylinder.

DAMPER A blade or series of blades inside a duct that may be adjusted in order to change the airflow rate.

DEAREATOR The tank in which boiler feedwater is brought into contact with steam in order to remove the air from the feedwater.

DEFROST CYCLE The automatic sequence that causes ice to melt off the evaporator, either by the use of an electric heating coil on the evaporator, or by the introduction of hot gas inside the evaporator tubes.

DEFROST TIMER The clock and switch assembly that controls the operation of the defrost cycle.

DEGREE DAY A measure of the amount of fuel required to heat a building, equal to the average number of degrees that the outside air was below 65°F for a one-day period.

DEHUMIDIFIER A refrigeration application in which room air is passed over an evaporator to remove moisture, and then over the condenser.

DEMAND The peak consumption rate of electricty, refrigeration, steam, hot water, etc.

DESICCANT A chemical that absorbs water in a refrigeration system.

DESIGN PRESSURE The maximum pressure at which a device may be expected to operated without sustaining physical damage.

DESIGN TEMPERATURE Usually the outdoor temperature (maximum or minimum) that has been assumed in order to design the capacity of a heating or air conditioning system. Also, the space design temperature.

DEW The moisture that is formed when outdoor air is cooled sufficiently to form condensate.

DEWPOINT The temperature to which air must be cooled in order for the moisture to begin to condense.

DIAPHRAGM A thin piece of material that can flex when a pressure difference exists from one side to the other.

DIELECTRIC Material used in capacitors with a high electrical resistance.

DIFFERENTIAL The difference betwen the cut-in and cut-out settings of an automatic switch.

DIFFUSER A duct fitting for delivering conditioned air to the occupied space.

DIRECT ACTING The action of a pneumatic controller that increases the branch pressure as the controlled variable increases.

DIVERTING VALVE A valve with one inlet and two outlets.

DRAFT GAUGE An air-pressure meter that measures the negative pressure available in a flue stack to remove products of combustion.

DRIP PAN The pan installed beneath an evaporator coil to collect the condensate formed on the outside surface of the coil.

DRY BULB The air temperature sensed by a standard thermometer that has not been wetted.

ECONOMIZER CYCLE A method of operating an air system to introduce extra outside air when it is cool outside.

EFFECTIVE AREA The area available in a grille or diffuser through which the air can actually flow.

ELECTRIC DEFROST A method of melting ice off an evaporator coil by using an electric heating element mounted on the coil.

ELECTRIC HEAT Heat that is produced by passing a current through a resistance coil. Resistance heat.

END BELL The end piece of an electric motor that contains the bearing or bushing.

END PLAY The movement of a shaft along its centerline.

ENTHALPY Heat.

EVACUATION The service procedure that removes all air and water vapor from inside a system by using a vacuum pump.

EVAPORATION The process in which a liquid absorbs heat and turns into a vapor.

EVAPORATIVE CONDENSER A condenser that uses evaporating water on the outside of tubes that carry condensing refrigerant.

EVAPORATIVE COOLER A unit that cools air through a water spray or pad and discharges it to an occupied space.

EVAPORATOR The component in a refrigeration system where the cold refrigerant is allowed to absorb heat from the cooled space.

EXCESS AIR A quantity of combustion air that is more than the amount theoretically required for complete combustion.

EXFILTRATION Air leaking out of a building.

EXPANSION TANK An open tank on a water system that provides room for expansion of the water.

EXPANSION VALVE A metering device.

EXPENDABLE REFRIGERANT SYSTEM A cooling system that allows a liquid refrigerant to flash to atmosphere.

FAHRENHEIT The temperature measurement system that defines freezing water as 32°F and boiling water as 212°F.

FARAD A unit of capacitance.

FEMALE THREAD A screw thread on the inside of a pipe or fitting.

FIN COMB A tool used to straighten bent fins on a condenser or evaporator coil.

FIRE DAMPER A temperature-actuated damper in a duct that closes during a fire. Prevents the spread of flames by the air system.

FIRESTAT A thermostat that senses temperature in a return-air duct and shuts down the fans in the event of a fire.

FLARE The enlarged end of a tube that is used with mechanical connectors.

FLASH GAS The vapor that is formed when a liquid is exposed to a pressure that is less than the saturation pressure of the liquid.

FLASH POINT The minimum temperature of a fluid that will support a flash flame.

FLOAT VALVE An automatic level valve that responds to an element that floats on a liquid.

FLOODED EVAPORATOR An evaporator that stores cold, liquid refrigerant.

FLUE GAS The products of combustion in the vent stack.

FLUID A liquid or a vapor.

FLUX The chemical paste applied to metals to be soldered or brazed to prevent the formation of an oxide coating.

FOAMING Formation of foam in oil that is caused by dissolved refrigerant boiling out of solution.

FOOT-POUND A unit of work that is required to lift a one pound mass one foot high.

FORCED DRAFT Airflow that is caused by a fan.

fpm Feet per minute.

FREEZING The change of state that involves removing the heat of fusion from a liquid to form a solid.

FREEZE-UP The formation of ice on the face of an evaporator coil that blocks the airflow. The formation of ice on the inside of evaporator tubes in a water chiller that causes the tubes to burst.

FREON Trade name for refrigerants manufactured by E. I. duPont de Nemours & Co., Inc.

FURNACE Equipment that burns a fuel to provide heated air to a space.

FUSE Device used in an electrical circuit that limits the maximum amount of current to which the wiring can be subjected.

FUSIBLE PLUG A plug used in a pressure-containing system that will melt out in the event that the system is exposed to high temperatures.

GAUGE PORT A ¼-male flare connection designed for the technician to use for attaching manifold gauges.

GAUGE PRESSURE The pressure reading on a gauge that starts at zero for atmospheric pressure.

gpm Gallons per minute.

GRADUAL SWITCH A pneumatic device that provides an output pressure that is selected by turning a knob.

GRAIN A unit of weight. Seven thousand grains are equal to one pound.

GROUND An electrical condition where the current-carrying wires have made contact with the exposed casing parts.

GROUND WIRE A wire at neutral potential that is used to carry current in the event of a short circuit and an open neutral line.

HALIDE TORCH A leak detector that uses a propane flame and a reactor plate.

HEAD PRESSURE The high-side pressure. Condensing pressure.

HEADER A pipe that distributes or collects liquid or vapor to or from a number of smaller tubes or pipes.

HEAT A form of energy that causes a change in temperature.

HEAT EXCHANGER A device that carries two different fluids for the purpose of transferring heat from the warmer stream to the cooler.

HEAT LOAD The rate at which heat enters a space that is being cooled.

HEAT OF COMPRESSION The heat equivalent of the work done by the compressor on a refrigerant.

HEAT OF RESPIRATION The heat released by the action of foods that are drying out while in cold storage.

HEAT PUMP A refrigeration system that cools outside air (or water) and rejects the heat into a space that is being heated.

HEATING COIL A finned-tube device that uses steam or hot water on the tube side to heat air. Also, an electric resistance heater.

HEATING VALUE The number of Btus released by a pound or a cubic foot of a fuel.

HERMETIC COMPRESSOR A compressor-motor assembly that is contained inside a sealed refrigerant circuit.

HERTZ Electrical cycles per second.

HIGH PRESSURE CUTOUT A switch that senses high-side pressure and opens when the pressure rises to a set point.

HIGH SIDE The portion of the refrigeration system between the compressor discharge valve and the metering device. The compressor and condenser assembly located outside the cooled space, where heat is to be rejected. A condensing unit.

HOLDBACK VALVE A crankcase pressure regulator.

HORSEPOWER A unit of power equivalent to 33,000 ft-lb per minute, or 746 kW.

HOT GAS The vapor being discharged from the compressor.

HOT GAS BYPASS A method of capacity control that sends hot gas from the compressor directly to the evaporator inlet or compressor inlet.

HOT GAS DEFROST A method of defrosting an evaporator or harvesting ice cubes in an ice maker. Hot gas is sent directly to the evaporator.

HUMIDIFIER A device used to add moisture to air to increase its humidity.

HUMIDISTAT A switch that senses relative humidity and closes when the sensed humidity falls below the set point.

HUMIDITY Relative humidity or absolute humidity.

HUNTING The cycling of a controlled variable above and below the set point.

HYDROCARBONS Materials whose molecules are comprised entirely of only carbon and hydrogen atoms.

HYDRONIC A system that circulates water as the heat transfer medium.

IGNITION TRANSFORMER Used on fuel oil heating systems to provide a high voltage spark between a pair of electrodes to ignite the fuel.

IMPEDENCE The resistance to electrical flow that is created by a coil in an ac circuit. Also, back EMF.

IMPELLER The rotating part of a centrifugal compressor or centrifugal water pump.

INCLINED MANOMETER A manometer that is sloped rather than vertical to increase its accuracy at low pressure differences.

INFRARED Having a wavelength outside the visible spectrum, just below the wavelength of the color red.

INSULATION *Electrical:* Coating on a wire that will not conduct electricity. *Heat transfer:* A material that exhibits a high resistance to the transmission of heat.

INTERMITTENT IGNITION A heating system that only

provides a source of ignition for the main flame just before the main flame is ignited.

IR DROP The change in voltage that occurs when current flows through a device.

KELVIN An absolute temperature scale that reads 273 degrees higher than the Celsius scale.

KILOWATT (kW) A unit of electrical power equal to 1000 watts.

KILOWATT-HOUR (kWh) A unit of electrical energy.

KING VALVE The service valve located on the receiver outlet.

LATENT HEAT Heat that has the effect of changing the state of a substance without changing its temperature.

LATENT HEAT OF CONDENSATION The amount of heat that must be removed from one pound of a saturated vapor to change it into a saturated liquid.

LATENT HEAT OF FUSION The amount of heat that must be removed from one pound of a substance at its freezing point to change it into a solid, at the same temperature.

LATENT HEAT OF MELTING The amount of heat that will be absorbed by one pound of a solid at its freezing temperature when it melts, at the same temperature.

LATENT HEAT OF VAPORIZATION The amount of heat required to change one pound of a saturated liquid into a saturated vapor.

LIMIT SWITCH An electrical safety switch, usually on a heater, that will open to turn off a heat source if the sensed temperature rises to the set point.

LIQUID LINE The tube that carries liquid refrigerant from the condenser or receiver to the metering device.

LOUVER Duct fitting on an outside air intake that permits air to enter but keeps rain out.

LOW PRESSURE CUTOUT An electric switch that senses low-side pressure. It opens on a drop in pressure, and closes on a rise in pressure.

LOW SIDE The part of the refrigeration system between the metering device outlet and the compressor. The components that include the metering device and the evaporator, located inside the conditioned space.

LOW-SIDE FLOAT A metering device that regulates the flow of refrigerant to maintain a constant level of refrigerant inside a flooded evaporator.

MAGNETIC CLUTCH A device used to transmit the motion of a belt to the air conditioning compressor in an automobile.

MAGNETIC STARTER A switch that provides voltage to a large motor. It is closed by the magnetic field

produced by a coil of wire, and it contains electrical overloads.

MAKE-UP AIR Air that is introduced to a building to replace air that has been exhausted to atmosphere.

MALE THREAD A screw thread on the outside of a pipe or fitting.

MANIFOLD GAUGES A set of two gauges and three hoses mounted on a manifold with two valves. May have additional gauges, hoses, or valves.

MANOMETER A device used to measure pressure by observing the difference in the height of liquid that the pressure can cause.

MBH Thousands of Btus per hour.

MASTER CONTROLLER A pneumatic controller that senses a variable and whose output pressure resets the set point of another controller.

MELTING POINT The temperature at which a solid will turn to liquid when it absorbs heat.

MICROFARAD A unit of capacitance equal to a millionth of a farad.

MICRON A thousandth of a meter.

MICRON GAUGE A pressure gauge used to accurately measure pressures just higher than that of a perfect vacuum.

MISCIBILITY The ability of two liquids to mix with each other.

MIXED AIR The mixture of outside air and return air entering an air handler.

MIXING VALVE A valve with two inlets and one outlet.

MODULATING Having the ability to adjust position to a continuous number of positions.

MOLLIER DIAGRAM A graph, usually of a refrigerant, that shows the relationship among many of the thermal properties.

MULLION The frame piece between the doors of a two-door refrigerator/freezer.

MULLION HEATER A resistance heater that prevents the mullion from sweating.

NATURAL CONVECTION The movement of air that occurs because of the buoyancy of heated air.

NOMINAL The approximate size or capacity of a part or system.

NONCONDENSIBLE A vapor that cannot be changed to a liquid by the removal of heat.

NONFERROUS Metals or alloys containing no iron. Includes copper, brass, and monel.

NORMALLY CLOSED A pneumatic device that opens as the pressure applied to the diaphragm increases. A solenoid valve that opens when a voltage is applied to the coil. A switch that is closed unless acted upon by an external force.

NORMALLY OPEN Opposite action from a normally closed device.

OHM (Ω) The unit used to measure electrical resistance.

OHM'S LAW The relationship between volts, ohms, and amps flowing in a circuit.

OIL SEPARATOR Device installed in the hot gas line to separate oil from the refrigerant and return it to the compressor.

OPEN CIRCUIT A circuit that does not have continuity.

OPEN COMPRESSOR A compressor that is driven from a power source that is located outside the refrigerant circuit.

OPERATING PRESSURE The pressure inside a system when it is operating normally.

ORIFICE A small, accurately sized opening.

OVERCURRENT A very high current flow resulting from a short circuit.

OVERLOAD A higher than normal current flow. The electrical switch that senses overload and opens when it occurs.

PACKAGED A cooling or heating system that is delivered from the factory already assembled as a complete system.

PACKING The soft material that is compressed around a valve stem or rotating shaft to create a seal.

PERMEANCE A measure of the ability of moisture to pass through a material.

pH A measurement of acidity (1–7) or alkalinity (7–14).

PILOT FLAME A small flame used to ignite a main flame.

PILOT GENERATOR A series of thermocouples together in one unit that generates the voltage required to operate a gas valve.

PISTON A moving part inside a cylinder.

PISTON PIN The part of a compressor that connects the connecting rod to the piston.

PISTON DISPLACEMENT The volume displaced by the motion of one piston completing one stroke.

PITOT TUBE A tube that separately senses total pressure and static pressure of air.

PNEUMATIC Operated by air pressure.

POA VALUE Pressure operated altitude valve used in automobiles to maintain a minimum allowable evaporator pressure.

POLYSTYRENE Styrofoam.

POTENTIAL A difference in electrical pressure.

POTENTIAL RELAY A starting relay that opens its contacts based on the back EMF produced by a start winding.

POUR POINT The lowest temperature at which a refrigerant oil will pour out of an open container.

POWER BURNER A fuel burner in which both the fuel and the air are supplied under pressure.

POWER ELEMENT The part of a thermostatic expansion valve that includes the charged bulb, diaphragm, and connecting tubing.

POWER FACTOR The ratio between actual power and calculated power, accounting for the difference in phase between volts and amps.

PRESSURE DROP The difference in pressure in a flowing air or water stream caused by friction.

PRESSURE REGULATOR A piping accessory that maintains a constant downstream pressure.

PRIMARY AIR The air that is mixed with air inside a gas burner before combustion takes place.

PRIMARY CONTROL The controller for an oil burner.

PRIMARY VOLTAGE The voltage supplied to a transformer.

PROCESS TUBE The compressor connection used by the manufacturer to add the initial charge of refrigerant to the system.

psi Pressure expressed in pounds per square inch.

psia Absolute pressure, expressed in pounds per square inch.

PSYCHROMETER A device consisting of a dry-bulb and wet-bulb thermometer.

PSYCHROMETRIC CHART A graph that shows the relationships among the thermodynamic properties of air.

PUMPDOWN Using the compressor to pump all the refrigerant out of the low side and into the high side.

PURGING Removing undesired vapors from an area or system.

PYROMETER An optical instrument used to measure surface temperatures.

RADIANT HEAT A method of heating that uses high temperature surfaces to heat people and objects directly, without heating the air between the heat source and the people.

RANKINE Absolute temperature scale that reads 460 degrees higher than the Fahrenheit scale.

RECEIVER The storage tank located in the liquid line.

RECEIVER CONTROLLER A pneumatic device that receives a signal from one or more transmitters and sends a branch signal to a controlled device.

RECIPROCATING Moving back and forth.

RECORDING THERMOMETER A thermometer that senses temperature and has a pen that records the sensed temperature on a moving chart.

REED VALVE A flat, springy part that bends easily,

used as a suction or discharge valve on a compressor cylinder.

REFRIGERANT Any fluid used to absorb heat from one location and release heat to a different location.

REGISTER A duct termination that has a decorative face and a volume damper behind.

RELATIVE HUMIDITY The percent of moisture in air compared to the maximum amount of moisture that can be held by air at that temperature.

RELIEF VALVE An automatic valve that senses pressure inside a system, and opens to relieve the pressure if it rises to a set point.

RETROFIT Replacement of parts in a system with parts of a newer, improved design.

RETURN AIR The room air that is returning to the air conditioning unit.

REVERSE ACTING The action of a pneumatic controller that causes the output pressure to decrease as the controlled variable increases.

REVERSING VALVE The four-way valve used in a heat pump to change the system between heating and cooling.

RUN WINDING The lower resistance winding in a single-phase compressor.

SADDLE VALVE A valve that clamps on a pressurized line and pierces it.

SAFETY CONTROL Any switch or device that is designed to operate only in the event of unusual (abnormal) operating pressures or temperatures.

SAFETY VALVE Relief valve.

SATURATION TEMPERATURE The temperature at which boiling or condensation can take place.

SCHRADER VALVE A refrigerant fitting that allows the attachment of a pressure gauge without loss of refrigerant.

SECONDARY VOLTAGE The output voltage from a transformer.

SEMIHERMETIC COMPRESSOR An accessible hermetic.

SENSIBLE HEAT Heat that causes a change in the temperature of a substance.

SENSITIVITY A measure of how quickly the output signal changes for a given change in input signal.

SERVICE VALVE A manually operated valve in a refrigeration system.

SERVICEABLE HERMETIC COMPRESSOR An accessible hermetic.

SHAFT SEAL The seal around the rotating shaft of an open compressor where the shaft passes through the compressor body.

SHORT CIRCUIT A circuit that provides a path between wires at different potential without passing through a load device.

SHORT CYCLING Turning on and off too frequently.

SIGHT GLASS A window in an operating system.

SILVER BRAZING Brazing with a filler material that contains silver.

SLING PSYCHROMETER A psychrometer that creates a velocity past the wet-bulb thermometer by swinging it in a circle.

SLIP The difference between synchronous speed and actual speed for a motor.

SLUGGING Condition that occurs when liquid refrigerant is allowed to enter the clearance space in a reciprocating compressor.

SOLDERING The process of joining metals together using a filler material that melts at a temperature below 800°F.

SOLENOID VALVE A valve that is operated by a coil that creates an electromagnet.

SPECIFIC GRAVITY Comparative weight of a liquid or vapor compared to water or air.

SPECIFIC HEAT The number of Btus required to change the temperature of a substance by 1°F, not involving a change of state.

SPECIFIC VOLUME The volume occupied by one pound of a substance.

SPLIT-PHASE MOTOR A motor that has a start and run stator winding, and the start winding is switched out of the circuit after start-up.

SPLIT SYSTEM A refrigeration system that has a condensing unit located remotely from the low side.

SPRING RANGE The range of air pressures required to completely stroke a pneumatic valve or damper actuator.

START WINDING The high resistance winding in a split-phase motor.

STARTING RELAY The switching mechanism used to take the start winding out of a circuit after the motor starts.

STATIC PRESSURE The outward push of air against the walls of a duct or of a liquid against a pipe.

STATOR The nonrotating winding in an electric motor.

STEAM-JET REFRIGERATION A system that uses steam through a venturi to create a low pressure.

STEAM TRAP A device that will allow condensate to pass, but will not allow steam through.

STRATIFICATION Accumulation of warmer air at the top of a space caused by its buoyancy.

SUBCOOLED A liquid that is at a temperature below its saturation temperature.

SUBMASTER CONTROLLER A controller whose set point is reset by a master controller.

SUCTION LINE The tube connecting the evaporator to the compressor.

SUPPLY AIR The conditioned air being supplied to the occupied space.

SUPERHEAT A vapor that is at a temperature higher than its saturation temperature.

SWAGE A tool used to expand the end of a tube so that it can fit snugly over the end of another tube.

SWAMP COOLER An evaporative cooler.

SWEATING (1) The formation of moisture on a surface that is colder than the dew point of the air. (2) Soldering.

SYNCHRONOUS SPEED The theoretical speed of a motor. 3600 rpm for a two-pole motor, 1800 rpm for a four-pole motor, and 1200 rpm for a six-pole motor.

TAIL COIL The last pass of refrigerant in an evaporator.

THERM A unit of heat equal to 100,000 Btu.

THERMISTER A device whose resistance changes dramatically with a change in temperature. A solid state switch.

THERMOCOUPLE A device that generates a voltage when exposed to high temperature. Used to prove a pilot flame, or as the sensor in electronic thermometers.

THERMODYNAMICS Science of the relationships between heat and work.

THERMOSTAT An automatic switch that opens or closes in response to temperature. A pneumatic temperature-sensing device that sends a signal to a controlled device.

THERMOSTATIC EXPANSION VALVE A metering device that maintains a constant number of degrees of superheat entering the compressor.

THREE-WAY VALVE A mixing valve or diverting valve.

THROTTLING RANGE The range of pressure or temperature over which a pneumatic control operates.

TON A measurement of refrigeration capacity equal to 12,000 Btu per hour.

TORQUE Twisting or turning force.

TRANSFORMER Electrical device used to step up or step down an input ac voltage.

TRANSMITTER Pneumatic device that senses temperature (or other variable) and responds to changes by changing the output pressure. There are no adjustments to sensitivity.

TURNDOWN RATIO The minimum percentage of full load to which a burner can be modulated.

TWO-WAY VALVE A valve with one inlet and one outlet.

ULTRAVIOLET Having a wavelength just outside the visible spectrum, just above the wavelength of the color violet.

VACUUM Any pressure lower than atmospheric pressure.

VALVE PLATE The part of a compressor between the cylinders and the head. It contains the suction and discharge valves.

VAPOR BARRIER A plastic or metallic barrier to prevent the migration of moisture.

VARIABLE PITCH A pulley that may be adjusted to different pitch diameters.

VELOCITY PRESSURE The pressure caused solely by the velocity of a fluid.

VIBRATION ELIMINATORS Flexible couplings around vibrating parts to prevent transmission of vibration to other parts of the system.

VIBRATION ISOLATORS Mounting pads for vibrating machinery that prevents the transmission of the vibration to other parts of the building.

VISCOSITY A measure of how well a liquid flows. Internal friction.

VOLTAGE Electrical pressure.

VOLTAGE RELAY A potential relay.

VOLUMETRIC EFFICIENCY Actual quantity of refrigerant pumped by a compressor compared to the theoretical flow rate.

WALK-IN BOX A large storage room for refrigerated or frozen foods.

WATT Unit of measurement for electrical power.

WET BULB The temperature registered on a thermometer that has been wetted.

WRIST PIN The pin that connects the connecting rod to the inside of the piston.

INDEX